Solar Capacitors and Batteries

Scrivener Publishing
100 Cummings Center, Suite 541J
Beverly, MA 01915-6106

Advances in Solar Cell Materials and Storage

Series Editor: Nurdan Demirci Sankir (ndsankir@gmail.com)
and Mehmet Sankir

Because the use of solar energy as a primary source of energy will exponentially increase for the foreseeable future, this recent series on *Advances in Solar Cell Materials and Storage* focuses on new and novel solar cell materials and their application for storage. The scope of this series covers the solution-based manufacturing methods, nanomaterials, organic solar cells, flexible solar cells, batteries and supercapacitors for solar energy storage, and solar cells for space.

Publishers at Scrivener
Martin Scrivener (martin@scrivenerpublishing.com)
Phillip Carmical (pcarmical@scrivenerpublishing.com)

Solar Capacitors and Batteries

Edited by

Nurdan Demirci Sankir

Department of Materials Science and Nanotechnology Engineering, TOBB University of Economics and Technology, Ankara, Turkey

and

Mehmet Sankir

Department of Materials Science and Nanotechnology Engineering, TOBB University of Economics and Technology, Ankara, Turkey

Scrivener Publishing

WILEY

This edition first published 2025 by John Wiley & Sons, Inc., 111 River Street, Hoboken, NJ 07030, USA and Scrivener Publishing LLC, 100 Cummings Center, Suite 541J, Beverly, MA 01915, USA

For more information about Scrivener publications please visit www.scrivenerpublishing.com.

Wiley Global Headquarters
111 River Street, Hoboken, NJ 07030, USA

For details of our global editorial offices, customer services, and more information about Wiley products visit us at www.wiley.com.

The manufacturer's authorized representative according to the EU General Product Safety Regulation is Wiley-VCH GmbH, Boschstr. 12, 69469 Weinheim, Germany, e-mail: Product_Safety@wiley.com.

Library of Congress Cataloging-in-Publication Data

ISBN 978-1-394-23376-2

Front cover images courtesy of Wikimedia Commons
Cover design by Russell Richardson

Set in size of 11pt and Minion Pro by Manila Typesetting Company, Makati, Philippines

Printed in the USA

10 9 8 7 6 5 4 3 2 1

Contents

Preface

Although the great potential of solar energy compared to other renewable energy sources, the minimization of the environmental damage caused by fossil fuels with photovoltaic conversion, and the achievability of the goal of zero carbon emissions in electricity production, it is an inevitable fact that solar energy needs to be stored for use when desired. Therefore, the storage of solar energy is as important as its harvesting, and so far, solar energy has been stored mostly thanks to electrochemistry-based batteries. Generally, batteries that can be charged with solar energy serve domestic and portable systems, especially in devices with low energy use. Devices that can be charged with solar energy can be grouped into two general categories. These are rechargeable batteries and capacitors. In addition to classical devices, innovative devices have been developed in recent years that provide both the conversion of solar energy and the storage of the obtained energy. The fourth volume of the ASCMS series covers all these devices that provide solar energy storage and the materials that reflect the new and cutting-edge technology developed for these devices.

The aim of the book is to introduce the reader to different systems that enable the conversion and storage of solar energy. In this context, systems in which photovoltaic cells are designed separately, storage units such as batteries and capacitors are separately addressed, and the structures and operating principles of integrated devices with conversion/storage functions are explained. In addition to device structures, working principles, and performance tests, the highly efficient and innovative materials used in these systems are also covered in the book. Moreover, cutting-edge new technologies for the future of solar energy conversion/storage systems are included in the book.

The book is structured in four parts for easier follow-up. In the first section (Part I), the reader is introduced to capacitors and photo-capacitors that can be charged with solar energy. In addition to the structures, working principles, and test methods of the 2, 3, and 4 terminal devices, detailed information about the metal oxide, metal-organic frameworks, carbon-based and polymeric, and other innovative materials used in these systems is presented in this section. Besides the various potential application areas of the photo-supercapacitors are introduced in Part I.

The second section is devoted to the battery type and hybrid solar rechargeable devices. Photo-rechargeable all-solid-state batteries based on photoelectrochemistry and solid-state ionics, novel hybrid perovskites and inorganic semiconductors for photo rechargeable lithium-ion battery photoelectrodes, and 2D materials for solar-assisted hybrid energy storage devices were explained in Part II.

The third section covers the integrated systems of solar cells and conventional energy storage devices for various application fields including electric vehicles, uninterruptible power supply systems, Internet of Things (IoT) devices, and solar cooking systems. Moreover, readers will find detailed information about the integration of the supercapacitors in photovoltaic cells and the future perspectives of the technology in Part III.

The last part, Part IV, focuses on the photoelectrochemical flow batteries and perovskite-based photo supercapacitors. This part covers the solar flow battery systems and the adaptation of cutting-edge semiconductor materials to electrochemical energy storage devices.

As a concluding remark, we would like to emphasize that the fourth volume of the Advances in Solar Cell Materials and Storage series provides an overall view of the new and highly promising photoactive materials and system designs for solar energy conversion and storage devices. Therefore, readers from diverse fields such as chemistry, physics, materials science and engineering, and mechanical and chemical engineering will take advantage of this book to comprehend the impacts of solar capacitors and batteries. We would like to thank all the valuable authors who contributed to the book, and Martin Scrivener and Scrivener Publishing for their support and efforts in publication.

Nurdan Demirci Sankir, PhD and Mehmet Sankir, PhD
Department of Materials Science and Nanotechnology Engineering,
TOBB University of Economics and Technology, Ankara, Turkey

Part 1

SOLAR RECHARGEABLE CAPACITORS AND PHOTO-SUPERCAPACITORS

1

Photosupercapacitor

Mohamad Mohsen Momeni* and Hossein Mohammadzadeh Aydisheh

Department of Chemistry, Isfahan University of Technology, Isfahan, Iran

Abstract

Energy is a fundamental pillar of human society, playing a pivotal role across various sectors, including industry, agriculture, and transportation. Over the past decades, the energy demand has steadily risen. However, our global energy supply still heavily relies on nonrenewable fossil fuels—coal, oil, and natural gas—which pose risks such as energy crises and the accumulation of greenhouse and toxic gases in the atmosphere, contributing to global environmental challenges. We must turn to cleaner and more sustainable energy sources to address these issues. Wind, hydro, tidal, geothermal, and solar energy are excellent examples of such alternatives. These renewable resources offer inexhaustible supplies and are increasingly competitive in today's energy landscape. Among the available energy sources, solar energy stands out as a renewable and abundant clean alternative to conventional fossil fuels. Leveraging the abundance of solar energy, storage devices like batteries and supercapacitors can serve as uninterrupted energy reservoirs when integrated with solar cells. However, the highly fluctuating nature of solar energy often leads to misalignment with actual energy demand. To address this challenge, direct conversion and electrochemical storage of solar energy are practical strategies. In this chapter, we delve into the principles governing integrated devices. We explore major types of essential integrated devices, emphasizing the development of high-performance systems based on solar energy conversion components, including photosupercapacitors. Additionally, we combine recent advancements and modern technologies in solar energy storage, with a keen focus on novel materials. Lastly, we address the challenges and opportunities for future development.

Keywords: Environmental challenges, renewable energy, solar energy, photosupercapacitor

**Corresponding author*: mm.momeni@cc.iut.ac.ir

Nurdan Demirci Sankir and Mehmet Sankir (eds.) Solar Capacitors and Batteries, (3–30) © 2025 Scrivener Publishing LLC

1.1 Introduction

The indiscriminate use of fossil fuels has led to the significant production of toxic and greenhouse gases (e.g., CO, CO_2, SO_x, NO_x), resulting in global warming and critical environmental and health issues. Factors such as population growth, fossil fuel depletion, and limitations in conventional energy sources have prompted a shift toward renewable energy alternatives, including wind, tidal, hydro, and solar energy (see Figure 1.1). Among these options, harnessing abundant solar energy for green electricity stands out due to its global availability, lack of pollution, and cost-effectiveness. This approach represents a promising pathway toward achieving a sustainable global civilization. Therefore, the development of solar energy conversion technologies and solar cells is of utmost importance. One of the major limitations the present solar electricity production technologies hold is that they are not capable of storing the solar energy directly within them. Unlike fossil fuels, which can be stored for later use, solar energy is typically converted into electricity for immediate consumption.

Figure 1.1 Various clean and renewable energy sources [1–7].

Thus, in practical terms for application, an energy storage system like a battery or supercapacitor is needed to store the energy for later use. Among these storage devices, supercapacitors show high power and energy densities against traditional batteries and capacitors. Supercapacitors have been considered in the past for storing solar energy using an external wire connection. This approach results in two systems with no integration: the solar cell system and the supercapacitor system. Unfortunately, this architecture had very complicated development steps, so the cell and supercapacitor designs were extremely large and space-consuming. Moreover, high external resistance was created, which induced high power losses because of the external interconnection among the cells and energy storage devices. To overcome these limitations, integrated energy harvesting and storage devices have emerged. These integrated systems, while efficient, pose challenges related to device packaging efficiency and ohmic transport losses. Additionally, the output voltage of photovoltaic cells often falls short of fully charging batteries or capacitors. Consequently, supplementary electronics are required, adding to the overall cost of these energy storage solutions [1–7]. In the realm of energy harvesting and storage, a critical imperative arises: the development of novel strategies that meet the demands of current and future wearable devices. Addressing this challenge has led to an empirical solution—integrating solar cells and energy storage components within a single module. This remarkable hybrid technology, known as the photosupercapacitor, seamlessly combines energy harvesting and storage functionalities in a single device. Its potential impact extends to various domains, including public transport, hybrid electric vehicles, military applications, and space exploration. As we look ahead, this integrated approach promises to be a significant boon for efficient energy utilization [2, 3]. In this chapter, we include the basic concepts and the latest developments on advanced integrated energy devices. We first give some basic concepts about these advanced devices, considering their integration modes and complex processes of solar energy conversion and storage. We then categorize two kinds of integrated devices into photovoltaic and photoelectrochemical rechargeable supercapacitors according to their two different device configurations, summarized in detail. Finally, we outline the most important challenges and opportunities for future research in developing new, state-of-the-art integrated devices for the conversion and storage of solar energy. We think that by entering this frontier, we will see transformational breakthroughs that will set a new trajectory in using sustainable energy in the future [4, 5].

1.2 Photosupercapacitors

Photosupercapacitors represent an innovative class of energy storage devices that combine the principles of supercapacitors with the characteristics of photovoltaic cells. These devices are designed by integrating elements from dye-sensitized perovskite or organic/silicon solar cells with various supercapacitor technologies into a unified module. The resulting bifunctional systems are capable of efficiently converting solar energy into electrical charge, thereby enhancing energy storage capabilities [6, 7]. When light interacts with the photoactive material in photosupercapacitors, it generates an electrical charge that is subsequently stored within the supercapacitor. This stored energy can be readily released as needed to power electronic devices. One of the classes of energy storage devices that has promise is photosupercapacitors, which elegantly merge concepts from supercapacitors and photovoltaic cells. The photoactive material, under illumination of light, generates electrons that migrate toward the transparent electrode. These electrons finally get stored as electrostatic charge in the supercapacitor until it is needed. Photosupercapacitors offer a number of advantages compared to traditional batteries. They are also highly efficient and long lasting in that they can charge and discharge fast and as many times as one desires without loss of performance. Another added advantage of their eco-friendliness is their reliance upon renewable energy sources and their cycle life, which permits them to undergo many charge–discharge cycles without degradation. Potential applications for photosupercapacitors include wearable electronics, sensors, and energy-harvesting devices. Moreover, they have huge potential for smart windows; they will store the solar energy and emit heat at night, which can help save energy in heating and cooling. Though photosupercapacitors offer such benefits, the process is still in an infant stage. Researchers constantly search for new materials and design innovations that will increase their efficiency, durability, and scalability since there are several problems within the development of this exciting field [8].

1.3 Designs and Principles of Photosupercapacitor

The generic functionality of a photosupercapacitor involves two main processes: light absorption and charge storage. In short, the mechanism entails that the incident light interfacing with the photovoltaic material within the device generates electron–hole pairs. These photogenerated electrons and holes are separated thereafter and moved toward the electrodes, generating accumulation of charge on the electrode surfaces. This stored charge is released, on demand, for the delivery of electrical power.

Basically, photosupercapacitors can be divided into two main groups: two-electrode and three-electrode systems. Their design for each type and the operational mechanism have been explained in this paper below [9].

1.3.1 Three-Electrode Systems

In laboratory, researchers employ three-electrode systems for researching and developing new materials for photosupercapacitor applications. These systems consist of the following components:

Photoelectrode: This electrode, made from conducting and optically transparent substrates, serves as the working electrode. Commonly used substrates include Sn-doped In_2O_3 (ITO), F-doped SnO_2 (FTO), Sb-doped SnO_2 (ATO), and Al-doped ZnO (AZO).

Reference electrode: The reference electrode provides a stable reference potential. Typically, Ag/AgCl or Hg/Hg_2Cl_2 electrodes are employed in the three-electrode configuration.

Counter electrode: To streamline our research, we often employ platinum foil or platinum mesh as the counter electrode due to its low overpotential for hydrogen and oxygen evolution reactions.

A solar simulator or xenon lamp, fitted with a filter to simulate the solar spectrum (typically calibrated at 100 mW/cm^2, AM 1.5G), is utilized as the

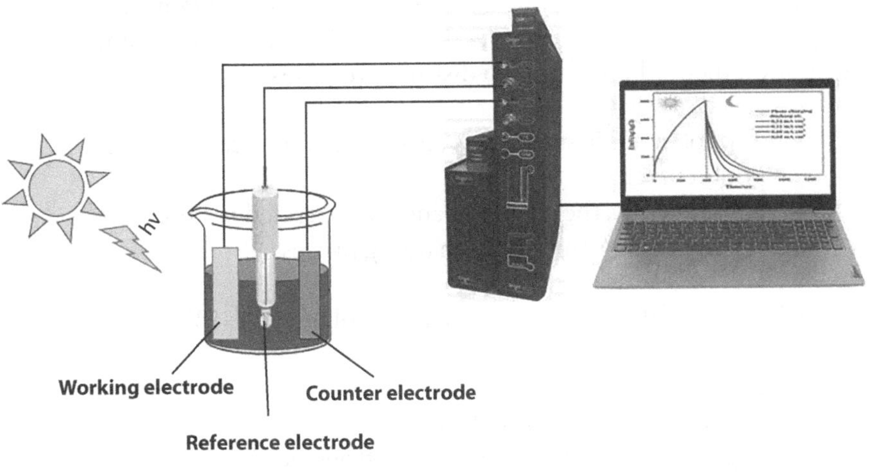

Figure 1.2 Working principle of the photoelectrode in three-electrode system.

light source [10]. In these systems, when the working electrode (photoelectrode) is immersed in an aqueous electrolyte, a space charge layer develops on the semiconductor surface. Upon illumination, the electrode material generates photo-induced electrons and holes. The electrons then travel through the external circuit to the counter electrode, where they facilitate the adsorption of cations. Meanwhile, the ho les move toward the electrode surface, where they interact with anions in the electrolyte (Figure 1.2).

1.3.2 Two-Electrode Systems (Photosupercapacitor Devices)

The integration methods between solar cells and photorechargeable supercapacitor devices can be categorized into two main types:

Type I: In this configuration, the solar cell and the energy storage system unit (tandem photosupercapacitors) are directly linked through an external circuit. Essentially, the solar cell and its associated energy storage unit (EES) function independently of one another.

Type II: In this configuration, the solar cell and the EES are integrated into a single device. Let us explore these two systems in detail.

1.3.2.1 Tandem Photosupercapacitors (Type I)

In this configuration of photosupercapacitors, the photoconversion component is directly linked to the energy storage component (supercapacitor) *via* an external circuit, forming a tandem arrangement (Figure 1.3). In this setup, each unit operates independently. Hybrid devices employing this integration method undergo two distinct processes:

- The solar cell first captures solar energy and converts it into photocurrent.
- Subsequently, the electrical energy generated is stored in the supercapacitor through a charging process.

The benefits of this hybrid system are as follows [11, 12]:

- Easy fabrication process
- Versatile design possibilities
- High overall energy conversion efficiencies ($\eta_{overcell}$)
- Established and thoroughly researched device type

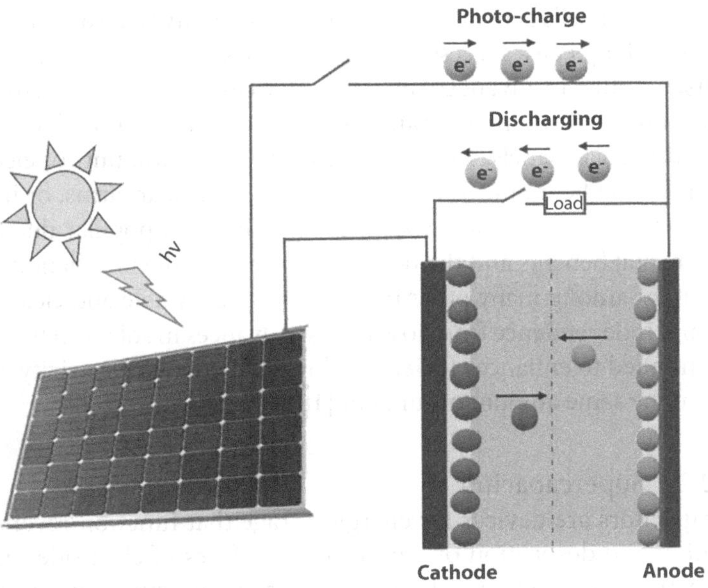

Figure 1.3 Working principle of the tandem photosupercapacitors.

And their disadvantages are as follows:

- Necessitates external electric circuits to link the two separate units, resulting in ohmic energy losses [13, 14].
- Takes up a relatively larger volume and weight because of the series arrangement.

1.3.2.2 Components of the Tandem Photosupercapacitors

1.3.2.2.1 Solar Cell Part

Solar cells are major constituent parts in photorechargeable integrated devices. They absorb solar radiation and, in turn, produce electricity. A solar cell is a device that transforms sunlight into electricity by utilizing the photovoltaic effect. The standard structure of a solar cell consists of a thin layer of silicon or any other semiconducting material, having two electrodes: one positive, the anode, and the other negative, the cathode. Light from the sun falls on the cell, transferring energy to the semiconductor material's electrons because of the photons. In response, it energizes them and causes an electron flow from the negative to the positive electrode, hence an electric current [15, 16]. The produced electricity from the solar cells may be stored in batteries, supercapacitors, or fed into the electrical grid for use straight away or for future use. Solar cells generate direct current electricity that supplies devices. In fact, there has been a successful integration of solar cells with supercapacitor units, resulting in advanced integrated devices [17]. These systems can be applied individually in small applications like calculators or outdoor lights or combined in solar panels, which can generate larger amounts of electricity. Solar panels can be mounted on rooftops, set up in solar farms, or incorporated into building materials. Solar cells have become popular due to their environmental benefits and the now-reduced cost of the installation of solar panels. They, among many other things, offer a renewable and clean energy source and independence from fossil fuels. Advances in solar cell technology have continued to enhance efficiency, allowing increased electricity production from the same amount of sunlight [18, 19].

1.3.2.2.2 Supercapacitor

Supercapacitors are devices for energy storage that function reversibly by adsorption and desorption of ions at the interfaces of electrode materials with electrolytes. Compared to a regular capacitor, supercapacitors have considerably higher capacitance but at much lower limits of voltage. The energy storage systems fill the gap between electrolytic capacitors and rechargeable batteries. A supercapacitor contains 10 to 100 times more energy per unit volume or mass when related to energy storage than capacitors. Apart from having this high energy storing capacity, they have the potential to accept rapid charge and deliver it back at an exceptionally fast rate that leaves the conventional batteries way behind. Moreover, supercapacitors can undergo thousands or even millions of charge and discharge cycles, which makes them more durable than rechargeable batteries. There are three major supercapacitor types, electrostatic double-layer capacitors (EDLCs), electrochemical pseudocapacitors, and hybrid supercapacitors (HSCs). EDLCs utilize electrodes made of carbon or its derivatives, offering significant capacitance [20]. An electrochemical pseudocapacitor utilizes metal oxide or conducting polymer electrodes. HSCs are constructed simply by combining electric double-layer capacitors with pseudocapacitance materials. Interestingly, the specific capacitance for HSCs is way higher compared to EDLCs or supercapacitors based on pseudocapacitance [21].

1.3.2.3 Type II: Photoelectrode and Supercapacitor Integrated Into One Single

In our previous discussions, we delved into the intricacies of tandem photosupercapacitors. However, these devices face a challenge due to the considerable distance between the solar cell and the EES. This spatial separation often results in bulky, heavy integrated systems and introduces ohmic energy losses. Consequently, their volume energy density and portability suffer. To address this issue, the development of highly integrated hybrid devices has become crucial for next-generation photosupercapacitors. One promising approach involves the introduction of integrated photosupercapacitors. In this configuration, the photoelectrode is seamlessly integrated into the supercapacitor's cathode side, akin to a simplified Type 1 system with external circuits removed (see Figure 1.4). The result is simultaneous conversion and storage of solar energy without intermediate steps [22–24]. During the photocharging process, electron–hole pairs form on the photoelectrode when exposed to light. The photoexcited electrons migrate to the supercapacitor's anode, where they reduce cation ions. Simultaneously, the holes move to the cathode, causing oxidation and enabling the photocharge function. It consists of two parts, one as the photoconversion electrode and another that acts as the energy storage electrode. In the case of photocharging, the counter electrode and photoelectrode are interconnected and the intermediate electrode is kept open. Upon incidence of light in the photoelectrode, photons get absorbed and charge separation takes place. The electrons flow from the conduction band (CB) of the photoelectrode to the counter electrode through the external circuit and get stored in the supercapacitor. During discharge, the intermediate electrode connects with the counter electrode, which allows the supercapacitor to discharge the

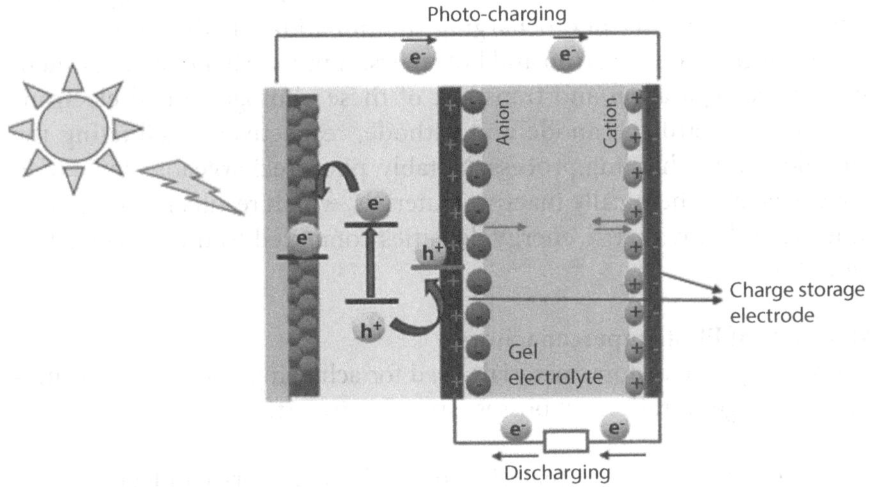

Figure 1.4 Schematic illustration of photosupercapacitor [9].

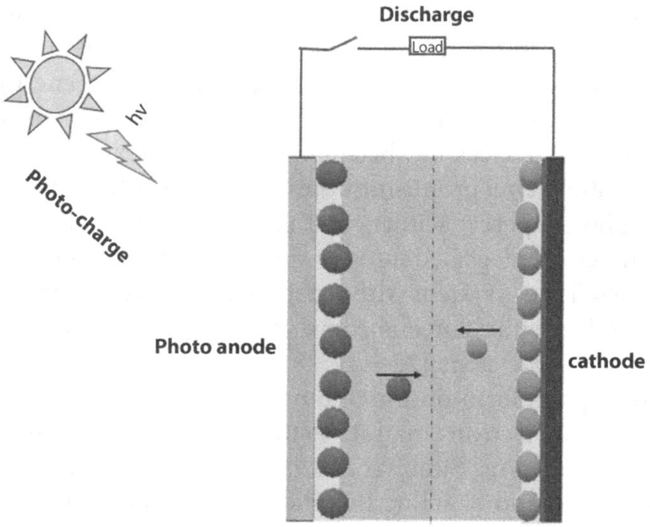

Figure 1.5 schematic illustration of photorechargeable capacitor [25].

stored energy. One of the key advantages of the integrated device is that it can overcome the large internal resistance by preventing the electrons from returning to the photoelectric electrode [25].

Another type of photorechargeable supercapacitor consists of a conventional electrode (either anode or cathode) paired with a photoelectrode (photo-anode or photocathode). In this configuration, the photo-active cathode (or anode) serves as the key component responsible for both solar energy conversion and electrochemical energy storage (Figure 1.5). During photocharging, illumination on the photo-active electrode generates electron and hole pairs. A precise bandgap alignment aids in the separation and transport of these photogenerated electrons and holes toward the anode and cathode, respectively, facilitating the intended photocharging process. Notably, photorechargeable devices do not use electrochemically inactive materials, which results in higher volumetric and gravimetric energy densities compared to the previous two types [26, 27].

Materials of Photosupercapacitors

In photosupercapacitors, materials used for achieving both photosensitivity and charge storage must possess specific properties:

Short bandgap: The material should have a short bandgap to efficiently absorb visible light.

Transparency: Transparency is essential for the material to allow light penetration.
Charge separation capability: The material must effectively separate photogenerated electrons and holes.
Long excited hole lifespan: Ensuring a long lifespan for excited holes is crucial.

Metal Oxides:
Metal oxides have been comprehensively researched to date for their potential implementation in supercapacitors, a very promising technology for energy storage with the capability to deliver both high power and cycle life. Recently, increasing interest has been shown in photosupercapacitors, which combine the energy storage ability of supercapacitors with the light harvesting ability. Metal oxides have been the most significant materials in the design of photosupercapacitors due to their unique electronic and electrochemical attributes. Investigation on titanium dioxide, zinc oxide, vanadium oxide, and tungsten trioxide has become more common for photosupercapacitor applications. These metal oxides have appropriate bandgaps to result in efficient visible light absorption and the generation of photoexcited charge carriers. These carriers can be harnessed for charge storage in supercapacitor electrodes. Enhancing the performance of metal oxide-based photosupercapacitors involves strategies such as optimizing the morphology and composition of metal oxide nanostructures, tuning the bandgap for improved light absorption, and incorporating co-catalysts or sensitizers to enhance charge separation and transfer. Additionally, the choice of electrolyte and device configuration significantly influences the overall performance of photosupercapacitors [28–30].

1. Titanium Dioxide (TiO_2)
   Among metal oxides, TiO_2 has been vastly studied for photosupercapacitor applications because of its inexpensive, biocompatible, highly stable, nonpoisonous, and inert nature. This oxide is available in large amounts and demonstrates appropriate alignment of energy levels with the redox potentials of water [31]. TiO_2 has three polymorphic phases: anatase, rutile, and brookite. Among them, anatase TiO_2 has been regarded as one of the more promising photoactive materials because it has a higher CB edge energy due to surface chemistry [32, 33]. The major limitation of TiO_2 is the wide bandgap (~3.2 eV) and weak absorption in the visible region. That is, although it efficiently absorbs ultraviolet light,

visible light absorption is poor. Additionally, in conventional photocatalysts based on TiO_2, fast electron–hole recombination also affects their photocatalytic efficiency adversely [34]. Hence, it is required to develop a visible light-sensitive photocatalyst. Next, it becomes very important to shift the optical response of TiO_2 from the UV to the visible spectral range so as to maximize the use of solar energy in photocatalytic reactions. As such, impressive efforts have gone into the development of visible-light active photocatalysts. Over time, a number of options have been assessed to help overcome this limitation. First, they hybridize wide bandgap photocatalysts with narrow bandgap semiconductors, forming heterostructures—an effective approach. Second, they develop new semiconductor materials with narrow bandgaps [35]. For example, Momeni *et al.* described a one-step anodization technique for co-doping titanium dioxide nanotubes with sulfur and tungsten. The resulting photoelectrodes demonstrated a remarkable average specific capacitance of 179 mF/cm^2 at a current density of 0.4 mA/cm^2, which is approximately five times higher than that of unmodified TiO_2 electrodes [36]. In a separate investigation, nickel–cobalt was electrodeposited on the tungsten-doped TiO_2 nanotube substrates, producing a photoelectrode that exhibited a specific capacitance of 252.4 mF/cm^2 at a current density of 0.71 mA/cm^2 in a 3.0 M aqueous KOH solution under light illumination. At a scan rate of 20 mV/s, an increase in areal capacitance of about 68.3% was recorded. The maximum photocharged areal capacitance attained was 75.2 mF/cm^2 at a current density of 0.04 mA/cm^2, with corresponding power density and areal energy values of 75.4 mW/cm^2 and 33.93 mWh/cm^2, respectively [37].

2. Zinc Oxide (ZnO)

   Zinc oxide (ZnO) is a frequently studied metal oxide for photosupercapacitors. It possesses a relatively large bandgap, allowing it to absorb UV and part of the visible light spectrum. However, 1D ZnO exhibits lower conversion efficiency compared to TiO_2 due to its limited internal surface area and high bandgap. Researchers have tackled this issue by synthesizing hierarchical ZnO nanostructures with multiscale configurations, including nanoflowers, dendritic nanowires, nanosheets, and tetrapod-like structures.

These diverse morphologies enhance the surface area and improve light absorption and charge transport, leading to better performance in various applications such as photocatalysis, sensing, and energy storage. Moreover, doping ZnO with various elements or combining it with other materials has been investigated to improve its efficacy in photosupercapacitors. Hierarchical ZnO nanostructures show great potential as photoelectrode candidates. For example, a hierarchical structure comprising ZnO nanowires and nanosheets in dye-sensitized solar cells (DSSCs) achieved a power conversion efficiency of 4.8%, nearly double that of a DSSC based solely on nanosheets. Additionally, the rapid synthesis of hierarchical nanostructures on ZnO nanowire and nanoneedle arrays using a room-temperature chemical bath deposition method demonstrated exceptional photovoltaic performance. The ZnO nanocactus array, in particular, exhibited superior electron transport characteristics, an elevated CB edge, and improved dye adsorption when compared to traditional one-dimensional ZnO nanostructure arrays. A flexible DSSC constructed with these nanocactus arrays achieved an efficiency of 4.43%, significantly outperforming hierarchical ZnO nanorod-nanosheet arrays (2.37%) and bare ZnO nanosheet arrays (1.24%) by margins of 2.06% and 3.19%, respectively. In their research, Cigdem *et al.* developed photosupercapacitors utilizing nanoscaled ZnO powders with two distinct morphologies: nanowires and nanoflowers. These ZnO nanopowders were synthesized using a hydrothermal method. The feasibility of pristine ZnO as a photoactive material was validated by fabricating photosupercapacitor devices through a simple drop-casting technique on both glass and plastic substrates, making them suitable for large-area applications. Remarkably, the nanowire-based photosupercapacitors achieved an energy density of 78.1 mWh/kg, indicating significant potential. Additionally, the nearly 100% capacitance retention observed over 3000 cycles demonstrated outstanding stability for photosupercapacitors made from these materials [38, 39]. In another study, Daniel *et al.* created an integrated photosupercapacitor utilizing ZnO nanorod arrays. They employed a passivated ZnS film combined with silver sulfide quantum dots, which were deposited on ZnO nanorods electrochemically grown on ZnO seed layers, forming the photoanode. The supercapacitor component consisted of an electrodeposited film of poly(3,4-ethylene-dioxythiophene) serving as the counter electrode, with an ionic liquid-based electrolyte applied between the two components using the dip-coating method. Electrochemical characterization indicated that this device could reach a potential of 0.33 V during photocharging and demonstrated a storage efficiency of 6.83% [40].

3. Vanadium oxide (V_2O_5)

A range of vanadium oxides, including V_2O_5, VO_2, V_2O_3, and VO, display exceptional and intriguing characteristics due to the stable oxidation states of vanadium, which span from V^{2+} to V^{5+}. Among these, V_2O_5, recognized as the most stable oxide, has attracted considerable attention [41, 42]. As an n-type semiconductor, V_2O_5 is synthesized using various techniques, allowing the creation of thin films on a range of substrates. This versatility positions V_2O_5 as a key candidate for photosupercapacitor applications. However, despite its theoretical specific capacitance of 2120 F/g and a broad bandgap of 2.2 eV, V_2O_5 encounters challenges, including low electrical conductivity and vanadium dissolution in electrolytes, both of which represent significant limitations.

To address these limitations, researchers have explored designing various V_2O_5 nanostructures or nanocomposites. The most effective approach thus far involves hybridizing V_2O_5 nanostructures with conductive materials such as carbon nanostructures, polyaniline, and polypyrrole [42–44]. In the work of Momeni *et al.*, vanadium pentoxide nanowires have been studied as an active material for photosupercapacitors. The nanowires serve as excellent electrode materials, demonstrating high electrochemical performance and long cycle life in photoresponsive supercapacitance applications. Their unique structure facilitates efficient charge storage and transfer, enhancing overall energy performance. As a result, these nanowires are promising candidates for integrating energy harvesting and storage technologies. To improve their photo-assisted supercapacitor performance, selenium and polypyrrole were incorporated. Capacitive performance has been improved to 375 mF/cm^2 at 0.25 mA/cm^2 for Se–V_2O_5–PPy. This value is more than 2.1 times that of pure V_2O_5, 1.5 times greater than the corresponding values for V_2O_5–PPy, 1.6 times higher than Se–V_2O_5, and 1.7 times greater than Se–PPy at identical current densities. In the case of Se–V_2O_5–PPy electrodes, areal capacitance values under light irradiation were 151 and 375 mF/cm^2 at a current density of 0.25 mA/cm^2 in dark and light conditions, respectively. Remarkably, the fabricated device shows good cycle stability of 91% even after 10,000 cycles, where the voltage window can be further extended up to 2.0 V and maximum volumetric energy densities reach as high as 16.8 Wh/kg at dark and 41.6 Wh/kg at light [45]. In another piece of related work, Momeni's group electrodeposited bismuth vanadate–vanadium oxide onto titania nanotubes and obtained photoelectrodes with a specific capacitance as high as 288 mF/cm^2. This was 15 times that of the pristine TiNT electrode in the dark under light irradiation. Besides, they achieved a high energy density of 40 mWh/cm^2 at a power density of 0.24 mW/cm^2. Specifically, at 0.12 mA/cm^2 current density, in light illumination, specific capacitance showed an increase of 65% from the value measured in the dark at the same current density that amounted to 100 mF/cm^2 [46].

4. Tungsten trioxide (WO_3)
   Tungsten trioxide (WO_3) is an intriguing metal oxide for photosupercapacitor applications due to its unique electrochromic properties. With a moderate bandgap, WO_3 efficiently absorbs a wide range of visible light (with a photon energy of approximately 2.5 eV), capturing nearly 12% of the visible light spectrum. WO_3 exhibits several advantages, including robust stability, extended lifespan, non-toxicity, affordability, and suitable band edges. Notably, it has shown significant stability in acidic environments, positioning it as an excellent material for treating organic acid-contaminated water [47, 48]. Its narrow bandgap (2.6–2.8 eV), non-toxic nature, and strong adaptability position WO_3 as an active material in photosupercapacitors [49, 50]. Photosupercapacitors based on WO_3 can simultaneously store energy and exhibit photoelectrochemical behavior, making them promising candidates for multifunctional devices. However, WO_3 shares some limitations with TiO_2, such as a wide bandgap and rapid electron–hole recombination. To enhance the photocatalytic activity of WO_3, various

strategies have been explored. These include modifying the WO_3 photoresponse value and improving the efficiency of photogenerated carrier recombination. Notably, effective approaches include surface modification, controlling morphology and size, creating synthetic heterojunctions, and element doping. Each of these strategies aims to optimize the performance of WO_3 by increasing light absorption, facilitating charge separation, and reducing recombination losses. For example, Zhang *et al.* produced WO_3 nanoflakes by employing etching and reduction techniques in a mildly acidic environment. The resulting photoelectrodes displayed an increased rough surface area due to the formation of oxygen vacancies during the etching process. These enhanced photoelectrodes achieved a photocurrent density of approximately 1.1 mA/cm^2, which is about twice that of unmodified WO_3 nanoflakes, measured against a 1.0 V Ag/AgCl reference electrode [51].

5. Metal sulfides

Metal sulfides have attracted considerable interest as electrode materials owing to their unique electrochemical characteristics, cost-effectiveness, abundant availability, stress-free production, and high theoretical capacities. These attributes position them as promising alternatives to noble metals across various fields, where their effectiveness is particularly noteworthy [52]. Metal sulfides, such as copper sulfide or zinc sulfide, are currently under investigation as photoactive materials for photosupercapacitors. These materials exhibit high efficiency in converting light into electrical charge and can be synthesized using simple and cost-effective methods. While numerous materials are being explored for their potential as photoactive components in photosupercapacitors, further research is needed to identify optimal materials and designs. Nevertheless, these metal sulfides hold promise for enhancing the performance and scalability of photosupercapacitors. Additionally, it is important to note that not all semiconductor materials are suitable for use as photoelectrodes in integrated devices; considerations such as band structure and reversible redox potential play a crucial role in material selection. Energy level alignment is a crucial factor in material selection. In particular, if the CB and valence band (VB) of a semiconductor are positioned within the redox potential window, photogenerated electron–hole pairs can effectively participate in the reduction or oxidation processes at the cathode or anode. Conversely, when the VB and CB lie outside the redox potential window, the material demonstrates significant redox activity. Complete photocharging is achieved only when the charging voltage is exactly 0.0 V.

Additionally, materials with narrower bandgaps exhibit lower absorption wavelengths, which can lead to reduced solar light absorption efficiency. Consequently, semiconductors with low bandgaps are well suited as photoanodes for achieving high photoconversion efficiency. In their study, Momeni *et al.* fabricated photoelectrodes using cobalt–manganese sulfide deposited on TiO_2 nanotubes, creating stable photoactive electrodes for energy-efficient photosupercapacitors. These photo-assisted supercapacitors (photo-ASCs) effectively capture light energy while concurrently storing electrical charge, resulting in improved charge storage performance when exposed to light compared to dark conditions. Exposure to light leads to prolonged discharge durations, with a capacity enhancement of approximately 70% achieved at 42.2 mF/cm^2 under illumination relative to dark conditions. The demonstrated light absorption properties and charge separation efficiency of transition metal sulfide–metal oxide-based photocathodes render them well suited for light-chargeable energy storage applications. Furthermore, the assembled photo-ASC exhibits stable charge retention over extended cycling, with only about a 4% reduction in capacity after 5000 cycles [53].

6. Carbon-based functional materials
   From a materials perspective, the integration of novel carbon-based materials can significantly enhance the efficiency and stability of hybrid solar energy systems. Carbon is among the most abundant elements found on the face of the earth, which delivers quite a few outstanding merits while being used in forms like multi-walled carbon nanotubes, single-walled carbon nanotubes, graphene, and graphene quantum dots. Such merits include high cyclic stability, low cost, non-toxicity, high conductivity, fast kinetics, and high-power density. Such is the high application potential of carbon-based materials in the integration of devices toward commercialization. Solar cells, as crucial components within

these integrated systems, directly impact the overall system performance. The incorporation of suitable carbon-based functional materials can greatly enhance the photoelectric properties of solar cells, creating a strong foundation for high-performance hybrid devices. For example, the addition of carbon nanotubes to the photoelectrode materials of solar cells can improve photovoltaic conversion efficiency by minimizing exciton recombination and promoting the rapid transport of photogenerated electrons. Moreover, graphene is recognized for its ability to inhibit back-transport reactions in solar cells, further enhancing energy conversion efficiency. These advancements not only elevate the performance of solar cells but also facilitate the development of more efficient and durable hybrid solar energy systems [53–55].

7. Metal-organic frameworks (MOFs)
   Research about MOFs as promising photoactive materials is aimed at improving energy storage in supercapacitors. MOFs are a class of porous materials consisting of metal ions or clusters linked by coordination bonds with organic ligands. A large surface area and adjustability of pore size render the MOF material an attractive candidate for photosupercapacitors. One of the main reasons for their demand is that MOFs possess a large-enough charge storage capacity and can be functionalized to enable them to cover a wide-enough range of light absorption, therefore making them efficient photoactive materials. On the other hand, they suffer from major stability and durability issues. MOFs have generally been rather sensitive to moisture and temperature changes, which may undermine their performance with time. In order to resolve these issues, researchers are working on novel synthesis methods and new encapsulation techniques that would increase the stability and durability of MOFs. One more problem that has been tagged with MOFs is that they are relatively poor electrical conductors when compared with a material like carbon nanotubes or graphene, which reduces their charge storage capacity and makes them inefficient as photoactive materials. To overcome this limitation, researchers are working on ways to improve the conductivity of MOFs—for example, by adding metal ions or embedding the MOFs in a conductive matrix. Lastly, it is the cost factor

that makes the production of MOFs really quite expensive and might affect their scalability and commercial viability. In fact, research in low-cost production techniques of MOFs is underway with alternative synthesis routes and scalable techniques.

Advantages of Photosupercapacitors compared to Traditional Batteries and other Energy Storage Devices

Photosupercapacitors, however, possess some unique benefits over other energy storage systems. They have high power density, thereby they are able to charge and discharge energy quickly at very high rates. In particular, where photobatteries could take hours to charge, a photosupercapacitor does this in seconds or minutes. Hence, they have useful applications in situations where sudden storage and release of energy is needed, such as in regenerative breaking of electric vehicles. Additionally, photosupercapacitors have an increased cycle life, unlike conventional batteries, which decrease in performance over time, and thus can be charged and discharged repetitively with minimal loss in performance. Finally, their eco-friendliness increases their bottled advantages. Photosupercapacitors utilize renewable energy sources, like sunlight, in generating electrical charge. This inherent dependence on renewables makes them environmentally friendly compared to traditional batteries, which are often prepared from non-renewable materials. Despite the obvious advantages of photosupercapacitors, several challenges need to be overcome before they become widely adopted. Notably, their energy density, the amount of energy stored per unit of volume, is currently lower compared with conventional batteries and hence might limit their applicability in certain uses. Much research in this line is still at work and thus holds much promise; we should expect more improvements in this technology.

Challenges of Photosupercapacitors

Some key challenges facing photosupercapacitors are that, although having high-power density, which makes them capable of charging and discharging very fast, their energy density is still relatively low compared with traditional batteries at present. Therefore, they might not be suitable for applications that require large amounts of energy to be stored, such as in an electric vehicle or grid-scale energy system. Other challenges that photosupercapacitors have are in terms of stability and durability. Over time, their interaction with light and water and temperature variations can lead to performance degradation and the resultant life reduction. Researchers work on discovering new materials and encapsulation technologies for more stability and durability. For the effective harnessing of photosupercapacitors, they need to be integrated into existing systems, whether power grids or electronic devices. This can be made possible through the development of complementary components and technologies. Safety is the most basic demand for any energy storage device and so for photosupercapacitors. From the very beginning of the design, safety considerations have to be taken into account in order to avoid overheating, short-circuiting, and other hazards that may be emulated out of these devices. New safety features of photosupercapacitors are being researched by scientists, such as self-healing materials and fail-safe mechanisms. This constitutes part of the complications, but continued research and development in this field succeeds in making headway toward the betterment of these issues. Therefore, provided sustained innovation and investment, photosupercapacitors will emerge to be one of the vastly adopted critical technologies for energy storage and conversion.

Different Applications of Photosupercapacitor

Even though photosupercapacitors are still in their infant stages, a number of their potent applications have currently appeared (see Figure 1.6). Herein, a few examples are given as illustration:

Energy harvesting: Photosupercapacitors act as efficient energy harvesters, which capture energy from the environment like sunlight or vibrations. This harvested energy gets stored within the supercapacitor and is used to run small electronic devices like sensors and wearables.

Smart windows: The windows, integrated with photosupercapacitors, will store the energy from the sun during the day. This energy is then released in the form of heat at night, thus considerably reducing the energy demand for heating and cooling.

Electric vehicles: Photosupercapacitors are a vital part of an electric vehicle since they store the energy obtained by regenerative braking. This stored energy can further be used to run the vehicle when it accelerates.

Consumer electronics: As they are being developed into energy sources, photosupercapacitors will find use in compact devices such as smartphones or laptops. Other than these uses, they can also be utilized in order to produce the energy needed for wearable gadgets like smartwatches and fitness trackers.

Military applications: Photosupercapacitors can be used for military applications by powering sensors or communication devices at far-off places.

Sensing and detection: Photosupercapacitors act as flexible devices in sensors and detectors. For instance, they may read pollutants or environmental conditions. For example, photosupercapacitors could measure air quality by quantifying pollutants, such as nitrogen oxides or volatile organic compounds.

Wearable electronics: Photosupercapacitors can perfectly fit into wearable devices like smartwatches or fitness trackers, providing power to the various sensors and other critical components. In this respect, their capability of fast and efficient charging offers a distinct advantage.

Environmental monitoring: Photosupercapacitors can power remote sensors and data loggers used in environmental monitoring. For instance, they may be powering temperature, humidity, or soil moisture sensors that are installed at a considerable distance from any nearby base.

Figure 1.6 Illustration of functions and application of integrated device [56–60].

Electrolytes for Photosupercapacitor

The selection of electrolytes plays a crucial role in determining the performance of photosupercapacitors. Important factors to consider include the voltage range, operating temperature, energy density, and specific attributes such as high ionic conductivity, stability, low resistance, and compatibility with photosensitive materials. There are three primary categories of electrolytes utilized in photosupercapacitors: liquid electrolytes, quasi-solid electrolytes (also known as gel electrolytes), and solid electrolytes.

1. Liquid electrolytes for photosupercapacitors
   Photosupercapacitors commonly employ two types of liquid electrolytes: aqueous and organic. Several electrolyte features, such as potential window, ion concentration, ion size, and operating temperature, significantly impact the performance of photosupercapacitors. For instance, adjusting the concentration and temperature of the electrolyte can reduce electrode material resistance in solar electrochemical capacitors, leading to increased specific capacitance due to enhanced ion diffusion within the electrolyte. Aqueous electrolytes used in photosupercapacitors fall into three categories: acidic, alkaline, and neutral. Among acid electrolytes, aqueous H_2SO_4 is commonly favored for its high ionic conductivity. Similarly, KOH and NaOH are widely employed as alkaline electrolytes due to their favorable ionic conductivity. It is essential to maintain an optimal concentration range for these electrolytes to ensure efficient ionic conductivity. In addition to acidic and alkaline aqueous electrolytes, neutral electrolytes are essential in solar electrochemical capacitors. Their benefits encompass a wide operating potential range, reduced corrosion, and the capability to provide high energy density. Specifically:

Potential window:

While acid and alkaline electrolytes are limited to a 1.3-V potential window for all electrode materials, neutral electrolytes extend this range to 2.2 V [56].

Organic electrolytes:

Some years ago, researchers explored organic electrolytes, such as iodide–triiodide (I^-/I_3^-), as alternatives to conventional aqueous electrolytes. These organic electrolytes exhibit favorable kinetic properties. For instance, I^- oxidizes rapidly at the electrolyte interface, facilitating efficient dye regeneration; I_3^- reduces slowly at the counter electrode interface, promoting high carrier collection. These organic electrolytes offer relative stability, cost-effectiveness, and ease of preparation [57]. Liquid electrolytes (including organic ones) are commonly used in electrochemical energy storage devices and solar cells to enhance ion mobility. However, their use may lead to leakage or evaporation, significantly impacting device performance. To address these challenges, photosupercapacitors often employ solid and quasi-solid-state electrolytes. These alternatives offer improved stability and safety.

2. Quasi-solid-state electrolytes (gel electrolyte)
   Challenges related to sealing and the long-term durability of liquid electrolytes pose significant limitations on the practical use of photosupercapacitors. As a result, researchers have investigated alternatives to liquid electrolytes, such as quasi-solid-state and gel electrolytes [58]. To mitigate issues of volatilization and leakage commonly associated with liquid electrolytes, ionic liquids and polymer gels containing redox couples are frequently utilized as quasi-solid-state or gel electrolyte options. A gel polymer electrolyte combines liquid-state components (such as aqueous, organic, or ionic electrolytes) with a solid-state polymer matrix, often referred to as a quasi-solid-state or gel polymer electrolyte. A variety of host polymers, including poly(vinyl alcohol), poly(vinylidene fluoride-co-hexafluoropropylene), poly(ethylene oxide), and poly(methyl methacrylate), can be utilized in gel polymer electrolytes. Among these, polyvinyl alcohol is a widely used host polymer for solar electrochemical capacitors, often combined with aqueous solutions such as strong acids (e.g., H_3PO_4, H_2SO_4), strong alkaline electrolytes (e.g., KOH), or neutral electrolytes (e.g., LiCl) to produce hydrogels. Despite the significant advantages and widespread use of gel electrolytes, they do have limitations. These include thermodynamic instability at high temperatures and the potential for solvent leakage. Consequently, meticulous sealing is necessary when employing gel electrolytes in high-temperature environments. As we look ahead, solid-state electrolytes will continue to be the primary focus of research and industrialization for photosupercapacitors.
3. Solid-state electrolytes
   Solid-state electrolytes have proven to be promising alternatives for the electronics industry, as they can be easily

packaged and produced in electrochemical solar capacitors. In particular, solid-state electrolytes composed of p-type semiconductor materials offer an effective solution to the challenges associated with photosupercapacitors based on liquid electrolytes. Solid-state electrolytes can be categorized into three main types: inorganic solid electrolytes, solid polymer electrolytes, and composite polymer electrolytes. At room temperature, solid-state electrolytes exhibit ionic conductivity within their solid structure. These electrolytes offer several advantages, including absolute safety, absence of toxic organic solvent leakage, low flammability, non-volatility, mechanical and thermal stability, ease of processing, low self-discharge, and the potential for higher achievable power density and cyclability. However, their primary limitation lies in their lower ionic conductivity compared to liquid counterparts. Commonly used inorganic hole-transfer materials with good conductivities include CuI, CuBr, $CsSnI_3$, and CuSCN [59].

Although solid-state-based photosupercapacitors demonstrate relatively low efficiency due to charge recombination occurring between the semiconductor and the hole-transfer material, they provide enhanced mechanical stability. In particular, CuI/CuSCN hole-transporting materials are known for their high hole mobility; however, their fast crystallization rates can result in insufficient filling within the photoanode films. This challenge contributes to a relatively low power conversion efficiency of 3.8% for these photosupercapacitors [60].

References

1. Poompavai, T. and Kowsalya, M., Control and energy management strategies applied for solar photovoltaic and wind energy fed water pumping system: a review. *Renew. Sust. Energ. Rev.*, 107, 108–122, 2019. https://renewablewatch.in/2016/11/07/emerging-renewables/.
2. Decarolis, J.F. and Keith, D.W., The real cost of wind energy. *Science*, 294, 1000–1001, 2001. https://ekbis.sindonews.com/berita/954868/34/alasan-chevron-mundur-dari-proyek-geothermal-ciremai.
3. Sun, H., You, X., Yang, Z., Deng, J., Peng, H., Winding ultrathin, transparent, and electrically conductive carbon nanotube sheets into high-performance

fiber-shaped dye-sensitized solar cells. *J. Mater. Chem. A.*, 1, 12422–12425, 2013. https://www.kenyaengineer.co.ke/water-energy-nexus-in-kenya/#google_vignette.

4. Carrillo, A.J., González-Aguilar, J., Romero, M., Coronado, J.M., Solar energy on demand: a review on high temperature thermochemical heat storage systems and materials. *Chem. Rev.*, 119, 4777–4816, 2019. https://group.intesasanpaolo.com/content/dam/portalgroup/repositorydocumenti/sostenibilt%C3%A0/inglese/Wind_Farm_Croatia.pdf.
5. Wang, L., Feng, J., Tong, Y., Liang, J.A., Reduced graphene oxide interface layer for improved power conversion efficiency of aqueous quantum dots sensitized solar cells. *Int. J. Hydrog. Energy*, 44, 128–135, 2019. https://www.ainonline.com/aviation-news/aerospace/2023-06-16/new-study-will-examine-sustainable-fuel-feedstocks?qt-latest_trending=0.
6. Jurado, U.T., Pu, S.H., White, N.M., Grid of hybrid nanogenerators for improving ocean wave impact energy harvesting self-powered applications. *Nano Energy*, 72, 104701, 2020. https://www.dreamstime.com/stock-photo-earth-isolated-white-background-soft-shadow-map-data-used-computer-generated-public-domain-www-image52804663.
7. Huang, S.P. and Liu, J.Q., Geothermal energy stuck between a rock and a hot place. *Nature*, 463, 293, 2010. https://www.wafaq.com.sa/solar-energy/.
8. Gong, J.L., Li, C., Wasielewski, M.R., Advances in solar energy conversion. *Chem. Soc. Rev.*, 48, 1862–1864, 2019.
9. Namsheer, K. and Rout, C.S., Photo-powered integrated supercapacitors: a review on recent developments, challenges and future perspectives. *J. Mater. Chem. A*, 9, 8248–8278, 2021.
10. Sankir, N.D. and Sankir, M. (Eds.), *Photoelectrochemical Solar Cells*, John Wiley & Sons, 2018.
11. Zhong, Y., Xia, X., Mai, W., Tu, J., Fan, H.J., Integration of energy harvesting and electrochemical storage devices. *Adv. Mater. Technol.*, 2, 1700182, 2017.
12. Luo, B., Ye, D.L., Wang, L.Z., Recent progress on integrated energy conversion and storage systems. *Adv. Sci.*, 4, 1700104, 2017.
13. Heo, Y.J. and Takeuchi, S., Towards smart tattoos: implantable biosensors for continuous glucose monitoring. *Adv. Healthc. Mater.*, 2, 43–56, 2013.
14. Chae, J.S., Park, S.K., Roh, K.C., Park, H.S., Electrode materials for biomedical patchable and implantable energy storage devices. *Energy Stor. Mater.*, 24, 113–128, 2020.
15. Kakiage, K., Aoyama, Y., Yano, T., Otsuka, T., Kyomen, T., Unno, M., Hanaya, M., An achievement of over 12 percent efficiency in an organic dye-sensitized solar cell. *Chem. Commun.*, 50, 6379–6381, 2014.
16. Graetzel, M., Janssen, R.A.J., Mitzi, D.B., Sargent, E.H., Materials interface engineering for solution-processed photovoltaics. *Nature*, 488, 304–312, 2012.

17. Rani, S., Suri, P., Shishodia, P.K., Mehra, R.M., Synthesis of nanocrystalline ZnO powder *via* sol-gel route for dye-sensitized solar cells. *Sol. Energy Mater. Sol. Cells*, 92, 1639–1645, 2008.
19. Bokhari, S.W., Siddique, A.H., Sherrell, P.C., Yue, X., Karumbaiah, K.M., Wei, S., Ellis, A.V., Gao, W., Advances in graphene-based supercapacitor electrodes. *Energy Rep.*, 6, 2768–2784, 2020.
20. Huang, X., Zhu, X., Luo, S., Li, R., Rajput, N., Chiesa, M., Liao, K., Chan, V., $MnO_{1.88}$/R-MnO_2/$Ti_3C_2(OH/F)_x$ composite electrodes for high-performance pseudo-supercapacitors prepared from reduced MXenes. *New J. Chem.*, 44, 6583–6588, 2020.
21. Reedijk, J., *Reference Module in Chemistry, Molecular Sciences and Chemical Engineering*, Elsevier, 2014.
22. Pal, B., Yang, S., Ramesh, S., Thangadurai, V., Jose, R., Electrolyte selection for supercapacitive devices: a critical review. *Nanoscale Adv.*, 10, 3807–3835, 2019.
23. Miyasaka, T. and Murakami, T.N., The photocapacitor: An efficient self-charging capacitor for direct storage of solar energy. *Appl. Phys. Lett.*, 85, 3932–3934, 2004.
24. Calandra, P., Calogero, G., Sinopoli, A., Gucciardi, P.G., Metal nanoparticles and carbon-based nanostructures as advanced materials for cathode application in dye-sensitized solar cells. *Int. J. Photoenergy*, 2010, 15, 2010.
25. Ng, C.H., Lim, H.N., Hayase, S., Harrison, I., Pandikumar, A., Huang, N.M., Potential active materials for photo-supercapacitor: A review. *J. Power Sources*, 296, 169–185, 2015.
26. Das, A., Deshagani, S., Kumar, R., Deepa, M., Bifunctional photo-supercapacitor with a new architecture converts and stores solar energy as charge. *ACS Appl. Mater. Interfaces*, 10, 35932–35945, 2018.
28. Gong, J., Liang, J., Sumathy, K., Review on dye-sensitized solar cells (DSSCs): Fundamental concepts and novel materials. *Renew. Sustain. Energy Rev.*, 16, 5848–5860, 2012.
29. Chang, Q., Ma, Z., Wang, J., Yan, Y., Shi, W., Chen, Q., Huang, Y., Yu, Q., Huang, L., Graphene nanosheets@ZnO nanorods as three-dimensional high efficient counter electrodes for dye sensitized solar cells. *Electrochim. Acta*, 151, 459–466, 2015.
30. Lee, Y., Chae, J., Kang, M., Comparison of the photovoltaic efficiency on DSSC for nanometer sized TiO_2 using a conventional sol–gel and solvothermal methods. *J. Ind. Eng. Chem.*, 16, 609–614, 2010.
33. Park, N.G., Van de Lagemaat, J., Frank, A.J., Comparison of dye-sensitized rutile- and anatase-based TiO_2 solar cells. *J. Phys. Chem. B*, 104, 8989–8994, 2000.
34. Fujishima, A. and Honda, K., Electrochemical photolysis of water at a semiconductor electrode. *Nature*, 238, 37–38, 1972.
35. Linsebigler, A.L., Lu, G., Yates, J.T., Photocatalysis on TiO_2 surfaces: principles, mechanisms, and selected results. *Chem. Rev.*, 95, 735–758, 1995.

36. Momeni, M.M., Aydisheh, H.M., Lee, B.K., Farrokhpour, H., Najafi, M., Preparation of photo-rechargeable asymmetric supercapacitors using S, W-codoped titania: Experimental and theoretical insights. *J. Alloys Compd.*, 960, 170722, 2023.
37. Momeni, M.M., Navandian, S., Aydisheh, H.M., Lee, B.K., Photo-assisted rechargeable supercapacitors based on nickel-cobalt-deposited tungsten-doped titania photoelectrodes: a novel self-powered supercapacitor. *J. Power Sources*, 557, 232588, 2023.
38. Lim, S.P., Pandikumar, A., Huang, N.M., Lim, H.N., Enhanced photovoltaic performance of silver@titania plasmonic photoanode in dye-sensitized solar cells. *RSC Adv.*, 4, 38111–38118, 2014.
39. Altaf, C.T., Coskun, O., Kumtepe, A., Rostas, A.M., Iatsunskyi, I., Coy, E., Erdem, E., Sankir, M., Sankir, N.D., Photo-supercapacitors based on nanoscaled ZnO. *Sci. Rep.*, *12*, 11487, 2022.
40. Solís-Cortés, D., Navarrete-Astorga, E., Schrebler, R., Peinado-Pérez, J.J., Martín, F., Ramos-Barrado, J.R., Dalchiele, E.A., A solid-state integrated photo-supercapacitor based on ZnO nanorod arrays decorated with Ag_2S quantum dots as the photoanode and a PEDOT charge storage counter-electrode. *RSC Adv.*, 10, 5712–5721, 2020.
41. Beke, S., A review of the growth of V_2O_5 films from 1885 to 2010. *Thin Solid Films*, 519, 1761–5171, 2011.
42. Sreedhara, M.B., Ghatak, J., Bharath, B., Rao, C.N.R., Atomic layer deposition of ultrathin crystalline epitaxial films of V_2O_5. *ACS Appl. Mater. Interfaces*, 9, 3178–3185, 2017.
43. Vijayakumar, Y., Reddy, K.N., Moholkar, A.V., Influence of the substrate temperature on the structural, optical and thermoelectric properties of sprayed V_2O_5 thin films. *Mater. Technol.*, 49, 371–376, 2015.
44. Beke, S., Korösi, L., Papp, S., Oszkó, A., Nánai, L., XRD and XPS analysis of laser treated vanadium oxide thin films. *Appl. Surf. Sci.*, 255, 9779–9782, 2009.
45. Aydisheh, H.M. and Momeni, M.M., Photoelectrodes based on selenium-polypyrrole-vanadium pentoxide nanowire films for high-performance lightweight symmetric photo-supercapacitors: a flexible photo-rechargeable electrical energy storage device. *Electrochim. Acta*, 467, 143066, 2023.
46. Renani, A.S., Momeni, M.M., Aydisheh, H.M., Lee, B.K., New photoelectrodes based on bismuth vanadate-V_2O_5@TiNT for photo-rechargeable supercapacitors. *J. Energy Storage*, 62, 106866, 2023.
47. Bhatt, M.D. and Lee, J.S., Recent theoretical progress in the development of photoanode materials for solar water splitting photoelectrochemical cells. *J. Mater. Chem. A*, 3, 10632–10659, 2015.
48. Dias, P., Lopes, T., Meda, L., Andrade, L., Mendes, A., Photoelectrochemical water splitting using WO_3 photoanodes: the substrate and temperature roles. *Phys. Chem. Chem. Phys.*, 18, 5232–5243, 2016.

49. Wang, P.Q., Bai, Y., Luo, P.Y., Liu, J.Y., Graphene-WO_3 nanobelt composite: elevated conduction band toward photocatalytic reduction of CO_2 into hydrocarbon fuels. *Catal. Commun.*, 38, 82–85, 2013.
50. Deng, S., Yang, Z., Lv, G., Zhu, Y., Li, H., Wang, F., Zhang, X., WO_3 nanosheets/g-C_3N_4 nanosheets nanocomposite as an effective photocatalyst for degradation of rhodamine B. *Appl. Phys. A*, 125, 44, 2019.
51. Zhang, T., Su, J., Guo, L., Hierarchical architecture of WO_3 nanosheets by self-assembly of nanorods for photoelectrochemical applications. *CrystEngComm.*, 18, 665–669, 2016.
52. Barik, R. and Ingole, P.P., Challenges and prospects of metal sulfide materials for supercapacitors. *Curr. Opin. Electrochem.*, 21, 327–334, 2020.
53. Wang, L.Q. and Feng, J.M., Improved energy conversion efficiency of CdS quantum dots sensitised solar cells using TiO_2/carbon nanotube nanocomposite photoanodes. *Mater. Res. Innov.*, 19, 41–46, 2015.
54. Habisreutinger, S.N., Nicholas, R.J., Snaith, H.J., Carbon nanotubes in perovskite solar cells. *Adv. Energy Mater.*, 7, 1601839, 2017.
55. Yun, S., Qin, Y., Uhl, A.R., Vlachopoulos, N., Yin, M., Li, D., Han, X., Hagfeldt, A., New-generation integrated devices based on dye-sensitized and perovskite solar cells. *Energy Environ. Sci.*, 11, 476–526, 2018.
56. de Freitas, J.N., Gonçalves, A.D.S., De Paoli, M.A., Durrant, J.R., Nogueira, A.F., The role of gel electrolyte composition in the kinetics and performance of dye-sensitized solar cells. *Electrochim. Acta*, 53, 7166–7172, 2008. https://no.words-sidekick.com/solar-plane-plans-ohio-detour-due-to-bad-weather5186#google_vignette.
57. An, H., Xue, B., Li, D., Li, H., Meng, Q., Guo, L., Chen, L., Environmentally friendly LiI/ethanol based gel electrolyte for dye-sensitized solar cells. *Electrochem. Commun.*, 8, 170–172, 2006. https://hop.bg/en/apple-products-117/apple-iphone-12-117%28id=56134%29.
58. Ivanova, S., Laguna, O.H., Centeno, M.A., Eleta, A., Montes, M., Odriozola, J.A., Microprocess technology for hydrogen purification, in: *Renewable. Hydrogen Technologies*, pp. 225–243, 2013. https://www.just-style.com/news/tommy-hilfiger-launches-solar-powered-jacket/.
59. Stergiopoulos, T., Rozi, E., Karagianni, C.S., Falaras, P., Influence of electrolyte co-additives on the performance of dye-sensitized solar cells. *Nanoscale Res. Lett.*, 6, 307, 2011. https://www.ecostars-europe.eu/en/reduce-transport-pollution/.
60. Li, B., Wang, L., Kang, B., Wang, P., Qiu, Y., Review of recent progress in solid-state dye-sensitized solar cells. *Sol. Energ. Mat. Sol. C*, 90, 549–573, 2006. https://www.amazon.co.uk/BELIYO-Fitbit-Versa-Sense-Replacement/dp/B0C3GCF6MP?th=1&psc=1.

2

Solar Rechargeable Capacitors and Photosupercapacitors

Nirmal Roy[1], Nandlal Pingua[2] and Rupam Sinha[2]*

[1]*Department of Electronics and Communication Engineering, Motilal Nehru National Institute of Technology Allahabad, Prayagraj, Uttar Pradesh, India*
[2]*Department of Chemical Engineering, National Institute of Technology Tiruchirappalli, Tamil Nadu, India*

Abstract
In the current era of rapid urbanization, the excessive use of nonrenewable energy sources like fossil fuels has turned out to be a major concern. Renewable and clean energy sources such as solar energy can be perceived as the greatest alternative to such energy crisis. Nonetheless, due to the inherent variability in solar energy, there is a huge mismatch between the actual energy consumption and the total solar energy production. This issue can be partially resolved by developing devices that can both convert as well as store solar energy effectively. One of the best ways to achieve this goal is by using photocapacitors and photo rechargeable supercapacitors, which can combine the characteristics of solar cells/photovoltaic cells and electrochemical energy storage systems like supercapacitors and batteries in a single device. Such a combination helps improve overall energy efficiency, specifically from the point of view of the rapid miniaturization of electronic devices. This chapter provides an overview of the current state of research on solar rechargeable capacitors and photosupercapacitors, highlighting their key features, potential applications, challenges, and future directions for research and development in this rapidly evolving field.

***Keywords*:** Photorechargeable, photocapacitor, solar energy, electrochemical energy storage, solar cells

**Corresponding author*: rupamsinha97@gmail.com

Nurdan Demirci Sankir and Mehmet Sankir (eds.) Solar Capacitors and Batteries, (31–54) © 2025 Scrivener Publishing LLC

2.1 Introduction

Nobel laureate in chemistry, Richard Errett Smalley, asserted that energy is the most crucial issue confronting humanity and represents the largest enterprise globally. This underscores the fundamental importance of energy sources in human society, with vital roles across various sectors including industry, agriculture, and transportation. However, the rapid industrialization and population growth have led to excessive energy consumption in recent decades. It is further intensified by the ever-increasing global energy demand, which is anticipated to increase by approximately 60% within the next decade and double by 2050 [1, 2]. Regrettably, global energy production and consumption still rely heavily on nonrenewable fossil fuels, including natural gas, coal, and oil. This is likely to result in an energy crisis in the future. Additionally, these fossil fuels release different types of toxic and greenhouse gases during their combustion (e.g., CO, CO_2, NO_x, SO_x, etc.), contributing to severe environmental problems like global warming, deterioration of water quality, and El Niño [3, 4].

A sustainable development strategy that combines renewable and clean energy sources is considered the most promising remedy for simultaneously addressing the energy crisis as well as environmental pollution. Energy and material science's advancement has led to a comprehensive investigation of numerous sustainable energy sources, including solar power, hydropower, wind power, ocean power, biomass power, and geothermal energy (Figure 2.1a) [2].

Solar energy emerges as a highly cost-effective and sustainable alternative to fossil fuels, making it one of the most innovative and competitive energy solutions available. Aside from offering multiple advantages, it is also a natural resource that can be used as a base for other types of renewable energy since it is abundant, is environmentally friendly, is easy to harvest, and has a wide range of uses [2, 5, 6]. The amount of solar radiation that reaches the Earth's surface each year is about 5.4×10^{21} kWh, which is nearly 5000 times the world's annual energy consumption. In other words, just 2% of this energy would be enough to supply all of the energy needs of an entire society on an annual basis. Additionally, the use of solar energy produces almost no greenhouse gases, which is a significant benefit for the environment. A major advantage of solar energy over traditional fossil fuels is that it is equally accessible throughout the world, making it a suitable alternative source of energy for remote locations (Figure 2.1b).

Despite its many benefits, solar energy does possess inherent limitations that greatly hamper its application on a large scale and over a long period

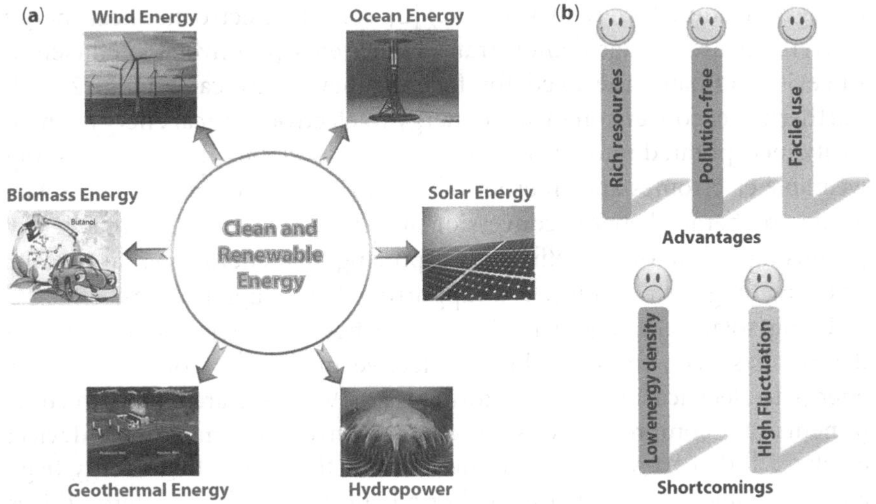

Figure 2.1 (a) Various sources of clean and renewable energy. (b) Solar energy advantages and disadvantages. Reprinted with permission (Licensed under CC-BY 4.0) [2].

of time [7, 8]. Despite its low energy density, the solar system only emits about 1 $kWcm^{-2}$ on its surface, which represents one of its drawbacks. In comparison with other various energy sources like wind energy, geothermal energy, and fossil fuels, this level of energy consumption is considerably lower. As a matter of fact, solar energy is inherently variable and unstable, as it is greatly affected by weather conditions and the natural day–night cycle (Figure 2.1b). The actual use of solar energy is severely hampered by such inadequate stability and energy density. The largest obstacle to the commercial use of solar energy has been thought to be its variable character, which typically results in a poor fit with the real energy demands of civilization [2, 9, 10]. Therefore, looking for practical solar energy storage technologies has become a research hotspot in order to establish a reliable energy supply.

Storing solar energy directly is challenging and impractical. Typically, solar energy is transformed into other forms before storage [2, 11]. Several prominent methods exist for converting and storing solar energy, including solar-to-chemical, solar-to-electrochemical, solar-to-thermal, and solar-to-biomass energy conversion and storage [2, 12–15]. The solar-to-electrochemical approach offers distinct benefits. Compared to the methods of solar-to-biomass and solar-to-thermal energy conversion, it provides a simpler and more cost-effective energy storage process. In contrast to solar-to-chemical technique (e.g., solar-powered hydrogen production, nitrogen fixation, CO_2 reduction, etc.) [2], solar-to-electrochemical energy conversion and storage demonstrates higher energy conversion efficiency, since it eliminates the need for forming new chemical bonds [2, 16]. Furthermore, conversion of solar energy to electrochemical energy can be easily incorporated into a variety of recently developed devices, including wearable electronics and Internet-of-Things (IoT) devices [17, 18]. Solar-to-electrochemical energy conversion and storage has emerged as a highly promising technology for efficiently capturing solar energy on a large scale, ensuring long-term storage, and supporting diverse applications.

Using solar cells together with electrochemical energy storage (EES) devices has been found to be an effective method of converting solar energy to electrical energy and storing it [2, 19]. There are two main components of a comprehensive solar energy conversion and storage device: a solar cell that harvests energy and a device that stores that energy (e.g., a rechargeable supercapacitor [SC] or a battery). A hybrid system utilizes solar cells to capture solar energy and convert it into electricity and store it in an EES. Photocharging is the process of combining these two processes together in order to acquire electrochemical energy. After this, the electrochemical energy is released during the discharge process of the EES unit to power other electrical devices, and this arrangement is termed as a photocapacitor (PC) or photosupercapacitor. Integrated solar energy conversion and storage devices, such as solar cells and rechargeable batteries, work efficiently and are easy to use. Because of these advantages, they have great potential in many emerging fields. An example would be combining compact and flexible integrated devices with wearables or implanted sensors to create a portable and self-powered electronic device.

With all the preceding context, this chapter provides a detailed exploration of the characteristics of PCs, and their applications, aiming to inspire further innovation in this emerging domain. With continued developments in PC technology, we will be able to accelerate the global shift to renewable energy while discovering new applications for them across a variety of industries, with the ultimate goal of contributing to a sustainable future.

2.2 Applications of Photorechargeable Capacitors and Photosupercapacitors

Despite their widespread use due to high energy density, lithium-ion batteries (LIBs) face challenges such as elevated costs, safety issues, limited durability, and environmental concerns associated with lithium and cobalt extraction [2]. Additionally, recycling difficulties result in billions of discarded units yearly. SCs, utilizing the electrical double layer (EDL) principle, present an alternative for rapid and secure energy storage, in spite of having lower energy density than LIBs [20]. Their quick charge and discharge capabilities make them suitable for high-power, brief-duration uses.

Combining SCs with photovoltaic (PV) cells results in PCs, enabling sustainable, self-reliant energy systems powered by solar or indoor illumination. PCs show potential for various applications, including wearable electronics, biosensors, IoT devices, and even power grids [21–25]. They surpass LIBs in stability and can operate small devices under ambient light, with indoor light sources such as LEDs providing adequate irradiance for low-power electronics.

Contemporary PCs are evolving with designs like the hybrid four-terminal (4-T) and integrated three- and two-terminal (3-T, 2-T) configurations (Figure 2.2). Integrated designs eliminate external wiring requirements and enhance compactness, making them ideal for future innovations. In IoT ecosystems, PCs can power microprocessors and sensors with minimal energy needs. For instance, indoor PVs under ambient light conditions achieve power conversion efficiencies (PCE) approaching 35% [26], sufficient to meet IoT device demands. Progress toward a PCE of 50% could further optimize PCs for ambient energy harvesting [25].

PCs provide a portable and reliable alternative to bulky battery systems. It is also possible to use them in combination with batteries in hybrid

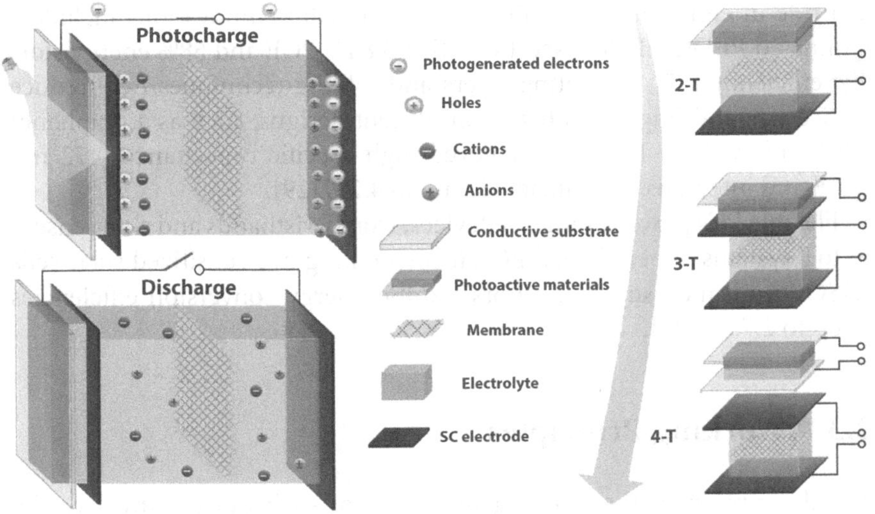

Figure 2.2 An illustration of a PC's photocharging and dark discharge and a PC integration into different terminal (T) configurations: 2-t, 3-t, 4-t. Reprinted with permission (Licensed under CC-BY 4.0) [25].

setups to enhance energy supply, particularly in the case of high-power systems. Ongoing research in material engineering and device integration will drive PC efficiency and adoption, aligning with the shift toward renewable energy and sustainable technologies.

In the present era, energy storage in the form of flexible PCs is a key area of research, providing lightweight and versatile solutions for diverse applications [25]. A significant advancement in flexible PV cells has resulted in the development of flexible PCs using materials, including polyethylene terephthalate, polyimide), and polyethylene naphthalate. These materials enable the creation of devices with the necessary flexibility for wearable electronics, smart textiles, and IoT applications. Despite their lower energy conversion and storage efficiency, flexible personal computers are on the verge of becoming more efficient due to innovative designs and manufacturing techniques.

Slot-die technology and roll-to-roll coating techniques are typically used in the manufacture of flexible computer monitors, as well as 3D, inkjet printing, and spray processing [25, 27]. This technique allows for the deposition of materials with tailored pore structures, improving charge storage and surface area [25]. While porous carbon and gel electrolytes have been successfully incorporated, developing appropriate inks for 3D printing remains challenging. Flash infrared annealing (FIRA) is a promising low-temperature method for processing components like TiO_2, addressing the limitations of high annealing temperatures incompatible with flexible substrates [25, 28]. Conductive polymers, such as PEDOT:PSS, play a crucial role in enhancing flexibility and functionality. Innovations include a laminated PEDOT:PSS-based PC with a 3.84% PCE and 58% energy storage efficiency [25]. Insulating layers and folding techniques help reduce device size and improve integration. Recent designs, such as a graphene/CNT hollow fiber PC, have achieved high specific capacitance (472 mF cm^{-2}) and integrated PV efficiencies up to 4.2% [29].

Flexible PCs power a range of devices, from wristbands and skin sensors to IoT systems. Fiber-shaped PCs, incorporating dye-sensitized solar cells (DSCs) and microsupercapacitors, exhibit energy conversion efficiencies of up to 2.73% [25].

2.3 Working Principles

The description of the working principles of such devices involves the understanding of the working mechanism of PV cells, SCs, and their integrated arrangements. This section discusses the working mechanism of each section in detail.

2.3.1 Photovoltaic Cells

The PV effect is one of the fundamental principles underlying solar or PV cell technologies in that it is capable of converting light into electricity. In semiconductor materials, when light is incident on them, photons with energies equal to or greater than their bandgap absorb the photons. This causes electrons or other particles to gain more energy and move to a higher level [30, 31]. The light creates electron–hole pairs, which then separate. The electrons flow as electric current through a circuit, while the holes move to the positive side [31]. Solar energy has long been dominated by technologies based on silicon, such as crystalline and amorphous silicon (c-Si and a-Si:H, respectively). The PCE of c-Si, which is 26.7% under standard illumination conditions, is the highest among all semiconductors, with a bandgap of 1.1 eV [31]. A-Si:H exhibits a lower efficiency of 11.9% due to its 1.6-eV bandgap; however, its simpler manufacturing process makes it a more cost-effective option [31]. For certain applications, a-Si:H, however, can deliver an impressive 21% PCE under less intense light (1000 lx).

The second generation of PV technologies including materials such as GaAs, CdTe, and CIGS has also emerged as a promising alternative. These materials offer higher efficiencies compared to their silicon-based counterparts, but their widespread adoption has been hindered by their higher costs and limited resource availability. Consequently, researchers have been exploring ways to improve the cost-effectiveness and scalability of these advanced PV technologies.

The landscape of PV technologies has evolved significantly, with the emergence of novel approaches that offer distinct advantages over traditional silicon-based solar cells. One such category, referred to as third-generation or emerging technologies, includes perovskite solar cells (PSCs), organic PVs, and DSCs [32]. The advantages of these technologies include that they can be prepared at a minimal cost, as they employ easily available materials, which makes them particularly suitable for industrial production on a large scale [31].

A prominent example of this third-generation technology is organic solar cells, which are composed of semiconductors (carbon-based) like conjugated polymers. When light hits organic semiconductor films, it creates tightly bound excitons that have a short distance to move. A benzodithiophene, difluorobenzothiadiazole, or fullerene-based organic semiconductor is an electron donor, while a fullerene-based semiconductor is an electron acceptor. In organic solar cells, the donor and acceptor molecules may be deposited in a planar structure in order to form a heterojunction, or the molecules could be mixed in order to reduce exciton separation [25]. As organic solar cells are readily available, nontoxic, and manufactured in low-cost, low-temperature processes, it is feasible to print on a larger surface area and on stretchy materials [33, 34]. Organic solar cells have been developed continuously over the past decade with single-junction efficiencies reaching up to 18.2% and 28.8% under one sun illumination [25] and under an LED lamp with 1000 lx illumination intensity (2600K), respectively [25].

PSCs have recently received considerable interest from academic researchers and industry professionals, primarily due to their straightforward manufacturing process based on simple solutions. These cells in tandem arrangements with silicon solar cells have achieved remarkable PCE of over 31%, compared with 25% in single junction configurations [25]. Perovskites, such as $CH_3NH_3PbI_3$, which are light-absorbing organic halides, have several distinctive characteristics, including high extinction coefficients, composition-dependent bandgap tunability, lower rates of carrier recombination, and micrometer-scale carrier diffusion lengths [35–37]. Typically, PSCs consist of metal contacts, a hole transport material (HTM), and an electron transport layer (ETL). By absorbing light, the ETL generates charge carriers that can be used as light-harvesting components. The electron–hole pair is then separated, and the hole carriers are collected by the HTM. Over the years, researchers have been focusing on a variety of architectural designs, including mesoscopic, planar, and inverted ones. Under indoor lighting conditions of 1000 lx, perovskite PVs can achieve PCE values exceeding 40% [25].

DSCs were the pioneers in the integration of energy storage devices into PCs, which was among the third generation of PV technologies [38–40]. Their appeal lies in their uncomplicated design, use of abundant and nontoxic materials, and straightforward manufacturing process [25], resulting in cost-effective production. With DSCs, substrates can be selected from rigid or flexible options such as lightweight plastics, wires, fibers, or metal foils. Furthermore, active materials can also be opted with adjustable properties like transparency and color [41, 42]. DSCs are composed of a semiconductor with a broad energy gap like TiO_2, coated with dye molecules (sensitizers), a redox mediator, and a counter electrode (CE). Photoelectrons are generated by dye molecules when exposed to light, and these electrons are then injected into the conduction band of TiO_2. Redox shuttles are important for the regeneration of dye and transfer of holes between the working electrode and the CE.

With recent advances in molecular systems, and new HTMs, PCEs have increased to 15% under AM1.5G [25]. Among the notable achievements is the 34.5% efficiency achieved under 1000-lx ambient light [25]. Since DSCs perform exceptionally well in low-light conditions, they have great potential for developing ultra-efficient PCs that can run electronic devices in ambient conditions [25].

Nanocrystalline semiconductor quantum dots (QDs) are used as photoactive materials or sensitizers in QD solar cells (QDSCs) [43–45]. QDs have special properties that change with their size, such as adjustable bandgaps. This allows them to absorb light across a wide range of wavelengths and generate multiple excitons from a single absorbed photon. These materials can be processed using simple solution-based methods, making them easy to work with. Additionally, they can be manufactured at relatively low temperatures, which helps reduce energy costs and simplifies production [46]. Studies have shown that QDSCs have reached a record efficiency of up to 18.1% by using lead-free and cadmium-free QD sensitizers [25].

2.3.2 Supercapacitors

Batteries, traditional capacitors, and SCs or ultracapacitors are the primary energy storage technologies utilized in various applications. Many devices can be powered by these technologies, such as power generation systems, household appliances, electronic devices, wireless gadgets, computers, chargers, stationary grid storage, electric vehicles, industrial equipment, and medical sensors. These technologies are used for a wide range of devices. A primary distinction between batteries and SCs can be found in their capabilities: batteries are capable of storing greater amounts of energy per unit weight, typically several hundred watt hours per kilogram, while SCs are capable of delivering a greater amount of power in a short period of time [47].

An SC stores and releases charge through the formation of EDLs [48]. An ionic electrolyte layer (Helmholtz layer) compacts at the electrode interface during charging, assembling counter charges. This is followed by a diffuse layer of charges located near the Helmholtz layer. At the interface between electrolyte and electrodes, faradaic charges accumulate pseudocapacitively due to redox reactions [49]. The rapid charging and discharging rates of SC make them ideal for regenerative brakes, as well as fast-charging backup power systems that rely on batteries to operate. One of the primary advantages of SCs is that they can provide short bursts of high power for a short period of time.

In contrast to this, batteries store charge when they undergo diffusion-controlled faradaic oxidation and reduction at a controlled rate [50]. A typical faradaic battery (FB) battery will take hours to charge and discharge due to ion diffusion across the active material layer, whereas a typical SC will charge and discharge within seconds. Batteries usually have a high energy density due to this difference. This makes them ideal for long-term power applications. Devices like mobile phones and laptops need a lot of energy, so they benefit from high-energy-density batteries.

An SC cannot use toxic or limited components like lithium or cobalt, both of which are a major concern in terms of a battery's sustainability. SCs can be assembled using materials that are less toxic or limiting. Furthermore, SCs require minimal maintenance and boast a very long operational life of over a million charge–discharge cycles, meaning that they can last for a very long time without deterioration. In addition, batteries can function safely across a wide temperature range. They work well from -40 to +60°C. There is no significant loss in performance within this range [51, 52]. The use of SCs with batteries is a very cost-effective way of enhancing and extending battery life, and in some cases, they can serve as an adequate replacement for batteries in certain situations. One possible approach is to pair an SC with a PV unit to create a PC. This device captures light, stores the generated charges, and works similarly to a battery.

Batteries and SCs can be characterized using a three-electrode electrochemical cell setup. This setup includes the active material as the working electrode. It also consists of a CE and a reference electrode. In symmetrical SCs, active materials can be deposited on two electrodes or asymmetric SCs can be tested by testing different materials inside each electrode. There is a wide range of capacitance values offered by activated carbon in commercial SCs, which range from 100 to 400 Fg^{-1} [53, 54]. Various materials, such as carbon nanotubes (CNTs), graphene, reduced graphene oxides, and transition metal oxides and chalcogenides (e.g., RuO_2, MnO_2, MoS_2), can enhance capacitance. However, they are still difficult to synthesize, costly to produce, and challenging to scale up for practical use. A carbon-based material has a very high surface area (up to several thousand m^2g^{-1}), as a result of which it can achieve a high capacitance and power [53, 54]. Conducting polymers like PEDOT, PANI, and PPy also exhibit pseudocapacitive characteristics [55–57], enabling reversible doping by electrolyte anions [25] and providing high electrical conductivity, chemical stability, and rapid charge–discharge kinetics. However, their tendency to swell and shrink during cycles hinders supercapacitance, impeding commercialization efforts.

Unlike symmetric SCs, hybrid SCs incorporate separate materials for the anode and cathode, each incorporating a different charge-storage process to achieve the desired output [25]. Combining pseudocapacitive and EDL electrodes improves performance and operating voltage. The EDL electrode offers high power capability, while the pseudocapacitor electrode provides high energy density, allowing for a longer discharge time.

The electrolytes also play a critical role in separating charges, and their selection has a significant impact on performance and operating window. In SCs, aqueous electrolytes, commonly used for their low cost, high ionic conductivity, and low toxicity, are restricted to 1.2 V due to the splitting of water. The voltage of asymmetric SCs can exceed 1.2 V even with aqueous electrolytes. Since organic electrolytes have reduced ionic conductivity, they have a voltage range of 2.5–3 V and a longer life cycle. However, they also have a lower specific capacitance due to the decreased voltage range. A potential window of 4.5 V can be extended for ionic liquid electrolytes.

An SC containing gel or solid electrolytes is typically preferred in design for its durability, structural stability, and flexibility during cycling [58]. Various adhesives and polymeric substances can be utilized to enhance the adhesion of hydrogels to electrode surfaces. A few of these materials are alginate, carboxymethyl cellulose, xanthan gum, chitosan, polyethylene glycol, polymethyl methacrylate, poly(acrylic acid), poly(amine-ester), and polyvinyl alcohol [25, 59].

An important component of any high-performance SC is the membrane, which serves as a barrier between electrodes to prevent recombination of stored charges and short circuits. Ion diffusion in the electrolyte should be minimized by the membrane by being nonconductive. Chemical and thermal stability, inertness to electrode materials, electrolyte components, or redox mediators, and thermal resilience are all important characteristics of these materials. To prevent shrinkage, it is essential to have a high degree of mechanical stability. Additionally, wettability determines the long-term cycling stability of the SC unit, which in turn affects internal resistance and, subsequently, the life of the system [25].

Depending on the size of the pores in the membrane, the conductivity and compatibility of the membrane with the electrolyte solvent can be affected. In order to achieve optimal swelling, pores should be approximately 1 micron in diameter, and overall porosity should be between 40% and 60% [25]. It has been recognized that there are many different types of SC membranes, including glass, various types of paper, ceramic materials, ion-exchange membranes, such as Nafion, polyvinylidene fluoride (PVDF), PVDF-high purity (PVDF-HP), polyethylene, polypropylene, crosslinked polymers, chitosan, cellulose, and chitin, among many others [25]. Although thin membrane films do not necessarily lack mechanical strength, they may have poor cyclability and a degradation of mechanical strength over time. The use of certain solid electrolytes previously mentioned can eliminate the need for membranes to serve as separators between electrodes in applications that use solid electrolytes.

2.3.3 Integrated Photosupercapacitor

The rapid advancement of portable and wearable electronics has driven an increasing need for efficient and adaptable energy harvesting and storage solutions. SCs have long been known for their high power density and long cycle life, while PCs offer the same light-harvesting capability as PVs [60–63].

The energy storage setup and the PV unit are the core components of PCs as shown in Figure 2.2, where light is converted into electrical energy [25]. Typically, these units are manufactured using either rigid or flexible platforms that feature transparent conductors such as indium tin oxide or fluorine-doped tin oxide. In PC applications, dyes, QDs, perovskites, and organic semiconductors have been deposited on such substrates [60]. The PV unit is then integrated with an SC section, where the generated electrical energy is stored.

SCs in PCs are typically constructed of "Janus"-type electrodes that are coated with capacitive materials (such as carbon, metal oxides, chalcogenides, and conducting polymers) on one side and serve as energy storage layers on the CE. A membrane separates the electrodes, preventing charge recombination, and the device is filled with an ion-conducting electrolyte that allows ions to flow between the electrodes.

The integration of PVs and SCs enables the harvesting of light energy and its immediate storage, addressing the intermittency of renewable energy sources. The high power density and long cycle life of SCs make them well suited for applications requiring rapid energy discharge and recovery, such as in portable electronics and hybrid vehicles. The PC is a revolutionary hybrid technology that combines PVs' light-harvesting capabilities with SCs' energy storage capabilities, ushering in a new era of solar energy conversion and storage. During photocharging, the PV module absorbs light, causing charge separation in the photoactive material (e.g., dye, organic semiconductors, QDs, or metal-halide perovskite). LUMO and conduction bands formed by dyes, QDs, and perovskites are injected into conduction bands formed by metal oxides, such as TiO_2, and transported to the current collector by photogenerated electrons. Photoactive materials contain holes that are simultaneously scavenged by the hole transport layer and reduced by the redox electrolyte species.

SCs use the electrons generated by PV electrodes to store charge, balancing it with cations in the electrolyte of the SC to allow PCs to work. A PC is usually discharged by disconnecting the PV unit from the SC to power an electronic device with the energy stored in the capacitor [64]. A rectifying diode is often placed between the PV unit and the external load. Its purpose is to stop charges from flowing back into the PV unit. This helps prevent charge recombination, which could reduce the system's performance. In spite of these challenges, many researchers are avoiding the use of the diode because of the voltage drop caused by it, which is typically several hundred millivolts, which could impede the overall efficiency of the circuit [65].

A novel material or design of PC with a diode-like terminal and minimal voltage loss is crucial to addressing this issue and to advancing the field of PCs [60, 64]. As the voltage drop during the dark discharge is as small as possible, the PC is able to maximize the energy extraction during the dark discharge, relying on the ability to minimize the voltage drop, which avoids the self-discharge of the SC into the electrolyte, thereby delivering the full power potential to the external electronic device at the same time.

A straightforward way to charge an energy storage system, such as batteries or SCs, is by connecting it to an energy generation system, such as solar cells or thermo-piezo-tribo-electric devices. This can be done using a 4-terminal (4-T) configuration, where the two separate devices are linked through an external wire. This setup allows efficient energy transfer from the generation system to the storage unit, ensuring effective power utilization. Although a 4-T configuration allows for independent enhancement of each unit, it has several disadvantages. The most significant of these are complicated packaging, energy losses due to the external resistance of wires, and reduced overall performance [25]. Furthermore, these externally connected systems are often bulky and inflexible, limiting their potential applications.

The use of tandem PVs is analogous to that of PCs, which can be integrated in a 2-terminal (2-T) arrangement, more commonly known as a monolithic arrangement. A combination of both PV and SC materials can be used in this setup, but they will be either stacked together or separated by a membrane [66]. The first 2-T PC was introduced by Miyasaka *et al.* [66] in 2004. They used a DSC as well as a carbon-based SC CE to design their device. With one sun illumination, the system was able to demonstrate a photocharged voltage of 0.45 V and a capacitance of 0.69 F cm^{-2}, both under photocharged light conditions.

It is advantageous to use a 3-T layout rather than a 2-T layout due to the fact that fewer materials and substrates are required. A two-terminal architecture is generally not preferred due to limitations including low charging voltages, high resistances, poor charge–discharge efficiency, self-discharge to the photoactive side of the device, and a high resistance [40].

Considering these constraints, it was necessary to develop a 3-terminal (3-T) architecture that would allow both PV and SC devices to share a single electrode yet function independently of each other [25]. Initially, a 2-T structure was used to demonstrate a PC using a DSC as the photogeneration device in 2004 [25, 67, 68]. Before the 3-T structure became increasingly effective. In the 3-T design, electrons and holes were transferred more efficiently during charge–discharge, resulting in photocharged voltages of 0.8 V and 47 μWh cm^{-2}. In the decades that have followed, a wide range of materials and configurations have been investigated [25, 67, 68].

In most cases, the voltage stored in the unit after photocharging is less than 1 V, which may not be enough for some devices like sensors or low-power electronics. To increase the voltage, effective solutions include connecting multiple PV units in series, building a system that combines PV modules and SCs, and using microsupercapacitors with a larger surface area for better energy storage. Using a 2-T configuration PC, Sun *et al.* [69] demonstrated an extremely high energy density (up to 32.3 Whcm-2) as well as an elevated output voltage of approximately 2 V. By combining the PSC with the micro-SC, it successfully powered an array of red and white LEDs, a micromotor, and a timer. A 3-T integrated system reported by Gao *et al.* [70] was capable of 1.8 V across the open-circuit voltage, 4.7 V across the open-circuit voltage, 0.18 Wh m-2 occupancy, and a 7.0% efficiency.

2.4 Techniques for Performance Analysis

The overall efficiency of the combined PC is determined by multiplying the individual efficiencies of the solar cell and SC components. The PCE is measured using standard current–voltage (I–V) characteristics, which are calculated using the following equation [25]:

$$PCE = \eta_{PV} = \frac{V_{OC} J_{SC} FF}{P_{in}} \times 100 \tag{2.1}$$

where V_{OC}, J_{SC}, fill factor (FF), and Pin stand for the open-circuit voltage, short-circuit current density, and incoming light power density, respectively. In Figure 2.3, V_{OC} is the voltage at which open-circuit conditions are present (at J = 0). The J_{SC} simultaneously displays the maximum current that the solar cells can produce at zero bias.

The capacitance of the SC is another important factor in influencing the integrated PC module's overall efficiency. The capacitance of a charged electrode can be estimated using the contributions of the Helmholtz double layer. According to the Stern model, ions are treated as point charges. These ions are attracted to the inner layer of the electrode by electrostatic forces.

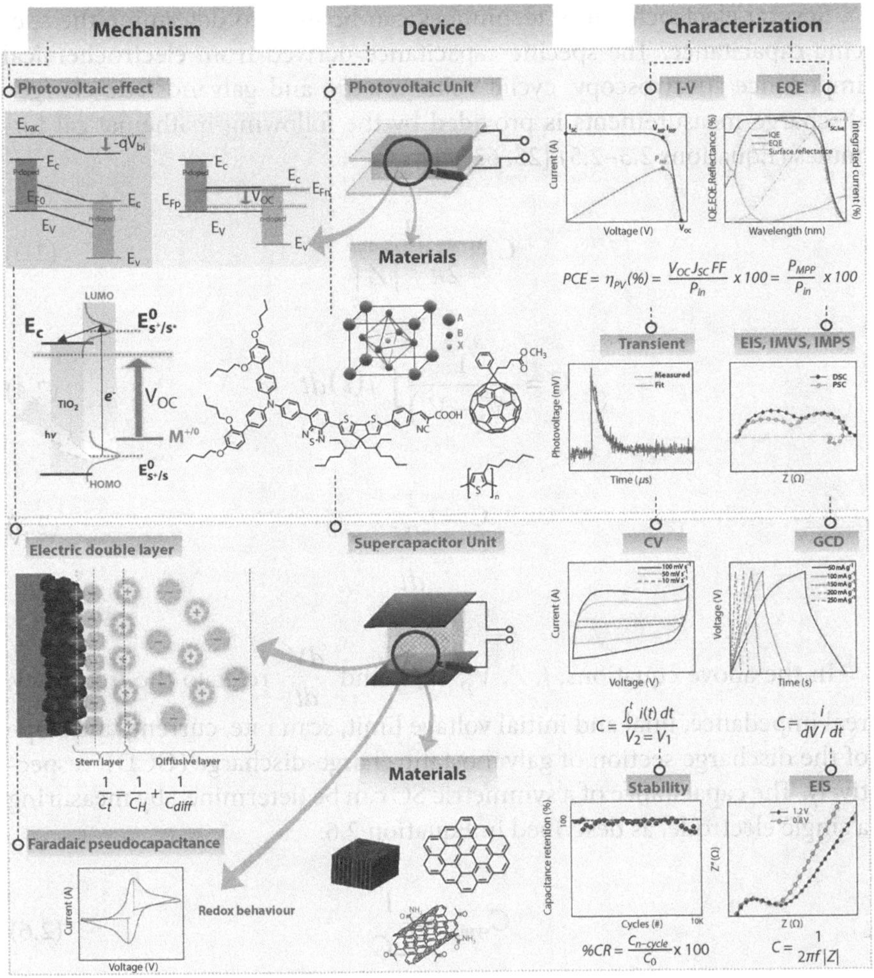

Figure 2.3 A schematic guideline illustrating the working mechanism, materials, and characterization methods of PV and SC devices. Reprinted with permission (Licensed under CC-BY 4.0) [25].

Near the electrode, there is a diffusive layer made up of solvated anions and cations. This layer helps to balance the charge on the electrode. The diffusive layer capacitance is represented by C_{diff} in Equation 2.2 [25] and the inner layer capacitance by C_H.

$$\frac{1}{C} = \frac{1}{C_H} + \frac{1}{C_{diff}} \tag{2.2}$$

Several electrochemical techniques can be used to determine the specific capacitance. The specific capacitance derived from electrochemical impedance spectroscopy, cyclic voltammetry, and galvanostatic charge–discharge measurements is provided by the following mathematical formulas (Equations 2.3–2.5) [25, 63, 71].

$$C = \frac{1}{2\pi f \,|Z|} \tag{2.3}$$

$$C = \frac{1}{(V_2 - V_1)} \int_0^t i(t)\,dt \tag{2.4}$$

$$C = \frac{i}{\frac{dV}{dt}} \tag{2.5}$$

In the above equations, f, Z, V_2, V_1, i, and $\frac{dV}{dt}$ refer to the frequency, real impedance, final and initial voltage limit, scan rate, current, and slope of the discharge section of galvanostatic charge-discharge (GCD), respectively. The capacitance of a symmetric SC can be determined by measuring a single electrode, as described in Equation 2.6:

$$C_{symm} = \frac{1}{2C} \tag{2.6}$$

The output energy (E_{Out}), energy density (E_A, in Wh g^{-1}) normalized by the mass loading, and power density (P_A, in W g^{-1}) of the SC during charging can all be determined using Equations 2.7–2.9 [25], where R stands for the SC's internal resistance and V is for the potential window.

$$E_{Out} = \frac{CV^2}{2} \tag{2.7}$$

$$E_A = \frac{1}{2}\frac{CV^2}{3600m} \tag{2.8}$$

$$P_A = \frac{E_A}{\Delta t} = \frac{V^2}{4R} \tag{2.9}$$

Using Equation 2.10, where A_{PV} represents the surface area of the solar cell and E_{in} represents the energy received while charging (Equation 2.11), the overall efficiency of the combined PC module can be determined. The efficiency is calculated using the following formula:

$$\eta_{Overall} = \frac{E_{Out}}{E_{in} A_{PV}} \times 100 \tag{2.10}$$

$$E_{in} = P_{in} t \tag{2.11}$$

The PCE, SC performance, and their interface, all of which are critical for sophisticated stacking systems, which determine a PC's overall efficiency. PCs exhibit a maximum efficiency that depends on their PV efficiency, assuming the SC module displays a constant capacitance. Therefore, if a PC device is to be effective, an efficient solar cell must also be used.

2.5 Future Prospect and Conclusions

The integration of solar cells and photoanodes with energy storage units has attracted global interest as an electrochemical method for converting and storing solar energy. These devices offer potential benefits such as high efficiency, diverse applications, and portability. Considerable research has been dedicated to developing and constructing these advanced integrated systems that merge solar energy harvesting and storage processes.

PVs, photoelectrochemicals, and SCs are among the photorechargeable integrated devices analyzed in this chapter. A significant contribution of these works is the conversion and storage of solar energy to electrochemical energy, as well as the miniaturization, portability, and wearability of integrated devices. In particular, carbon-based functional materials have been found to play important roles in cutting-edge hybrid solar energy systems as conductive additives, active materials, electrode stabilizers, flexible substrates, as well as shared electrodes that provide both environmental and energy benefits.

Despite these accomplishments, research in this field remains in its early stages. Many current studies focus primarily on designing and demonstrating integrated device prototypes, often neglecting efficiency improvements and practicality enhancements, which are essential for real-world applications. To progress from laboratory to commercial use, efforts should simultaneously address efficiency improvement and practicality enhancement. The following are a few recommendations for developing photorechargeable integrated devices in the future:

(i) Device coupling: Currently, the low overall efficiency of the devices taking advantage of solar energy is mainly due to the mismatch between the parameters of the conversion components and the energy storage components. When solar cells or photoelectrodes generate photocurrent and photovoltage, the working conditions they provide may not be adequate for energy storage units during photocharging, resulting in underutilization. In the future, research should take into account performance coupling between the components of solar energy and those of energy storage systems. A smart chip design that adjusts the outputs of the solar cells/photoelectrodes is one of the most effective ways to meet the electrochemical requirements of energy storage units.

(ii) Appropriate applications: There are many different types of integrated devices, each with its own application. A PSC-, organic solar cells (OSC)-, or dye-sensitized solar cell (DSSC)-based battery or SC, for instance, can be miniaturized, made lightweight, or even flexible. These features make them perfect for portable and wearable devices. They can be used in self-powered mobile phones and biosensors that can be worn or implanted. On the other hand, a silicon-based solar cell and LIB hybrid system provides high energy efficiency and safety. This makes it a good choice for the growing "Internet of Things". It can be used to power electric vehicles and industrial robots that operate autonomously. In essence, it is important to place more emphasis on practical applications and finding out which types of integrated devices are best suited for which usages.

(iii) Life-cycle stability: It is crucial for the commercialization of devices to ensure their durability, particularly their stability throughout their entire life cycle. Therefore, efforts need to continue to be made in this regard to enhance the stability of components within integrated devices in order to make them more stable. In addition to this, sophisticated encapsulation techniques play a very important part in ensuring that the overall performance of the integrated system is as stable as possible.

(iv) Multi-integration: A piezoelectric effect or a triboelectric effect makes it possible to convert mechanical energy into electrical energy and store it in EES units [2, 72]. Moreover, waste heat from industries can be used as an energy source and can be converted and stored as thermal, electrical, or electrochemical energy [2, 73]. The development of multi-integrated devices that are capable of utilizing solar, mechanical, and industrial waste energies synergistically is going to establish the basis for a major change in the energy industry.

Furthermore, alongside the aforementioned points, there is also a need to pay attention to other emerging technologies in order to develop high-performance integrated devices by combining them with relevant existing technologies. Fortunately, tandem solar cells (much like perovskite/silicon tandem solar cells) have the advantage of being able to convert solar energy into electricity with a higher efficiency than conventional cells. The benefits of Zn-ion batteries make them a promising option for future energy storage systems. These include the characteristics of high safety and availability, as well as environmental friendliness [2]. Thus, integrating these devices into advanced solar energy hybrid systems has a great deal of potential as a means of making these systems more practical and competitive. A single PE-based integrated device is thought to be a promising approach for converting, storing, and utilizing solar energy efficiently and reliably. The electrode materials in these devices can absorb sunlight directly from the sky without using solar cells or other special part, thus eliminating the need for solar cells/PEs and EES components to be compatible.

In conclusion, it can be concluded that photorechargeable integrated devices, i.e., PCs, are likely to be of significant value in practical solar energy applications. The emergence of state-of-the-art hybrid solar energy systems has been hampered by a number of significant challenges, but we believe that these groundbreaking solutions will pave the way for a new era in the development of solar energy systems. As a matter of fact, we forecast that carbon-based functional materials are going to play an increasingly crucial role in the development of photorechargeable capacitors and PCs, especially for green energy applications in the near future.

References

1. Guo, L., Sun, J., Wei, J., Liu, Y., Hou, L., Yuan, C., Conductive metal-organic frameworks: Recent advances in electrochemical energy-related applications and perspectives. *Carbon Energy*, 2, 203, 2020.
2. Wang, L., Wen, L., Tong, Y., Wang, S., Hou, X., An, X., Dou, S.X., Liang, J., Photo-rechargeable batteries and supercapacitors: Critical roles of carbon-based functional materials. *Carbon Energy*, 3, 225, 2021.
3. Zhou, P., Yu, J., Jaroniec, M., All-Solid-State Z-Scheme Photocatalytic Systems. *Adv. Mater.*, 26, 4920, 2014.
4. Ong, W.-J., Tan, L.-L., Ng, Y.H., Yong, S.-T., Chai, S.-P., Graphitic Carbon Nitride (g-C3N4)-Based Photocatalysts for Artificial Photosynthesis and Environmental Remediation: Are We a Step Closer To Achieving Sustainability? *Chem. Rev.*, 116, 7159, 2016.
5. He, R., Huang, X., Chee, M., Hao, F., Dong, P., Carbon-based perovskite solar cells: From single-junction to modules. *Carbon Energy*, 1, 109, 2019.
6. Gong, J., Li, C., Wasielewski, M.R., Advances in solar energy conversion. *Chem. Soc. Rev.*, 48, 1862, 2019.
7. Xu, B., Li, P., Chan, C., Application of phase change materials for thermal energy storage in concentrated solar thermal power plants: A review to recent developments. *Appl. Energy*, 160, 286, 2015.
8. Guo, Y., Li, H., Ma, W., Shi, W., Zhu, Y., Choi, W., Photocatalytic activity enhanced *via* surface hybridization. *Carbon Energy*, 2, 308, 2020.
9. Luo, B., Ye, D., Wang, L., Recent Progress on Integrated Energy Conversion and Storage Systems. *Adv. Sci.*, 4, 1700104, 2017.
10. Zhang, P., Wang, F., Yu, M., Zhuang, X., Feng, X., Two-dimensional materials for miniaturized energy storage devices: from individual devices to smart integrated systems. *Chem. Soc. Rev.*, 47, 7426, 2018.
11. Yu, S., Wilson, A.J., Kumari, G., Zhang, X., Jain, P.K., Opportunities and Challenges of Solar-Energy-Driven Carbon Dioxide to Fuel Conversion with Plasmonic Catalysts. *ACS Energy Lett.*, 2, 2058, 2017.
12. Dahash, A., Ochs, F., Janetti, M.B., Streicher, W., Advances in seasonal thermal energy storage for solar district heating applications: A critical review on large-scale hot-water tank and pit thermal energy storage systems. *Appl. Energy*, 239, 296, 2019.
13. Bentsen, N.S. and Møller, I.M., Solar energy conserved in biomass: Sustainable bioenergy use and reduction of land use change. *Renew. Sustain. Energy Rev.*, 71, 954, 2017.
14. Bie, C., Zhu, B., Xu, F., Zhang, L., Yu, J., *In Situ* Grown Monolayer N-Doped Graphene on CdS Hollow Spheres with Seamless Contact for Photocatalytic CO2 Reduction. *Adv. Mater.*, 31, 1902868, 2019.
15. Meng, H., Pang, S., Cui, G., Photo-Supercapacitors Based on Third-Generation Solar Cells. *ChemSusChem*, 12, 3431, 2019.

16. Vega-Garita, V., Ramirez-Elizondo, L., Narayan, N., Bauer, P., Integrating a photovoltaic storage system in one device: A critical review. *Prog. Photovoltaics Res. Appl.*, 27, 346, 2019.
17. Shu, X., Chen, S., Chen, S., Pan, W., Zhang, J., Cobalt nitride embedded holey N-doped graphene as advanced bifunctional electrocatalysts for Zn-Air batteries and overall water splitting. *Carbon*, 157, 234, 2020.
18. Liu, R., Liu, Y., Zou, H., Song, T., Sun, B., Integrated solar capacitors for energy conversion and storage. *Nano Res.*, 10, 1545, 2017.
19. Yang, Z., Deng, J., Sun, H., Ren, J., Pan, S., Peng, H., Self-Powered Energy Fiber: Energy Conversion in the Sheath and Storage in the Core. *Adv. Mater.*, 26, 7038, 2014.
20. Shao, Y., El-Kady, M.F., Sun, J., Li, Y., Zhang, Q., Zhu, M., Wang, H., Dunn, B., Kaner, R.B., Design and Mechanisms of Asymmetric Supercapacitors. *Chem. Rev.*, 118, 9233, 2018.
21. K, N. and Rout, C.S., Photo-powered integrated supercapacitors: a review on recent developments, challenges and future perspectives. *J. Mater. Chem. A*, 9, 8248, 2021.
22. Jahandar, M., Kim, S., Lim, D.C., Indoor Organic Photovoltaics for Self-Sustaining IoT Devices: Progress, Challenges and Practicalization. *ChemSusChem*, 14, 3449, 2021.
23. Jin, W.-Y., Ovhal, M.M., Lee, H.B., Tyagi, B., Kang, J.-W., Scalable, All-Printed Photocapacitor Fibers and Modules based on Metal-Embedded Flexible Transparent Conductive Electrodes for Self-Charging Wearable Applications. *Adv. Energy Mater.*, 11, 2003509, 2021.
24. Li, C., Cong, S., Tian, Z., Song, Y., Yu, L., Lu, C., Shao, Y., Li, J., Zou, G., Rümmeli, M.H., Dou, S., Sun, J., Liu, Z., Flexible perovskite solar cell-driven photo-rechargeable lithium-ion capacitor for self-powered wearable strain sensors. *Nano Energy*, 60, 247, 2019.
25. Flores-Diaz, N., De Rossi, F., Das, A., Deepa, M., Brunetti, F., Freitag, M., Progress of Photocapacitors. *Chem. Rev.*, 123, 9327, 2023.
26. Zhang, D., Stojanovic, M., Ren, Y., Cao, Y., Eickemeyer, F.T., Socie, E., Vlachopoulos, N., Moser, J.-E., Zakeeruddin, S.M., Hagfeldt, A., Grätzel, M., A molecular photosensitizer achieves a Voc of 1.24 V enabling highly efficient and stable dye-sensitized solar cells with copper(II/I)-based electrolyte. *Nat. Commun.*, 12, 1777, 2021.
27. Shen, Y., Tian, Z., Lu, D., Huang, J., Zhang, Z., Li, X., Li, J., Impact of pneumonia and lung cancer on mortality of women with hypertension. *Sci. Rep.*, 6, 20, 2016.
28. Sánchez, S., Vallés-Pelarda, M., Alberola-Borràs, J.-A., Vidal, R., Jerónimo-Rendón, J.J., Saliba, M., Boix, P.P., Mora-Seró, I., Flash infrared annealing as a cost-effective and low environmental impact processing method for planar perovskite solar cells. *Mater. Today*, 31, 39, 2019.

29. Liu, K., Chen, Z., Lv, T., Yao, Y., Li, N., Li, H., Chen, T., A Self-supported Graphene/Carbon Nanotube Hollow Fiber for Integrated Energy Conversion and Storage. *Nano Micro Lett.*, 12, 64, 2020.
30. Ma, D., Pan, Y., Yang, S., Solar power generation technology and its development prospect. *AIP Conf. Proc.*, 2122, 020016, 2019.
31. Das, S.K. and Frank, D., Assessment of various parametric effects on the performance of a commercial grade photovoltaic solar panel. *E3S Web Conf.*, 122, 02002, 2019.
32. Rajeswari, R., Mrinalini, M., Prasanthkumar, S., Giribabu, L., Emerging of Inorganic Hole Transporting Materials For Perovskite Solar Cells. *Chem. Rec.*, 17, 681, 2017.
33. Zhang, G., Zhao, J., Chow, P.C.Y., Jiang, K., Zhang, J., Zhu, Z., Zhang, J., Huang, F., Yan, H., Nonfullerene Acceptor Molecules for Bulk Heterojunction Organic Solar Cells. *Chem. Rev.*, 118, 3447, 2018.
34. Tong, Y., Xiao, Z., Du, X., Zuo, C., Li, Y., Lv, M., Yuan, Y., Yi, C., Hao, F., Hua, Y., Lei, T., Lin, Q., Sun, K., Zhao, D., Duan, C., Shao, X., Li, W., Yip, H.-L., Xiao, Z., Zhang, B., Bian, Q., Cheng, Y., Liu, S., Cheng, M., Jin, Z., Yang, S., Ding, L., Progress of the key materials for organic solar cells. *Sci. China Chem.*, 63, 758, 2020.
35. Ansari, M.I.H., Qurashi, A., Nazeeruddin, M.K., Frontiers, opportunities, and challenges in perovskite solar cells: A critical review. *J. Photochem. Photobiol., C*, 35, 1S, 2018.
36. Correa-Baena, J.P. and Saliba, M., Promises and challenges of perovskite solar cells. *Science*, 358, 739, 2017.
37. Kim, J.Y., Lee, J.-W., Jung, H.S., Shin, H., Park, N.-G., High-Efficiency Perovskite Solar Cells. *Chem. Rev.*, 120, 7867, 2020.
38. Miyasaka, T. and Murakami, T.N., Response to "Comment on 'The photocapacitor: An efficient self-charging capacitor for direct storage of solar energy'" [Appl. Phys. Lett. 86, 196101 (2005)]. *Appl. Phys. Lett.*, 86, 196101, 2005.
39. Murakami, T.N., Kawashima, N., Miyasaka, T., A high-voltage dye-sensitized photocapacitor of a three-electrode system. *Chem. Commun.*, 26, 3346, 2005.
40. Miyasaka, T. and Murakami, T.N., The photocapacitor: An efficient self-charging capacitor for direct storage of solar energy. *Appl. Phys. Lett.*, 85, 3932, 2004.
41. Muñoz-García, A.B., Benesperi, I., Boschloo, G., Concepcion, J.J., Delcamp, J.H., Gibson, E.A., Meyer, G.J., Pavone, M., Pettersson, H., Hagfeldt, A., Freitag, M., Dye-sensitized solar cells strike back. *Chem. Soc. Rev.*, 50, 12450, 2021.
42. Kokkonen, M., Talebi, P., Zhou, J., Asgari, S., Soomro, S.A., Elsehrawy, F., Halme, J., Ahmad, S., Hagfeldt, A., Hashmi, S.G., Advanced research trends in dye-sensitized solar cells. *J. Mater. Chem. A*, 9, 10527, 2021.
43. Pan, Z., Rao, H., Mora-Seró, I., Bisquert, J., Zhong, X., Quantum dot-sensitized solar cells. *Chem. Soc. Rev.*, 47, 7659, 2018.

44. Kouhnavard, M., Ikeda, S., Ludin, N.A., Ahmad Khairudin, N.B., Ghaffari, B.V., Mat-Teridi, M.A., Ibrahim, M.A., Sepeai, S., Sopian, K., A review of semiconductor materials as sensitizers for quantum dot-sensitized solar cells. *Renew. Sustain. Energy Rev.*, 37, 397, 2014.
45. Mora-Seró, I., Current Challenges in the Development of Quantum Dot Sensitized Solar Cells. *Adv. Energy Mater.*, 10, 2001774, 2020.
46. Singh, S., Khan, Z.H., Khan, M.B., Kumar, P., Kumar, P., Quantum dots-sensitized solar cells: a review on strategic developments. *Bull. Mater. Sci.*, 45, 81, 2022.
47. Winter, M., Brodd, R.J., Batteries, W.A., Fuel Cells, and Supercapacitors? *Chem. Rev.*, 104, 4245, 2004.
48. Wu, J., Understanding the Electric Double-Layer Structure, Capacitance, and Charging Dynamics. *Chem. Rev.*, 122, 10821, 2022.
49. Simon, P. and Gogotsi, Y., Materials for electrochemical capacitors. *Nat. Mater.*, 7, 845, 2008.
50. Eftekhari, A., The mechanism of ultrafast supercapacitors. *J. Mater. Chem. A*, 6, 2866, 2018.
51. Simon, P., Gogotsi, Y., Dunn, B., Where Do Batteries End and Supercapacitors Begin? *Sci. (New York N.Y.)*, 343, 1210, 2014.
52. Huang, S., Zhu, X., Sarkar, S., Zhao, Y., Challenges and opportunities for supercapacitors. *APL Mater.*, 7, 100901, 2019.
53. Dubey, R. and Guruviah, V., Review of carbon-based electrode materials for supercapacitor energy storage. *Ionics*, 25, 1419, 2019.
54. Liu, C., Yu, Z., Neff, D., Zhamu, A., Jang, B.Z., Graphene-Based Supercapacitor with an Ultrahigh Energy Density. *Nano Lett.*, 10, 4863, 2010.
55. Rajesh, M., Raj, C.J., Manikandan, R., Kim, B.C., Park, S.Y., Yu, K.H., A high performance PEDOT/PEDOT symmetric supercapacitor by facile *in-situ* hydrothermal polymerization of PEDOT nanostructures on flexible carbon fibre cloth electrodes. *Mater. Today Energy*, 6, 96, 2017.
56. Zhao, J., Wu, J., Li, B., Du, W., Huang, Q., Zheng, M., Xue, H., Pang, H., Facile synthesis of polypyrrole nanowires for high-performance supercapacitor electrode materials. *Prog. Nat. Sci.: Mater. Int.*, 26, 237, 2016.
57. Huang, Z., Ji, Z., Feng, Y., Wang, P., Huang, Y., Flexible and stretchable polyaniline supercapacitor with a high rate capability. *Polym. Int.*, 70, 437, 2021.
58. Yang, Y., Zhu, T., Shen, L., Liu, Y., Zhang, D., Zheng, B., Gong, K., Zheng, J., Gong, X., Recent progress in the all-solid-state flexible supercapacitors. *SmartMat*, 3, 349, 2022.
59. Chen, H., Ling, M., Hencz, L., Ling, H.Y., Li, G., Lin, Z., Liu, G., Zhang, S., Chemical, E., Mechanical, and Electrical Functionalities of Binders for Advanced Energy-Storage Devices. *Chem. Rev.*, 118, 8936, 2018.
60. Sheng, X., Wang, C., Liu, Y., Lee, H.G., Chang, N., Yang, H., A high-efficiency dual-channel photovoltaic power system for nonvolatile sensor nodes, in: *2014 IEEE Non-Volatile Memory Systems and Applications Symposium (NVMSA)*, pp. 1–2, 2014.

61. Zhang, L.L., Lei, Z., Zhang, J., Tian, X., Zhao, X.S., Supercapacitors: Electrode Materials Aspects, in: *Encyclopedia of Inorganic Chemistry*, pp. 1–5, 2005.
62. Kumar, A., Sol gel synthesis of zinc oxide nanoparticles and their application as nano-composite electrode material for supercapacitor. *J. Mol. Struct.*, 1220, 128654, 2020.
63. Sinha, R., Roy, N., Mandal, T.K., N-Doped Carbon Dots and ZnO Conglomerated Electrodes for Optically Responsive Supercapacitor Applications. *Langmuir*, 39, 4518, 2023.
64. Wang, Y. and Xia, Y., Recent Progress in Supercapacitors: From Materials Design to System Construction. *Adv. Mater.*, 25, 5336, 2013.
65. Pan, M., Zeng, W., Quan, H., Cui, J., Guo, Y., Wang, Y., Chen, D., Low-crystalline Ni/Co-oxyhydroxides nanoarrays on carbon cloth with high mass loading and hierarchical structure as cathode for supercapacitors. *Electrochim. Acta*, 357, 136886, 2020.
66. Todorov, T., Gunawan, O., Guha, S., A road towards 25% efficiency and beyond: perovskite tandem solar cells. *Mol. Syst. Des. Eng.*, 1, 370, 2016.
67. Shi, C., Dong, H., Zhu, R., Li, H., Sun, Y., Xu, D., Zhao, Q., Yu, D., An "all-in-one" mesh-typed integrated energy unit for both photoelectric conversion and energy storage in uniform electrochemical system. *Nano Energy*, 13, 670, 2015.
68. Maddala, G., Ambapuram, M., Tankasala, V., Mitty, R., Optimal Dye Sensitized Solar Cell and Photocapacitor Performance with Efficient Electrocatalytic SWCNH Assisted Carbon Electrode. *ACS Appl. Energy Mater.*, 4, 11225, 2021.
69. Sun, J., Gao, K., Lin, X., Gao, C., Ti, D., Zhang, Z., Laser-Assisted Fabrication of Microphotocapacitors with High Energy Density and Output Voltage. *ACS Appl. Mater. Interfaces*, 13, 419, 2021.
70. Gao, K., Ti, D., Zhang, Z., A photocapacitor with high working voltage and energy density. *Sustain. Energy Fuels*, 3, 1937, 2019.
71. Sinha, R., Roy, N., Mandal, T.K., SWCNT/ZnO nanocomposite decorated with carbon dots for photoresponsive supercapacitor applications. *Chem. Eng. J.*, 431, 133915, 2022.
72. Pu, X., Li, L., Song, H., Du, C., Zhao, Z., Jiang, C., Cao, G., Hu, W., Wang, Z.L., A Self-Charging Power Unit by Integration of a Textile Triboelectric Nanogenerator and a Flexible Lithium-Ion Battery for Wearable Electronics. *Adv. Mater.*, 27, 2472, 2015.
73. Zhang, F., Liu, J., Yang, W., Logan, B.E., A thermally regenerative ammonia-based battery for efficient harvesting of low-grade thermal energy as electrical power. *Energy Environ. Sci.*, 8, 343, 2015.

3

Role of Photoactive Materials in Photo-Supercapacitors

Esakkimuthu Shanmugasundaram[1], Suganya Bharathi Balakrishnan[2], Amos Ravi[3] and Stalin Thambusamy[1]*

[1]Department of Industrial Chemistry, Alagappa University, Karaikudi, Tamil Nadu, India
[2]Department of Chemistry, School of Engineering, Dayananda Sagar University, Ramanagara, Karnataka, India
[3]Department of Materials Science, Central University of Tamil Nadu, Thiruvarur, Tamil Nadu, India

Abstract
Photo-supercapacitors have much more attention in wearable and portable energy devices due to their dual-functional combination of solar cell energy conversion (photovoltaic effect) and supercapacitor energy storage capability. The performance of the photo-supercapacitors mainly depends on the photoactive material used in the devices. The photoactive metal oxides, conducting polymers, dye molecules, and redox electrolytes, are commonly used photoactive materials in photo-supercapacitors. These materials contribute to the energy conversion process and enhance the performance of the photo-supercapacitors. In this chapter, firstly, the working principle and the components of photo-supercapacitors are discussed. Then, the types of photoactive materials and their performance in photo-supercapacitors are clearly to be explained. Finally, the future trends for photo-supercapacitors based on potent photoactive materials are summarized.

***Keywords*:** Photoactive materials, energy conversion, energy storage, photo-supercapacitors

**Corresponding author*: drstalin76@gmail.com

Nurdan Demirci Sankir and Mehmet Sankir (eds.) Solar Capacitors and Batteries, (55–82) © 2025 Scrivener Publishing LLC

Abbreviations

DSSCs	Dye-sensitized solar cells
OSCs	Organic solar cells
SCs	Supercapacitors
CPs	Conducting polymers
PANI	Polyaniline
PPy	Polypyrrole
PTh	Polythiophene
PEDOT	Poly (3,4-ethylenedioxythiophene)
MnO_2	Manganese dioxide
Fe_2O_3	Iron oxide
NiO	Nickel oxide
Co_3O_4	Cobalt oxide
RuO_2	Ruthenium oxide
HOMO	Highest occupied molecular orbital
LUMO	Lowest unoccupied molecular orbital
TiO_2	Titanium dioxide
ZnO	Zinc oxide
ETL	Electron transport layer
V_{oc}	Photovoltage
Ag	Silver
1D	One-dimensional
CeO_2	Cerium oxide
SnO_2	Tin oxide
WS_2	Tungsten disulfide
CuI	Copper iodide
CuBr	Copper bromide
CuSCN	Copper thiocyanate
EDLC	Electrical double-layer capacitor
CNTs	Carbon nanotubes
H_2SO_4	Sulfuric acid
HCl	Hydrochloric acid
Na_2SO_4	Sodium sulfate
TMDs	Transition metal dichalcogenides
MoS_2	Molybdenum disulfide
GM	Galvanostatic method
PGM	Pulse galvanostatic method
$NaNO_3$	Sodium nitrate
C_3N_4	Carbon nitride

PVA	Poly (vinyl alcohol)
GO	Graphene oxide
2D	Two-dimensional
$Ti_3C_2T_x$	Titanium carbide
J_{sc}	Photocurrent
FF	Fill factor
η	Efficiency of the cell
P_{in}	Incident light power
$FeSi_2$	Iron disilicide
$CH_3NH_3PbI_3$	Methylammonium lead iodide
CdSe	Cadmium selenide
CdS	Cadmium sulfide
Ag_2S	Silver sulfide

3.1 Introduction

To regulate the global temperature and protect the environment, energy is produced from natural energy sources like solar, wind, hydrogen, and tidal energy [1–3]. Solar energy is a renewable and environmentally friendly energy source that delivers 3 × 1024 J per year from the sun. Solar energy is 10^4 times more abundant than human usage, so researchers are developing various types of solar cells to convert light energy into electrical energy [4–7]. A key component of solar cell technology is the photovoltaic effect, which harvests solar energy to generate electricity [8]. The first-generation solar cell was generated by BELL Laboratory in 1954 [9]. After that, new materials, architecture, and technologies developed in emerging ways and produced second-generation (thin film-based solar cells) and third-generation dye-sensitized solar cells (DSSCs), organic solar cells (OSCs), and perovskite solar cells were introduced [10, 11]. Third-generation solar cells have substantial advantages such as being lightweight, flexible, and transparent and having working capability. Photoactive materials play a vital role in solar cell performance because they can enhance the efficiency of solar cell devices with the help of their light-harvesting properties [12–14]. Based on the types, the photoactive materials are differing; for instance, dyes are a photoactive material in DSSCs, whereas polymers or small molecules are used as photoactive material in OSCs [15–17]. A brief timeline showing the history of photo-supercapacitors (SCs) is given in Figure 3.1.

Energy storage systems like batteries or SCs store the generated electrical energy. The solar system's storage devices are chosen based on usage; batteries are used for long-term use, while SCs are utilized for short-term

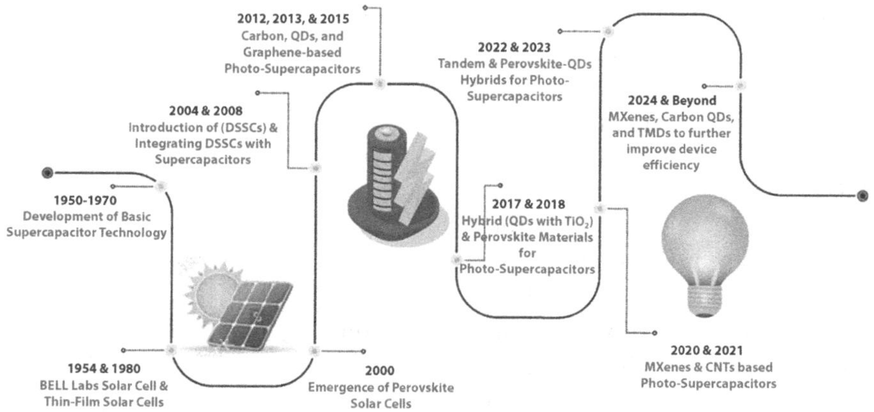

Figure 3.1 A brief timeline of the history of photo-supercapacitors.

use [18]. Batteries store energy by the redox reactions, and the SCs store energy by electrical double-layer and pseudo-capacitor nature between the electrode–electrolyte interfaces. Compared to batteries, SCs give rapid energy release, which is the reason SCs are suitable for quick power output systems [19, 20].

In the early stage, the solar cell was connected to the SCs with an external wire connection. In this method, the architecture needs heavy space and size consumption, which is a complex fabrication. Moreover, the interconnection between the solar cell and storage device causes power loss owing to the external resistance of the wire [21, 22]. Researchers try to develop a combined device in which the solar cell and energy devices are coupled in a single module to solve this issue. The combination of solar cells (photovoltaics) and energy storage (SCs) is called a photo-SC, which can capture light energy and convert electrical energy and store it. The stored energy is released rapidly when needed [23, 24]. Several types of combined conversion-storage modules such as triboelectric nanogenerators, piezoelectric nanogenerators, and pyroelectric nanogenerators are reported before photo-supercapacitors [25]. These modules generate electricity from mechanical or thermal action. These modules have some drawbacks such as insufficient energy sources, low power density, and poor cyclability [26]. Photo-supercapacitor overcomes the above-mentioned issues because it has higher power density and sufficient energy sources from the sun. When the solar cell involves the photovoltaic effect, the SCs are quickly charged. Then, the SCs deliver the energy when solar power is scarce and the load is properly maintained. In solar cells, the cell voltage reaches a higher range when the solar cell comes in contact with light; the voltage rises and then drops to zero in the absence of light. When the solar cell is interconnected with the SCs, the cell voltage does not reach zero immediately. That is the reason why photo-supercapacitors deliver higher power output than other modules [27–29].

The light source is a hub of the photo-supercapacitor so the photoactive materials are vital in photo-supercapacitors. Photoactive materials involve the photovoltaic effect (solar cells) with the presence of light, which produces the electrons that are stored in SCs [30]. Dyes, conducting polymers (CPs), metal oxides, and quantum dots are commonly used photoactive materials for photo-supercapacitors [31]. Dye molecules in DSSCs play an important role in photo-supercapacitor systems by combining solar energy harvesting and electrochemical energy storage. In the photo-supercapacitor system, dye molecules in DSSCs absorb sunlight and convert it into electrical energy, which is then stored and released on demand, all within a single integrated device [22, 30, 32]. The combination of DSSCs and an SC allows the system to store solar energy in real time, ensuring that the energy generated by sunlight is immediately available for storage. DSSCs' architecture is compatible with SCs since they can be coupled *via* a shared electrode. Moreover, the hybridization of a DSSC with the SC has provided a 43% internal resistance reduction that can increase the power delivery of the device [33, 34]. CPs stand out among the many materials utilized for photo-supercapacitor electrodes because of their high conductivity, sufficient light-harvesting properties, pseudocapacitance, environmental stability, and ease of fabrication [35]. In OSCs, CPs such as polyaniline (PANI), polypyrrole (PPy), polythiophene (PTh), and poly (3,4-ethylenedioxythiophene) (PEDOT) are considered promising photoactive materials due to their sufficient light-harvesting properties and suitable band-gap structure. During charging and discharging, CPs in SCs undergo redox reactions, which aid in faradaic processes that enhance the capacitance of the SC device [36, 37]. Some metal oxides have light-harvesting properties like manganese dioxide (MnO_2), iron oxide (Fe_2O_3), nickel oxide (NiO), cobalt oxide (Co_3O_4), and ruthenium oxide (RuO_2), which are used as a photoactive material for photo-supercapacitors [38–40]. Nowadays, quantum dots are used as a photoactive material in photo-supercapacitors owing to their superior absorbance and emission behavior [33]. This chapter summarizes the working principle and components of photo-supercapacitors. Following that, the various types of photoactive materials and their performance in photo-supercapacitors are thoroughly explained. Finally, new trends in photo-supercapacitors based on powerful photoactive materials are discussed. Figure 3.1 displays a brief timeline of the history of SCs.

3.2 Working Principle of Photo-Supercapacitors

A photo-supercapacitor is an integrated device that combines energy conversion and storage. The electron is excited from the photoactive material's conduction band in the energy conversion device (solar cell) and stored in the SC energy storage device, also known as the reservoir [29, 34].

For instance, the DSSC-SC-based photo-supercapacitors contain photoanode, counter anode, dye, and redox electrolytes in the energy conversion (DSSC) device. Moreover, the SCs constructed with the cathode and anode material with the electrolyte are stacked in a sandwich type [41, 42]. These two devices are commonly connected with the common counter electrode whose connection completed the circuit. DSSC is called an electron donor because it donates the electron to the SC device. The absorption of light by dye molecules excites electrons from the dye's highest occupied molecular orbital (HOMO), which are stimulated and migrated to the lowest unoccupied molecular orbital (LUMO) of a photoactive metal oxide, such as titanium dioxide (TiO_2) or zinc oxide (ZnO). The metal oxides act as efficient electron pathways, allowing electrons to be transported from the DSSC compartments to the SC device. The produced electrons are stored in a shared counter electrode [40, 43]. In the absence of light, the SC device uses a discharge mechanism in which the storage electrons generate electricity, which is then employed for energy-related purposes. The same method is used by different types of photo-supercapacitors, such as organic, perovskite, and quantum dot-based solar cells that contain SCs; however, the materials used to build the solar cells and SCs vary depending on the type of photo-supercapacitor [33, 44]. A photo-supercapacitor is schematically depicted in Figure 3.2.

3.3 Basic Components of Photo-Supercapacitors

The photo-supercapacitor consists of two major components: energy conversion (solar cells) and energy storage (SC). Solar cells are made up of three key elements: the photoanode, the photoactive material, and the electrolyte, whereas the SC is made up of a counter electrode, a collecting electrode, and an electrolyte. The counter electrode is a common electrode component used in solar cells and SCs [34, 42]. The components are addressed using the materials indicated below.

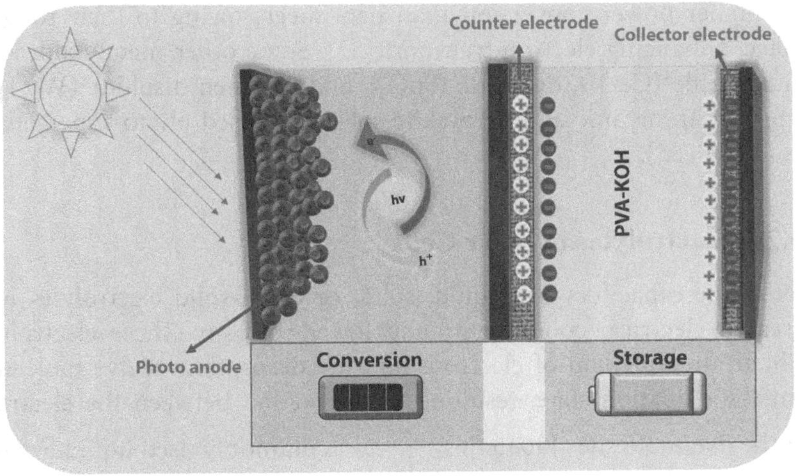

Figure 3.2 Schematic representation of a photo-supercapacitor.

3.3.1 Photoanode Materials in Solar Cells

Photoanode is a transport material of the photo-supercapacitor device because the material can collect the electrons from the LUMO of the photoactive material and transport them to the counter electrode of the SC device. The photoanode materials are also called electron transport layer (ETL) in organic and perovskite solar cells [41, 45, 46]. TiO_2 is commonly used as a potential photoanode material owing to its biocompatibility, non-toxicity, high stability, and low cost. It has three types such as brookite, anatase, and rutile. TiO_2 has a band gap of 3.2 eV, which permits efficient absorption of visible light. Researchers reported that the rutile TiO_2 has sufficient effort to transport the electron to the counter electrode from the dye. Rutile TiO_2 gave lower V_{oc} (photovoltage) because it has higher positive conduction band edge potential so that the electrons easily transport to the counter electrode. Moreover, TiO_2 doped metal oxide photoanode delivers higher efficiency than undoped TiO_2 [47, 48]. For example, Ag-doped TiO_2 exhibits increased photo-conversion efficiency due to the synergistic effect between silver (Ag) and TiO_2. ZnO is also used as an efficient transport material for photo-supercapacitors, particularly 1D ZnO can reduce electron recombination and quickly transport the electron to the counter electrodes [49, 50]. Different nanostructure variants of ZnO, such as nanowires, nanoflowers, nanofibers, and nanosheets, are considered effective photoelectrode materials due to their increased power conversion efficiency [51]. Particularly, ZnO nanowire-based DSSC devices have higher power conversion than nanosheets owing to their rougher surface, enhancing electron transport [52]. Some other metal oxides like cerium oxide (CeO_2), tin oxide (SnO_2), and tungsten disulfide (WS_2) are using ETL in organic and perovskite solar cell-based photo-supercapacitors [39, 40, 53].

3.3.2 Electrolytes in Solar Cells

Photo-supercapacitors use liquid, solid, or quasi-solid electrolytes with increased electrical conductivity and lower viscosity. These electrolytes facilitate the diffusion of electrons and the desorption of dye molecules from the oxidation state, resulting in interaction between the electrode and the dye molecules [30, 54]. I_3^- / I^- is a commonly used liquid electrolyte in DSSC-based photo-supercapacitors. However, the liquid electrolyte has a leakage problem; hence, researchers developed solid-state electrolytes known as hole-transfer materials such as copper iodide (CuI), copper bromide (CuBr), and copper thiocyanate (CuSCN). These materials offer superior conductivity properties and high mechanical stability, resulting in a minimal recombination effect and improved transport behavior [55, 56].

3.3.3 Counter and Collector Materials in SCs

Carbon-based materials, such as graphene, graphite, carbon nanotubes (CNTs), carbon aerogels, and carbon nanofibers, are being widely investigated as electrode materials for electrical double-layer capacitors (EDLCs). They have a high specific area, stability, variable pore size, and great electrical conductivity [57–59]. The second category focuses on recent advancements in metal oxide materials used in high-performance SCs. Metal oxide electrode materials are differentiated by their strong redox activity, exceptional conductivity, and chemical stability, making them ideal for solar cell applications [39, 40]. CPs can also improve photo-supercapacitor efficiency due to their inherent conductivity, high pseudocapacitance, and flexibility [37]. Metal oxides and polymers are the most frequent electrode materials utilized in photo-supercapacitor applications. Figure 3.3 shows the numerous advantages of photo-supercapacitors.

3.3.3.1 Metal Oxide-Based Electrodes

Metal oxide-based electrode materials have a larger specific capacitance, which is the amount of charge they can hold per unit volume or mass. Due

Figure 3.3 Advantages of photo-supercapacitors.

to the high energy density and specific capacitance of 2D materials compared to carbonaceous materials and conductive polymers, most of current studies concentrate on metal oxides and their composites with carbon and non-carbon-based materials [60]. Certain metal oxides, such as MnO_2, Co_3O_4, NiO, and RuO_2, have been suggested to be suitable electrode materials for SCs. These metal oxide-based materials have good cycling stability, allowing SCs to sustain many charges and discharges without damage. This results in higher capacitance and energy storage, which improves the overall performance and feasibility of metal oxide-based SCs [61, 62].

3.3.3.1.1 Manganese Dioxide

MnO_2 has strong pseudocapacitive characteristics, making it a common electrode material for SCs. MnO_2 may store energy through quick and reversible faradaic processes, resulting in high capacitance. Huang *et al.* fabricated MnO_2-decorated graphene nanosheets, which gave 280 F/g specific capacitance with sufficient cyclic stability [63]. MnO_2 is commonly coupled with TiO_2 to form hybrid photo-supercapacitors. TiO_2 enhances light absorption, while MnO_2 increases energy storage capacity. MnO_2 has excellent pseudo-capacitance, but its low conductivity limits its rate capability. To address this constraint, hybridization with conductive materials (e.g., graphene or CNTs) is a common approach [64].

3.3.3.1.2 Iron Oxide

Iron oxide (Fe_2O_3), particularly α-Fe_2O_3 (hematite), is an abundant material with photocatalytic characteristics, making it a viable photoactive electrode material for SCs [65, 66]. Hematite's narrow band gap (~2.1 eV) can absorb visible light, which makes it a better material for solar-powered applications than ZnO and TiO_2. Munichandraiah *et al.* prepared porous α-Fe_2O_3, which demonstrated good cyclic stability and a specific capacitance of 300 F/g. Fe_3O_4 is frequently hybridized with carbon materials or other metal oxides (e.g., TiO_2) to enhance its performance in photo-supercapacitors because of its low electrical conductivity [67].

3.3.3.1.3 Nickel Oxide

NiO's narrow band gap (~3.7 eV) allows it to function as a pseudocapacitive electrode and photoactive material in photo-supercapacitors. Lee *et al.* designed a NiO nanoparticle-based electrochromic-SC, whose capacitance range is 1386 F/g, and demonstrated consistent cycling performance. While NiO has outstanding electrochemical properties, its band gap is rather broad, limiting its capacity to capture visible light. NiO is typically coupled with other semiconducting materials, such as TiO_2 or graphene, to enhance its photoelectrochemical performance and extend its absorption into the visible spectrum [68].

3.3.3.1.4 Cobalt Oxide

Co_3O_4-based SCs reveal extensive research in the field due to cobalt's high theoretical capacitance and multiple oxidation states. Co_3O_4 has pseudocapacitance, which is caused by faradaic redox reactions involving cobalt ions (Co^{2+}/Co^{3+}) during the charge/discharge cycles. Studies have demonstrated that Co_3O_4 undergoes multiple electron transfer reactions, enhancing the capacitance compared to carbon-based EDLCs [69, 70]. In a comprehensive study by Xu *et al.*, Co_3O_4 nanoparticles were synthesized as electrode materials for SC applications, resulting in a high specific capacitance of 970 F/g at a current density of 1 A/g. One of the primary challenges in SC fabrication is enhancing the surface area for improved ion diffusion and charge storage [68]. To solve this, researchers have focused on developing nanostructured Co_3O_4. These include nanowires, nanorods, nanoflakes, and mesoporous structures.

3.3.3.1.5 Ruthenium Oxide

RuO_2 is a well-studied material for SCs for its high capacitance, rapid and reversible redox processes (Ru^{3+}/Ru^{4+}), and efficient charge storage and release. RuO_2 can improve photo-supercapacitor efficiency by increasing catalytic activity and facilitating efficient electron transport in hybrid systems [71]. Arunachalam *et al.* developed a RuO_2 electrode for SC applications with enhanced capacitance, which is 1724 F/g. Furthermore, graphene's high conductivity and vast surface area make it ideal for enhancing electrochemical performance when combined with RuO_2 materials [72]. CNTs are also mixed with RuO_2 to prepare hybrid electrodes that improve conductivity and mechanical stability [73].

3.3.3.2 Polymer-Based Electrodes

CPs stand out among the many materials utilized for SC electrodes owing to their high conductivity, pseudocapacitance, environmental stability, and simple fabrication methods. During charging/discharging, CPs in SCs undergo redox reactions, which aid in faradaic processes that enhance the capacitance overall [36, 37]. Higher energy densities for CP-based SCs result from this characteristic, contrasting the pure electrostatic charge storage in EDLCs. PANI, PPy, PTh, and PEDOT CPs have high specific capacitance and numerous active sites. Furthermore, researchers have explored various techniques to enhance the conductive properties of CPs and their nanocomposite materials like graphene and CNTs to accomplish unique functionality [74, 75].

3.3.3.2.1 Polyaniline

PANI's electrochemical characteristics significantly depend on its doping level, morphology, and structural configuration (emeraldine, leucoemeraldine, and pernigraniline forms). The emeraldine form is the most conductive and is often used in SC electrodes [76, 77]. PANI's pseudocapacitive characteristic is linked to redox-active regions in the polymer backbone, which undergoes faradaic reactions, increasing charge storage capacity. Pure PANI has high specific capacitance values, depending on its shape and production method. Specific capacitance varies significantly with different electrolytes, commonly sulfuric acid (H_2SO_4), hydrochloric acid (HCl), and sodium sulfate (Na_2SO_4); acidic electrolytes typically result in higher capacitance [78, 79]. Lokhande *et al.* employed nanostructured PANI as an electrode material for SC applications, yielding a specific capacitance of 503 F/g. However, as an electrode material, PANI has low mechanical resistance and poor film-forming characteristics [80]. Liu *et al.* developed graphene-wrapped PANI hollow spheres as a hybrid electrode material for SC applications. They demonstrated that the hybrids' specific capacitance may reach 614 F/g with a current density of 1 A/g, demonstrating good cycling stability for 500 charging/discharging cycles [81].

To overcome cycling stability constraints and increase capacitance, PANI is frequently coupled with carbon-based materials, metal oxides, or transition metal dichalcogenides (TMDs) [82]. TMDs resemble graphene due to their two-dimensional layered structure. The weak van der Waals interactions between layers facilitate ion intercalation, which is advantageous for quick charge and discharge cycles. This large surface area maximizes ion access to active sites, which increases capacitance. Materials like molybdenum disulfide (MoS_2) and WS_2 exhibit excellent pseudocapacitive characteristics, leading to higher energy density in SCs compared to standard carbon-based electrodes [83, 84]. PPy/PANI@MoS_2 ternary composite material was prepared by Dai *et al.* using simple hydrothermal and *in situ* polymerization procedures. In three-electrode systems, the produced composites have exceptional electrochemical properties, such as a high capacitance of 1171 F/g [85].

3.3.3.2.2 Polypyrrole

PPy has received significant attention as an electrode material for SCs due to its high conductivity, electrochemical activity, and flexibility. It possesses high electrical conductivity when compared to other polymers, which is crucial for effective charge storage and transfer in SCs. This conductivity allows for faster electron transport, which enables greater charge and discharge rates. PPy is frequently used in conjunction with other materials, such as CNT, graphene, or metal oxides, to form hybrid electrodes with improved characteristics [86, 87]. These composites combine the high conductivity and stability of the additional materials with PPy's pseudocapacitive characteristic, resulting in SCs with higher energy density, conductivity, and cycle life. Furthermore, PPy is biocompatible, making it ideal for SCs in biomedical applications such as implanted devices, which require bio-safe materials [88, 89]. Kang *et al.* used the pulse galvanostatic method (PGM) and the galvanostatic method (GM) to prepare PPy films. The findings revealed that PGM-PPy films outperformed GM-PPy films in terms of capacitance and good rate capability making them more attractive for energy storage applications. PGM-PPy films had a capacitance of 403 F g^{-1} [90]. Dubal *et al.* found an excellent specific capacitance of 533 F/g in a 0.5 M H_2SO_4 solution with a potential range of -0.4 to 0.6 V. The PPy electrode showed 93% charge/discharge efficiency and 83% capacitance retention at a current density of 10 mA cm^{-2}, indicating high electrochemical reversibility and rate capability [91]. Pereira *et al.* highlighted that the full-organic composite composed of graphitic-carbon nitride (g-C_3N_4) and PPy delivers higher capacitance range than pristine PPy. The prepared material was completely described using both theoretical and experimental methods [92].

3.3.3.2.3 Polythiophene

PTh is a conjugated polymer with a thiophene ring as the repeating unit. Its structure gives great electrical conductivity and flexibility, making it an attractive material for SCs. PTh's conjugated π-electron system allows for efficient electron transfer. The redox reactions associated with PTh entail the doping and de-doping of ions from the electrolyte into the polymer matrix, which allows for rapid charge transfer [93]. This pseudocapacitive activity often leads to higher specific capacitance than EDLC materials such as activated carbon. PTh's charge storage mechanism combines faradaic processes with some degree of EDLC formation at the interface, resulting in hybrid capacitance properties. Patil *et al.* used the chemical bath deposition method at room temperature with oxidative polymerization to generate PTh thin films. The produced PTh thin film has a maximal specific capacitance of 300 F/g [94].

Until now, flexible-wire-shaped devices have been in high demand due to their exceptional properties, such as stretchability, lightweight design, and ability to be woven into textile fibers for future portable and wearable electronics. These wearable devices have lately demonstrated significant potential in SCs that can be incorporated into smart textiles [95, 96]. Ambade *et al.* established a symmetric wire-shaped device that is based on an electropolymerization technique based on conducting PTh on titanium wire. It delivers 1357.31 mF/g capacitance and 23.11 μWh cm^{-2} energy density [95, 96].

3.3.3.2.4 Poly (3,4-Ethylenedioxythiophene)

PEDOT, as a member of the PTh family, is among the most conductive polymers known. The polymer is electron-rich, with a consistent conductivity that enables effective electron transport within the electrode. PEDOT has demonstrated better stability under various environmental conditions than many other polymers. This stability becomes crucial for practical SCs, particularly in outdoor or harsh-environment applications where the device may be subjected to temperature changes and humidity [97, 98]. Kurungol *et al.* reported a unique method for fabricating a high-conductive PEDOT on flexible cellulose paper. The flexible solid-state SC has a volumetric energy density of 1 mWh/cm^3. The PEDOT system has a specific capacitance of 115 F/g and a volumetric capacitance of 145 F/cm^3. These flexible devices exhibit high stability during charge/discharge cycling under twisting and bending conditions for more than 3800 cycles [99]. CP nanofibers made of poly (vinyl alcohol) (PVA) and graphene oxide (GO) coated with PEDOT for SC applications were prepared by Sulaiman *et al.* using integrated techniques such as electrospinning and electrodeposition. They disclosed that the nanocomposite has the highest specific capacitance of 224.27 F/g. PVA–GO/PEDOT nanocomposite demonstrated 1.8-V wide working potential windows, which considerably improved its capacitance [100].

MXenes, a two-dimensional (2D) transition metal carbide, nitride, and carbonitride, have shown promising results as electrode materials for SCs. MXenes' redox-active surfaces can undergo reversible Faradaic reactions, resulting in pseudocapacitive behavior. These reactions help to store charge in addition to simple electrical double-layer capacitance [101, 102]. Transition metals, such as titanium in titanium carbide ($Ti_3C_2T_x$), allow MXenes to engage in surface redox processes, resulting in higher specific capacitance and energy density compared to EDLC-based materials. Lu *et al.* developed Ti_3C_2Tx@PEDOT (1:10) composites using a cost-effective *in situ* polymerization technique. The Ti_3C_2Tx@PEDOT (1:10) composite gave 564 F/g capacitance with 96.5% capacitance retention. The surface redox process and synergistic effects of Ti_3C_2Tx@PEDOT enhance electrochemical performance and stability [103].

3.3.4 Photoactive Materials in Solar Cell

Photoactive materials use the photovoltaic effect to transform light energy into electrical energy; they act as a sensitizer for the photo-supercapacitor. Photoactive materials are often made of materials with exceptional optical and electrical properties as well as band gaps appropriate for the solar spectrum [6]. In DSSC-based photo-supercapacitors, dye molecules absorb light energy and generate electrons, whereas organic, perovskite, and quantum dot solar cell-based photo-supercapacitors use polymers, perovskite, and quantum dot materials as photoactive material sensitizers [12].

3.3.4.1 Dyes as a Photoactive Material

Synthetic and organic dyes are popular sensitizers for DSSC-based photo-supercapacitors. When sunlight enters the solar cell through the transparent photoanode, the dyes absorb photons and transfer electrons from the ground to the excited states. This leads electrons to transition to the photoanode's conduction band [12, 16, 22, 30]. The electron then moves to the SC device's counter electrode. In the meantime, the dye oxidizes the electrolyte to tri-iodide form by absorbing electrons from the iodine. The tri-iodide restores the electron from the counter electrode after passing the external circuit. Thus, the entire operation is based on two redox processes: dye and electrolyte. The entire reaction is detailed below.

$$TiO_2 + Dye + h\nu \rightarrow TiO_2 + Dye^{*} \text{ (Excitation)}$$

$$TiO_2 + Dye^{*} \rightarrow TiO_2 + e^{-} + Dye^{+} \text{ ((Transfer the electron to photoanode)}$$

$$TiO_2 + Dye^{+} + I^{-} \rightarrow TiO_2 + Dye + I_3^{-} \text{ (Dye collect the electron from electrolyte)}$$

$$TiO_2 + Dye + I_3^{-} + 2e^{-} \text{ (Counter electrode)} \rightarrow \text{Counter electrode} + 3I^{-} \text{ (Electrolyte reduction)}$$

The photoanode and redox potential of the electrolyte determines the photovoltage (V_{oc}), whereas the light-harvesting efficiency and electron production and collection determine the photocurrent (J_{sc}). The total efficiency of the photo-supercapacitor device is estimated by the fill factor (FF) and efficiency of the cell (η)

$$FF = \frac{V_{max} \times J_{max}}{V_{OC} \times J_{SC}}$$

$$\eta = \frac{V_{OC} \times J_{SC} \times FF}{P_{in} \times 100\%}$$

where J_{max} and V_{max} correspond to current and voltage values, respectively; V_{oc}—open-circuit voltage (V), J_{SC}—short-circuit current density (mA cm^{-2}), and P_{in}—the incident light power.

Carboxy-anchor organic dye-based photo-supercapacitor achieved 4.9% power conversion efficiency, and the capacitance and energy densities are 32 F/g and 2.3 Wh kg^{-1}, respectively. Das *et al.* found that Rose bengal dye-based photo-supercapacitor had an 8.25% power conversion efficiency and 79.2 m F cm^{-2} specific capacitance at 0.07 mA cm^{-2} [104]. N719 dye molecule-based capacitors have 172.8 F/g specific capacitance. The organic sensitizer dye molecules like porphyrin and ruthenium have delivered >14% power conversion efficiency under standard illumination. Compared to metal-based sensitizers, metal-free organic sensitizers have

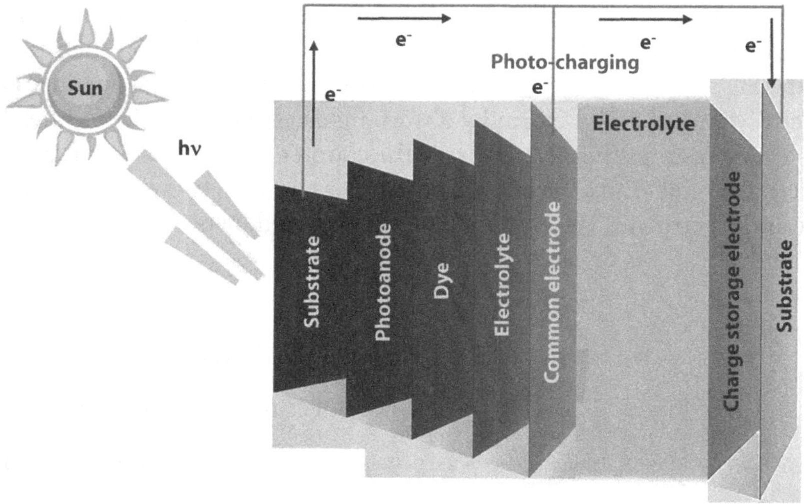

Figure 3.4 Construction of a photo-supercapacitor based on DSSC.

much more attention in photo-supercapacitors owing to their unique properties such as low toxicity, environmentally friendly, flexibility, and facile synthesis. This type of molecule easily binds with the semiconductor surface and easily transports to the excited electrons. Carbazole, phenothiazine, and triphenylamine are organic sensitizers that can easily harvest light energy and convert electrical energy with the help of their low band-gap nature. Nowadays, researchers synthesized and reported novel low band-gap dye molecules that give higher power conversion efficiency than natural dyes. The synthesized dye molecules' band-gap structures are easily tuned by the doping or functional group modification methods. Duvva *et al.* prepared three types of dyes such as tetrathia-fulvalene donor, benzo-thiadiazolethiophene, and rhodanine-3-acetic acid. Among these, rhodanine-3-acetic acid delivers higher efficiency, which is 6.0% [105]. Han *et al.* synthesized three benzocarbazole dyes with benzene, furan, and thiophene bridge units that deliver 5.10%, 7.67%, and 6.58% power conversion efficiency, respectively [106]. Figure 3.4 shows the construction of a photo-supercapacitor based on DSSC.

3.3.4.2 Polymers as a Photoactive Material

OSCs have much more attention in photo-supercapacitors owing to their significant properties such as flexibility, low cost, and easy fabrication processes. Polymers are photoactive materials for OSCs that band gap easily tuned by doping or functionalization methods. This tunability behavior of OSCs can enhance the performance of the photo-supercapacitors. OSCs significantly enhance the photo-supercapacitor owing to their sufficient light-harvesting polymers presented in the solar device, which can generate more electrons and transfer to the storage electrodes. For instance, PBDB-T-2F:Y_6:$PC_{71}BM$ polymer OSCs deliver 17% power conversion efficiency, open-circuit voltage is 0.84 V, and J_{SC} is 26 mA cm^{-2} [107]. PM6:Y6:C8-DTC polymer OSCs deliver 17.52% power conversion efficiency, whereas p-BTP-PhC_6 polymer OSCs give 17.7% power conversion efficiency [108]. The results suggest that OSCs are a potential candidate for the conversion of light energy into electrical energy for photo-supercapacitor applications. Liu *et al.* prepared the PBDTTTOFT: $PC_{71}BM$ polymer solution-based photo-supercapacitor and reported a high capacitance retention of 96% with 100 charge/discharge cycles for one week [109]. PCDTBT: PC71BM-based photo-supercapacitor delivers 6.2% power conversion efficiency. Moreover, the device delivered 13.3 mWh/Kg and 2.8 mW/Kg energy and power density, respectively, which is a higher energy storage performance than other photoactive polymers [110].

3.3.4.3 Perovskite Materials as a Photoactive Material

Perovskite solar cells have gained attention in the field of photo-supercapacitors due to their good light-harvesting behavior and sufficient stability. Moreover, these kinds of solar cells are mainly employed in flexible and wearable electronics so it is preferable for flexible photo-supercapacitors. These perovskite materials have a flexible nature that has easily changed their band gap by the doping processes. This character can easily absorb the solar spectrum photons without loss and enhance the power conversion efficiency. Rugut *et al.* reported that the pseudo-cubic $CH_3NH_3PbI_3$ band gap is 1.49 eV, which is suitable for solar light harvesting in perovskite solar cell applications. Even though perovskite materials have higher power conversion efficiency, the material is easily degraded by oxygen, moisture, and high temperatures. Moreover, lead can seriously affect the environment and human health. However, the photo-supercapacitor encounters issues because the energy storage can cover the solar cell from the environment. That is the reason the perovskite-based photo-supercapacitors have much attention from researchers for development in future energy applications. $FA_{0.75}Cs_{0.25}Pb(I_{0.8}Br_{0.2})_3$ perovskite solar cell-based photo-supercapacitor delivers 11.5% and the SC is 92% [30]. Methylammonium lead iodide ($CH_3NH_3PbI_3$) photoactive material-based photo-supercapacitor delivers 22.9% power conversion efficiency with higher specific capacitance and

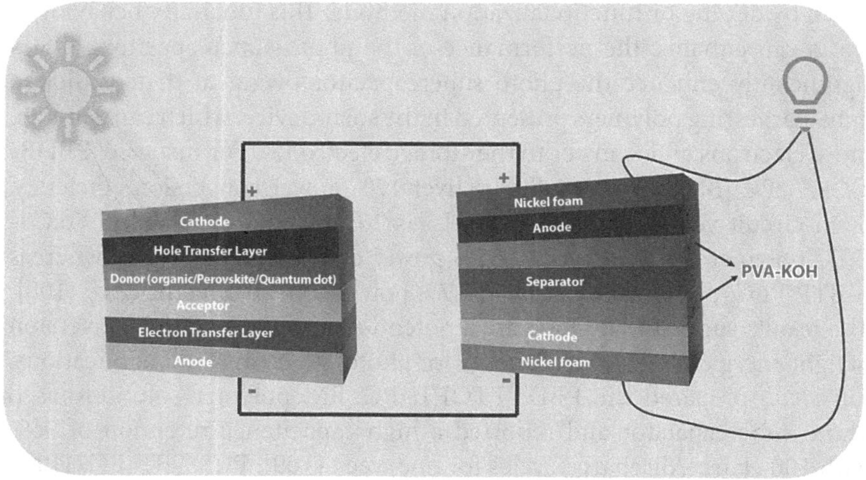

Figure 3.5 Schematic diagram of a photo-supercapacitor system based on organics, perovskites, and quantum dots.

superior energy density. Perovskite material combined with a spiro-OMe-TAD-based photo-supercapacitor gave 20% power conversion efficiency with high power energy density. Ng *et al.* reported that cesium lead halide inorganic-based photo-supercapacitor shows a high areal capacitance of 150 mF cm^{-2} with 4% cyclic capacitance reduction after 5000 cyclic stability. Figure 3.5 shows a schematic diagram of a photo-supercapacitor system based on organics, perovskites, and quantum dots.

3.3.4.4 Quantum Dot Materials as a Photoactive Material

Quantum dots are zero-dimensional materials, which are 2-10 nm. These dots have quantum confinement in which the electrons possess oscillation strength in all directions. The quantum dots contain some significant properties such as low cost, easy synthesis, superior optical and electrical properties, and effortless band-gap tunability. In addition, the quantum dots emit higher fluorescence behavior, which is the reason why it is used in optical applications like sensors, bio-imaging, wearable, portable electronics, and photocatalytic applications, owing to their light-absorbing property, which is used in solar cell applications. The quantum dot solar cells are promising third-generation solar cells that can harvest solar light and electrical energy with the help of a suitable band-gap nature with the solar spectrum. Two types of quantum dots are used in solar cells such as metal-based quantum dots and carbon quantum dots. Xiao *et al.* reported

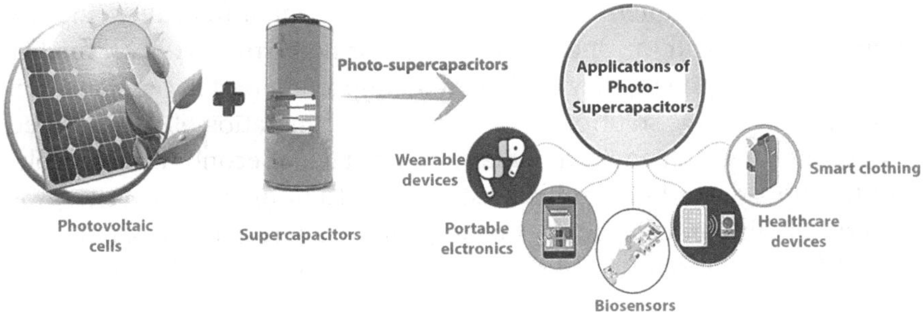

Figure 3.6 Different applications of photo-supercapacitors.

that the acetonitrile-based SnO_2-PbS quantum dot solar cell has delivered 10.49% power conversion efficiency, which is a higher methanol-based SnO_2-PbS quantum dot solar cell device [111]. Zheng *et al.* reported that cadmium sulfide (CdS)/cadmium selenide (CdSe) quantum dot-based photo-supercapacitor delivered 2.66% overall efficiency, and the capacitance is 132.83 mF/cm^2 [112]. The device delivers 76.7% cyclic stability after 100 cycles. Silver sulfide quantum dots with ZnO nanorod-based photo-supercapacitors reach 0.33-V potential during photocurrent, and the charge capacitance is 6.83% [113]. Liu *et al.* discovered that the black phosphorous quantum dot photo-supercapacitor enhances the energy storage capacity, which is 138.4 mA h g^{-1}. Moreover, the energy density is 44.4 Wh kg^{-1}, and the power density is 800 W kg^{-1}. The capacitance retention of the material is 104.8% after 13,000 cycles [114]. Figure 3.6 depicts different applications of photo-supercapacitors.

3.4 Conclusions

Photo-supercapacitors represent a promising and novel innovation in energy storage and renewable energy technologies. Photo-supercapacitors, which integrate the principles of solar energy harvesting and high-capacity energy storage, have the potential to deliver continuous, sustainable power for a broad spectrum of applications, from portable devices to off-grid systems. This technology bridges a major gap in energy systems, enabling devices to capture and effectively store sunlight for later use. The fundamental principles and operating mechanisms of photo-supercapacitors have been examined, specifically how they differ from regular SCs by including light-harvesting processes. Moreover, the role of photoactive materials like conductive polymers, carbon materials, metal oxides, and perovskites is discussed in the chapter. These materials are essential for enhancing the photosensitivity and the photo-supercapacitor performance. With further study and development, they could play an important part in the transition to sustainable energy solutions, providing a new generation of self-powered devices that are both environmentally beneficial and economically viable. This technology marks a significant step toward a future in which energy is both sustainable and accessible, making photo-supercapacitors a potential subject for further scientific research and technological development.

References

1. Indrajith, B., Gunawardane, K., Jayasinghe, H., A review: Dc microgrids for sustainable power delivery in offshore industries. *Proc. Int. Conf. Offshore Mech. Arct. Eng. - OMAE*, 86922, V010T13A019, 2023.
2. Awasthi, D.K. and Panday, J., Eco-Friendly technology in energy production. *Biochem. Cell. Arch.*, 16, 12–15, 2016.
3. Lodewijks, P. and Duerinck, J., Long Term Energy System Planning: to a 100 %. *Renewable Energy Syst. By 2050*, 1, EEPP3088, 2019.
4. Kumar, V., Shrivastava, R.L., Untawale, S.P., Solar Energy: Review of Potential Green & Clean Energy for Coastal and Offshore Applications. *Aquat. Procedia*, 4, 473–480, 2015.
5. Obaideen, K., Olabi, A.G., Al Swailmeen, Y., Shehata, N., Abdelkareem, M.A., Alami, A.H., Rodriguez, C., Sayed, E.T., Solar Energy: Applications, Trends Analysis, Bibliometric Analysis and Research Contribution to Sustainable Development Goals (SDGs). *Sustain.*, 15, 1418, 2023.
6. Bist, A., Pant, B., Ojha, G.P., Acharya, J., Park, M., Saud, P.S., Novel Materials in Perovskite Solar Cells: Efficiency, Stability, and Future Perspectives. *Nanomaterials*, 13, 1724, 2023.
7. Xie, Z., Peng, Y.P., Yu, L., Xing, C., Qiu, M., Hu, J., Zhang, H., Solar-Inspired Water Purification Based on Emerging 2D Materials: Status and Challenges. *Sol. RRL*, 4, 19004002020, 2020.
8. Ameen, S., Rub, M.A., Kosa, S.A., Alamry, K.A., Akhtar, M.S., Shin, H.S., Seo, H.K., Asiri, A.M., Nazeeruddin, M.K., Perovskite Solar Cells: Influence of Hole Transporting Materials on Power Conversion Efficiency. *ChemSusChem.*, 9, 10–27, 2016.
9. Jardine, C.N., Solar photovoltaic panels. *Domest. Microgeneration Renew. Distrib. Energy Technol. Policies Econ.*, 1, 163–18, 22015.
10. Goswami, R., Three Generations of Solar Cells. *Adv. Mater. Res.*, 1165, 113–130, 2021.
11. Seroka, N.S., Taziwa, R., Khotseng, L., Solar Energy Materials-Evolution and Niche Applications: A Literature Review. *Mater. (Basel)*, 15, 5338, 2022.

12. Saravanan, V., Ganesan, S., Rajakumar, P., Synthesis and DSSC application of BODIPY decorated triazole bridged and benzene nucleus cored conjugated dendrimers. *RSC Adv.*, 10, 18390–18399, 2020.
13. Jaiswal, R., Kumar, A., Yadav, A., Nanomaterials based solar cells. *Nanotechnol. Automot. Ind.*, 1, 467–484, 2022.
14. Hosseinnezhad, M. and Ranjbar, Z., A review on flexible dye-sensitized solar cells as new sustainable energy resources. *Pigm. Resin Technol.*, 52, 310–320, 2023.
15. Alajlani, Y., Alaswad, A., Placido, F., Gibson, D., Diyaf, A., Inorganic Thin Film Materials for Solar Cell Applications, in: *Encycl. Smart Mater*, 2021.
16. Higashino, T. and Imahori, H., Emergence of Copper (I/II) Complexes as Third-Generation Redox Shuttles for Dye-Sensitized Solar Cells. *ACS Energy Lett.*, 7, 1926–1938, 2022.
17. Meyer, E.L., Mbese, J.Z., Agoro, M.A., The frontiers of nanomaterials (SnS, PbS and CuS) for dye-sensitized solar cell applications: An exciting new infrared material. *Molecules*, 24, 4223, 2019.
18. Kim, S.L., Hsu, J.H., Yu, C., Intercalated graphene oxide for flexible and practically large thermoelectric voltage generation and simultaneous energy storage. *Nano Energy*, 48, 582–589, 2018.
19. Hong, J. and Jo, C., Study on Morphology Control and N-Doping of Porous Carbon Cathode for Zinc-Ion Hybrid Supercapacitor. *ECS Meet. Abstr.*, 244, 689–689, 2023.
20. Wang, X., Li, Q., Zhang, L., Hu, Z., Yu, L., Jiang, T., Lu, C., Yan, C., Sun, J., Liu, Z., Caging Nb_2O_5 Nanowires in PECVD-Derived Graphene Capsules toward Bendable Sodium-Ion Hybrid Supercapacitors. *Adv. Mater.*, 30, 1800963, 2018.
21. Varma, S.J., Sambath Kumar, K., Seal, S., Rajaraman, S., Thomas, J., Cells, F.-T.S., Nanogenerators, Batteries, and Supercapacitors for Wearable Applications. *Adv. Sci.*, 5, 1800340, 2018.
22. Rahman, M.M., A Comprehensive Review on Perovskite Solar Cells Integrated Photo-supercapacitors and Perovskites-Based Electrochemical Supercapacitors. *Chem. Rec.*, 24, 202300183, 2024.
23. Keppetipola, N.M., Olivier, C., Toupance, T., Cojocaru, L., Biomass-derived carbon electrodes for supercapacitors and hybrid solar cells: Towards sustainable photo-supercapacitors. *Sustain. Energy Fuels*, 5, 4784–4806, 2021.
24. Chen, X., Ding, J., Chen, X., Liu, X., Zhuang, G., Zhang, Z., Porous tremella-like $NiCo_2S_4$ networks electrodes for high-performance dye-sensitized solar cells and supercapacitors. *Sol. Energy*, 176, 762–770, 2018.
25. Zhao, K., Wang, Y., Han, L., Wang, Y., Luo, X., Zhang, Z., Yang, Y., Nanogenerator-Based Self-Charging Energy Storage Devices. *Nano Micro Lett.*, 11, 1–19, 2019.
26. Jang, Y., Jo, J., Lee, S.H., Fabric Supercapacitors Fabrication for Smart Textiles By Using All Solution Process. *ECS Meet. Abstr.*, 3, 142–142, 2019.

27. Berestok, T., Diestel, C., Ortlieb, N., Glunz, S.W., Fischer, A., A Monolithic Silicon-Mesoporous Carbon Photosupercapacitor with High Overall Photoconversion Efficiency. *Adv. Mater. Technol.*, 7, 2200237, 2022.
28. Arjunkumar, B., Mangalaraja, R.V., Ragupathi, C., Ramalingam, G., Alam, M.M., Al-Sehemi, A.G., Anbukumaran, S.K., Green carbon nanospheres for integrated photo-supercapacitors device with improved efficiency. *Diam. Relat. Mater.*, 150, 111699, 2024.
29. Keerthana, T., Kumar, B.A., Ramalingam, G., Vetrivelan, V., Ragupathi, C., Mangalaraja, R.V., Alam, M.M., Anbukumaran, S.K., Nanoarchitectonics of MoS2 petals with 0D@ 2D Carbon Derivatives for Integrated Photo Capacitor Application. *J. Alloys Compd.*, 1010, 177622, 2024.
30. Berestok, T., Diestel, C., Ortlieb, N., Buettner, J., Matthews, J., Schulze, P.S.C., Goldschmidt, J.C., Glunz, S.W., Fischer, A., High-Efficiency Monolithic Photosupercapacitors: Smart Integration of a Perovskite Solar Cell with a Mesoporous Carbon Double-Layer Capacitor. *Sol. RRL*, 5, 2100662, 2021.
31. Nandi, A.K. and Chatterjee, D.P., Hybrid polymer gels for energy applications. *J. Mater. Chem. A*, 11, 12593–12642, 2023.
32. Nordin, N.A., Ansari, M.N.M., Nomanbhay, S.M., Hamid, N.A., Tan, N.M.L., Yahya, Z., Abdullah, I., Integrating photovoltaic (Pv) solar cells and supercapacitors for sustainable energy devices: A review. *Energies*, 14, 7211, 2021.
33. Kumar, B.A., Ran, F., Maram, P.S., Sangaraju, S., Progressive horizons of energy generation and storage: Nook and cranny of photo-supercapacitors. *J. Energy Storage*, 97, 112876, 2024.
34. Meng, H., Pang, S., Cui, G., Photo-Supercapacitors Based on Third-Generation Solar Cells. *ChemSusChem*, 12, 3431–3447, 2019.
35. Sun, K., Zhang, S., Li, P., Xia, Y., Zhang, X., Du, D., Isikgor, F.H., Ouyang, J., Review on application of PEDOTs and PEDOT:PSS in energy conversion and storage devices. *J. Mater. Sci. - Mater. Electron.*, 26, 4438–4462, 2015.
36. Wang, W., Cao, J., Yu, J., Tian, F., Luo, X., Hao, Y., Huang, J., Wang, F., Zhou, W., Xu, J., Liu, X., Yang, H., Flexible Supercapacitors Based on Stretchable Conducting Polymer Electrodes. *Polym. (Basel)*, 15, 1856, 2023.
37. Han, Y. and Dai, L., Conducting Polymers for Flexible Supercapacitors. *Macromol. Chem. Phys.*, 220, 1800355, 2019.
38. Liang, R., Du, Y., Xiao, P., Cheng, J., Yuan, S., Chen, Y., Yuan, J., Chen, J., Transition metal oxide electrode materials for supercapacitors: A review of recent developments. *Nanomaterials*, 11, 1248, 2021.
39. Wang, R., Li, X., Nie, Z., Zhao, Y., Wang, H., Metal/Metal Oxide Nanoparticles-Composited Porous Carbon for High-Performance Supercapacitors. *J. Energy Storage*, 38, 102479, 2021.
40. An, C., Zhang, Y., Guo, H., Wang, Y., Metal oxide-based supercapacitors: progress and prospectives. *Nanoscale Adv.*, 1, 4644–4658, 2019.
41. Maheshwaran, G., Pandi, P., Suganya, S., Kumar, B.A., Ramalingam, G., Prabhu, M.R., Sudhahar, S., Fabrication of self charging supercapacitor based on two dimensional bismuthene-graphitic carbon nitride nanocomposite

powered by dye sensitized solar cells. *J. Energy Storage*, 56105900, 2022, 2022.

42. Ng, C.H., Lim, H.N., Hayase, S., Harrison, I., Pandikumar, A., Huang, N.M., Potential active materials for photo-supercapacitor: A review. *J. Power Sources*, 296, 169–185, 2015.
43. Wang, Q., Tian, X., Gao, Q., Zhang, X., Rong, K., Cao, M., Han, L., Fan, W., Diversified constructions and electrochemical cycling stability of metal oxide fiber supercapacitors. *J. Mater. Res. Technol.*, 24, 909–917, 2023.
44. Kumar, R., Kumar, A., Shukla, P.S., Das Varma, G., Venkataraman, D., Bag, M., Photorechargeable Hybrid Halide Perovskite Supercapacitors. *ACS Appl. Mater. Interfaces*, 14, 35592–35599, 2022.
45. Narayanan, R., Kumar, P.N., Deepa, M., Srivastava, A.K., Combining Energy Conversion and Storage: A Solar Powered Supercapacitor. *Electrochim. Acta*, 178, 113–126, 2015.
46. Das, A., Deshagani, S., Kumar, R., Deepa, M., Bifunctional Photo-Supercapacitor with a New Architecture Converts and Stores Solar Energy as Charge. *ACS Appl. Mater. Interfaces*, 10, 35932–35945, 2018.
47. Ates, M., Kuzgun, O., Candan, I., Supercapacitor performances of titanium-polymeric nanocomposites: a review study. *Iran. Polym. J.*, 31, 1–27, 2022.
48. Wang, Q., Li, M., Wang, Z., Supercapacitive performance of TiO_2 boosted by a unique porous TiO_2/Ti network and activated Ti^{3+}. *RSC Adv.*, 9, 7811–7817, 2019.
49. Cui, J., Cao, L., Zeng, D., Wang, X., Li, W., Lin, Z., Zhang, P., Surface Characteristic Effect of Ag/TiO_2 Nanoarray Composite Structure on Supercapacitor Electrode Properties. *Scanning*, 2018, 2464981, 2018.
50. Kumar, A., Mittal, H., Khanuja, M., Facile synthesis of 2D acid-etched g-C_3N_4 nanosheets with 1D ZnO nanorods as a promising electrode material for supercapacitor. *J. Energy Storage*, 67, 107496, 2023.
51. Kong, L.B., Que, W., Liu, L., Boey, F.Y.C., Xu, Z.J., Zhou, K., Li, S., Zhang, T., Wang, C., Oxide based supercapacitors: II-other oxides, in: *Nanomater. Supercapacitors*, pp. 277–421, CRC Press, Boca Raton, FL, 2017.
52. Angaiah, S., Arunachalam, S., Murugadoss, V., Vijayakumar, G., A Facile Polyvinylpyrrolidone Assisted Solvothermal Synthesis of Zinc Oxide Nanowires and Nanoparticles and Their Influence on the Photovoltaic Performance of Dye Sensitized Solar Cell. *ES Energy Environ.*, 4, 59–65, 2019.
53. Janaky, C., Rajeshwar, K., Samu, G.F., (Photo) Electrochemically Prepared Organic/Inorganic Hybrid Assemblies for Energy Conversion and Storage. *ECS Meet. Abstr.*, 227, 934–934, 2015.
54. Gallegos-Perez, W.R., Corpus-Mendoza, A.N., Lobato-Peralta, D.R., Cuentas-Gallegos, A.K., Camacho-Caceres, J., Arias-Ramos, C.F., Hu, H., A Simplified Equivalent Circuit Model for the Photo-Charging Process of

Carbon-Based Quasi-Solid Photosupercapacitors. *ECS J. Solid State Sci. Technol.*, 12, 115003, 2023.
55. Ezealigo, B.N., Nwanya, A.C., Simo, A., Osuji, R.U., Bucher, R., Maaza, M., Ezema, F.I., Optical and electrochemical capacitive properties of copper (I) iodide thin film deposited by SILAR method. *Arab. J. Chem.*, 12, 5380–5391, 2019.
56. Ma, S., Xu, X., Wu, W., Dual-Aspect Expanding Electrochemical Characteristics of Flexible Supercapacitors via an Ionic Liquid and a High-Boiling-Point Solvent. *ACS Appl. Energy Mater.*, 5, 3401–3408, 2022.
57. Zhang, L. and Zhao, X.S., Carbon-based materials as supercapacitor electrodes. *Chem. Soc. Rev.*, 38, 2520–2531, 2009.
58. Borenstein, A., Hanna, O., Attias, R., Luski, S., Brousse, T., Aurbach, D., Carbon-based composite materials for supercapacitor electrodes: A review. *J. Mater. Chem. A*, 5, 12653–12672, 2017.
59. Ciszewski, M., Koszorek, A., Radko, T., Szatkowski, P., Janas, D., Review of the Selected Carbon-Based Materials for Symmetric Supercapacitor Application. *J. Electron. Mater.*, 48, 717–744, 2019.
60. Salunkhe, R.R., Kaneti, Y.V., Yamauchi, Y., Metal-Organic Framework-Derived Nanoporous Metal Oxides toward Supercapacitor Applications: Progress and Prospects. *ACS Nano*, 11, 5293–5308, 2017.
61. Sayyed, S.G., Mahadik, M.A., Shaikh, A.V., Jang, J.S., Pathan, H.M., Nano-Metal Oxide Based Supercapacitor via Electrochemical Deposition. *ES Energy Environ.*, 3, 25–44, 2019.
62. Lamba, P., Singh, P., Singh, P., Singh, P., Bharti, Kumar, A., Gupta, M., Kumar, Y., Recent advancements in supercapacitors based on different electrode materials: Classifications, synthesis methods and comparative performance. *J. Energy Storage*, 48, 103871, 2022.
63. Unnikrishnan, B., Wu, C.W., Chen, I.W.P., Chang, H.T., Lin, C.H., Huang, C.C., Carbon Dot-Mediated Synthesis of Manganese Oxide Decorated Graphene Nanosheets for Supercapacitor Application. *ACS Sustain. Chem. Eng.*, 4, 3008–3016, 2016.
64. Rajagopal, R. and Ryu, K.S., Homogeneous MnO_2@TiO_2 core-shell nanostructure for high performance supercapacitor and Li-ion battery applications. *J. Electroanal. Chem.*, 856, 113669, 2020.
65. Guo, G., Su, Q., Zhou, W., Wei, M., Wang, Y., Cycling stability of Fe_2O_3 nanosheets as supercapacitor sheet electrodes enhanced by $MgFe_2O_4$ nanoparticles. *RSC Adv.*, 13, 3643–3651, 2023.
66. Wu, H., Li, Y., Xiao, W., Tian, L., Song, J., Preparation and electrochemical properties of mesoporous α-Fe_2O_3 nanowires for supercapacitor application. *J. Mater. Sci. - Mater. Electron.*, 34, 1098, 2023.
67. Shivakumara, S., Penki, T.R., Munichandraiah, N., High specific surface area α-Fe_2O_3 nanostructures as high performance electrode material for supercapacitors. *Mater. Lett.*, 131, 100–103, 2014.

68. Cai, G., Wang, X., Cui, M., Darmawan, P., Wang, J., Eh, A.L.S., Lee, P.S., Electrochromo-supercapacitor based on direct growth of NiO nanoparticles. *Nano Energy*, 12, 258–267, 2015.
69. Lu, C., Liu, L., Yang, Y., Ma, Y., Luo, Q., Zhu, M., Recent Progress in Co_3O_4-Based Nanomaterials for Supercapacitors. *ChemNanoMat*, 9, e202200537, 2023.
70. Akhtar, M.A., Chowdhury, A., Chandra, A., Addition of redox additives - Synergic strategy for enhancing the electrochemical activity of spinel Co_3O_4 based supercapacitors. *J. Phys. D. Appl. Phys.*, 52, 155501, 2019.
71. Brousse, K., Pinaud, S., Nguyen, S., Fazzini, P.F., Makarem, R., Josse, C., Thimont, Y., Chaudret, B., Taberna, P.L., Respaud, M., Simon, P., Facile and Scalable Preparation of Ruthenium Oxide-Based Flexible Micro-Supercapacitors. *Adv. Energy Mater.*, 10, 1903136, 2020.
72. Arunachalam, R., Prataap, R.K.V., Pavul Raj, R., Mohan, S., Vijayakumar, J., Peter, L., Al Ahmad, M., Pulse electrodeposited RuO_2 electrodes for high-performance supercapacitor applications. *Surf. Eng.*, 35, 102–108, 2019.
73. Zhu, D., Zhou, Q., Liang, A., Zhou, W., Chang, Y., Li, D., Wu, J., Ye, G., Xu, J., Ren, Y., Two-step preparation of carbon nanotubes/RuO_2/polyindole ternary nanocomposites and their application as high-performance supercapacitors. *Front. Mater. Sci.*, 14, 109–119, 2020.
74. Varghese, A., Devi, S.K.R., Kausar, F., Pinheiro, D., Evaluative study on supercapacitance behavior of polyaniline/polypyrrole – metal oxide based composites electrodes: a review. *Mater. Today Chem.*, 29, 101424, 2023.
75. Afzal, A., Abuilaiwi, F.A., Habib, A., Awais, M., Waje, S.B., Atieh, M.A., Polypyrrole/carbon nanotube supercapacitors technological advances and challenges. *J. Power Sources*, 352, 174–186, 2017.
76. Eftekhari, A., Li, L., Yang, Y., Polyaniline supercapacitors. *J. Power Sources*, 347, 86–107, 2017.
77. Inci Yesilyurt, E., Pionteck, J., Simon, F., Voit, B., Fabrication of PANI/MWCNT supercapacitors based on a chitosan binder and aqueous electrolyte for enhanced energy storage. *RSC Appl. Polym.*, 1, 97–110, 2023.
78. Dmitry, R., Patel, M., Runa, M.J., Wimmer, M.A., Jacobs, J.J., Hallab, N.J., Mathew, T., Fretting-corrosion behavior of CoCrMo-CoCrMo couple in modular hip junctions: a parametric and mechanistic study. *J. Orthop. Res.*, 62, 570–587, 2016.
79. Shim, Y., Jung, Y., Kim, H.J., Graphene-based supercapacitors: A computer simulation study. *J. Phys. Chem. C*, 115, 23574–23583, 2011.
80. Dhawale, D.S., Vinu, A., Lokhande, C.D., Stable nanostructured polyaniline electrode for supercapacitor application. *Electrochim. Acta*, 56, 9482–9487, 2011.
81. Fan, W., Zhang, C., Tjiu, W.W., Pramoda, K.P., He, C., Liu, T., Graphene-wrapped polyaniline hollow spheres as novel hybrid electrode materials for supercapacitor applications. *ACS Appl. Mater. Interfaces*, 5, 3382–3391, 2013.

82. Crisci, M., Boll, F., Merola, L., Pflug, J.J., Liu, Z., Gallego, J., Lamberti, F., Gatti, T., Nanostructured 2D WS_2@PANI nanohybrids for electrochemical energy storage. *Front. Chem.*, 10, 1000910, 2022.
83. Rahman, R., Rani, P., Ghosh, S., Midya, A., Pathak, A., Nath, T.K., Engineering Hierarchical Self-Assembled PANI-/1T@2H MoS_2 Nanostructure toward Ultrahigh Performance Supercapacitor Electrodes. *ACS Appl. Energy Mater.*, 6, 11012–11029, 2023.
84. Kuttan, S.S., Girija, N., Devaki, S.J., Nair, B.N., Jose, R., Rajaram, N., Mohamed, A.P., Saraswathy Hareesh, U.N., Modulating Electrochemical Performance of Interfacially Polymerized, MoS_2 Decorated Polyaniline Composites for Electrochemical Capacitor Applications. *ACS Appl. Energy Mater.*, 5, 8510–8525, 2022.
85. Luo, L., Wang, X., Yang, S., Dai, J., Li, D., Xia, L., Chi, C., Cabot, A., Xu, Y., Dai, L., PPy/PANI@MoS_2 Composites with a Dual-Channel Architecture for Advanced Asymmetric Supercapacitors. *ACS Appl. Energy Mater.*, 6, 5940–5951, 2023.
86. Huang, Y., Li, H., Wang, Z., Zhu, M., Pei, Z., Xue, Q., Huang, Y., Zhi, C., Nanostructured Polypyrrole as a flexible electrode material of supercapacitor. *Nano Energy*, 22, 422–438, 2016.
87. Huang, Y., Zhu, M., Pei, Z., Huang, Y., Geng, H., Zhi, C., Extremely Stable Polypyrrole Achieved via Molecular Ordering for Highly Flexible Supercapacitors. *ACS Appl. Mater. Interfaces*, 8, 2435–2440, 2016.
88. Kashani, H., Chen, L., Ito, Y., Han, J., Hirata, A., Chen, M., Bicontinuous nanotubular graphene-polypyrrole hybrid for high-performance flexible supercapacitors. *Nano Energy*, 19, 391–400, 2016.
89. Ullah, R., Khan, N., Khattak, R., Khan, M., Khan, M.S., Ali, O.M., Preparation of Electrochemical Supercapacitor Based on Polypyrrole/Gum Arabic Composites. *Polym. (Basel)*, 14, 242, 2022.
90. Zhang, J., Bin Kong, L., Li, H., Luo, Y.C., Kang, L., Synthesis of polypyrrole film by pulse galvanostatic method and its application as supercapacitor electrode materials. *J. Mater. Sci.*, 45, 1947–1954, 2010.
91. Dubal, D.P., Patil, S.V., Jagadale, A.D., Lokhande, C.D., Two step novel chemical synthesis of polypyrrole nanoplates for supercapacitor application. *J. Alloys Compd.*, 509, 8183–8188, 2011.
92. Gonçalves, R., Paiva, R.S., Lima, T.M., Paixao, M.W., Pereira, E.C., Carbon nitride/polypyrrole composite supercapacitor: Boosting performance and stability. *Electrochim. Acta*, 368, 137570, 2021.
93. Vijeth, H., Ashokkumar, S.P., Yesappa, L., Niranjana, M., Vandana, M., Devendrappa, H., Flexible and high energy density solid-state asymmetric supercapacitor based on polythiophene nanocomposites and charcoal. *RSC Adv.*, 8, 31414–31426, 2018.
94. Patil, B.H., Patil, S.J., Lokhande, C.D., Electrochemical characterization of chemically synthesized polythiophene thin films: Performance of asymmetric supercapacitor device. *Electroanalysis*, 26, 2023–2032, 2014.

95. Xue, Q., Sun, J., Huang, Y., Zhu, M., Pei, Z., Li, H., Wang, Y., Li, N., Zhang, H., Zhi, C., Recent Progress on Flexible and Wearable Supercapacitors. *Small*, 13, 1701827, 2017.
96. Tadesse, M.G. and Lubben, J.F., Review on Hydrogel-Based Flexible Supercapacitors for Wearable Applications. *Gels*, 9, 106, 2023.
97. Wang, H., Diao, Y., Lu, Y., Yang, H., Zhou, Q., Chrulski, K., D'Arcy, J.M., Energy storing bricks for stationary PEDOT supercapacitors. *Nat. Commun.*, 11, 1–9, 2020.
98. Bi, W., Wu, Y., Liu, C., Wang, J., Du, Y., Gao, G., Wu, G., Cao, G., Gradient oxygen vacancies in V_2O_5/PEDOT nanocables for high-performance supercapacitors. *ACS Appl. Energy Mater.*, 2, 668–677, 2019.
99. Anothumakkool, B., Soni, R., Bhange, S.N., Kurungot, S., Novel scalable synthesis of highly conducting and robust PEDOT paper for a high performance flexible solid supercapacitor. *Energy Environ. Sci.*, 8, 1339–1347, 2015.
100. Mohd Abdah, M.A.A., Zubair, N.A., Azman, N.H.N., Sulaiman, Y., Fabrication of PEDOT coated PVA-GO nanofiber for supercapacitor. *Mater. Chem. Phys.*, 192, 161–169, 2017.
101. Qin, L., Jiang, J., Tao, Q., Wang, C., Persson, I., Fahlman, M., Persson, P.O.A., Hou, L., Rosen, J., Zhang, F., A flexible semitransparent photovoltaic supercapacitor based on water-processed MXene electrodes. *J. Mater. Chem. A*, 8, 5467–5475, 2020.
102. Akhter, R. and Maktedar, S.S., MXenes: A comprehensive review of synthesis, properties, and progress in supercapacitor applications. *J. Mater.*, 9, 1196–1241, 2023.
103. Liu, Z., Wang, L., Xu, Y., Guo, J., Zhang, S., Lu, Y., A Ti_3C_2TX@PEDOT composite for electrode materials of supercapacitors. *J. Electroanal. Chem.*, 881, 114958, 2021.
104. Das, A., Deshagani, S., Ghosal, P., Deepa, M., Redox active and electrically conducting cobalt telluride Nanorods/Poly(1-aminoanthraquinone) composite and photoactive Rose Bengal dye based photo-supercapacitor. *Appl. Mater. Today*, 19, 100592, 2020.
105. Duvva, N., Reddy, G., Singh, S.P., Chowdhury, T.H., Bedja, I., Islam, A., Giribabu, L., Functional π-conjugated tetrathiafulvalene decorated with benzothiadiazole organic sensitizers for dye sensitized solar cells. *New J. Chem.*, 43, 8919–8929, 2019.
106. Han, L., Meng, X., Ke, Y., Ye, H., Cui, Y., Novel D-π-A benzocarbazole dyes with simple structures for efficient dye-sensitized solar cells. *J. Photochem. Photobiol. A Chem.*, 376, 127–134, 2019.
107. Liu, R., Takakuwa, M., Li, A., Inoue, D., Hashizume, D., Yu, K., Umezu, S., Fukuda, K., Someya, T., An Efficient Ultra-Flexible Photo-Charging System Integrating Organic Photovoltaics and Supercapacitors. *Adv. Energy Mater.*, 10, 2000523, 2020.

108. Ma, Q., Jia, Z., Meng, L., Zhang, J., Zhang, H., Huang, W., Yuan, J., Gao, F., Wan, Y., Zhang, Z., Li, Y., Promoting charge separation resulting in ternary organic solar cells efficiency over 17.5 %. *Nano Energy*, 78, 105272, 2020.
109. Lin, Y., Adilbekova, B., Firdaus, Y., Yengel, E., Faber, H., Sajjad, M., Zheng, X., Yarali, E., Seitkhan, A., Bakr, O.M., El-Labban, A., Schwingenschlogl, U., Tung, V., McCulloch, I., Laquai, F., Anthopoulos, T.D., 17 % Efficient Organic Solar Cells Based on Liquid Exfoliated WS_2 as a Replacement for PEDOT:PSS. *Adv. Mater.*, 31, 1902965, 2019.
110. Zhu, Y., Yang, L., Zhao, S., Huang, Y., Xu, Z., Yang, Q., Wang, P., Li, Y., Xu, X., Improved performances of PCDTBT:PC71BM BHJ solar cells through incorporating small molecule donor. *Phys. Chem. Chem. Phys.*, 17, 26777–26782, 2015.
111. Ng, C.H., Lim, H.N., Hayase, S., Zainal, Z., Shafie, S., Lee, H.W., Huang, N.M., Cesium Lead Halide Inorganic-Based Perovskite-Sensitized Solar Cell for Photo-Supercapacitor Application under High Humidity Condition. *ACS Appl. Energy Mater.*, 1, 692–699, 2018.
112. Song, Z., Wu, J., Sun, L., Zhu, T., Deng, C., Wang, X., Li, G., Du, Y., Chen, Q., Sun, W., Fan, L., Chen, H., Lin, J., Lan, Z., Photocapacitor integrating perovskite solar cell and symmetrical supercapacitor generating a conversion storage efficiency over 20 %. *Nano Energy*, 100, 107501, 2022.
113. Solís-Cortes, D., Navarrete-Astorga, E., Schrebler, R., Peinado-Perez, J.J., Martin, F., Ramos-Barrado, J.R., Dalchiele, E.A., A solid-state integrated photo-supercapacitor based on ZnO nanorod arrays decorated with Ag_2S quantum dots as the photoanode and a PEDOT charge storage counter-electrode. *RSC Adv.*, 10, 5712–5721, 2020.
114. Liu, Y., Chen, X., Mao, B., Ying, Y., Li, L., Shi, W., Black phosphorus quantum dots enabled photo-assisted supercapacitor with boosted volumetric charge storage capability. *J. Mater. Sci. Technol.*, 191, 80–88, 2024.

Part 2

SOLAR RECHARGEABLE BATTERIES AND HYBRID DEVICES

4

Photo-Rechargeable All-Solid-State Batteries Based on Photoelectrochemistry and Solid-State Ionics

Kenta Watanabe[1]* and Masaaki Hirayama[1,2]

[1]*Department of Chemical Science and Engineering, School of Materials and Chemical Technology, Institute of Science Tokyo, Yokohama, Japan*
[2]*Research Center for All-Solid-State Battery, Institute of Integrated Research, Institute of Science Tokyo, Yokohama, Japan*

Abstract

All charge and discharge reactions of batteries including all-solid-state batteries (ASSBs) and Li-ion batteries (LIBs) consist of electrochemical redox reactions. Therefore, photoelectrochemical redox reactions must proceed so as to charge batteries directly with energies of irradiated light. Photoelectrochemical reactions generally utilize semiconductor materials as electrodes. n- and p-type semiconductors are active for photoelectrochemical oxidation and reduction, respectively. Photoelectrochemical reactions proceed even without external voltages when n- and p-type semiconductors are applied to each electrode. This means that charge reactions can photoelectrochemically proceed without external voltages by using n- and p-type semiconductor electrodes. However, photoelectrochemical charge reactions without external voltages do not proceed in all combination between n- and p-type semiconductors. A photo-responsive potential region is a key factor. In photo-rechargeable ASSBs (PR-ASSBs), Li^+-(de)intercalation abilities are also required to electrode materials, which should be studied with based on solid-state ionics. In this chapter, we first introduce general photoelectrochemical reactions using n- and p-type semiconductor electrodes in cells using liquid electrolytes. Then, bulk and thin-film type ASSBs are explained. Finally, PR-ASSBs and their principles are introduced.

**Corresponding author*: watanabe.k.c173@m.isct.ac.jp

Nurdan Demirci Sankir and Mehmet Sankir (eds.) Solar Capacitors and Batteries, (85–104) © 2025 Scrivener Publishing LLC

***Keywords*:** Photoelectrochemistry, semiconductor, flat band, band bending, solid-state ionics, Li^+-(de)intercalation

4.1 Introduction: Problems of Research on Photo-Rechargeable Batteries

Both primary and secondary batteries work based on electrochemistry. Considering from this viewpoint, photo-rechargeable batteries (PRBs) should photoelectrochemically function. This is why all PRBs must be studied based on photoelectrochemistry. However, there are few reports about PRBs studied from the viewpoint of photoelectrochemistry. This is because photoelectrochemistry has been developed in the research field of not batteries but photocatalysis and artificial photosynthesis [1–7]. Photoelectrochemistry started to be studied from finding Honda-Fujishima effect [8, 9]. The reports of Honda-Fujishima effect focused on photoelectrochemical water splitting into hydrogen and oxygen. Then, many researchers have studied about photoelectrochemistry as parts of photocatalysis and artificial photosynthesis toward solar energy conversion. This is why photoelectrochemistry is hardly related to research fields of batteries. Nevertheless, almost all PRBs are studied by researchers in research fields of batteries, resulting in no consideration from the viewpoint of photoelectrochemistry [10–23]. This leads to no basic understanding of photo-rechargeabilities. To accelerate basic understanding of PRBs, they should be studied based on photoelectrochemistry, as usual batteries based on electrochemistry. In the cases of giving photo-rechargeabilities to all-solid-state batteries (ASSBs), solid-state ionics also should be considered because ionic conduction is required in all parts of cathodes, solid electrolytes, and anodes. From these points, the authors have studied photo-rechargeable ASSBs (PR-ASSBs) based on both photoelectrochemistry and solid-state ionics [24, 25]. In this chapter, the authors first explain the basic principles about photoelectrochemical reactions. Then, ASSBs are introduced. Finally, photo-rechargeabilities in ASSBs are explained.

4.2 General Principles of Photoelectrochemical Reactions Using Semiconductor Electrodes

In photoelectrochemical reactions, electrons and holes must be generated by photoexcitation using incident photons. Therefore, most of photoelectrodes consist of semiconductor materials. When incident photons possess energies larger than band gaps (BGs) of semiconductors, electrons are photoexcited from their valence bands (VBs) to the conduction bands (CBs), resulting in generation of holes in the VB. These photogenerated electrons and holes are utilized in photoelectrochemical reactions. Semiconductor properties decide which carrier is applied to photoelectrochemical reactions. The photogenerated electrons and holes are utilized in p- and n-type semiconductor electrodes, respectively. Detailed mechanisms are described in the following sections.

4.2.1 Under Dark Conditions

Before explanation about photoelectrochemical reactions, we have to understand semiconductors themselves and mechanisms of usual (= dark) electrochemical reactions using the semiconductor electrodes. Differences between p- and n-type semiconductors are mainly kinds of majority carriers and potentials of fermi levels (E_f). Fermi levels in intrinsic (i-type) semiconductors are located at the middle of their BG, as shown in Figure 4.1(a). This is because their CBs never include free electrons and because their VBs are completely filled with electrons, respectively. Electrons in the VB are thermally excited to the CB with increasing temperature, resulting in electronic conductivities. In the case of n-type semiconductors (Figure 4.1(b)), donor levels (DLs) are formed at slightly more positive potentials than their conduction band minima (CBM). Electrons in the donor levels are thermally excited to the CBs even at room temperatures. The electrons excited from the DL function as majority carriers for electric conductivities. Then, E_f is located near the CBM. In contrast, acceptor levels (ALs) are formed at slightly more negative potentials than valence band maxima (VBM) of p-type semiconductors, as shown in Figure 4.1(c). Electrons in the VB are thermally excited to the AL even at room temperatures, resulting in generation of holes in the VB. In the p-type semiconductors, the holes generated in the VB work as majority carriers. Then, E_f is close to the

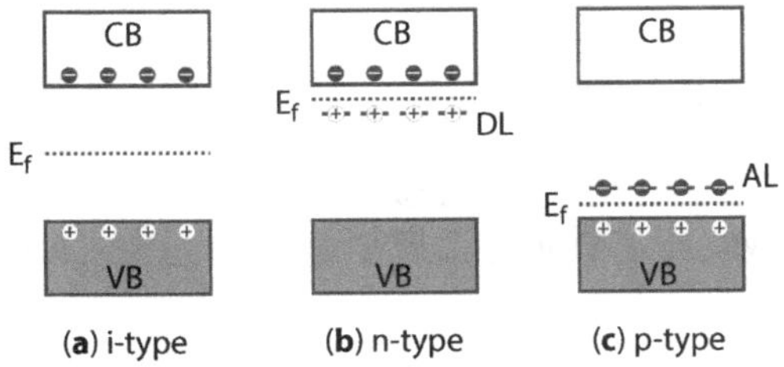

Figure 4.1 Band structures of i-, p-, and n-type semiconductors.

VBM. Thus, p-type semiconductors possess opposite properties to n-type ones in majority carriers and E_f.

Opposite electrochemical properties are also shown by p- and n-type semiconductor electrodes. As a premise, band edges of semiconductors bend toward negative or positive potentials with keeping the edge potentials by applied voltages and/or contacts with other materials, as shown in Figure 4.2. The kept potentials of the band edges correspond to flat band potentials (E_{FB}). When n-type semiconductor electrodes are located at more positive potentials than E_{FB} (Figure 4.2(a)), electrons of majority carriers cannot migrate to electrolytes because of energy barriers by the band edges bent toward negative potentials. Consequently, the electrons cannot be used for reduction, and hence, any currents do not flow. The energy barriers forbid carriers from migrating also from the electrolytes to the electrodes. At more negative potentials than E_{FB} (Figure 4.2(b)), E_f overlaps into CB because of the band bending toward positive potentials, resulting in the appearance of metallic electron conductivities. Additionally, the band bending toward positive potentials does not prohibit electrons from migration to electrolytes because of no energy

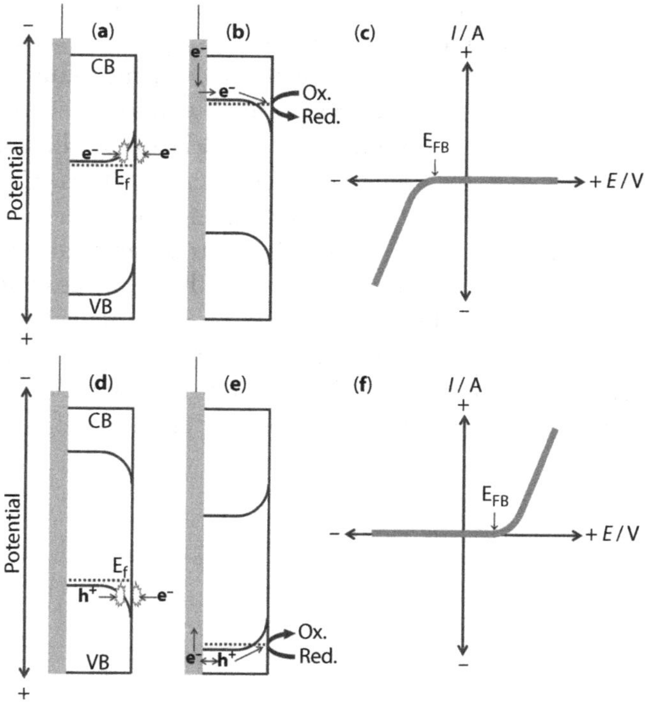

Figure 4.2 Electrochemical properties of n- and p-type semiconductor electrodes under dark conditions.

barriers. Both the metallic electron conductivities and the no energy barriers enable the electrons to reduce oxidants. Thus, when n-type semiconductors are utilized as electrodes, only cathodic currents flow at more negative potentials than E_{FB} (Figure 4.2(c)). p-type semiconductor electrodes show opposite electrochemical properties to n-type ones. When p-type semiconductors are located at more negative potentials than E_{FB} (Figure 4.2(d)), holes in the VB cannot migrate to electrolytes owing to energy barriers by the band bending toward positive potentials. The energy barriers prohibit carriers from migrating also from the electrolytes to the electrodes. At more positive potentials than E_{FB} (Figure 4.2(e)), the holes can oxidize reductants due to metallic electron conductivities and no energy barriers. Therefore, only anodic currents flow at more positive potentials than E_{FB} (Figure 4.2(f)). These are basic electrochemical properties of p- and n-type semiconductors.

4.2.2 Under Light Irradiation

It is the most different point under light irradiation from dark conditions to consider both photogenerated electrons and holes. When n-type semiconductor electrodes are located at more positive potentials than E_{FB}, electrons in CB cannot migrate to electrolytes under dark conditions, resulting in no reactions and no currents, as mentioned in the previous section. However, photogenerated holes can migrate to electrolytes under light irradiation, as shown in Figure 4.3(a). Then, if the VBM (or quasi fermi levels of holes (q-E_{fh})) is more positive than redox potentials, oxidation can proceed. The original and photoexcited electrons in the CB flow to external circuits corresponding to the amount of the reacted holes. At more negative potentials than E_{FB} (Figure 4.3(b)), n-type semiconductors are in the states showing metallic electron conductivities, whose carrier density is higher than those of the photogenerated carriers. Of course, materials with many free electrons are used as current collectors, meaning that they have also metallic electron conductivities. Thus, both the photogenerated electrons and holes cannot contribute to electrochemical properties [26]. For these reasons, n-type semiconductor electrodes exhibit anodic photocurrents at more positive potentials than E_{FB} under light irradiation, as shown in Figure 4.3(c). This is why photoresponsive n-type semiconductor electrodes are called photoanodes. Also cathodic currents flow at more negative potentials than E_{FB} under light irradiation, but these currents are almost the same as those under dark conditions. In the case of p-type semiconductor electrodes, photogenerated electrons can migrate to electrolytes at more negative potentials than E_{FB}, even holes cannot migrate due to the energy barriers by band bending toward positive potentials (Figure 4.3(d)).

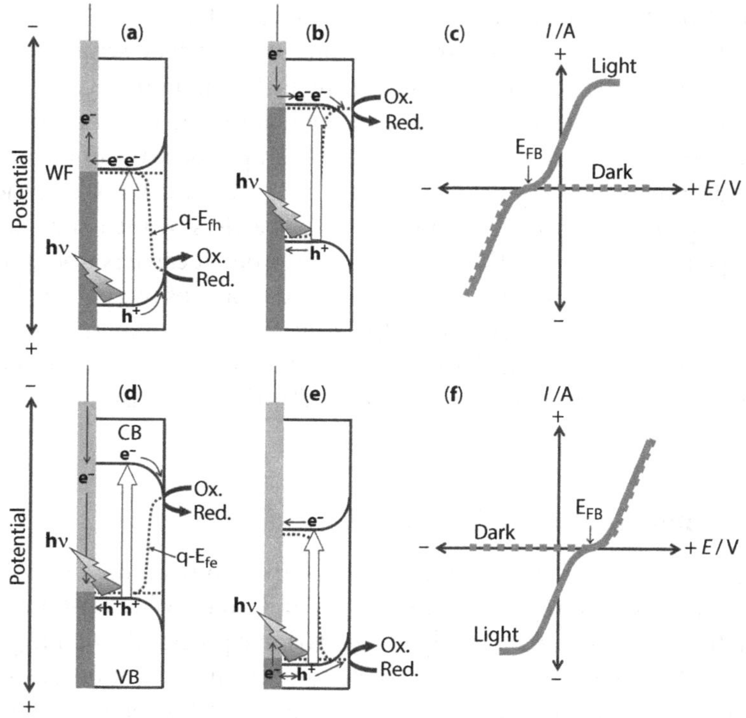

Figure 4.3 Photoelectrochemical properties of n- and p-type semiconductor electrodes under light irradiation.

Then, if the CBM (or quasi fermi levels of electrons (q-E_{fe})) is more negative than redox potentials, reduction can proceed. The original and photogenerated holes in the VB flow to external circuits corresponding to the amount of the reacted electrons. When p-type semiconductors are located at more positive potentials than E_{FB} (Figure 4.3(e)), electrochemical properties are hardly changed by light irradiation like n-type semiconductors at more negative potentials than E_{FB}. This is also because the photogenerated carriers contribute lower to the electrochemical properties than the original carriers in the state showing metallic electron conductivities. For these reasons, p-type semiconductor electrodes exhibit cathodic photocurrents at more negative potentials than E_{FB} under light irradiation, as shown in Figure 4.3(f). This is why photo-responsive p-type semiconductor electrodes are called photocathodes. Also anodic currents flow at more positive potentials than E_{FB} under light irradiation, but these currents are almost the same as those under dark conditions.

4.2.3 Photoelectrochemical Reactions without External Voltages

Electrochemical reactions consist of combinations between reduction at cathodes and oxidation at anodes. p- and n-type semiconductors are active for photoelectrochemical reduction and oxidation as photocathodes and photoanodes, respectively. Thus, photoelectrochemical reactions are expected to proceed without external voltages by combining photocathodes and photoanodes. In fact, there are reports of photoelectrochemical water splitting into H_2 and O_2 without external voltages by combining photocathodes and photoanodes [27–35]. However, photoelectrochemical reactions without external voltages do not proceed in all combination between photocathodes and photoanodes even if the photoelectrodes show high activities. In this section, photoelectrochemical reactions without external voltages are explained in detail.

Photoelectrochemical reactions without external voltages mean the reactions in short circuits. In short circuits, all of cathodes, anodes, and electrolytes (redox couples) try to be in the same potentials as each other, meaning the entire short circuits get equilibrium. Therefore, both cathodes and anodes should be photoelectrochemically active at the same potentials as each other. When photocathodes or photoanodes are connected to non-photo-responsive counter electrodes such as Pt and carbon (Figure 4.4(a,b)), redox reactions attempt to proceed only at the photoelectrodes due to photogenerated carriers possessing high potential energies. The counter electrodes are in equilibrium with electrolytes including redox couples, resulting in no driving force for redox reactions. Consequently, photoelectrochemical reactions hardly proceed in the short circuits connecting the photoelectrodes with the non-photo-responsive counter electrodes.

Next, Figure 4.5 shows cases of short circuits connecting photocathodes and photoanodes showing high activities in utilizing separately. These short circuits should possess driving force for reduction and oxidation derived from photogenerated carriers, respectively. However, this is not enough for progress of photoelectrochemical reactions without external voltages. The photo-responsive potential regions of photocathodes need to overlap with those of photoanodes. When photo-responsive potential regions of photocathodes do not overlap with those of photoanodes like Figure 4.5(a), one electrode cannot photoelectrochemically function even if the other electrode can do, wherever equilibrium potential is. These situations are almost the same as those of the short circuit connecting photoelectrodes

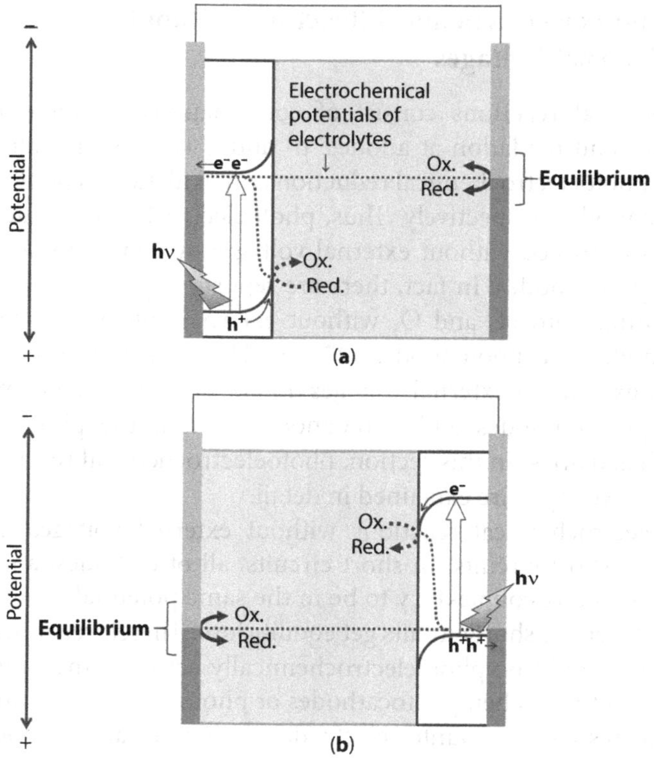

Figure 4.4 Short circuits consisting of n- or p-type semiconductor photoelectrodes and non-photo-responsive counter electrodes.

and non-photo-responsive counter electrodes. Additionally, band edges of photocathodes and/or photoanodes bend toward negative and/or positive potentials outside their photo-responsive potential regions, respectively (Figure 4.5(b-d)). This is why photogenerated electrons in photocathodes and holes in photoanodes cannot migrate to electrolytes, resulting in no redox reactions. Then, photoelectrochemical reactions do not proceed without external voltages. Of course, when both photocathodes and photoanodes are located at potentials outside their photo-responsive regions, photoelectrochemical reactions never proceed because of equilibrium at both the photoelectrodes (Figure 4.5(d)).

When photo-responsive potential regions of photocathodes obviously overlap with those of photoanodes like Figure 4.6(a), both the photocathodes and the photoanodes can photoelectrochemically work in the overlapped potential regions. In these situations, photoelectrochemical reactions

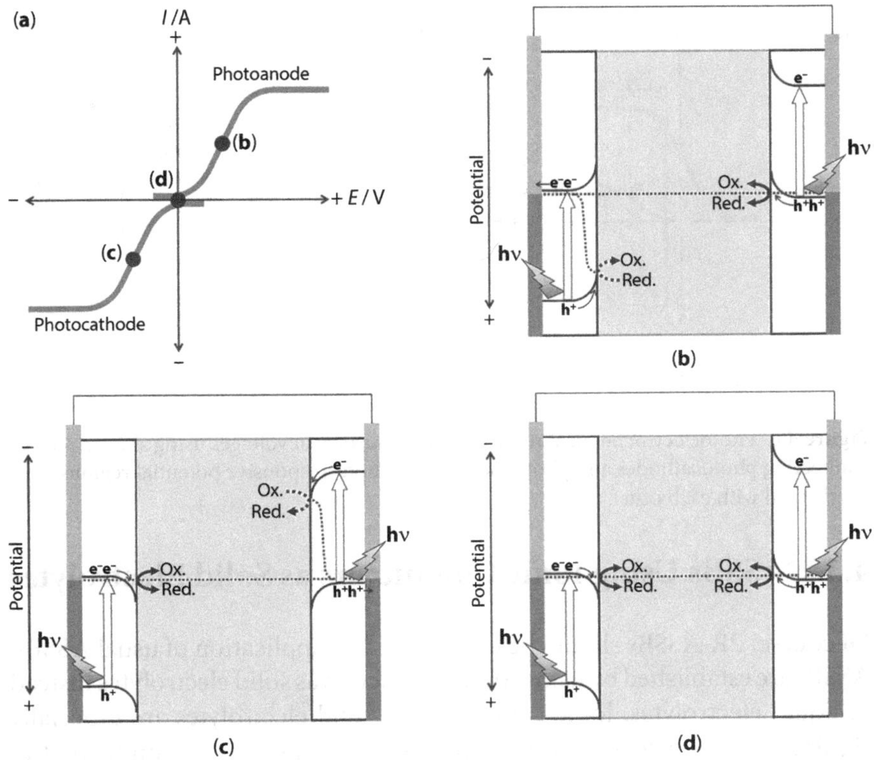

Figure 4.5 Short circuits connecting photocathodes and photoanodes not sharing photo-responsive potential regions.

proceed even without external voltages (Figure 4.6(b)). The amounts of electrons reacted at cathodes must be equal to those at anodes in electrochemical reactions. This means that cathodic photocurrents must agree with anodic photocurrents. In addition, potentials of photocathodes should be same as those of photoanodes in short circuits. From these points, both photocathodes and photoanodes function at the potentials where both of them show the same current in separate measurements of each other, in short circuit cells. Actually, it is reported that the photocurrent in the short circuit cell is consistent with the photocurrent displayed in both the separately measured photocathode and photoanode at the same potential as each other [34]. Thus, the overlap of photo-responsive potential regions is necessary to achieve photoelectrochemical reactions without external voltages. The principle of the photoelectrochemical reactions without external voltages can be applied to photo-charge of batteries without electric power sources.

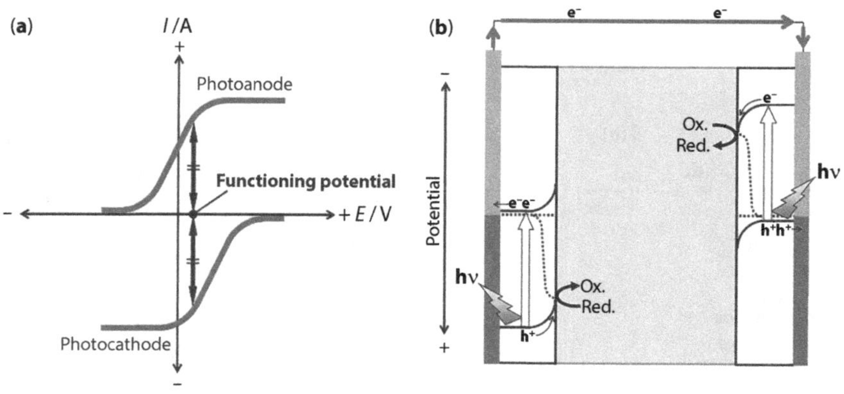

Figure 4.6 Photoelectrochemical reactions without external voltages using short circuits connecting photocathodes and photoanodes with photo-responsive potential regions overlapped with each other.

4.3 ASSBs Using Ionic Conductors as Solid Electrolytes

Of course, PR-ASSBs should be constructed by application of usual ASSBs. ASSBs are established by using ionic conductors as solid electrolytes instead of liquid electrolytes. Ionic conductors as solid electrolytes are especially significant owing to the large contribution of their ionic conductivities to battery performance. Many kinds of super ionic conductors have been developed to be applied as solid electrolytes into ASSBs [36–43]. The development of super ionic conductors has promoted research on ASSBs. Li^+ are often selected as carrier ions of ASSBs and their solid electrolytes because there are many kinds of lithium ionic super conductors. Therefore, materials and knowledges of lithium ion batteries (LIBs) are often applied to ASSBs. Figure 4.7 shows a conceptual sketch of a representative ASSB.

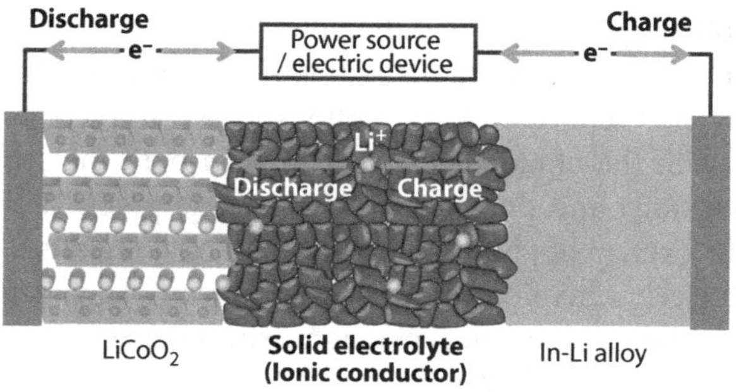

Figure 4.7 A conceptual sketch of a representative ASSB.

$LiCoO_2$ is used as a cathode active material like LIBs [44, 45]. Solid electrolytes tend to be metal phosphorus sulfides showing high Li^+-conductivities (10^{-3}–10^{-2} S cm^{-1}) such as LGPS-type [38, 43, 46], argyrodite-type [39, 42, 47], and $Li_7P_3S_{11}$ [48]. In–Li alloy [49–52] is applied to an anode because of the high electrochemical stability at interfaces between it and sulfide solid electrolytes. ASSBs are mainly divided into bulk and thin-film types.

4.3.1 Bulk Type

Bulk-type ASSBs are generally and widely studied by many researchers. The bulk-type ASSBs usually consist of powdered cathode materials, solid electrolytes, and anode materials. The powdered materials are pelletized with foils (+ meshes) as current collectors by pressing under several hundred MPa, as shown in Figure 4.8. It is relatively easy and low-cost to fabricate the bulk-type ASSBs in laboratory-scale experiments, compared with thin-film-type ASSBs. The bulk-type ASSBs can easily utilize and investigate many kinds of materials. Furthermore, various shapes and sizes can be realized by changing molds for pelletizing. This is advantageous in large-scale and -amount utilization. However, the bulk-type ASSBs have disadvantages in fully bringing out theoretical abilities of utilized materials. Charge and discharge reactions in ASSBs proceed *via* interfaces among particles such as cathode active material/cathode active material, cathode active material/solid electrolyte, solid electrolyte/solid electrolyte, solid electrolyte/anode active material, and anode active material/anode active material. This is why

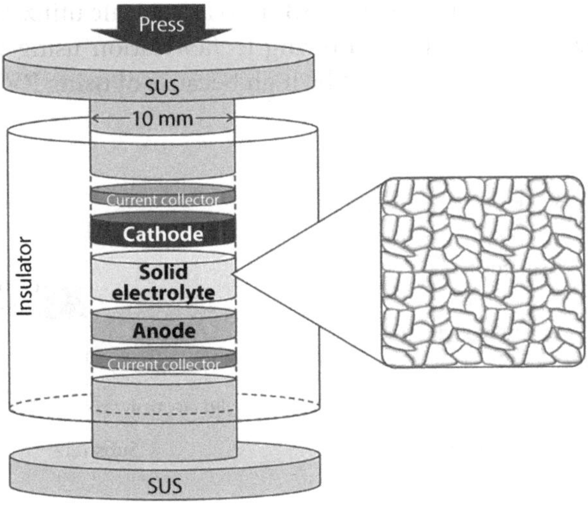

Figure 4.8 A schematic of bulk type ASSBs.

contacts among the particles are quite significant for performances of ASSBs. Contacts between active materials and solid electrolytes are especially severe because the solid electrolytes cannot penetrate into tiny spaces among the particles of the active materials unlike liquid electrolytes. Actually, the bulk-type ASSBs generally keep being pressed to maintain contacts among particles during charge–discharge tests. Moreover, composites comprising active materials and solid electrolytes are often utilized for electrodes to ensure contacts between the active materials and the solid electrolytes.

4.3.2 Thin-Film Type

Thin-film-type ASSBs are fabricated by stacking cathode materials, solid electrolytes, and anode materials layer by layer in order on substrates using various physical vapor deposition (PVD) methods, as shown Figure 4.9 [53]. In the thin-film-type ASSBs, resistances of electronic and ionic conductions in each layer can be decreased by decreasing thickness. For example, amorphous Li_3PO_4 solid electrolyte, which shows 7×10^{-8} S cm^{-1} of the ionic conductivity [54, 55], is generally utilized due to easy deposition. Additionally, each layer is uniformly deposited in two dimensions, resulting in steadily dense contacts at interfaces between the layers without pressures. Furthermore, charge–discharge reactions can be promoted by exposing crystal facets advantageous in Li^+-(de)intercalation to solid electrolytes using epitaxially grown active materials. From these points, the thin-film-type ASSBs easily show theoretical abilities [56, 57]. The thin-film-type ASSBs are suitable for small electric devices. In contrast, it is difficult to apply the thin-film-type ASSBs into large-scale utilization because increasing sizes of them is hard owing to fabrication using PVD methods. Fabrication costs also tend to be high because of using PVD methods.

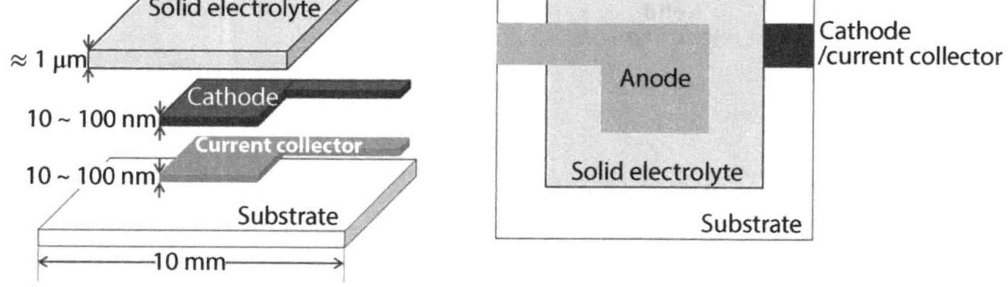

Figure 4.9 Schematics of thin-film-type ASSBs.

It is also difficult to accurately synthesize objective materials. However, the thin-film-type ASSBs are suitable for various analyses because of their simple compositions and structures. The authors have actually analyzed thin-film-type ASSBs *in situ* using surface X-ray diffractometry, neutron reflectometry, and hard X-ray photoelectron spectroscopy [57–61].

4.4 Photo-Rechargeable ASSBs

PRBs require pathways to working electrodes for incident light. The bulk-type ASSBs are surrounded with dies and pins, as shown in Figure 4.8. It is difficult to ensure pathways for incident light in dies or pins. In the thin-film-type ASSBs, materials shutting out incident light are only substrates. The substrates also do not interrupt incident light when they are transparent. Moreover, incident light can reach to the entire electrodes because of thinness. For these reasons, the thin-film-type ASSBs are more suitable for PR-ASSBs than the bulk-type ones. The authors have actually studied photo-rechargeabilities using thin-film-type ASSBs [24, 25]. Even in ASSBs, photoelectrochemical reactions should proceed based on the principles same as those mentioned in Section 4.2. The differences of ASSBs from usual cells using liquid electrolytes are that active materials themselves are reduced and oxidized and that Li^+ ions are (de)intercalated from/into the active materials. Then, the principles of photoelectrochemical reactions in ASSBs should be as discussed below.

When $LiMO_2$ (M: metal cation) of an n-type semiconductor possessing a Li^+-(de)intercalation ability is used as a working electrode (Figure 4.10(a)), photoelectrochemical oxidation proceeds at more positive potentials than E_{FB} of $LiMO_2$ following the reactions below.

Working electrode (n-type semiconductor)

$$x\text{h}\nu \rightarrow xe^- \text{ (CB)} + xh^+ \text{ (VB)} \tag{4.1}$$

$$xh^+ + LiM^{3+}O_2 \rightarrow Li_{1-x}M^{3+}{}_{1-x}M^{4+}{}_xO_2 + xLi^+ \tag{4.2}$$

xe^- in reaction (4.1) and xLi^+ in reaction (4.2) migrate from the working electrode to the counter electrodes (metallic Li) through the external circuit and the solid electrolyte, respectively. Then, the following reaction proceeds.

Counter electrode

$$xe^- + xLi^+ \rightarrow xLi \tag{4.3}$$

The authors have experimentally demonstrated that reactions (4.1) and (4.2) photoelectrochemically proceed at more positive potentials than E_{FB} of Nb-doped anatase-Li_xTiO_2 electrode (a-Li_xTiO_2:Nb) [24, 25]. This demonstration also indicates that photoelectrochemical reactions can progress even in all-solid-state cells based on the same principles as those in usual cells using liquid electrolytes.

Considering both the demonstration using a-Li_xTiO_2:Nb and the usual photoelectrochemical reduction described in Figure 4.3, below, photoelectrochemical reduction should progress at more negative potentials than E_{FB} of p-type semiconductors with Li^+-(de)intercalation abilities (MO_2), as shown in Figure 4.10(b).

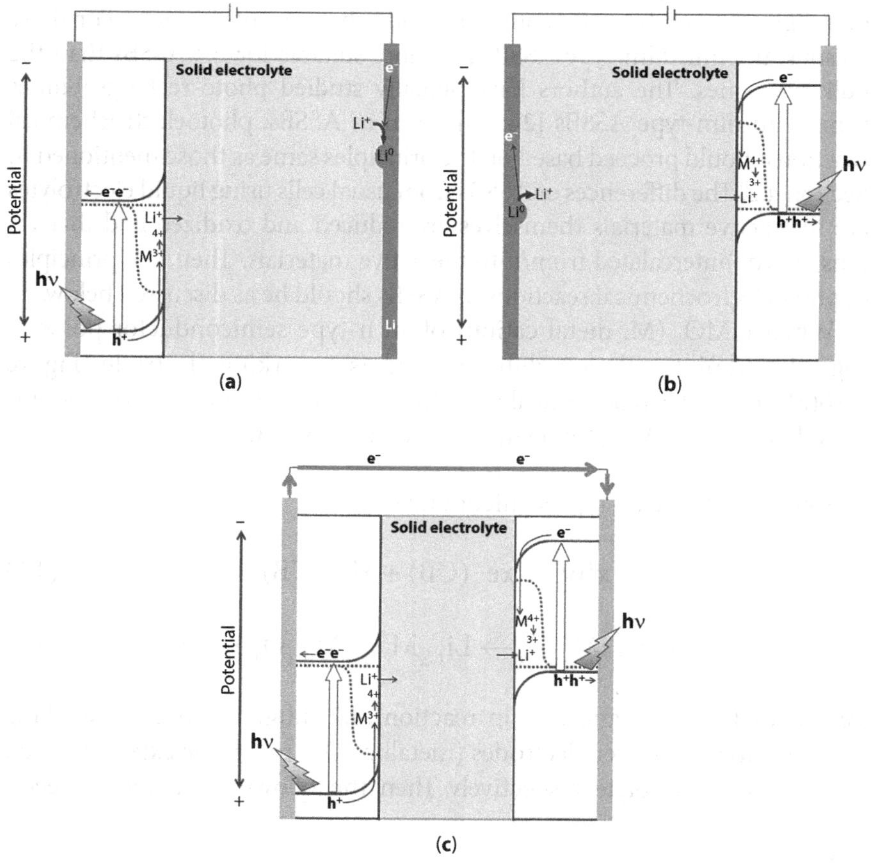

Figure 4.10 Photoelectrochemical reactions using n- and p- semiconductor photoelectrodes possessing Li^+-(de)intercalation abilities in all-solid-state cells.

Working electrode (p-type semiconductor)

$$x\text{h}\nu \rightarrow xe^- \text{ (CB)} + x\text{h}^+ \text{ (VB)} \tag{4.4}$$

$$xe^- + x\text{Li}^+ + \text{M}^{4+}\text{O}_2 \rightarrow \text{Li}_x\text{M}^{4+}{}_{1-x}\text{M}^{3+}{}_x\text{O}_2 \tag{4.5}$$

$x\text{Li}^+$ in reaction (4.5) comes from the counter electrode through the solid electrolyte. $x\text{h}^+$ in reaction (4.4) migrates from the working electrode to the external circuit. Then, the following reaction proceeds, corresponding to the amount of h^+ migrating to the external circuit.

Counter electrode

$$x\text{Li} \rightarrow x\text{Li}^+ + xe^- \tag{4.6}$$

The produced $x\text{Li}^+$ and xe^- migrate to the working electrode through the solid electrolyte and the external circuit, respectively.

Finally, when the above p- and n-type semiconductors are connected to each other, photoelectrochemical reactions without external voltages proceed in the overlapped photo-responsive potential region, as shown in Figure 4.10(c). Reactions (4.1), (4.2) and (4.4), (4.5) progress at the n- and p-type semiconductor photoelectrodes, respectively. The Li^+ ions deintercalated from the n-type semiconductor photoelectrodes migrate to the p-type semiconductor photoelectrodes through the solid electrolytes and are subsequently intercalated again. e^- and h^+ in n- and p-type semiconductor photoelectrodes migrate to the external circuits, respectively, resulting in photocurrents. The progress of these photoelectrochemical reactions means photo-charge without electric power sources. Thus, it is necessary to develop both p- and n-type semiconductors possessing Li^+-(de)intercalation abilities and overlapped photo-responsive potential regions so as to construct PR-ASSBs.

References

1. Iwase, A. and Kudo, A., Photoelectrochemical Water Splitting Using Visible-Light-Responsive $BiVO_4$ Fine Particles Prepared in an Aqueous Acetic Acid Solution. *J. Mater. Chem.*, 20, 7536, 2010.
2. Abe, R., Recent Progress on Photocatalytic and Photoelectrochemical Water Splitting under Visible Light Irradiation. *J. Photochem. Photobiol. C-Photochem. Rev.*, 11, 179, 2010.

3. Pinaud, B.A., Benck, J.D., Seitz, L.C., Forman, A.J., Chen, Z., Deutsch, T.G., James, B.D., Baum, K.N., Baum, G.N., Ardo, S., Wang, H., Miller, E., Jaramillo, T.F., Technical and Economic Feasibility of Centralized Facilities for Solar Hydrogen Production via Photocatalysis and Photoelectrochemistry. *Energy Environ. Sci.*, 6, 1983, 2013.
4. Hisatomi, T., Kubota, J., Domen, K., Recent Advances in Semiconductors for Photocatalytic and Photoelectrochemical Water Splitting. *Chem. Soc. Rev.*, 43, 7520, 2014.
5. Ager, J.W., Shaner, M.R., Walczak, K.A., Sharp, I.D., Ardo, S., Experimental Demonstrations of Spontaneous, Solar-Driven Photoelectrochemical Water Splitting. *Energy Environ. Sci.*, 8, 2811, 2015.
6. Maeda, K., Metal-Complex/Semiconductor Hybrid Photocatalysts and Photoelectrodes for CO_2 Reduction Driven by Visible Light. *Adv. Mater.*, 31, 1808205, 2019.
7. Yoshino, S., Takayama, T., Yamaguchi, Y., Iwase, A., Kudo, A., CO_2 Reduction Using Water as an Electron Donor over Heterogeneous Photocatalysts Aiming at Artificial Photosynthesis. *Acc. Chem. Res.*, 55, 966, 2022.
8. Fujishima, A. and Honda, K., Electrochemical Evidence for the Mechanism of the Primary Stage of Photosynthesis. *Bull. Chem. Soc. Jpn.*, 44, 1148, 1971.
9. Fujishima, A. and Honda, K., Electrochemical Photolysis of Water at a Semiconductor Electrode. *Nature*, *238*, 37, 1972.
10. Kanbara, T., Takada, K., Yamamura, Y., Kondo, S., Photo-Rechargeable Solid State Battery. *Solid State Ion*, 40–41, 955, 1990.
11. Zou, X., Maesako, N., Nomiyama, T., Horie, Y., Miyazaki, T., Photo-Rechargeable Battery with TiO_2/Carbon Fiber Electrodes Prepared by Laser Deposition. *Sol. Energy Mater. Sol. Cells*, 62, 133, 2000.
12. Usui, H., Miyamoto, O., Nomiyama, T., Horie, Y., Miyazaki, T., Photo-Rechargeability of TiO_2 Film Electrodes Prepared by Pulsed Laser Deposition. *Sol. Energy Mater. Sol. Cells*, 86, 123, 2005.
13. Paolella, A., Faure, C., Bertoni, G., Marras, S., Guerfi, A., Darwiche, A., Hovington, P., Commarieu, B., Wang, Z., Prato, M., Colombo, M., Monaco, S., Zhu, W., Feng, Z., Vijh, A., George, C., Demopoulos, G.P., Armand, M., Zaghib, K., Light-Assisted Delithiation of Lithium Iron Phosphate Nanocrystals towards Photo-Rechargeable Lithium Ion Batteries. *Nat. Commun.*, 8, 14643, 2017.
14. Ahmad, S., George, C., Beesley, D.J., Baumberg, J.J., De Volder, M., Photo-Rechargeable Organo-Halide Perovskite Batteries. *Nano Lett.*, 18, 1856, 2018.
15. Boruah, B.D., Mathieson, A., Wen, B., Jo, C., Deschler, F., De Volder, M., Photo-Rechargeable Zinc-Ion Capacitor Using 2D Graphitic Carbon Nitride. *Nano Lett.*, 20, 5967, 2020.
16. Boruah, B.D., Wen, B., De Volder, M., Light Rechargeable Lithium-Ion Batteries Using V_2O_5 Cathodes. *Nano Lett.*, 21, 3527, 2021.

17. Kim, G.Y., Senocrate, A., Wang, Y.-R., Moia, D., Maier, J., Photo-Effect on Ion Transport in Mixed Cation and Halide Perovskites and Implications for Photo-Demixing**. *Angew. Chem.-Int. Edit.*, 60, 820, 2021.
18. Usui, H., Kojima, H., Domi, Y., Sakaguchi, H., Regeneration of Nicotinamide Adenine Dinucleotide Phosphate by a Chlorophyll A-Coated TiO_2 Film Electrode. *ACS Appl. Bio Mater.*, 4, 5975, 2021.
19. Shimokawa, K., Matsubara, S., Okamoto, A., Ichitsubo, T., Light-Induced Li Extraction from $LiMn_2O_4/TiO_2$ in a Water-in-Salt Electrolyte for Photo-Rechargeable Batteries. *Chem. Commun.*, 58, 9634, 2022.
20. Wang, P., Xue, M., Jiang, D., Yang, Y., Zhang, J., Dong, H., Sun, G., Yao, Y., Luo, W., Zou, Z., Photovoltage Memory Effect in a Portable Faradaic Junction Solar Rechargeable Device. *Nat. Commun.*, 13, 2544, 2022.
21. Pujari, A., Kim, B.-M., Sayed, F.N., Sanders, K., Dose, W.M., Mathieson, A., Grey, C.P., Greenham, N.C., De Volder, M., Does Heat Play a Role in the Observed Behavior of Aqueous Photobatteries? *ACS Energy Lett.*, 8, 4625, 2023.
22. Shimokawa, K., Matsubara, S., Kawaguchi, T., Okamoto, A., Ichitsubo, T., Optimizing $LiMn_{1.5}M_{0.5}O_4$ Cathode Materials for Aqueous Photo-Rechargeable Batteries. *Chem. Commun.*, 59, 7947, 2023.
23. Zhang, R., Chen, Z., Ma, J., Zhang, P., Liu, M., Li, X., Zhao, R., Tang, J., Ren, Z., Li, S., An All-Solid-State Photo-Rechargeable Battery Based on $Cs_3Bi_2I_9$. *Chem. Commun.*, 59, 2911, 2023.
24. Watanabe, K., Horisawa, Y., Yoshimoto, M., Tamura, K., Suzuki, K., Kanno, R., Hirayama, M., Stable Photoelectrochemical Reactions at Solid/Solid Interfaces toward Solar Energy Conversion and Storage. *Nano Lett.*, 24, 1916, 2024.
25. Yoshimoto, M., Tamura, K., Watanabe, K., Shimizu, K., Horisawa, Y., Kobayashi, T., Tsurita, H., Suzuki, K., Kanno, R., Hirayama, M., Intercalative and Non-Intercalative Photo-Recharge Using All-Solid-State Cells for Solar Energy Conversion and Storage. *Sustain. Energy Fuels*, 8, 1236, 2024.
26. Kawasaki, S., Nakatsuji, K., Yoshinobu, J., Komori, F., Takahashi, R., Lippmaa, M., Mase, K., Kudo, A., Epitaxial Rh-Doped $SrTiO_3$ Thin Film Photocathode for Water Splitting under Visible Light Irradiation. *Appl. Phys. Lett.*, 101, 033910, 2012.
27. Jia, Q., Iwashina, K., Kudo, A., Facile Fabrication of an Efficient $BiVO_4$ Thin Film Electrode for Water Splitting under Visible Light Irradiation. *PNAS*, 109, 11564, 2012.
28. Iwase, A., Ng, Y.H., Amal, R., Kudo, A., Solar Hydrogen Evolution Using a $CuGaS_2$ Photocathode Improved by Incorporating Reduced Graphene Oxide. *J. Mater. Chem. A*, 3, 8566, 2015.
29. Kaga, H., Tsutsui, Y., Nagane, A., Iwase, A., Kudo, A., An Effect of Ag(I)-Substitution at Cu Sites in $CuGaS_2$ on Photocatalytic and Photoelectrochemical Properties for Solar Hydrogen Evolution. *J. Mater. Chem. A*, 3, 21815, 2015.

30. Kato, T., Hakari, Y., Ikeda, S., Jia, Q., Iwase, A., Kudo, A., Utilization of Metal Sulfide Material of $(CuGa)_{1-x}Zn_{2x}S_2$ Solid Solution with Visible Light Response in Photocatalytic and Photoelectrochemical Solar Water Splitting Systems. *J. Phys. Chem. Lett.*, 6, 1042, 2015.
31. Goto, Y., Minegishi, T., Kageshima, Y., Higashi, T., Kaneko, H., Kuang, Y., Nakabayashi, M., Shibata, N., Ishihara, H., Hayashi, T., Kudo, A., Yamada, T., Domen, K., A Particulate $(ZnSe)_{0.85}(CuIn_{0.7}Ga_{0.3}Se_2)_{0.15}$ Photocathode Modified with CdS and ZnS for Sunlight-Driven Overall Water Splitting. *J. Mater. Chem. A*, 5, 21242, 2017.
32. Higashi, T., Shinohara, Y., Ohnishi, A., Liu, J., Ueda, K., Okamura, S., Hisatomi, T., Katayama, M., Nishiyama, H., Yamada, T., Minegishi, T., Domen, K., Sunlight-Driven Overall Water Splitting by the Combination of Surface-Modified $La_5Ti_2Cu_{0.9}Ag_{0.1}S_5O_7$ and $BaTaO_2N$ Photoelectrodes. *ChemPhotoChem*, 1, 167, 2017.
33. Hayashi, T., Niishiro, R., Ishihara, H., Yamaguchi, M., Jia, Q., Kuang, Y., Higashi, T., Iwase, A., Minegishi, T., Yamada, T., Domen, K., Kudo, A., Powder-Based $(CuGa_{1-y}In_y)_{1-x}Zn_{2x}S_2$ Solid Solution Photocathodes with a Largely Positive Onset Potential for Solar Water Splitting. *Sustain. Energ. Fuels*, 2, 2016, 2018.
34. Kobayashi, H., Sato, N., Orita, M., Kuang, Y., Kaneko, H., Minegishi, T., Yamada, T., Domen, K., Development of Highly Efficient $CuIn_{0.5}Ga_{0.5}Se_2$-Based Photocathode and Application to Overall Solar Driven Water Splitting. *Energy Environ. Sci.*, 11, 3003, 2018.
35. Takayama, T., Iwase, A., Kudo, A., Enhancing Photocathodic Performances of Particulate-$CuGaS_2$-Based Photoelectrodes via Conjugation with Conductive Organic Polymers for Efficient Solar-Driven Hydrogen Production and CO_2 Reduction. *ACS Appl. Mater. Interfaces*, 16, 36423, 2024.
36. Zheng, Y., Yao, Y., Ou, J., Li, M., Luo, D., Dou, H., Li, Z., Amine, K., Yu, A., Chen, Z., A Review of Composite Solid-State Electrolytes for Lithium Batteries: Fundamentals, Key Materials and Advanced Structures. *Chem. Soc. Rev.*, 49, 8790, 2020.
37. Wang, C., Fu, K., Kammampata, S.P., McOwen, D.W., Samson, A.J., Zhang, L., Hitz, G.T., Nolan, A.M., Wachsman, E.D., Mo, Y., Thangadurai, V., Hu, L., Garnet-Type Solid-State Electrolytes: Materials, Interfaces, and Batteries. *Chem. Rev.*, 120, 4257, 2020.
38. Kato, Y., Hori, S., Kanno, R., $Li_{10}GeP_2S_{12}$-Type Superionic Conductors: Synthesis, Structure, and Ionic Transportation. *Adv. Energy Mater.*, 10, 2002153, 2020.
39. Bai, X., Duan, Y., Zhuang, W., Yang, R., Wang, J., Research Progress in Li-Argyrodite-Based Solid-State Electrolytes. *J. Mater. Chem. A*, 8, 25663, 2020.
40. Kim, K.J., Balaish, M., Wadaguchi, M., Kong, L., Rupp, J.L.M., Solid-State Li–Metal Batteries: Challenges and Horizons of Oxide and Sulfide Solid Electrolytes and Their Interfaces. *Adv. Energy Mater.*, 11, 2002689, 2021.

41. Bouibes, A., Takenaka, N., Kubota, K., Komaba, S., Nagaoka, M., Development of Advanced Electrolytes in Na-Ion Batteries: Application of the Red Moon Method for Molecular Structure Design of the SEI Layer. *RSC Adv.*, 12, 971, 2021.
42. Zhou, L., Minafra, N., Zeier, W.G., Nazar, L.F., Innovative Approaches to Li-Argyrodite Solid Electrolytes for All-Solid-State Lithium Batteries. *Acc. Chem. Res.*, 54, 2717, 2021.
43. Kanno, R., Developments in New Materials for Electrochemistry and Energy Storage Devices. *Electrochemistry*, 91, 102001, 2023.
44. Mizushima, K., Jones, P.C., Wiseman, P.J., Goodenough, J.B., Li_xCoO_2 ($0<x\leq1$): A New Cathode Material for Batteries of High Energy Density. *Mater. Res. Bull.*, 15, 783, 1980.
45. Goodenough, J.B. and Kim, Y., Challenges for Rechargeable Li Batteries. *Chem. Mater.*, 22, 587, 2010.
46. Kamaya, N., Homma, K., Yamakawa, Y., Hirayama, M., Kanno, R., Yonemura, M., Kamiyama, T., Kato, Y., Hama, S., Kawamoto, K., Mitsui, A., A Lithium Superionic Conductor. *Nat. Mater.*, 10, 682, 2011.
47. Deiseroth, H.-J., Kong, S.-T., Eckert, H., Vannahme, J., Reiner, C., Zaiß, T., Schlosser, M., Li_6PS_5X: A Class of Crystalline Li-Rich Solids With an Unusually High Li^+ Mobility. *Angew. Chem.-Int. Edit.*, 47, 755, 2008.
48. Seino, Y., Ota, T., Takada, K., Hayashi, A., Tatsumisago, M., A Sulphide Lithium Super Ion Conductor Is Superior to Liquid Ion Conductors for Use in Rechargeable Batteries. *Energy Environ. Sci.*, 7, 627, 2014.
49. Wen, C.J. and Huggins, R.A., Thermodynamic and Mass Transport Properties of "LiIn". *Mater. Res. Bull.*, 15, 1225, 1980.
50. Takada, K., Aotani, N., Iwamoto, K., Kondo, S., Solid State Lithium Battery with Oxysulfide Glass. *Solid State Ion*, 86–88, 877, 1996.
51. Nagao, M., Hayashi, A., Tatsumisago, M., Bulk-Type Lithium Metal Secondary Battery with Indium Thin Layer at Interface between Li Electrode and Li_2S-P_2S_5 Solid Electrolyte. *Electrochemistry*, 80, 734, 2012.
52. Webb, S.A., Baggetto, L., Bridges, C.A., Veith, G.M., The Electrochemical Reactions of Pure Indium with Li and Na: Anomalous Electrolyte Decomposition, Benefits of FEC Additive, Phase Transitions and Electrode Performance. *J. Power Sources*, 248, 1105, 2014.
53. Ohnishi, T., Fabrication of Thin-Film Batteries Composed of $LiCoO_2$, Li_3PO_4, and Li Layers. *J. Solid State Electrochem.*, 28, 4355, 2024.
54. Bates, J.B., Dudney, N.J., Gruzalski, G.R., Zuhr, R.A., Choudhury, A., Luck, C.F., Robertson, J.D., Electrical Properties of Amorphous Lithium Electrolyte Thin Films. *Solid State Ion.*, 53–56, 647, 1992.
55. Yu, X., Bates, J.B., Jellison, G.E., Hart, F.X., A Stable Thin-Film Lithium Electrolyte: Lithium Phosphorus Oxynitride. *J. Electrochem. Soc.*, 144, 524, 1997.

56. Hikima, K., Suzuki, K., Taminato, S., Hirayama, M., Yasuno, S., Kanno, R., Thin Film All-Solid-State Battery Using Li_2MnO_3 Epitaxial Film Electrode. *Chem. Lett.*, 48, 192, 2019.
57. Ito, K., Tamura, K., Shimizu, K., Yamada, N.L., Watanabe, K., Suzuki, K., Kanno, R., Hirayama, M., Degradation of a Lithium Cobalt Oxide Cathode under High Voltage Operation at an Interface with an Oxide Solid Electrolyte. *RSC Appl. Interfaces*, 1, 790, 2024.
58. Hikima, K., Hinuma, Y., Shimizu, K., Suzuki, K., Taminato, S., Hirayama, M., Masuda, T., Tamura, K., Kanno, R., Reactions of the Li_2MnO_3 Cathode in an All-Solid-State Thin-Film Battery during Cycling. *ACS Appl. Mater. Interfaces*, 13, 7650, 2021.
59. Hikima, K., Shimizu, K., Kiuchi, H., Hinuma, Y., Suzuki, K., Hirayama, M., Matsubara, E., Kanno, R., Operando Analysis of Electronic Band Structure in an All-Solid-State Thin-Film Battery. *Commun. Chem.*, 5, 1, 2022.
60. Hikima, K., Shimizu, K., Kiuchi, H., Hinuma, Y., Suzuki, K., Hirayama, M., Matsubara, E., Kanno, R., Reaction Mechanism of Li_2MnO_3 Electrodes in an All-Solid-State Thin-Film Battery Analyzed by Operando Hard X-Ray Photoelectron Spectroscopy. *J. Am. Chem. Soc.*, 144, 236, 2022.
61. Asano, S., Hata, J., Watanabe, K., Shimizu, K., Matsui, N., Yamada, N.L., Suzuki, K., Kanno, R., Hirayama, M., Formation Processes of a Solid Electrolyte Interphase at a Silicon/Sulfide Electrolyte Interface in a Model All-Solid-State Li-Ion Battery. *ACS Appl. Mater. Interfaces*, 16, 7189, 2024.

5

Novel Hybrid Perovskites and Inorganic Semiconductors for Photorechargeable Li-Ion Battery Photoelectrodes

Shubham Chamola[1*], Rashid M. Ansari[1] and Shahab Ahmad[1,2†]

[1]*Advanced Energy Materials Lab, Department of Physics, Indian Institute of Technology Jodhpur, Jodhpur, Rajasthan, India*
[2]*Rishabh Centre for Research and Innovation in Clean Energy, Indian Institute of Technology Jodhpur, Jodhpur, Rajasthan, India*

Abstract

Among various recent advancements in solar energy harvesting, photorechargeable batteries (PRBs) provide a compact and low-cost solution to supply power at remote and off-grid locations, which is otherwise bulky and expensive with the existing solar cell and battery technology. PRB is a device that can harvest solar energy and store it in the form of electrochemical energy simultaneously; thus, mitigating concerns related to the intermittent nature of sun light as well as the associated ohmic losses between the two devices. However, merging solar energy harvesting and storage functionalities in a single material is challenging and thus requires novel materials to accomplish the task. This chapter focuses on the recent development in the domain of two-electrode Li-ion-based PRBs using emerging photoactive materials ranging from hybrid semiconducting materials such as emerging metal halide perovskites (MHPs) to inorganic materials such as transition metal oxides and metal chalcogenides as active materials. Two-electrode Li-PRBs using a single active material 2D lead iodide-based organo MHP demonstrated the simultaneous solar energy harvesting and storage resulting in first demonstration of single active materials based two-electrode PRB. Notably, innovative PRB design, based upon two-electrode configuration, enables seamless integration of solar energy conversion and Li-ion storage functionalities at a single active material level. However, these reported MHP-based PRBs suffered from

**Corresponding author*: shubham.2@iitj.ac.in
†*Corresponding author*: shahab@iitj.ac.in

Nurdan Demirci Sankir and Mehmet Sankir (eds.) Solar Capacitors and Batteries, (105–142) © 2025 Scrivener Publishing LLC

capacity fading after a few cycles due to material instability in liquid electrolyte mediums. Still, despite such limitations, these initial demonstrations opened up avenues to explore new materials and mechanisms for PRB applications. The chapter further discusses about the use of quasi-2D MHPs and MoS_2 nanoflake heterojunction-based photoelectrodes for Li-PRBs, providing insights about stable and efficient MHP-based Li-PRBs. The second part of this book chapter explores the recently developed inorganic active nanomaterials (Fe_2O_3 and Sb_2S_3) based on two-electrode Li-PRBs. Under illuminated conditions, these PRBs demonstrate enhanced electrochemical performance, such as low impedance, high capacity, higher diffusion constants, etc. Fe_2O_3-based PRB has shown high enhancements and stable performance using conversion mechanisms, but it suffers from capacity degradation when operated at a high C rate. Sb_2S_3-based Li-PRB has demonstrated very stable performance under prolonged light exposure (~99% retention) and a galvanostatic photoconversion and storage efficiency of ≈24.53% at 2200 mA g^{-1}. To further improve the performance and endurance of PRBs, it is crucial to explore the novel material systems and understand the underlying mechanisms to achieve compatibility and synergy between energy harvesting and storage properties of active material.

Keywords: Li-ion photobatteries, perovskite photobattery, inorganic nanomaterials, solar energy harvesting, energy storage, IoT devices

5.1 Introduction

Energy is a vital factor for human survival. Throughout human history, our progress has been driven by an unwavering pursuit of energy, starting from the earliest days of civilization to the intricate challenges of the present era. Electricity energizes our residences, powers our automobiles and all the consumer portable electronic devices. In the modern era of global interconnectivity, the importance of energy goes beyond simple lighting. It is the push that propels invention, fuels growth in the economy, and allows communities to prosper. Traditional fossil fuels, including coal, oil, and natural gas, have historically served as the fundamental basis for worldwide energy consumption. However, their combustion releases carbon dioxide and other greenhouse gases, contributing significantly to the rising global carbon emissions leading to climate change. According to reports, global carbon dioxide emissions reached 37.01 billion metric tons in 2023 and rose to a record high of 41.6 in 2024 [1]. Today, due to technological advancements and population growth, energy requirements are reaching unprecedented levels, and most of the energy demand is still being fulfilled by traditional fossil fuels.

In order to limit the global temperature, increase to 2°C or 1.5°C, global annual CO_2 emissions must reach net-zero or become net-negative by the end of this century [2, 3]. In light of the pressing need to tackle energy poverty in a sustainable manner, the UN's Sustainable Development Goals (SDGs) highlight the crucial role of renewable energy sources. In June 1992, at the Earth Summit in Rio de Janeiro, Brazil, more than 178 countries pledged to participate in a plan of action to build a global partnership for sustainable development for improving human lives by protecting the environment using SDGs [4]. These SDGs prioritize the utilization of solar energy, wind energy, hydropower, biomass energy, geothermal energy, tidal energy, and wave energy, aiming to encourage a worldwide transition toward cleaner and greener energy. Solar energy possesses significant potential among all renewable energy sources. The Earth is exposed to an immense amount of solar energy each year, specifically 173,000 TW, which is more than sufficient to fulfil our worldwide energy requirements multiple times. Undoubtedly, solar energy has made a significant contribution in recent years and reached the efficiency of 47.1% for multi-junction concentrator solar cells [5, 6]. Currently, solar PVs make much less (~4.5%) of total global electricity generation [7]. Unfortunately, solar energy is prone to intermittency caused by factors such as day–night cycles and cloudy weather. However, despite immense research on photovoltaic (PV) and historically low prices, the total share of solar power production is expected to reach 20% by 2050 [8].

Conventionally, solar energy is utilized by integrating solar cells with energy storage devices where solar cells harvest the solar energy in daylight hours, and energy storage devices store this energy for later use. This method is still the most popular for solar energy utilization today. However, this physical merger of solar cells and batteries makes the whole system bulky and costly, and at the same time, this device suffers from ohmic losses, duplicate components, and high costs. Lithium-ion batteries (LIBs) have become the most popular energy storage device in the past few years, powering a wide range of applications from portable consumer electronics to electric vehicles (EVs). Exceptional properties of LIBs, including a high energy density (120–170 Wh kg^{-1}), lightweight, compact design, high open-circuit potentials, minimal memory effects, fast charging capability, low self-discharge rates, and relative environmental friendliness, have placed them at the top of the energy sector [9]. However, most LIBs rely on the conventional power supply to charge and thus pose a significant barrier to achieving net-zero carbon emissions, demanding the urgent need for sustainable and renewable energy integration within LIB systems.

Nowadays, LIBs have been used to store solar energy by connecting to solar cells; however, the merger of these two devices is unsuitable for urban populated cities and devices requiring compact and low-weight power supply.

5.1.1 Photorechargeable Battery

Researchers have put forth great strides to address the abovementioned issues arising from the conventional solar cell–battery merger. These attempts resulted in the development of an innovative device that integrates the abilities of solar cells and batteries in one device. This integrated system is known as the photorechargeable battery (PRB) or photo battery. PRBs are a class of energy storage devices that can harvest solar energy during daylight hours and simultaneously store that energy in the form of electrochemical energy within a single device. The first kind of such device was introduced in 1976 by Hodes *et al.*, and the device was coined as a "photoelectrochemical storage cell" [10]. This cell utilized a Cd–Se polycrystalline chalcogenide photoelectrode and a porous Ag counter electrode immersed in an S^{2-}/S redox electrolyte. Subsequently, despite a few additional innovative ideas, progress in this field slowed down, particularly in the wake of significant advancements in Li-ion battery technology that revolutionized the world with its extraordinary capabilities. Nevertheless, there has been a notable increase in interest in these devices in recent decades, primarily because of the urgent need for renewable energy sources.

Figure 5.1 shows the traditional solar energy approach, which involves a solar panel connected to a secondary battery. This conventional approach has been used since the initial days of solar harvesting by solar panels and is still the most popular way to harvest solar energy. This physical combination of solar cells and batteries requires additional electronic components to regulate the current and voltage to recharge the battery. Conversely, the PRB provides an effective solution to these issues by directly storing solar energy within a single device utilizing the multifunctional electrode, which can power various low-power Internet of Things (IoT) devices 24/7. PRBs can broadly be classified into two main categories: three-electrode-based PRBs and two-electrode PRBs. The three-electrode PRB employs a direct integration technique, which combines rechargeable batteries with a third photosensitive electrode in a three-electrode system (Figure 5.2a) [11–13]. Upon exposure to light, the photoelectrode generates excited electrons and holes, which are then transported to the battery electrodes by suitable transport layers and external circuits. To achieve maximum performance, precise tuning of charge transport layers is critical to facilitate efficient

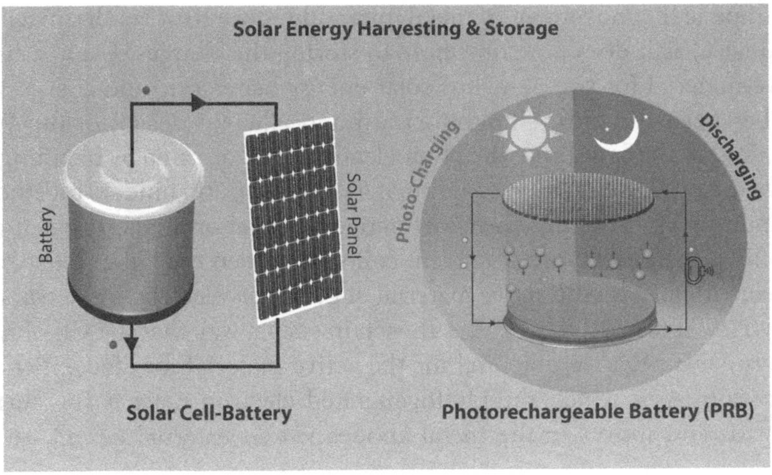

Figure 5.1 Schematic representation showing the conventional combination of solar cell and battery, and the photorechargeable battery (PRB) as a standalone device.

charge transfer. While this design does improve efficiency and compactness to a certain degree, the inclusion of the photoelectrode and transport layers unavoidably increases the complexity and cost of battery manufacture. Moreover, in setups where a metal electrode Li^+/Li is employed, it becomes difficult to charge the battery using only light. This would require a photo absorber active material with a greater negative potential of conduction band minima (CBM). A simplified solution in this context is the use of two-electrode PRBs. This configuration consists of a bifunctional electrode, which is photosensitive, and a counter electrode composed of a metal ion such as Li, Na, or Zn. Bifunctional electrodes commonly consist of one or multiple materials, allowing the electrode to possess the dual functionality of light absorption and charge storage on a single electrode [14]. The two-electrode design can be further divided into dual active material-based PRB and single active material-based PRB. Dual active material-based PRBs employ two distinct active materials blended together [15]. One of these materials is photoactive and responsible for collecting solar energy, while the second material stores metal ions by utilizing this energy. For such a system, it is crucial to choose the appropriate materials carefully, considering the efficient transfer of energy and the stability of these materials in the electrolyte. The limitation of this configuration is that only one material can harness solar energy. If the storage material has a large surface area, it might block sunlight. Conversely, the photoactive material can impede the movement of metal ions and reduce the overall capacity of the device, as it does not contribute to storing the charges. Using a single active material for simultaneous solar energy harvesting and charge storage has been the most straightforward and efficient design of the PRBs [16]. In this configuration, the photoelectrode is fabricated by blending the active material, electron transport material (ETM), and binders together in an appropriate ratio, or sometimes active material or a heterostructure is directly grown on a porous current collector. When the light of a suitable wavelength falls on the active material, it generates electron–hole pairs. As shown in Figure 5.2, the ETM is chosen in such a way that the CB of ETM always has a negative potential for the active material for efficient charge extraction. As a result, the photogenerated electron reaches the current collector and moves to the metal anodes *via* an external circuit. At the

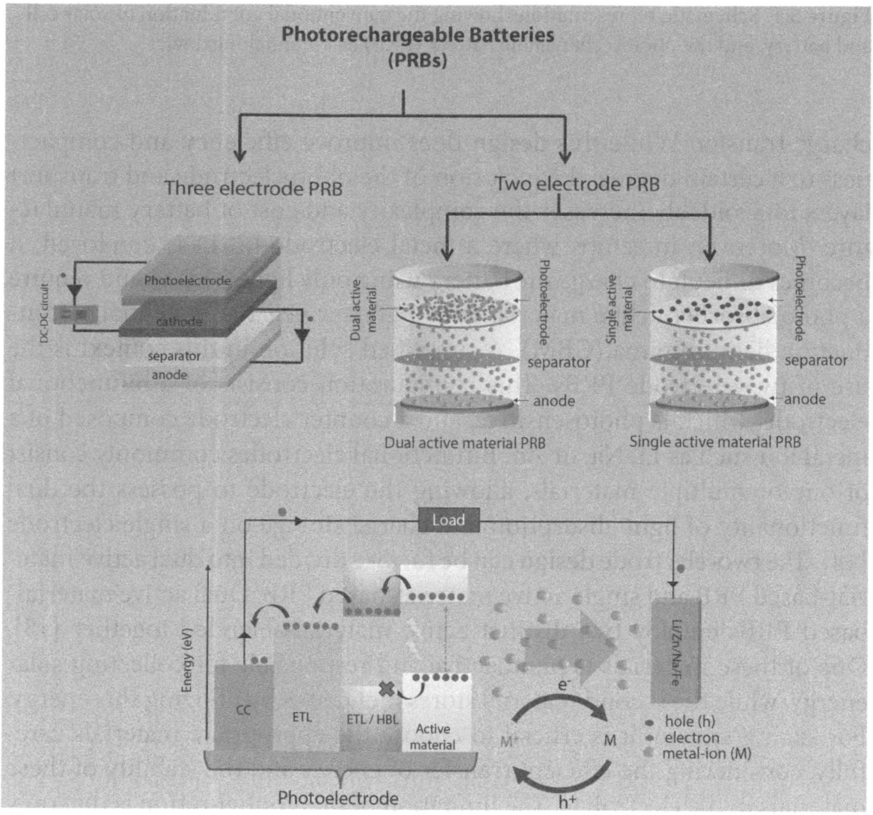

Figure 5.2 Schematics showing the classification of PRBs based on the number of electrodes, the type of active material used, and the working mechanism of single active materials-based PRBs.

same time, photo holes initiate the redox couple reactions by changing the oxidation state of active material and creating enough positive charges, and as a result, Li-ions migrate toward the Li metal.

In some cases, a hole-blocking material (HBM) is also employed to block the holes so that they do not recombine with electrons and repel the Li-ions toward the anode. In such cases, the CB of HBM has a more negative potential, and the valence band maxima (VBM) has a more positive potential with respect to the active material. It is essential to understand that these band positions are generally taken for pristine semiconductors while charging/discharging due to Li-ion intercalation/conversion of active material to a new species, which may alter the band positions. Under a potentiostatic charging/discharging operation, these PRBs exhibit enhanced capacities. These enhanced capacities are attributed to the constant generation of photocharge carriers, which contribute to the discharging current. As a result, less current is drawn from the battery, thus exhibiting a capacity over time. Additionally, due to the generation of electron–hole pairs, the active material's oxidation states get changed and result in the formation of a built-in electric field within the electrode, which allows the inactivated sites of the electrode to participate in the electrochemical reaction and exhibit enhanced capacities. It is evident that Li-ion battery technology has revolutionized the electronics industry and transformed how we power our gadgets over the last few years. High electro positivity, a low standard reduction potential, a high energy density of 150 to 250 watt-hours per kilogram, and lightweight are just a few of its remarkable electrochemical properties that have been instrumental in enabling the development of consumer electronics and extending the capabilities and range of EVs [17–22]. Photo-rechargeable lithium-ion batteries (Li-PRBs) present an innovative technique for storing renewable energy using the excellent electrochemical features of LIBs. Li-PRBs take advantage of the inherent benefits of LIBs, such as high energy density, lightweight design, and efficient charge–discharge cycles, and combine them with photocharging mechanisms to improve their overall functioning [23, 24]. Since single active material-based two-electrode Li-PRBs allow for the simultaneous storage of solar energy harvesting metal ions, they have received much attention in recent years. The progress in this field becomes apparent when comparing the initial development of Li-PRBs using 2D perovskite to current inorganic transition metal oxide (TMO) materials designed utilizing conversion mechanisms and heterostructures. These advancements have significantly enhanced photo conversion and storage efficiency (PC-SE), reaching a milestone of 2.6% from the first development of 0.03% [25, 26]. Although current findings may not yet match the standards required for commercial use, the rapid advancements being made indicate the possibility of developing highly effective devices in the future. Single active material-based PRBs have the potential to offer significant advantages in diverse applications that require an uninterrupted and compact power source for various off-grid sensors, solar cameras, and IoT devices. To evaluate the performance enhancements in Li-PRBs, the following protocols can be employed:

1. Optical absorption studies: Investigation of the optical absorbance spectrum of the active material to analyze its light-harvesting capability in the visible to near-infrared range (400–1100 nm) and bandgap estimations.
2. Electronic structure analysis: Utilizing ultraviolet photoelectron spectroscopy to examine the band positions and electronic states of the active material.
3. Charge carrier dynamics: Perform photoluminescence (PL) and time-resolved photoluminescence (TRPL) spectroscopy to analyze charge recombination kinetics in the presence of a suitable ETM.
4. Photocurrent generation studies: Fabrication of photodetectors (PDs) to evaluate the photocurrent response and charge extraction efficiency of the active material.
5. Electrochemical performance tests: Conducting cyclic voltammetry (CV), galvanostatic charge–discharge (GCD), and electrochemical impedance spectroscopy (EIS) to assess intrinsic electrochemical behavior for LIBs.
6. Photoelectrochemical testing: Performing chronoamperometry (CA), CV, GCD, and EIS of Li-PRB under dark and illuminated conditions to quantify the impact of light on electrochemical performance.
7. Temperature-dependent electrochemical analysis: Investigating the effect of temperature variations on electrochemical performance to isolate and quantify the contribution of photogenerated charge carriers.
8. Photoconversion and storage efficiency (PC-SE) calculation: Implementation of a precise formulation to separate illumination and thermal contributions to determine the photo-enhanced charge storage efficiency accurately.
9. *In situ* and *ex situ* spectroscopic and structural analyses: Performing operando PL, absorbance and Raman spectroscopies, as well as X-ray photoelectron spectroscopy (XPS),

and X-ray diffraction (XRD) under illumination at various voltages to quantify their effects on the structural and optical properties of active material.

5.2 MHPs and TMOs as Active Materials for PRBs

Researchers have explored various active materials for Li-PRBs in the last few years, such as organic MHPs and inorganic materials from TMOs. In the upcoming sections, we will discuss the various PRB systems that employ MHPs and TMO as active materials, providing insights on the development of these PRBs in terms of device configuration, optical and electrochemical stability, synthesis of active materials, their performance, the present trend in PRBs, and outlook on their future prospective.

5.2.1 Metal Halide Perovskites for PRBs

Like 3D perovskite, 2D MHPs are emerging as a potential material for optoelectronic and photoelectrochemical applications such as solar cells [27–32], PDs [33–38], light-emitting diodes (LEDs) [39–42], lasers [43–45], solar–water splitting [46–48], photo-batteries [25, 49], and so on because their Li-ion possesses nature, tunable [50, 51] energy bandgap (3.39 to 1.83 eV) [31, 52], exciton binding energy [39, 42], large absorption coefficient (~10^4 cm^{-1}) [53], charge carrier mobility (~9 cm^2/V-sec) [54], charge carriers lifetimes (~30 ns) [54], diffusion coefficients (~ 7–14 μm) [55], ease of synthesis and, solution-processable nature. Their quantum confinement nature due to their inorganic–organic layers provides them with relatively better structural stability and moisture-resistant nature [56–60]. The general formula for 2D MHPs are $(R\text{-}NH_3)_2(A)_{n-1}M_nX_{3n+1}$ for n=1-4, ..., n<∞ [61, 62]. Where R-NH_3 is organic moiety and called organic spacer cation. It can be linear such as C_mH_{2m+1}-NH_3, where m ≥4 (e.g., butylamine, C_4H_9–NH_3; pentylamine, C_5H_{11}–NH_3, etc.) as well as cyclic, such as benzylammonium, phenylethylammonium, etc. [63–65]. A is caged between inorganic $[MX_6]^{-4}$ octahedra layers, and it can be MA (CH_3NH_3), FA ($NH_2CH_3NH_3^+$), Cs^+, etc. M denotes divalent cations such as Pb^{2+}, Sn^{2+}, Ge^{2+}, etc. X denotes halides such as Cl^-, Br^-, and I^-. During the synthesis, R-NH_3 and A were chosen in such a way that the size of R-NH_3 should be larger than the size of A [66, 67]. Moreover, n denotes the number of inorganic octahedra layers. For instance, n=1 [$(R\text{-}NH_3)_2MX_4$] based MHP has only one inorganic octahedra layer, whereas n=4 [$(R\text{-}NH_3)_2A_3M_4X_{13}$] based MHP has four inorganic octahedra layers, and so on. These single or a few consecutive inorganic octahedra layers are separated by inorganic spacer cation and bonded together with the help of H-bonds [68].

Additionally, MHPs have also been explored as electrode materials in energy storage applications: Mathieson *et al.* investigated bromine and iodine based-Ruddlesden Popper (RP) phase MHPs series of the type $(BA)_2(MA)_{n-1}Pb_nX_{3n+1}$ (where BA is butylammonium as a linear organic spacer cation, MA is methylammonium, and n = 4) for LIB anodes [69]. Here, researchers reported stabilized gravimetric specific capacities of 55 mAh g^{-1} (for n=1), 60 mAh g^{-1} (for n=2), 70 mAh g^{-1} (for n=3), and 92 mAh g^{-1} (for n=4) achieved after 60 cycles at the current density of 30 mA g^{-1} in the voltage range of 2.85–0.1 V. Such increased specific capacities in reported LIBs from n=1 to n=4 are due to the increase in the stoichiometric amount of inorganic octahedra layers. These findings report that there is no lithiation or delithiation within the organic spacer cations as well. Furthermore, they also reported that bromide-based RP perovskite exhibits higher gravimetric specific capacities than iodide-based RP perovskite because bromide-based RP perovskite makes more active sites for Li-ion intercalation due to smaller octahedra formation. Moreover, $n = 4$ bromide-based perovskites exhibited improved and stable cyclic stability of the device by the introduction of 5 M of bistrifluoromethanesulfonimide lithium (LiTFSI) electrolyte (in ethylene carbonate/propylene carbonate) as compared to 1 M LiPF6 (in ethylene carbonate/dimethyl carbonate).

5.2.1.1 2D Perovskite of Type $(C_6H_9C_2H_4NH_3)_2PbI_4$ and Double Perovskite of Type $Cs_2Bi_2I_9$ Nanosheets for Li-PRBs

In 2018, Shahab *et al.* demonstrated the utilization of a single active material for simultaneous solar energy harvesting and Li-ion storage in a two-electrode-based Li-PRB [24]. 2D MHP of type $(C_6H_9C_2H_4NH_3)_2PbI_4$ (CHPI) was used as a single active material-based photocathode. The schematic crystal structure for the reported active material is shown in Figure 5.3a, where an aromatic organic spacer cation is bonded to a monolayer of inorganic $[PbI_6]^{-4}$ octahedra layers with the help of an H-bond. For photocathode fabrication, CHPI, PCBM/rGO, and PVDF were taken in a ratio of 85:1:5, respectively, with 1 mL of N-methy-2-pyrrolidone (NMP). The slurry was casted over precleaned and UV-ozone-treated FTO substrates (1.0 x 1.5 cm^2). The FTO substrates were extended with Cu-foil from one edge to take out the electrical contact. The photocathodes were dried overnight at 45°C on a hot plate inside the glove box. The scanning electrode microscope (SEM) imaging revealed vertically aligned nanoplatelets of ~320-nm thickness and 8–10 μm height. For fabricating the PRB device, Al-metal foil was employed as a current collector for the counter electrode (Li-metal). To separate the photocathode and anode, Whatman borosilicate paper (soaked in 1 M LiPF6, EC/DEC=1/1 electrolyte) was used (see Figure 5.3b). The other side of the device was sealed with glass. For this

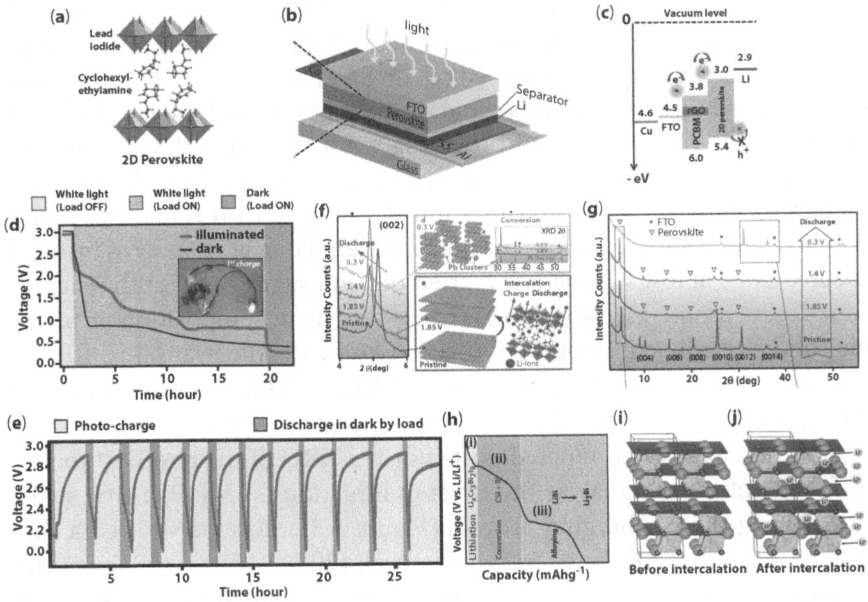

Figure 5.3 Schematic illustration for (a) crystal structure of 2D RP-perovskite CHPI (n=1) and (b) device configuration of 2D perovskite-based li-PRB. (c) Energy level diagram showing photogenerated charge extraction and transport mechanism in photocathode. (d) Discharge curves of li-PRBs under light (green curve) and dark (black curve). Inset camera image shows the prototype of li-PRB powering a commercial 3V white LED. (e) Long-cycle performance test of li-PRB under photocharging (λex ~ 420–650 nm, pin ~ 100 mW cm^{-2}) and dark discharge with a 21.5 kΩ load. (f) *Ex situ* XRD spectra recorded at different voltage states. (g) Close-up view of the characteristic (002) peak of CHPI, exhibiting a shift toward lower diffraction angles upon lithiation. (h) Illustration of intercalation, conversion, and alloying mechanisms taking place during first discharge at different voltages. (i, j) Crystal structure illustrating the intercalation of li-ions, accompanied by an increase in d-spacing between the (*0012*) plane peaks (highlighted in blue). Reproduced from ref [75] with copyright permission from the american chemical society.

work, the authors employed PCBM or reduced graphene oxide (rGO) as ETM to extract the photogenerated electrons from MHPs (see Figure 5.3c). To investigate the effect of light on the performance of Li-PRBs in details, first Li-PRBs were photocharged up to 3.0 V using white light LED (λ_{ex} ~ 420–650 nm, P_{in} ~ 100 mW cm^{-2}) and then discharged through an electrical resistor of 21.5 kΩ under illumination and dark condition (see Figure 5.3d). When discharged under light illumination, Li-PRB maintained a higher voltage (~3.0 to ~1.4 V), providing almost twice the discharge time **(orange zone)** as compared to the discharge time observed under dark conditions. The higher discharging time was attributed to the occurrence of simultaneous photo charging provided by the photogenerated charge carriers. When the light was turned off after 19 hours, a fast drop in voltage was observed (gray zone). This rapid drop demonstrated the effect of light on stabilizing the voltage and providing extra discharge time. A notable decrease in photocharging voltage was observed when Li-PRB was photocharged between 3.0 and 0.4 V and discharged with a load in a dark condition for multiple cycles. The post-mortem Fourier transform infrared spectroscopy at low voltage was carried out and suggested that this significant decay in potential was due to the electrolyte decomposition or unstable solid electrolyte interphase formation [70–72]. Afterward, Li-PRB was photocharged and discharged between 2.95 and 2.00 V, and no potential voltage decay was observed even after 10 cycles (see Figure 5.3e).

Also, the authors conducted post-mortem XRD of the photocathode at different voltages to analyze the effect of lithiation on MHP crystal structure during the discharging process (Figure 5.3f). Results showed a shift in the CHPI characteristic (002) plane peak (2θ ~ 5.2°) toward lower diffraction angles (2θ ~ 4.9°) for photocathode discharged to 1.85 or 1.40 V (Figure 5.3g). Such shift in XRD peaks was attributed to the increase in the atomic interlayer d-spacing from 17 Å for pristine photocathode to 18 Å when discharged to 1.85 or 1.40 V. Such increased d-spacing evidenced the Li-ion intercalation between inorganic octahedra layers (see bottom schematic of Figure 5.3g) [73]. Based on these facts, the authors suggested that photocharging process occurs within the perovskite rather than on the surface metal cluster formation when Li-PRB was discharged to deep potential levels (0.3 V) (see top schematic of Figure 5.3g) [74]. However, the specific capacity of the reported Li-PRB degraded in a liquid electrolytic medium due to its dissolution, but, still, it has shown a discharge specific capacity of ~100 mAh g^{-1} with a coulombic efficiency of ~17% for the first discharge and consecutive charge cycle. The reported PC-SE was 0.034% for the CHPI-based Li-PRB.

Subsequently, Neha *et al.*, in 2021, employed lead-free perovskite (hexagonal $Cs_2Bi_2I_9$ nanosheets) photocathode for Li-PRB [73]. PCBM was used as ETM, and the photocathode consisted of CHPI, PCBM, and PVDF in an 80:10:10 ratio. The Li-PRB was fabricated similarly to the previously discussed work and also with a modified current collector of carbon felt (CF) as a current collector. The CF based Li-PRB was fabricated in a modified 2032 coin cell, where a 8mm diameter hole was punched in the casing to allow the light exposure. The compound perovskite $Cs_3Bi_2I_9$ belongs to a member of the zero-dimensional Bi-based perovskite family. It possesses a hexagonal close packing structure within the space group $P6_3/mmc$. This structure consists of Bi–I octahedral pairs that are arranged in a face-sharing arrangement, forming isolated $[Bi_2I_9]^{3-}$ clusters. These clusters are surrounded by Cs^+ ions [76]. The top view SEM imaging has revealed the presence of $Cs_3Bi_2I_9$ in the electrode, appearing as thin hexagonal sheets with diameters ranging from 10 to 50 nm. To investigate the performance, Li-PRB was photocharged under standard 1 sun illumination (100 mW cm^{-2}) and discharged galvanostatically under dark conditions at a current density of 100 mA g^{-1} for the voltage range between 2.3 and 0.9 V. These measurements indicated that there was no decrease in peak voltage despite a drop in capacity when Li-PRB was operated for more than nine cycles, although working windows were wider as compared to previously reported work. These findings showed that real photocharging has taken place with significant discharge capacity even after several hours of photocharging.

In order to examine the valence states of different elements in the perovskite, XPS *ex situ* studies were performed over photocathodes at various states of charging and discharging. The pristine photocathode exhibited two peaks at 159.2 and 164.9 eV, indicating the oxidized states of bismuth (Bi^{+3}). As the reaction progresses toward lithiation, the appearance of two other peaks at 156.9 and 162.3 eV was attributed to the presence of bismuth metal (Bi^0). This indicated the conversion of Bi^{3+} to Bi^0 within the perovskite structure. This newly formed Bi^0 then reacts with Li ions to form Li_xBi alloys. However, these peaks remain even after delithiation, indicating the irreversible conversion of certain amounts of Bi^{3+} to Bi^0 [74]. This irreversible conversion caused a change in surface morphology and capacity fading in successive cycles. Furthermore, to examine the charge storage mechanism in perovskite-based Li-PRB, XRD was performed at different states of voltages in the discharging cycle. According to reported research articles, lithiation, conversion, and alloying mechanisms were involved during discharging (see Figure 5.3h). At high voltage, lithiation occurred where the crystal structure remained unchanged, which was confirmed by the XRD (see Figure 5.3i, j). The authors suggested that within the voltage range of 1.8–1.0 V, a conversion mechanism ($Bi^{3+} \rightarrow Bi^0$) and a below 0.8-V alloying mechanism occur, which was confirmed by an XPS analysis at these voltages. The mechanism was confirmed by the presence of additional peaks at $2\theta \sim 21.7°$ and 26.3°, which were associated with the formation of LiBi alloy and Bi^0 metal. Additionally, the characteristic XRD peaks at $2\theta \sim 7°$, 17°, and 25.5° disappeared, and there was a shift in the prominent peak at $2\theta = 51.8°$ to 50.4° (see (ii) and (iii) curves in Figure 5.3h) [77–79]. The observed shift in the peak signifies the expansion of the lattice to allow for the intercalation of Li-ions into the perovskite structure, as depicted in Figure 5.3d, e. Finally, to compare the performance of Li-PRBs fabricated over FTO and CF current collector, photoconversion and storage efficiency (PC-SE) were measured for Li-PRBs. The Li-PRB fabricated on the CF current collector exhibited a PC-SE of 0.428% for the first photocharge/discharge cycle, whereas the Li-PRB fabricated over the FTO current collector had a PC-SE of 0.071%. The substantial difference observed in PC-SE can be attributed to the significant variation in resistance exhibited by both current collectors. Furthermore, the PC-SE for the 5th cycle is 0.073% and 0.028%, correspondingly. The decrease in performance capacity and cycle stability of the positive electrode in LIBs was attributed to the deterioration of the active material, which necessitates additional enhancement.

5.2.1.2 *Quasi 2D RP Perovskite and MoS_2-Based Hybrid Heterojunction for Li-PRBs*

To tackle the electrochemical instability of MHP-based Li-PRb, Rashid *et al.* has studied the heterojunction of quasi 2D RP perovskite of type $(BA)_2(MA)_3Pb_4I_{13}$ and a few layers of MoS_2 nanoflakes (Type II; staggered type) [51]. Such heterojunction formation has improved the charge transport from RP perovskite to MoS2 by forming more decay channels. Moreover, the presence of MoS_2 nanoflakes in the RP perovskite precursor solution has triggered heterogeneous nucleation during the crystallization process, which has resulted in improved top-view surface morphology of hybrid drop-casted films as compared to its pristine counterpart (Figure 5.4a). Additionally, the cross-sectional view of the hybrid films revealed vertically aligned crystals of RP perovskite having an average size of ~65 μm, which was quite similar to what was observed in the top-view imaging

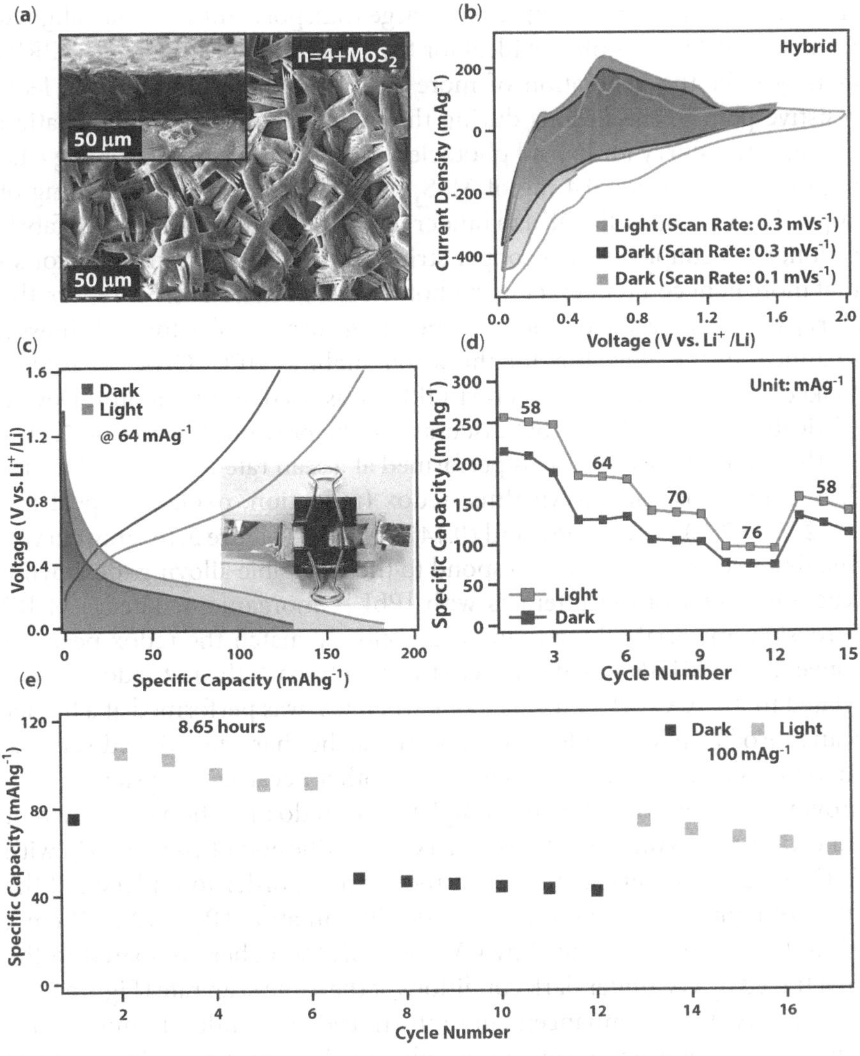

Figure 5.4 (a) Top view FESEM image of $(BA)_2(MA)_3Pb_4I_{13}$–MoS_2 drop-casted film; inset shows the cross-sectional view of the film. (b) CV of hybrid li-PRBs under dark and illumination at different scan rates. (c) GCD profile of hybrid li-PRB at 64 mA g^{-1} under dark and illumination. (d) Rate cyclability test of hybrid li-PRB at different current densities from 58 mA g^{-1} to 76 mA g^{-1} under dark and illumination. (e) Prolong cyclability test of hybrid li-PRB at a current density of 100 mA g^{-1} under dark and illumination. reproduced from ref [51] with copyright permission from John Wiley & Sons.

(Figure 5.4a; inset). Such improved charge transport and vertically aligned crystals of RP perovskite are ideal for the fabrication of LIBs and Li-PRBs as they offer the generation of more electron–holes pairs and the least resistive path to the Li-ions during the intercalation and deintercalation process. The slurry for hybrid photoelectrode was made by dissolving 84.6 mg of RP perovskite, 0.4 mg of MoS_2 nanoflakes, 5 mg of rGO, 3 mg of Super-P, and 7 mg of PVDF as a binder in 1-ml NMP. For the Li-PRB fabrication, FTO coupled with a copper strip was used as a current collector so that more light could interact with photoelectrode materials. Moreover, the slurry was drop-casted on the corresponding current collector and allowed to anneal in the glove box for the whole night at 70°C. Glass microfiber soaked with 30 µl electrolyte (1 M $LiPF_6$) was used as a separator. Device fabrication was similar to that discussed in the previous Subsection 2.11.

The CV for hybrid Li-PRBs performed at a scan rate of 0.1 mVs^{-1} under dark conditions had shown three redox (reduction, oxidation) peaks at (0.52 V, 0.69 V), (0.34, 0.56), and (0.14 V, 0.24 V) (Figure 5.4b, gray curve). The first two redox peaks correspond to the reversible alloying–dealloying reactions when Li-ion interacts with $[PbI_6]^{-4}$ inorganic octahedra of RP perovskite material. These redox peaks closely match the redox peaks of conventional MHPs-based LIBs. On the counterpart, the last redox peak is related to the rGO [51]. Furthermore, when CV was performed at a higher scan rate of 0.3 mVs^{-1} under dark conditions, the characteristic redox peaks RP perovskite material merged with each other because the system fails to provide sufficient time for the completion of redox reactions and current collection due to high resistivity of FTO and adhesion of copper strip with FTO (Figure 5.4b, black curve). Furthermore, in order to understand the effect of light, CV was performed under illumination (P_{in} ~ 12 mW cm^{-2}, white LED). The area traced by CV was ~27.8% higher compared to the area traced by CV under dark conditions at the same scan rate (Figure 5.4b, yellow curve). Such enhancement in the traced area under illumination is due to the generation of a greater number of electron–hole pairs and their contribution to store more Li-ion. Under illumination, the following electrochemical reactions (5.1) and (5.2) occur at photoelectrode:

$$Li_{3.2}Pb + 0.2h^{+} \rightarrow Li_3Pb + 0.2Li^{+} \ at \ 0.56V \tag{5.1}$$

$$Li_3Pb + 2h^{+} \rightarrow Li_3Pb + 2Li^{+} \ at \ 0.69V \tag{5.2}$$

The photogenerated electrons at the counter electrode will reduce the Li-ions into Li, as demonstrated by the electrochemical reaction (5.3):

$$Li^{+} + e^{-} \rightarrow Li \tag{5.3}$$

The GCD profile for the hybrid Li-PRB (Figure 5.4c, inset) at a current density of 64 mA g^{-1} is displayed, where the specific capacity for demonstrated Li-PRB has enhanced to 180.83 mAh g^{-1} from 129.77 mAh g^{-1} (photo-enhancement: 39.34%) under illumination (12.0 mW cm^{-2}, white LED). Furthermore, rate cycling was performed for the voltage range of 0.02 to 1.60 V at different current densities from 58 to 76 mA g^{-1} stepwise and then return to a lower current density of 58 mA g^{-1} under dark and illumination to investigate the effect of light on the specific capacities (Figure 5.4d). As anticipated with CV, the hybrid Li-PRB had shown enhanced average discharge specific capacity from 202.61 to 251.75 mAh g^{-1} (capacity retention under dark: 87.31%; capacity retention under illumination: 96.23%; photo-enhancement: 24.25% for 1–3 cycles) at a current density of 58 mA g^{-1}. Likewise, the average specific discharge capacity increased from 128.66 to 180.67 mAh g^{-1} (capacity retention under dark: 96.56%; capacity retention under illumination: 97.39%; photo-enhancement: 40.42% for 4–6 cycles) at a current density of 64 mA g^{-1}. As the current density has increased to 76 mA g^{-1} after testing at a current density of 70 mA g^{-1}, the hybrid Li-PRB has enhanced the average discharge specific capacity from 74.15 to 94.82 mAh g^{-1} (capacity retention under dark: 97.21%; capacity retention under illumination: 97.51%; photo-enhancement: 27.86% for 10–12 cycles). Lastly, when the current density has reversed back to 58 mA g^{-1} after being tested to 76 mA g-1, the Li-PRB still showed an average discharge specific capacity of 123.97 mAh g^{-1}, which further enhanced to 148.39 mAh g^{-1} (capacity retention under dark: 84.49%; capacity retention under illumination: 89.59%; photo-enhancement: 19.70% for 13–15 cycles). Furthermore, to examine the photoelectrochemical performance of hybrid Li-PRB, the GCD was performed under dark and illumination at a current density of 100 mA g^{-1} for the voltage rage of 0.5 to 1.6 V (Figure 5.4e). It was found that the discharge-specific capacity was enhanced to 105 mAh g^{-1} from 75.3 mAh g^{-1} under illuminations for the initial cycles, and even after 12 cycles, the discharge-specific capacity of Li-PRB has enhanced to 73 from 41 mAh g^{-1} under illumination. Even after multiple cycles, such enhanced discharge-specific capacity indicated that Li-PRB remains responsive to light and maintains stable performance at high current rates. Finally, this hybrid Li-PRB has shown PC-SE of 0.52% under 1 sun illumination, which is ~15 times higher than the first reported RP perovskite (n = 1) based Li-PRB [51]. Notably, this RP perovskite–MoS_2 hybrid photoelectrode-based Li-PRBs have demonstrated better performance and high stability when operated under varying current densities in comparison to the first reported RP perovskite (n = 1) based Li-PRB.

5.2.2 Inorganic Photoactive Materials for Li-PRBs

After realizing that a single active material can achieve bifunctional properties, researchers were primarily concerned with the stability of active materials in liquid electrolyte environments and non-reversible structural changes occurring during cycling. The quest to look for more stable material had directed the researchers toward inorganic semiconducting materials. Among various inorganic materials, TMOs, transition metal dichalcogenides, and other metal chalcogenides have been an essential part of materials science due to their unique and tunable properties in the fields of electronics, energy storage, photocatalysts, etc. [22, 80–86]. TMOs are primarily employed as a cathode material in LIBs such as lithium cobalt oxide ($LiCoO_2$), lithium nickel cobalt manganese oxide [$Li(NiCoMn)O_2$ or NMC], lithium manganese oxide ($LiMn_2O_4$ or LMO), and lithium titanate oxide ($Li_4Ti_5O_{12}$ or LTO) [87–89]. These materials have been widely used as cathodes in LIBs since the beginning of the LIB technology and are still the most popular choice due to their inherent stability, abundance, and substantial lithium storage capabilities. Unfortunately, most of these TMOs cannot be utilized as a single active material for the bifunctional electrode due to their poor photon absorption capabilities and low electronic conductivities.

Among the conventional battery cathode materials derived from TMOs, LMO stands out as a capable active material for harvesting solar radiation in the visible spectrum due to its suitable bandgap of 1–2 eV [90, 91]. Lee *et al.* in 2019 investigated the effect of illumination on the LMO-based photocathode [24]. The authors observed that the LNMO-based LIB showed enhanced capacity under illumination. The findings brought the hope of employing unconventional photovoltaic materials to achieve bifunctionality in Li-PRBs. A conventional slurry-making process was employed to prepare the slurry by mixing $LiMn_2O_4$, carbon conductive additive (acetylene carbon black), and Teflon binder in a weight ratio of 75:5:20. The free-standing electrode was placed onto an Al mesh and used as a photocathode, and Li metal was employed as an anode. A glass fiber separator soaked with an electrolyte ($LiPF_6$ in EC/EMC) was used as a separator. Customized 2032-coin cell casing, with a punched hole of 8-mm diameter in the top casing, was used as a cell assembly. The optical window was sealed with Kapton tape to avoid contact with oxygen and moisture and allow access to the light. After illuminating the Li-PRB with a 300-W standard Xenon lamp (1 SUN), the discharge capacity of Li-PRB was doubled compared to the dark discharge capacity, whereas the charging rate was also increased by a factor of 3.4 under illumination. However, these LMO-based PRBs have practical constraints due to their susceptibility to LMO deterioration when subjected to high temperatures (60°C) or with prolonged light exposure. Additionally, LMO exhibits a limited theoretical capacity and operates within a narrow voltage range, which can limit its long-term effectiveness for PRB applications. However, this study confirmed that solar radiation can be used to accelerate the charging of LMO-based PRBs and opened a way to look for more inorganic materials for PRB applications.

Later on in 2022, Boruah and colleagues presented a study on layered vanadium pentoxide (V_2O_5) nanofiber-based photocathodes for Li-ion PRBs [26]. Compared to LMO, V_2O_5 is a more favorable choice for a photocathode due to its broader absorption spectrum, high absorption coefficient, and high theoretical capacity of ~441 mAh g^{-1}. In this work, authors employed poly(3-hexylthiophene-2,5-diyl) as a hole-blocking layer so that holes can promote the oxidation of vanadium and thus release Li ions more efficiently. V_2O_5 NF-based Li-PRB showed a PC-SE of approximately 2.6% under 455-nm illumination and around 0.22% under 1 sun illumination.

With a similar idea, Wang *et al.* employed γ-LiV_2O_5 (LVO) nanorod (NR) bundles as an active material for the photocathode in Li-PRB [92]. LVO is a better choice of photocathode material due to its high conductivity and mobility than pristine V_2O_5 [93]. Up to this discussion, most discussed PRBs utilize an intercalation mechanism to store the Li-ions in the host active material. Intercalation mechanism-based Li-PRBs are stable but accommodate much smaller Li-ions, limiting their capacities [94].

5.2.2.1 *Fe_2O_3-Based Li-PRBs*

As a common but beneficial compound, iron oxide can be found widely in nature and can be synthesized on a large scale in a laboratory using low-cost techniques. Fe_2O_3 possesses a direct bandgap in the visible spectra (2.0–2.5 eV) and has been explored in many optoelectronic and energy storage applications, which include PDs, water splitting, ETM for solar cells, anodes for LIBs, etc. [95–98]. These properties make α-Fe_2O_3 a suitable active material for a photocathode in PRBs. In 2022, Shubham *et al.* investigated the applicability of alpha phase Fe_2O_3, also known as hematite, for Li-PRB photoelectrode [99]. α-Fe_2O_3 employs a conversion mechanism to store Li-ions, with a high lithium extraction rate of 6 mol Li per 1 mol of α-Fe_2O_3, thus resulting in a high theoretical capacity of 1006 mAh g^{-1} [95–97, 100].

A low-cost solution processing method was used to prepare nanoporous α-Fe_2O_3, demonstrating a bandgap of 2.1 eV, allowing Fe_2O_3 to effectively capture 40% of the solar spectrum (Figure 5.5a). Before utilizing α-Fe_2O_3 for photocathode, it is imperative to mitigate the volume expansion occurring during lithiation/delithiation. For the smooth photocharge transport and to provide the buffer matrix to Fe_2O_3, multi-walled carbon nanotubes (MWCNTs) and phenyl-C61-butyric acid methyl ester (PCBM) were employed as ETMs due to their well-aligned CBM and VBM levels with respect to Fe_2O_3. TRPL measurements showed a charge lifetime of ≈130 ps for pristine α-Fe_2O_3 and ≈106 ps for the Fe_2O_3–PCBM–MWCNT composite film (Figure 5.5b). The shorter life can be attributed to the faster charge extraction at the Fe_2O_3–PCBM–MWCNT interface due to the reduced

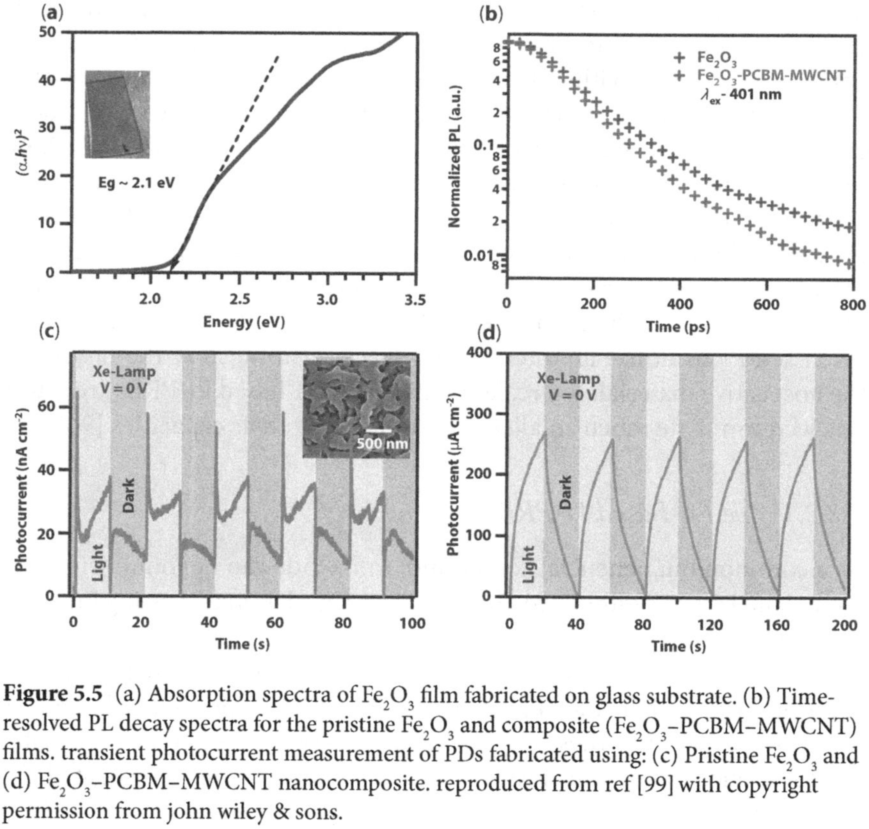

Figure 5.5 (a) Absorption spectra of Fe_2O_3 film fabricated on glass substrate. (b) Time-resolved PL decay spectra for the pristine Fe_2O_3 and composite (Fe_2O_3–PCBM–MWCNT) films. transient photocurrent measurement of PDs fabricated using: (c) Pristine Fe_2O_3 and (d) Fe_2O_3–PCBM–MWCNT nanocomposite. reproduced from ref [99] with copyright permission from john wiley & sons.

recombination rate and smooth extraction of charge carriers. In-depth optoelectronics properties of these materials were studied by fabricating the PDs and carrying out photocurrent studies. The transient photocurrent response revealed a photocurrent of 64 nA cm^{-2} for the pristine PD and an increased 270 nA cm^{-2} photocurrent for the composite PD (Figure 5.5c-d).

Adding functionalized MWCNTs with PCBM in α-Fe_2O_3–NMP solution resulted in the growth of Fe_2O_3 NRs on CNTs, leading to the formation of a highly porous structure illustrated in Figure 5.6a. Porous structures promote the interaction of light due to their large surface area, and they also exhibit superior electrochemical performance due to a better interface between the electrode and the electrolyte [101, 102]. The nanocomposite was further tested for electrochemical stability by fabricating coin cell-based Li-PRB. The LIB showed a specific capacity of 650 mAh g^{-1} after 200 cycles at a current rate of 1000 mA g^{-1}, confirming the stability and capability of Fe_2O_3 nanocomposite to store the Li-ion efficiently. For PRB fabrication, punched 2032-coin cell casings with 8-mm-diameter windows were employed (a similar configuration has been discussed in Figure 5.2). Whatman borosilicate paper was used as a separator, 1M $LiPF_6$ was employed as an electrolyte, and Li metal was employed as the counter electrode. The band alignment and working mechanism of Fe_2O_3-based Li-PRB are shown in Figure 5.6b. Upon light exposure, Fe_2O_3 NRs absorb light (E_g ~ 2.1 eV) and create photocharge carriers at the photocathode. The photoelectrons can easily be transported to the CF current collector due to the favorable energy paths provided by PCBM and MWCNTs and further reach the anode *via* an external circuit. At the same time, photoholes in the Fe_2O_3 NRs cause the oxidation of Fe^0 to Fe^{3+}. The author suggested that the oxidation process further facilitates the movement of Li^+ ions toward the Li-metal anode *via* the electrolyte. The oxidation of Fe^0 to Fe^{3+} resulted in the formation of Fe ions with a positive charge, which generated a repulsive force against Li-ions. This repulsion effectively directed Li-ions toward the anode, where they undergo reduction to produce Li-metal and photocharging of the PRB takes place. The following reaction mechanisms can be ascribed for photocharging of the Fe_2O_3-based Li-PRB under closed circuit condition (Equations 5.4–5.6).

At the photocathode:

$$2Fe^0 + 6h^+ \rightarrow 2Fe^{3+} \tag{5.4}$$

$$2Fe^{3+} + 3Li_2O \rightarrow Fe_2O_3 + 6Li^+ \tag{5.5}$$

At the anode, the photogenerated electrons will neutralize the Li-ions.

$$6Li^{+} + 6e^{-} \rightarrow 6Li \tag{5.6}$$

A high transient photocurrent of ~2.3 mA cm^{-2} was observed for PRB under the standard Xe lamp (100 mW cm^{-2}) light source within 50 s, which showed its ability to photocharge without external biasing. The authors studied *in situ* charge tracking of Li-PRB, where the OCV of Li-PRB was tracked under different operating conditions, as shown in Figure 5.6c. During the initial 10 minutes under dark conditions, OCV was found to be stable; however, when white LED (12 mW cm^{-2}) was turned on, OCV increased from 2.14 to 2.54 V in 1 hour (photocharging: shown by the orange curve). Subsequently, a fixed resistance of 21 kΩ was connected to discharge the PRB while keeping the light on. An abrupt decrease in voltage to approximately 2.35 V was observed. Interestingly, the OCV exhibited a slow increase after this sudden fall, reaching at 2.44 V. Authors attributed this voltage rise to the simultaneous photocharging process, which counterbalanced the discharging current. When this load was changed to 15 kΩ, the OCV experienced a gradual decline to 2.35 V in 1 hour, indicating that now the discharging process dominates the photo-charging process. As soon as the light was turned off, the OCV dropped to 2.0 V within 24 min. These voltage tracking measurements demonstrated the ability of Li-PRB to simultaneously photocharge while powering the consumer electronic device, which is crucial for powering devices that require an uninterrupted power supply 24×7. The author further recorded the CV under dark and illumination, which showed ~94% enhancement in the swept area under the Xe lamp, which was reduced to 41.3% under blue light illumination (λex ≈470 nm LED, 11.8 mW cm^{-2}). The rise in the redox peaks and enhancement in the CV area signified the early completion of redox reactions than anticipated under illumination. GCD measurements aligned perfectly with the CV measurements, where the specific capacity was increased by approximately 71.34% from 276 to 550 mAh g^{-1} within a voltage range of 3–0.5 V (see Figure 5.6d). During the cycling, a notable increase in coulombic efficiency from 99.28% under dark conditions to 113.51% under illumination was observed, which was attributed to the effect caused by the simultaneous photocharging process during the cycling. Similarly, at current densities of 750, 1000, 1500, and 2000 mA g^{-1}, consistent enhancement in the capacity was recorded to be 70.1%, 69.52%, 72.64%, and 92.96%, respectively (see Figure 5.6e). The authors further tested the PRB for temperature-dependent analysis to investigate

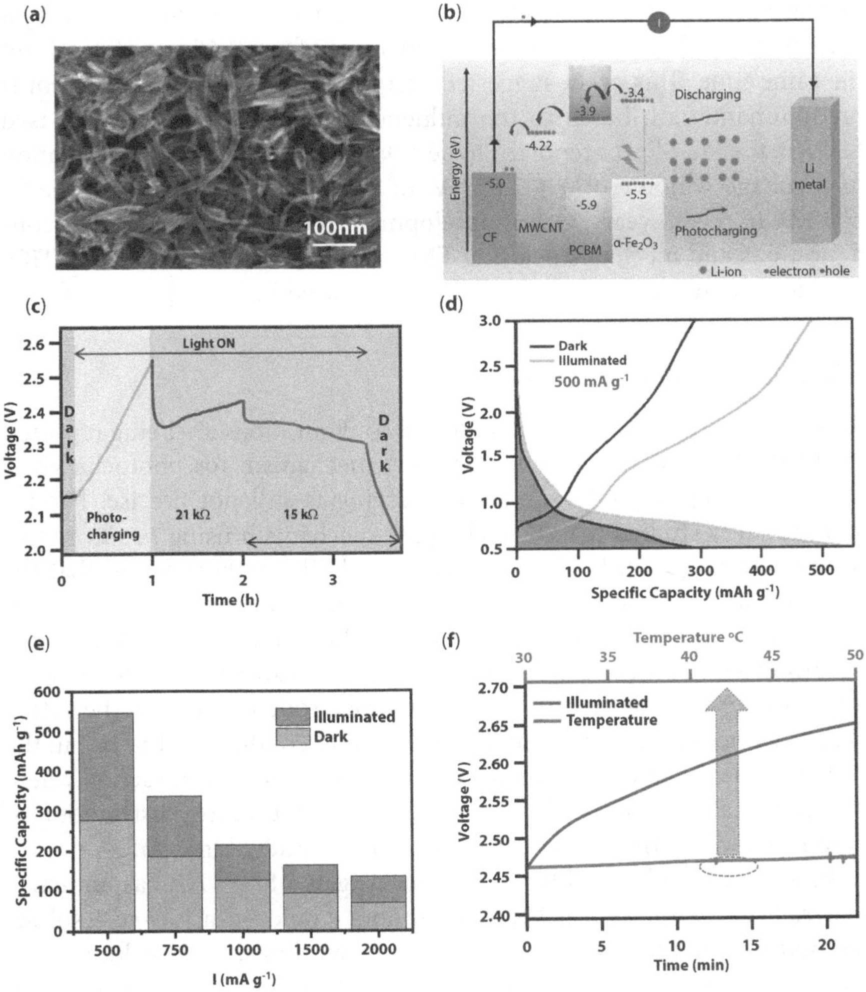

Figure 5.6 (a) FESEM image of the α-Fe_2O_3 photocathode showing the growth of Fe_2O_3 NRs on MWCNT bundles. (b) Energy band diagram and working mechanism of α -Fe_2O_3-based li-PRB. (c) Voltage monitoring of Fe_2O_3-based li-PRB during photo-enhanced charging and discharging under different loads in illuminated and dark conditions. (d) GCD cycle of Fe_2O_3-based li-PRB at 500 mA g^{-1} under dark and illumination. (e) Comparison in the specific capacity at various current rates under dark and illumination. (f) Comparison of open circuit voltage (OCV) under illumination and elevated temperature. reproduced from ref [99] with copyright permission from John Wiley & Sons.

the impact of heat on the OCV of the PRB. The photocathode was subjected to external heating, with the temperature gradually rising from 30 to 50°C in increments of 5°C, and each temperature was held for 5 minutes (see Figure 5.6f). The measurement did not show any appreciable increase during high temperatures. However, under exposure to blue light (11.8 mW cm^{-2}), the OCV of the PRB increased from 2.46 to 2.65 V within the same time. This event suggested that the observed OCV increment is predominantly attributed to the influence of light. For Fe_2O_3 NR-based Li-PRB, the PC-SE was reported to be 1.98% for the blue LED illumination in a voltage window of 0.8–2.574 V. In addition to the discussed Fe_2O_3 Li-PRB, in recent years, lots of development took place with similar configurations and mechanisms using TMOs such as CuO, Cu_2O, SnO_2, TiO_2 etc. based active material for two-electrode-based Li-PRBs [103–110].

5.2.2.2 *Sb_2S_3-Based Li-PRBs*

Every day, a new mechanism and device configurations are being reported for PRBs; however, the charge transport mechanism for photocharging/OCV charging or photo-enhanced charging is still not precise. Pandya *et al.* recently studied the OCV charging mechanism using *in situ* reflection spectroscopy on $Li_xV_2O_5$ material [111]. The authors found that the electrons and Li-ions produced under illumination on an open circuit remain confined to the electrode surface and provide higher voltage without any external medium. It was suggested that because there is no medium for Li-ion to migrate toward the counter electrode (Li-metal), the actual charging does not take place under open circuit conditions. The rise in the voltage under open circuit voltage is entirely a surface phenomenon where Li-ion migrates on the surface and causes the nonuniform distribution of Li ions. However, the phenomenon needs to be studied in detail.

In a recent study, Shubham *et al.* investigated Sb_2S_3 NRs as an active material for Li-ion PRB [112]. The authors employed a low-cost sol-gel method for the Sb_2S_3–rGO nanocomposite synthesis. Sb_2S_3 is known for its high absorption coefficient and low toxicity and is employed as an active material for solar cells [113, 114]. Additionally, Sb_2S_3 exhibits a very high specific capacity of 946 mAh g^{-1} when used as a LIB anode [115]. The sol-gel synthesized NRs exhibited a bandgap of 1.8 eV. Furthermore, rGO was employed due to its well-aligned band position with respect to the conduction band of Sb_2S_3 and its potential to minimize the volume expansion and reduce polysulfide shuttling in Sb_2S_3. Before conducting photoelectrochemical measurements, PDs were fabricated and transient photocurrent was recorded at 0.0 V bias to ensure the smooth transport of electrons from Sb_2S_3 to rGO. The Sb_2S_3–rGO nanocomposite-based PD demonstrated twice photocurrent as compared to pristine Sb_2S_3 PD. After ensuring the optoelectronics properties, the authors conducted electrochemical tests for coin cell-based LIBs. Furthermore, Cu mesh as a current collector was employed for the PRB fabrication. The porous structure of Cu-mesh allowed simultaneous light and electrolyte interaction with the active material (Figure 5.7a, b).

Electrochemical tests revealed that the Cu-mesh performed similarly to the Cu-foil current collector. The Cu-mesh current collector provided a new approach for a coin cell-based Li-PRB, as most of the coin cell-based PRBs either use FTO or carbon paper as a current collector. FTO suffers from high resistance, and carbon paper is costly, interfering with

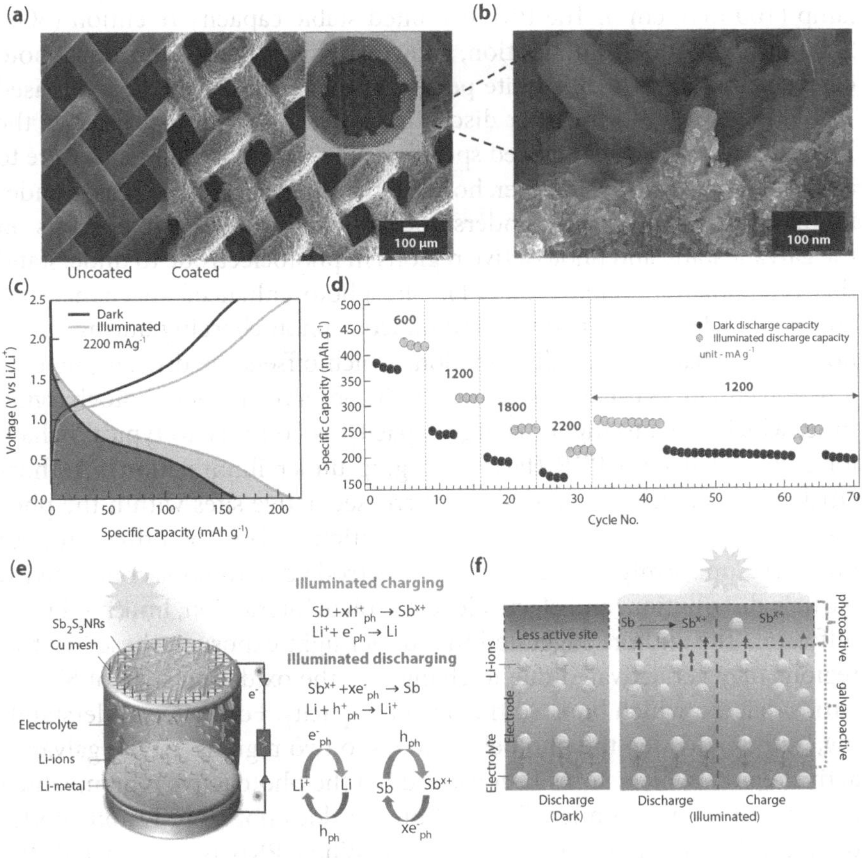

Figure 5.7 FESEM image of cu-mesh current collector: (a) Before coating (left) and after coating (right); inset shows the camera image of cu-mesh current collector. (b) Magnified FESEM image of coated cu-mesh with Sb_2S_3-rGO nanocomposite. (c) GCD cycles at 2200 mA g^{-1} under dark and illumination. (d) Rate cyclability test under dark and illumination. (e) Schematic representation of li-PRB working under illumination. (f) Schematic illustration of photoelectrode under illumination. reproduced from ref [112] with copyright permission from John Wiley & Sons.

electrochemical properties during cycling due to its ability to store Li-ions at low voltages. Coin cell-based PRB (similar to that discussed earlier) was employed for the study, and a 12-mW cm^{-2} LED was used as an illumination source. The GCD curves of the PRB under dark and illumination at 2200 mA g^{-1} showed an increase in specific capacity from ≈165 mAh g^{-1} (dark) to ≈212.72 mAh g^{-1} (illuminated), achieving a 28.94% photo-enhancement due to photogenerated charge carriers (Figure 5.7c). Considering the high applied current rate, which completes the fast charging and discharging, and low power LED, the authors ensured that heat plays a minimum role in enhancing the specific capacity of PRBs, which may happen with high-power Xe lamp (100 mW cm^{-2}). The PRB exhibited stable capacity retention (600–2200 mA g^{-1}) under illumination, showcasing its stability and non-photo corrosive nature, which is quite persistent in the perovskite or dye-based PRBs (Figure 5.7d). In earlier discussions, we have seen that most of the PRBs discussed show enhanced specific capacities under light exposure to a suitable wavelength. However, how the charging capacity increases under illumination is still poorly understood. The authors put a hypothesis on the galvanostatic and photoactive regions in photoelectrode to understand this phenomenon (Figure 5.7e, f). The photo-enhanced mechanism of Sb_2S_3 photoelectrodes involves simultaneous photocharging during galvanostatic discharge under illumination, which offsets the discharging current and enhances capacities. However, the observed longer photocharging time, which leads to higher charge capacity, is contrary to typical behavior expected from a PRB (faster charging under illumination). Authors attributed this phenomenon to the increased active sites within the photoelectrode under illumination. In conventional LIBs, the lithiation does not occur uniformly on the complete electrode due to poor conductivity, particle distribution, and electrode–electrolyte interaction, limiting Li-ion diffusion. The authors suggested that under light exposure, these inactive regions may get activated due to changes in the oxidation state of Sb and lattice defects, which boosted the overall capacity. For better understanding, authors divided the photoelectrode into two regions: (i) the galvanoactive region, which stores Li-ions, and (ii) the photoactive region, which can absorb photons under illumination but does not store Li-ions under dark conditions or ordinary conditions. When PRB is illuminated, the light exposure on photoelectrode modifies the oxidation state of Sb, activating the photoactive region, which then participates in the galvanostatic process, making it available for Li-ion storage, thus improving capacity. This photo-induced charge inhomogeneity drives Li-ions to inactive sites, enabling conversion-alloying reactions and resulting in a higher discharge capacity. The charging process under light involves the release of trapped

Li-ions from the photoactive region, resulting in higher charge time and capacity. The authors further validated their arguments by conducting *ex situ* XPS and Raman investigations where the oxidation state of Sb was found to be modified under light exposure, suggesting a change in the electronic state of Sb under illumination. The Sb_2S_3-based Li-PRB powered a digital hygrometer for 1 week, demonstrating its highly stable nature and capability to power low-power IoT devices [112]. Additionally, a galvanostatic PC-SE of ≈24.53% at 2200 mA g^{-1} was recorded under white LED illumination (12 mW cm^{-2}) by using the following formula (Equation 5.7). The galvanostatic PC-SE can be referred to as the energy-saving efficiency in photo-enhanced batteries. The quantity signifies the amount of energy saved with the help of illumination for a particular galvanostatic charge and discharge cycle.

$$\eta(\%) = \frac{E_{op(light)} - E_{op(dark)}}{(E_{inp})optical} \times 100 \tag{5.7}$$

where $E_{op\ (light)}$ and $E_{op\ (dark)}$ are the energy output during galvanostatic discharge under light and dark conditions, respectively; thus, $E_{op(light)} - E_{op(dark)} = V \times I \times (t_{dis}(light) - t_{dis}(dark))$. (E_{inp}) *optical* is the input energy due to illumination; thus, $(E_{inp})\ optical = P_{inp} \times t_{il} \times A$. P_{in} is the input illuminated power, t_{il} is the illumination time, and A is the active surface area of the PRB.

5.3 Conclusions

PRBs have emerged as a promising solution for efficient solar energy storage by directly converting and storing solar energy in a single device. This book chapter has focused on the recent reports on PRBs employing organic and inorganic materials such as MHP, MHP–MoS_2 heterostructure, TMOs, and Sb_2S_3 as an active material, highlighting their synthesis method, device configuration, operation, advantages, and issues associated with these PRBs. MHPs have been very attractive due to their well-known PV properties and have shown significant promise as an active material in PRBs. However, MHP suffers from structural instability post-lithiation and degradation in a liquid electrolyte medium. MoS_2-2D perovskite heterostructure demonstrated increased electrochemical stability by forming an interconnected network with improved morphology and crystallinity. These MoS_2-2D perovskite hybrid-based Li-PRBs have achieved an average dark discharge specific capacities of 128.66 $mAhg^{-1}$, which enhanced to 180.67 $mAhg^{-1}$ under illumination resulting in photo-enhancement of 40.42% at 64 mAg^{-1}. However, these MHP-based PRBs would require further attention to improve the long-term cyclability and compatibility with a wide range of electrolytes. On the other hand, inorganic semiconductors have shown significantly improved performance due to their high electrochemical stability under a liquid electrolyte medium as well as improved optoelectronic properties. Fe_2O_3-based PRB demonstrated a low-cost synthesis method and high enhancement in the specific capacity, but it would still require significant improvements for real-time applications. As MHP and Fe_2O_3 have distinct properties, MHP is known for its rich optical and optoelectronic properties but suffers from degradation. Fe_2O_3-based photoelectrode exhibits a high specific capacity (307.5 $mAhg^{-1}$ at 500 mAg^{-1} in dark conditions, which enhanced to 522.4 $mAhg^{-1}$ under illumination) but lower photoconversion ability than MHPs. For the development of PRBs, it is imperative to look for stable electrochemistry without losing its optical activities, which is one of the biggest concerns. Sb_2S_3 is known for its high specific capacity and for being employed in solar cell active material. Hence, Sb_2S_3 is investigated to address the above-discussed concerns and attain optical and electrochemical synergy in the Sb2S3 NRs demonstrated stable photocurrent generation capability when used for PD. Sb_2S_3-based PRB has shown a significant photo-enhancement of 28.94% in specific capacity at a very high current rate of 2200 mA g^{-1} with ~99% capacity retention under illumination. The approach of using Cu mesh as a current collector facilitated efficient light and electrolyte interaction, leading to stable cycling performance and minimal photo-corrosion. The findings suggest that light activates inactive regions in the photoelectrode, improving lithium-ion storage and overall capacity. In addition, the *ex situ* XPS and Raman spectroscopy analysis indicates that light modifies the oxidation state of Sb and thus promotes the redox activities, thus showing higher discharging capacity. This advancement supports the application of Sb_2S_3-based PRBs in powering low-power IoT devices. The smaller steps taking place in the PRB research domain may lead to future efficient PRBs. Currently, PRBs are in an early stage of development and thus require a well-established protocol for studying their detailed working mechanisms for a wide range of metal-ions and the corresponding performance. While significant progress has been made, several challenges remain, such as improving photoconversion efficiency, addressing stability issues, and developing scalable and cost-effective fabrication methods. Continued research and development efforts are crucial to overcome these challenges and realize the full potential of PRBs for a sustainable energy future.

Acknowledgments

S.A. acknowledges financial support from IITJ-SEED Grant (SA/20230015) and SERB-CRG (CRG/2023/003139) research grants.

References

1. *Annual carbon dioxide (CO2) emissions worldwide from 1940 to 2024 (in billion metric tons)*, Statista, & Global Carbon Budget, November 13, 2024, https://www.statista.com/statistics/276629/global-co2-emissions/.
2. DeAngelo, J., Azevedo, I., Bistline, J., Clarke, L., Luderer, G., Byers, E., Davis, S.J., Energy systems in scenarios at net-zero CO2 emissions. *Nat. Commun.*, 12, 1, 6096, 2021.
3. Matthews, H.D. and Caldeira, K., Stabilizing climate requires near-zero emissions. *Geophys. Res. Lett.*, 35, L04705, 2008.
4. *1992 United Nations Conference on Environment and Development*, Rio de Janeiro, Brazil, 3-14 June 1992, https://www.un.org/en/conferences/environment/rio.
5. Geisz, J.F., France, R.M., Schulte, K.L., Steiner, M.A., Norman, A.G., Guthrey, H.L., Young, M.R., Song, T., Moriarty, T., Six-junction III–V solar cells with 47.1% conversion efficiency under 143 Suns concentration. *Nat. Energy*, 5, 4, 326–335, 2020.
6. Nijsse, F.J.M.M., Mercure, J.-F., Ameli, N., Larosa, F., Kothari, S., Rickman, J., Vercoulen, P., Pollitt, H., The momentum of the solar energy transition. *Nat. Commun.*, 14, 1, 6542, 2023.
7. Saxena, A., Brown, C., Arneth, A., Rounsevell, M., Advanced photovoltaic technology can reduce land requirements and climate impact on energy generation. *Commun. Earth Environ.*, 5, 1, 586, 2024.
8. International Energy Agency, in: *World Energy Outlook 2021*, 2021, www.iea.org/weo.
9. Chamola, S. and Ahmad, S., Emerging opto-operando techniques to shed light on lithium-ion batteries. *Next Energy*, 7, 100239, 2025.
10. Hodes, G., Manassen, J., Cahen, D., Photoelectrochemical energy conversion and storage using polycrystalline chalcogenide electrodes. *Nature*, 261, 5559, 403–404, 1976.
11. Guo, W., Xue, X., Wang, S., Lin, C., Wang, Z.L., An Integrated Power Pack of Dye-Sensitized Solar Cell and Li Battery Based on Double-Sided TiO_2 Nanotube Arrays. *Nano Lett.*, 12, 5, 2520–2523, 2012.
12. Chen, T., Qiu, L., Yang, Z., Cai, Z., Ren, J., Li, H., Lin, H., Sun, X., Peng, H., An Integrated "Energy Wire" for both Photoelectric Conversion and Energy Storage. *Angew. Chem. Int. Ed.*, 51, 48, 11977–11980, 2012.
13. Gurung, A., Reza, K.M., Mabrouk, S., Bahrami, B., Pathak, R., Lamsal, B.S., Rahman, S.I., Ghimire, N., Bobba, R.S., Chen, K., Pokharel, J., Baniya, A.,

Laskar, M.A.R., Liang, M., Zhang, W., Zhang, W., Yang, S., Xu, K., Qiao, Q., Rear-Illuminated Perovskite Photorechargeable Lithium Battery. *Adv. Funct. Mater.*, 30, 2001865, 2020.

14. Salunke, A.D., Chamola, S., Mathieson, A., Boruah, B.D., de Volder, M., Ahmad, S., Photo-Rechargeable Li-Ion Batteries: Device Configurations, Mechanisms, and Materials. *ACS Appl. Energy Mater.*, 5, 7, 7891–7912, 2022.
15. Paolella, A., Faure, C., Bertoni, G., Marras, S., Guerfi, A., Darwiche, A., Hovington, P., Commarieu, B., Wang, Z., Prato, M., Colombo, M., Monaco, S., Zhu, W., Feng, Z., Vijh, A., George, C., Demopoulos, G.P., Armand, M., Zaghib, K., Light-assisted delithiation of lithium iron phosphate nanocrystals towards photo-rechargeable lithium ion batteries. *Nat. Commun.*, 8, 1, 14643, 2017.
16. Gouder, A. and Lotsch, B.V., Integrated Solar Batteries: Design and Device Concepts. *ACS Energy Lett.*, 8, 8, 3343–3355, 2023.
17. Kim, T., Song, W., Son, D.Y., Ono, L.K., Qi, Y., Lithium-ion batteries: outlook on present, future, and hybridized technologies. *J. Mater. Chem. A Mater.*, 7, 7, 2942–2964, 2019.
18. Grey, C.P. and Hall, D.S., Prospects for lithium-ion batteries and beyond—a 2030 vision. *Nat. Commun.*, 11, 6279, 2020.
19. Ahmad, S., Copic, D., George, C., De Volder, M., Hierarchical Assemblies of Carbon Nanotubes for Ultraflexible Li-Ion Batteries. *Adv. Mater.*, 28, 31, 6705–6710, 2016.
20. Goodenough, J.B., Electrochemical energy storage in a sustainable modern society. *Energy Environ. Sci.*, 7, 1, 14–18, 2014.
21. Manthiram, A., An Outlook on Lithium-Ion Battery Technology. *ACS Cent. Sci.*, 3, 10, 1063–1069, 2017.
22. Mondal, B., Azam, A., Ahmad, S., PCBM Functionalized WS_2 Hybrid Nanostructures for High Performance Li-Ion Battery Anodes: Toward Binder-Free Electrodes. *Energy Fuels*, 37, 20, 16105–16118, 2023.
23. Kato, K., Puthirath, A.B., Mojibpour, A., Miroshnikov, M., Satapathy, S., Thangavel, N.K., Mahankali, K., Dong, L., Arava, L.M.R., John, G., Bharadwaj, P., Babu, G., Ajayan, P.M., Light-Assisted Rechargeable Lithium Batteries: Organic Molecules for Simultaneous Energy Harvesting and Storage. *Nano Lett.*, 21, 2, 907–913, 2021.
24. Lee, A., Vörös, M., Dose, W.M., Niklas, J., Poluektov, O., Schaller, R.D., Iddir, H., Maroni, V.A., Lee, E., Ingram, B., Curtiss, L.A., Johnson, C.S., Photo-accelerated fast charging of lithium-ion batteries. *Nat. Commun.*, 10, 1, 4946, 2019.
25. Ahmad, S., George, C., Beesley, D.J., Baumberg, J.J., De Volder, M., Photo-Rechargeable Organo-Halide Perovskite Batteries. *Nano Lett.*, 18, 3, 1856–1862, 2018.
26. Boruah, B.D., Wen, B., De Volder, M., Light Rechargeable Lithium-Ion Batteries Using V2O5Cathodes. *Nano Lett.*, 21, 8, 3527–3532, 2021.

27. Shao, M., Bie, T., Yang, L., Gao, Y., Jin, X., He, F., Zheng, N., Yu, Y., Zhang, X., Over 21% Efficiency Stable 2D Perovskite Solar Cells. *Adv. Mater.*, 34, 1, 1–10, 2022.
28. Caiazzo, A. and Janssen, R.A.J., High Efficiency Quasi-2D Ruddlesden–Popper Perovskite Solar Cells. *Adv. Energy Mater.*, 12, 2202830, 2022.
29. Zhao, X., Liu, T., Loo, Y.L., Advancing 2D Perovskites for Efficient and Stable Solar Cells: Challenges and Opportunities. *Adv. Mater.*, 34, 3, 1–19, 2022.
30. Tsai, H., Nie, W., Blancon, J.C., Stoumpos, C.C., Asadpour, R., Harutyunyan, B., Neukirch, A.J., Verduzco, R., Crochet, J.J., Tretiak, S., Pedesseau, L., Even, J., Alam, M.A., Gupta, G., Lou, J., Ajayan, P.M., Bedzyk, M.J., Kanatzidis, M.G., Mohite, A.D., High-efficiency two-dimensional ruddlesden-popper perovskite solar cells. *Nature*, 536, 7616, 312–317, 2016.
31. Stoumpos, C.C., Soe, C.M.M., Tsai, H., Nie, W., Blancon, J.C., Cao, D.H., Liu, F., Traoré, B., Katan, C., Even, J., Mohite, A.D., Kanatzidis, M.G., High Members of the 2D Ruddlesden-Popper Halide Perovskites: Synthesis, Optical Properties, and Solar Cells of $(CH_3(CH_2)_3NH_3)_2(CH_3NH_3)_4Pb_5I_{16}$. *Chem.*, 2, 3, 427–440, 2017.
32. Cao, D.H., Stoumpos, C.C., Farha, O.K., Hupp, J.T., Kanatzidis, M.G., 2D Homologous Perovskites as Light-Absorbing Materials for Solar Cell Applications. *J. Am. Chem. Soc.*, 137, 24, 7843–7850, 2015.
33. Kang, D.H., Pae, S.R., Shim, J., Yoo, G., Jeon, J., Leem, J.W., Yu, J.S., Lee, S., Shin, B., Park, J.H., An Ultrahigh-Performance Photodetector based on a Perovskite–Transition-Metal-Dichalcogenide Hybrid Structure. *Adv. Mater.*, 28, 35, 7799–7806, 2016.
34. Wang, Y., Fullon, R., Acerce, M., Petoukhoff, C.E., Yang, J., Chen, C., Du, S., Lai, S.K., Lau, S.P., Voiry, D., O'Carroll, D., Gupta, G., Mohite, A.D., Zhang, S., Zhou, H., Chhowalla, M., Solution-Processed MoS_2/Organolead Trihalide Perovskite Photodetectors. *Adv. Mater.*, 29, 4, 1–7, 2017.
35. Rahil, M., Ansari, R.M., Ahmad, S., Islam, S.S., Nanostructured Ruddlesden–Popper-Layered Lead Bromide Perovskites with Stable and Selected Wavelength for Photodetection Applications. *ACS Appl. Nano Mater.*, 6, 7, 5187–5199, 2023.
36. Ahmad, S., Kanaujia, P.K., Beeson, H.J., Abate, A., Deschler, F., Credgington, D., Steiner, U., Prakash, G.V., Baumberg, J.J., Strong Photocurrent from Two-Dimensional Excitons in Solution-Processed Stacked Perovskite Semiconductor Sheets. *ACS Appl. Mater. Interfaces*, 7, 45, 25227–25236, 2015.
37. Rahil, M., Ansari, R.M., Prakash, C., Islam, S.S., Dixit, A., Ahmad, S., Ruddlesden–Popper 2D perovskites of type $(C_6H_9C_2H_4NH_3)_2(CH_3NH_3)_{n-1}Pb_n I3_{n+1}$ (n=1–4) for optoelectronic applications. *Sci. Rep.*, 12, 1, 1–13, 2022.
38. Yadav, A., Rahil, M., Ahmad, S. Facile and Effective Band Gap Engineering of 2D Ruddlesden–Popper Perovskites with Improved Structural and Optoelectronic Properties. *ACS Appl. Electron. Mater.*, 5, 2, 1024–1034, 2023.

39. Rahil, M., Mushtaq, N., Ansari, H.R., Alam, M.K., Ahmad, S., Ahmad, S., Exciton tunability in mixed halide hybrid layered perovskites. *AIP Conf. Proc.*, 2276, 020043, October, 2020.
40. Rahil, M., Rajput, P., Ghosh, D., Ahmad, S., Highly tunable single-phase excitons in mixed halide layered perovskites. *ACS Appl. Electron. Mater.*, 2, 10, 3199–3210, 2020.
41. Sadhanala, A., Ahmad, S., Zhao, B., Giesbrecht, N., Pearce, P.M., Deschler, F., Hoye, R.L.Z., Gödel, K.C., Bein, T., Docampo, P., Dutton, S.E., De Volder, M.F.L., Friend, R.H., Blue-Green Color Tunable Solution Processable Organolead Chloride-Bromide Mixed Halide Perovskites for Optoelectronic Applications. *Nano Lett.*, 15, 9, 6095–6101, 2015.
42. Ahmad, S., Baumberg, J.J., Vijaya Prakash, G., Structural tunability and switchable exciton emission in inorganic-organic hybrids with mixed halides. *J. Appl. Phys.*, 114, 233511, 2013.
43. Liang, S.Y., Liu, Y.F., Wang, S.Y., Ji, Z.K., Xia, H., Bai, B.F., Sun, H.B., High-Resolution Patterning of 2D Perovskite Films through Femtosecond Laser Direct Writing. *Adv. Funct. Mater.*, 2204957, 1–8, 2022.
44. Li, G., Lin, K., Zhao, K., Huang, Y., Ji, T., Shi, L., Hao, Y., Xiong, Q., Zheng, K., Pullerits, T., Cui, Y., Localized Bound Multiexcitons in Engineered Quasi-2D Perovskites Grains at Room Temperature for Efficient Lasers. *Adv. Mater.*, 35, 20, 1–9, 2023.
45. Lin, Y.K., Chen, C.H., Wang, Y.Y., Yu, M.H., Yang, J.W., Ni, I.C., Lin, B.H., Zhidkov, I.S., Kurmaev, E.Z., Lu, Y.J., Chueh, C.C., Realizing High Brightness Quasi-2D Perovskite Light-Emitting Diodes with Reduced Efficiency Roll-Off *via* Multifunctional Interface Engineering. *Adv. Sci.*, 10, 26, 1–10, 2023.
46. Lu, Y., Zhang, H., Wang, Y., Zhu, X., Xiao, W., Xu, H., Li, G., Li, Y., Fan, D., Zeng, H., Chen, Z., Yang, X., Solar-Driven Interfacial Evaporation Accelerated Electrocatalytic Water Splitting on 2D Perovskite Oxide/MXene Heterostructure. *Adv. Funct. Mater.*, 33, 21, 1–10, 2023.
47. Ahmad, S., Sadhanala, A., Hoye, R.L.Z., Andrei, V., Modarres, M.H., Zhao, B., Rongé, J., Friend, R., De Volder, M., Triple-Cation-Based Perovskite Photocathodes with AZO Protective Layer for Hydrogen Production Applications. *ACS Appl. Mater. Interfaces*, 11, 26, 23198–23206, 2019.
48. Andrei, V., Hoye, R.L.Z., Crespo-Quesada, M., Bajada, M., Ahmad, S., De Volder, M., Friend, R., Reisner, E., Scalable Triple Cation Mixed Halide Perovskite–BiVO4 Tandems for Bias-Free Water Splitting. *Adv. Energy Mater.*, 8, 25, 1–14, 2018.
49. Chen, Y., Chen, Z., Zhang, X., Chen, J., Wang, Y., An organic-halide perovskite-based photo-assisted Li-ion battery for photoelectrochemical storage. *Nanoscale*, 14, 30, 10903–10909, 2022.
50. Yadav, A. and Ahmad, S., Single Crystal Ruddlesden–Popper and Dion–Jacobson Metal Halide Perovskites for Visible Light Photodetectors: Present Status and Future Perspectives. *ACS Appl. Mater. Interfaces*, 16, 33, 43134–43155, 2024.

51. Ansari, R.M., Chamola, S., Ahmad, S., Ruddlesden–Popper 2D Perovskite-MoS2 Hybrid Heterojunction Photocathodes for Efficient and Scalable Photo-Rechargeable Li-Ion Batteries. *Small*, 20, 2401350, 2024.
52. Wang, S., Li, L., Weng, W., Ji, C., Liu, X., Sun, Z., Lin, W., Hong, M., Luo, J., Trilayered Lead Chloride Perovskite Ferroelectric Affording Self-Powered Visible-Blind Ultraviolet Photodetection with Large Zero-Bias Photocurrent. *J. Am. Chem. Soc.*, 142, 1, 55–59, 2020.
53. Li, M., Gao, Q., Liu, P., Liao, Q., Zhang, H., Yao, J., Hu, W., Wu, Y., Fu, H., Amplified Spontaneous Emission Based on 2D Ruddlesden–Popper Perovskites. *Adv. Funct. Mater.*, 28, 17, 1–9, 2018.
54. Zhang, F., Kim, D.H., Lu, H., Park, J.S., Larson, B.W., Hu, J., Gao, L., Xiao, C., Reid, O.G., Chen, X., Zhao, Q., Ndione, P.F., Berry, J.J., You, W., Walsh, A., Beard, M.C., Zhu, K., Enhanced Charge Transport in 2D Perovskites *via* Fluorination of Organic Cation. *J. Am. Chem. Soc.*, 141, 14, 5972–5979, 2019.
55. Shrestha, S., Li, X., Tsai, H., Hou, C.-H., Huang, H.-H., Ghosh, D., Shyue, J.-J., Wang, L., Tretiak, S., Ma, X., Nie, W., Long carrier diffusion length in two-dimensional lead halide perovskite single crystals. *Chem.*, 8, 4, 1107–1120, 2022.
56. Wang, K., Park, J.Y., Akriti, Dou, L., Two-dimensional halide perovskite quantum-well emitters: A critical review. *EcoMat.*, 3, 3, 1–13, 2021.
57. Guan, M., Xie, Y., Zhang, Y., Gu, Z., Qiu, L., He, Z., Ye, B., Suwardi, A., Dai, Z., Li, G., Hu, G., Moisture-Tailored 2D Dion–Jacobson Perovskites for Reconfigurable Optoelectronics. *Adv. Mater.*, 35, 22, 1–11, 2023.
58. Ye, J., Zheng, H., Zhu, L., Liu, G., Zhang, X., Hayat, T., Pan, X., Dai, S., Enhanced Moisture Stability of Perovskite Solar Cells With Mixed-Dimensional and Mixed-Compositional Light-Absorbing Materials. *Sol. RRL*, 1, 11, 1–8, 2017.
59. Xu, Y., Wang, M., Lei, Y., Ci, Z., Jin, Z., Crystallization Kinetics in 2D Perovskite Solar Cells. *Adv. Energy Mater.*, 10, 43, 1–18, 2020.
60. Yu, H., Xie, Y., Zhang, J., Duan, J., Chen, X., Liang, Y., Wang, K., Xu, L., Thermal and Humidity Stability of Mixed Spacer Cations 2D Perovskite Solar Cells. *Adv. Sci.*, 8, 12, 1–10, 2021.
61. Liu, R., Hu, X., Xu, M., Ren, H., Yu, H., Layered Low-Dimensional Ruddlesden-Popper and Dion-Jacobson Perovskites: From Material Properties to Photovoltaic Device Performance. *ChemSusChem*, 16, e202300736, 2023.
62. Gao, X., Zhang, X., Yin, W., Wang, H., Hu, Y., Zhang, Q., Shi, Z., Colvin, V.L., Yu, W.W., Zhang, Y., Ruddlesden–Popper Perovskites: Synthesis and Optical Properties for Optoelectronic Applications. *Adv. Sci.*, 6, 1900941, 2019.
63. Hu, J., Liu, J., Chen, H., Ran, X., Zhang, H., Li, P., Song, L., Müller-Buschbaum, P., Xia, Y., Chen, Y., A bromide-induced highly oriented low-dimensional Ruddlesden-Popper phase for efficient and stable perovskite solar cells. *J. Mater. Chem. A Mater.*, 9, 26, 15068–15075, 2021.
64. Zhang, Y., Siegler, T.D., Thomas, C.J., Abney, M.K., Shah, T., De Gorostiza, A., Greene, R.M., Korgel, B.A., A 'tips and Tricks' Practical Guide to the

Synthesis of Metal Halide Perovskite Nanocrystals. *Chem. Mater.*, 32, 13, 5410–5423, 2020.

65. Spanopoulos, I., Hadar, I., Ke, W., Tu, Q., Chen, M., Tsai, H., He, Y., Shekhawat, G., Dravid, V.P., Wasielewski, M.R., Mohite, A.D., Stoumpos, C.C., Kanatzidis, M.G., Uniaxial Expansion of the 2D Ruddlesden-Popper Perovskite Family for Improved Environmental Stability. *J. Am. Chem. Soc.*, 141, 13, 5518–5534, 2019.
66. Kong, L., Luo, Y., Turyanska, L., Zhang, T., Zhang, Z., Xing, G., Yang, Y., Zhang, C., Yang, X., A Spacer Cation Assisted Nucleation and Growth Strategy Enables Efficient and High-Luminance Quasi-2D Perovskite LEDs. *Adv. Funct. Mater.*, 33, 2209186, 2023.
67. Bailey, C.G., Gillan, L.V., Lee, M., Sloane, N., Liu, X., Soufiani, A.M., Hao, X., McCamey, D.R., Influence of Organic Spacer Cation on Dark Excitons in 2D Perovskites. *Adv. Funct. Mater.*, 34, 2308095, 2024.
68. Chu, Z., Chu, X., Zhao, Y., Ye, Q., Jiang, J., Zhang, X., You, J., Emerging Low-Dimensional Crystal Structure of Metal Halide Perovskite Optoelectronic Materials and Devices. *Small Struct.*, 2, 6, 1–21, 2021.
69. Mathieson, A., Rahil, M., Zhang, Y., Dose, W.M., Lee, J.T., Deschler, F., Ahmad, S., De Volder, M., Ruddlesden Popper 2D perovskites as Li-ion battery electrodes. *Mater. Adv.*, 2, 10, 3370–3377, 2021.
70. Jiang, Q., Chen, M., Li, J., Wang, M., Zeng, X., Besara, T., Lu, J., Xin, Y., Shan, X., Pan, B., Wang, C., Lin, S., Siegrist, T., Xiao, Q., Yu, Z., Electrochemical Doping of Halide Perovskites with Ion Intercalation. *ACS Nano*, 11, 1, 1073–1079, 2017.
71. Cabana, J., Monconduit, L., Larcher, D., Palacín, M.R., Beyond intercalation-based Li-ion batteries: The state of the art and challenges of electrode materials reacting through conversion reactions. *Adv. Mater.*, 22, 35, 170–192, 2010.
72. Zhuang, G.V., Yang, H., Blizanac, B., Ross, P.N., A study of electrochemical reduction of ethylene and propylene carbonate electrolytes on graphite using ATR-FTIR spectroscopy. *Electrochem. Solid-State Lett.*, 8, 9, 441–445, 2005.
73. Billing, D.G. and Lemmerer, A., Synthesis, characterization and phase transitions of the inorganic-organic layered perovskite-type hybrids $[(C_nH_{2n+1}NH_3)_2PbI_4]$ (n = 12, 14, 16 and 18). *New J. Chem.*, 32, 10, 1736–1746, 2008.
74. Dawson, J.A., Naylor, A.J., Eames, C., Roberts, M., Zhang, W., Snaith, H.J., Bruce, P.G., Saiful Islam, M., Mechanisms of Lithium Intercalation and Conversion Processes in Organic-Inorganic Halide Perovskites. *ACS Energy Lett.*, 2, 8, 1818–1824, 2017.
75. Tewari, N., Shivarudraiah, S.B., Halpert, J.E., Photorechargeable Lead-Free Perovskite Lithium-Ion Batteries Using Hexagonal Cs3Bi2I9Nanosheets. *Nano Lett.*, 21, 13, 5578–5585, 2021.
76. Lehner, A.J., Fabini, D.H., Evans, H.A., Hébert, C.A., Smock, S.R., Hu, J., Wang, H., Zwanziger, J.W., Chabinyc, M.L., Seshadri, R., Crystal and Electronic Structures of Complex Bismuth Iodides $A_3Bi_2I_9$ (A = K, Rb, Cs)

Related to Perovskite: Aiding the Rational Design of Photovoltaics. *Chem. Mater.*, 27, 20, 7137–7148, 2015.
77. Vicente, N. and Garcia-Belmonte, G., Methylammonium Lead Bromide Perovskite Battery Anodes Reversibly Host High Li-Ion Concentrations. *J. Phys. Chem. Lett.*, 8, 7, 1371–1374, 2017.
78. Vicente, N., Bresser, D., Passerini, S., Garcia-Belmonte, G., Probing the 3-step Lithium Storage Mechanism in $CH_3NH_3PbBr_3$ Perovskite Electrode by Operando-XRD Analysis. *ChemElectroChem*, 6, 2, 456–460, 2019.
79. López, L.T., Ramírez, D., Jaramillo, F., Calderón, J.A., Novel hybrid organic-inorganic $CH_3NH_3NiCl_3$ active material for high-capacity and sustainable lithium-ion batteries. *Electrochim. Acta*, 357, 136882, 2020.
80. Lany, S., Semiconducting transition metal oxides. *J. Phys.: Condens. Matter*, 27, 28, 283203, 2015.
81. Song, F., Bai, L., Moysiadou, A., Lee, S., Hu, C., Liardet, L., Hu, X., Transition Metal Oxides as Electrocatalysts for the Oxygen Evolution Reaction in Alkaline Solutions: An Application-Inspired Renaissance. *J. Am. Chem. Soc.*, 140, 25, 7748–7759, 2018.
82. Zhu, X., Recent advances of transition metal oxides and chalcogenides in pseudo-capacitors and hybrid capacitors: A review of structures, synthetic strategies, and mechanism studies. *J. Energy Storage*, 49, 104148, 2022.
83. Yuan, S., Duan, X., Liu, J., Ye, Y., Lv, F., Liu, T., Wang, Q., Zhang, X., Recent progress on transition metal oxides as advanced materials for energy conversion and storage. *Energy Storage Mater.*, 42, 317–369, 2021.
84. Lei, Z., Lee, J.M., Singh, G., Sathish, C.I., Chu, X., Al-Muhtaseb, A.H., Vinu, A., Yi, J., Recent advances of layered-transition metal oxides for energy-related applications. *Energy Storage Mater.*, 36, 514–550, 2021.
85. Lim, J., Lee, J.-I., Wang, Y., Gauriot, N., Sebastian, E., Chhowalla, M., Schnedermann, C., Rao, A., Photoredox phase engineering of transition metal dichalcogenides. *Nature*, 633, 8028, 83–89, 2024.
86. Lei, T., Chen, W., Huang, J., Yan, C., Sun, H., Wang, C., Zhang, W., Li, Y., Xiong, J., Multifunctional Layered WS2 Nanosheets for Enhancing the Performance of Lithium–Sulfur Batteries. *Adv. Energy Mater.*, 7, 4, 1–8, 2017.
87. Kanwade, A., Gupta, S., Kankane, A., Tiwari, M.K., Srivastava, A., Kumar Satrughna, J.A., Chand Yadav, S., Shirage, P.M., Transition metal oxides as a cathode for indispensable Na-ion batteries. *RSC Adv.*, 12, 36, 23284–23310, 2022.
88. Thackeray, M.M., Johnson, P.J., de Picciotto, L.A., Bruce, P.G., Goodenough, J.B., Electrochemical extraction of lithium from $LiMn_2O_4$. *Mater. Res. Bull.*, 19, 2, 179–187, 1984.
89. Mizushima, K., Jones, P.C., Wiseman, P.J., Goodenough, J.B., Li_xCoO_2: A new cathode material for batteries of high energy density. *Mater. Res. Bull.*, 15, 6, 783–789, 1980.

90. Joshi, Y., Hadjixenophontos, E., Nowak, S., Lawitzki, R., Ghosh, P.K., Schmitz, G., Modulation of the Optical Properties of Lithium Manganese Oxide *via* Li-Ion De/Intercalation. *Adv. Opt. Mater.*, 6, 1701362, 2018.
91. Chakraborty, A., Dixit, M., Aurbach, D., Major, D.T., Predicting accurate cathode properties of layered oxide materials using the SCAN meta-GGA density functional. *npj Comput. Mater.*, 4, 1, 60, 2018.
92. Wang, J., Wang, Y., Zhu, C., Liu, B., Photoinduced Rechargeable Lithium-Ion Battery. *ACS Appl. Mater. Interfaces*, 14, 3, 4071–4078, 2022.
93. Wang, W., Wang, H., Liu, S., Huang, J., Synthesis of γ-LiV_2O_5 nanorods as a high-performance cathode for Li ion battery. *J. Solid State Electrochem.*, 16, 7, 2555–2561, 2012.
94. Fang, S., Bresser, D., Passerini, S., Transition Metal Oxide Anodes for Electrochemical Energy Storage in Lithium- and Sodium-Ion Batteries. *Adv. Energy Mater.*, 10, 1902485, 2020.
95. Mishra, M. and Chun, D.M., α-Fe_2O_3 as a photocatalytic material: A review. *Appl. Catal. A Gen.*, 498, 126–141, 2015.
96. Keppeler, M., Shen, N., Nageswaran, S., Srinivasan, M., Synthesis of α-Fe_2O_3/carbon nanocomposites as high capacity electrodes for next generation lithium ion batteries: a review. *J. Mater. Chem. A Mater.*, 4, 47, 18223–18239, 2016.
97. Xue, Y. and Wang, Y., A review of the α-Fe_2O_3 (hematite) nanotube structure: recent advances in synthesis, characterization, and applications. *Nanoscale*, 12, 20, 10912–10932, 2020.
98. Chamola, S. and Ahmad, S., Photo-Enhanced Li-Ion Batteries Based on Conversion Type Hematite Phase Iron Oxide Nanostructures, in: *Advances in Sustainability Science and Technology*, A. Dixit, V.K. Singh, S. Ahmad (Eds.), pp. 71–79, Springer, Singapore, 20242024.
99. Chamola, S. and Ahmad, S., High Performance Photorechargeable Li-Ion Batteries Based on Nanoporous Fe_2O_3 Photocathodes. *Adv. Sustain Syst.*, 7, 2300043, 2023.
100. Yan, N., Zhou, X., Li, Y., Wang, F., Zhong, H., Wang, H., Chen, Q., Fe_2O_3 nanoparticles wrapped in multi-walled carbon nanotubes with enhanced lithium storage capability. *Sci. Rep.*, 3, 3392, 2013.
101. Geng, B., Tao, B., Li, X., Wei, W., Ni2+/surfactant-assisted route to porous α-Fe_2O_3 nanoarchitectures. *Nanoscale*, 4, 5, 1671–1676, 2012.
102. Raypah, M.E., Ahmed, A.A., Omar, A.F., Application of porous materials and structures for improving optical and thermal performance of inorganic and organic light-emitting diodes: A review. *Sens. Actuators A Phys.*, 347, 113966, 2022.
103. Zhang, Q., Wei, M., Dong, Q., Gao, Q., Cai, X., Zhang, S., Yuan, T., Peng, F., Fang, Y., Yang, S., Photoinduced Cu^+/Cu^{2+} interconversion for enhancing energy conversion and storage performances of CuO based Li-ion battery. *J. Energy Chem.*, 79, 83–91, 2023.

104. Ciria-Ramos, I., Juarez-Perez, E.J., Haro, M., Solar Energy Storage Using a Cu_2O-TiO_2 Photocathode in a Lithium Battery. *Small*, 19, 2301244, 2023.
105. Yan, W., Wang, J., Hu, Q., Fu, J., Albolkany, M.K., Zhang, T., Lu, X., Ye, F., Liu, B., Approaching the theoretical capacity of TiO_2 anode in a photo-rechargeable lithium-ion battery. *Nano Res.*, 17, 2655–2662, 2024.
106. Wang, S., Liao, W., Su, H., Pang, S., Yang, C., Fu, Y., Zhang, Y., Review on the Application of Semiconductor Heterostructures in Photocatalytic Hydrogen Evolution: State-of-the-Art and Outlook. *Energy Fuels*, 37, 3, 1633–1656, 2023.
107. Pham, P.V., Bodepudi, S.C., Shehzad, K., Liu, Y., Xu, Y., Yu, B., Duan, X., 2D Heterostructures for Ubiquitous Electronics and Optoelectronics: Principles, Opportunities, and Challenges. *Chem. Rev.*, 122, 6, 6514–6613, 2022.
108. Heinz, T.F., Optical Properties of van der Waals Heterostructures, in: *Frontiers in Optics 2015*, 2015, LW4H.1.
109. Wang, S., Zhao, S., Guo, X., Wang, G., 2D Material-Based Heterostructures for Rechargeable Batteries. *Adv. Energy Mater.*, 12, 2100864, 2022.
110. Hu, C., Chen, L., Hu, Y., Chen, A., Chen, L., Jiang, H., Li, C., Light-Motivated SnO_2/TiO_2 Heterojunctions Enabling the Breakthrough in Energy Density for Lithium-Ion Batteries. *Adv. Mater.*, 33, 2103558, 2021.
111. Pandya, R., Mathieson, A., Boruah, B.D., de Aguiar, H.B., de Volder, M., Interrogating the Light-Induced Charging Mechanism in Li-Ion Batteries Using *Operando* Optical Microscopy. *Nano Lett.*, 23, 16, 7288–7296, 2023.
112. Chamola, S., Zhou, R., Bakr, O.M., Ahmad, S., Sb_2S_3 Nanorods for Photoenhanced Li-Ion Batteries: A Synergetic Balance Between Energy Harvesting and Storage. *Adv. Funct. Mater.*, 35, 2414913, 2025.
113. Kondrotas, R., Chen, C., Tang, J., Sb2S3 Solar Cells. *Joule*, 2, 5, 857–878, 2018.
114. Chen, J., Li, G., Xu, Z., Xu, C., Naveed, F., Liu, B., Zhang, Y., Zhou, R., Chen, C., Wang, M., Xu, J., Li, L., Recent Advances and Prospects of Solution-Processed Efficient Sb2S3 Solar Cells. *Adv. Funct. Mater.*, 34, 18, 2313676, 2024.
115. Luo, W., Ao, X., Li, Z., Lv, L., Li, J., Hong, G., Wu, Q.-H., Wang, C., Imbedding ultrafine Sb_2S_3 nanoparticles in mesoporous carbon sphere for high-performance lithium-ion battery. *Electrochim. Acta*, 290, 185–192, 2018.

6

2D Materials for Solar-Assisted Hybrid Energy Storage Devices: Photo-Supercapacitors

Yasar Ozkan Yesilbag[1]*, Fatma Nur Tuzluca Yesilbag[1†], Ahmad Huseyin[2], Ahmed Jalal Salih Salih[2] and Mehmet Ertugrul[3,4]

[1]*Department of Physics, Erzincan Binali Yıldırım University, Erzincan, Turkey*
[2]*Department of Physics, Erzincan Binali Yıldırım University Graduate School of Natural and Applied Sciences, Erzincan, Turkey*
[3]*Department of Metallurgical and Materials Engineering, Faculty of Engineering, Karadeniz Technical University, Trabzon, Turkey*
[4]*Institute of Nanoscience & Nanotechnology (ION2), Universiti Putra Malaysia, Serdang, Selangor, Malaysia*

Abstract

Photo-supercapacitors (PSCs) are gaining recognition as a feasible option for future electronic devices and have attracted considerable research attention in recent years. These devices, designed for storing light energy, are viewed as a promising energy source for smart electronics, flexible technologies, and wearable devices. PSCs integrate solar cells with supercapacitors, facilitating the conversion of light energy into electrical energy, which is subsequently stored in chemical energy. Their exceptional light conversion efficiency and outstanding energy storage capabilities distinguish PSCs as key players in light-operated energy storage solutions. Recent studies have extensively explored the application of nanomaterials (1D, 2D, and 3D) as potential electrode materials in these devices. Furthermore, there has been a notable shift toward research to decrease dependence on traditional energy sources. This chapter delves into PSCs' structure, mechanisms, and operational principles alongside an overview of the available types. It concludes with significant recommendations for improving PSC performance by innovating new

**Corresponding author*: ftuzluca@gmail.com; ftuzluca@erzincan.edu.tr
†*Corresponding author*: oyesilbag@erzincan.edu.tr

Nurdan Demirci Sankir and Mehmet Sankir (eds.) Solar Capacitors and Batteries, (143–174) © 2025 Scrivener Publishing LLC

materials that exhibit optical and capacitive properties. Additionally, it addresses PSC's challenges and issues and proposes appropriate solutions.

***Keywords*:** Photo-supercapacitors, solar-assisted energy storage, 2D materials, graphene, MXene, hybrid energy devices

6.1 Introduction

Solar energy is considered one of the leading renewable energy sources, along with wind, hydropower, biomass, and geothermal energy. These sources naturally replenish themselves and do not release harmful environmental emissions. Solar energy can be efficiently converted into usable energy for daily applications at economically viable costs. The conversion of solar energy into electricity has garnered significant interest among researchers. The concept of converting sunlight into electrical energy was first introduced in 1839 by scientist Edmond Becquerel, who discovered that the voltage generated by two electrodes immersed in an electrolyte depended on the intensity of light incident on the electrodes. This foundational research culminated in 1956 with the development of the first solar cell, which utilized crystalline silicon, at Bell Laboratories. This breakthrough paved the way for extensive research and development in solar cell technology, particularly in production methods, the structure and design of electrode materials, and conversion efficiency. Significant efforts have been directed toward developing integrated systems that facilitate the cost-efficient generation, storage, and utilization of renewable energy sources in response to demand. One of the significant challenges associated with renewable energy is its intermittency; for example, solar energy is not continuously available. Therefore, optimizing solar energy utilization during daylight hours and ensuring efficient energy storage for nighttime or periods of insufficient sunlight (such as cloudy or rainy conditions) is crucial.

Among energy storage systems, batteries and supercapacitors (SCs) are widely utilized for storing solar energy. Batteries, characterized by their high energy density and low power density, can supply energy over extended periods. This capability arises from their energy storage mechanism, which involves oxidation–reduction reactions leading to phase changes in the electrode materials. In contrast, SCs, known for their high power density, deliver energy over shorter durations. Their charge storage mechanism relies on either the formation of an electric double layer at the interface between electrode materials and electrolytes or rapid surface redox reactions. Initially, solar cells and energy storage devices operated as separate components connected through external wiring. However, this approach resulted in energy losses due to complex designs, increased spatial requirements, and inefficiencies associated with cable connections. Integrating solar cells with energy storage devices into a single unit has proven to be an effective solution for overcoming these challenges. Due to the intermittent nature of sunlight, there is an increasing need for energy storage devices capable of rapid charging and reliable power delivery under low-light conditions. In this context, photo-supercapacitors (PSCs) have gained attention as promising solar energy systems. Their advantages include compact size, lightweight design, rapid charging capabilities, and seamless integration with solar cells, making them highly suitable for next-generation solar energy applications [1–6].

PSCs are emerging as a promising and efficient approach for directly converting solar energy into electrical energy [7, 8]. The primary challenge in this field is the development of photovoltaic SCs that are both cost-effective and efficient in terms of energy conversion and storage. In recent years, substantial efforts have focused on developing systems that can harvest and store solar energy. In these designs, the electrodes play a dual role, functioning as solar energy absorbers and energy storage components. PSCs are based on the integration of solar cells and SCs. Although both technologies have been extensively studied independently, considerable advancements have been achieved in synthesizing nanomaterials used as electrodes in these devices. In solar cells, the primary focus is light absorption to generate electron–hole pairs, enhancing the stored energy. In SCs, the goal is to optimize the electrode structure to improve specific capacitance (Cp) and energy density (Ed). This can be achieved using two-dimensional (2D) nanomaterials with a high surface area, such as graphene and MXene, or one-dimensional (1D) materials with high conductivity, such as carbon nanotubes (CNTs). This chapter will discuss recent advancements in PSC technology, emphasizing the integration of SCs and solar cells into a single device. Key topics will include the operating mechanisms, materials employed, and the electrode architecture of integrated PSC systems. Furthermore, we will explore the classification of PSCs based on different solar cell types and their operating principles, including dye-sensitized solar cells (DSSCs), quantum dot-sensitized solar cells (QDSSCs), organic solar cells (OSCs), and perovskite solar cells (PeSCs). Lastly, the modern applications of PSCs will be analyzed, focusing on their potential to drive advancements in next-generation energy storage and conversion technologies.

6.2 Fundamental Components of PSCs

6.2.1 Solar Cells

Solar cells harness sunlight and convert it into electrical energy *via* the photovoltaic effect. When photons interact with the surface of a semiconductor material, they excite electrons, resulting in the generation of an electric current that can either be used immediately or stored for later use. The scientific foundation of this process dates back to 1839 when French physicist Edmond Becquerel discovered the photovoltaic effect. Later, in 1873, Willoughby Smith identified the light sensitivity of selenium, which further fueled interest in solar energy research. In 1883, Charles Fritts developed the first solar cell using selenium as a semiconductor coated with a thin layer of gold. However, the breakthrough in high-efficiency solar technology occurred in 1954 when scientists at Bell Laboratories created the first silicon-based solar cell, achieving an efficiency of approximately 6%, paving the way for commercial solar applications.

The functionality of solar cells is based on a light-absorbing electrode material. When photons strike the surface of a solar cell composed of semiconductor materials such as silicon, the light energy is absorbed. This absorption excites electrons within the semiconductor, liberating them from their atomic structure. The most commonly used semiconductor material is silicon, which has two primary forms: phosphorus-doped silicon (n-type) and boron-doped silicon (p-type). Upon absorbing photon energy, an electric field is established, causing the n-type region to acquire a negative charge while the p-type region attains a positive charge. This separation of charges produces a voltage difference, facilitating the flow of electrons when the cell is connected to an electrical circuit, thereby generating electrical energy. Solar cells are manufactured using various materials, each exhibiting distinct advantages and limitations. The most widely used categories include the following.

A-) Silicon-Based Solar Cells

Silicon is the predominant material in solar cell production, accounting for approximately 90% of all commercial solar panels. The two main types of silicon-based solar cells are as follows:

- *Monocrystalline silicon (Mono-Si)*: This type exhibits high efficiency, typically exceeding 20%, due to its high crystal purity. It performs well under low-light conditions and has a long lifespan. However, its production cost is relatively high.

Polycrystalline silicon (Poly-Si): Although more cost-effective than monocrystalline silicon, it has lower efficiency, generally between 15% and 17%.

B-) Thin-Film Solar Cells

Thin-film solar cells are semiconductor layers deposited onto flexible substrates such as plastic or glass. Although these cells are generally more affordable than silicon-based alternatives, they may contain toxic elements. Notable thin-film technologies include the following:

- *Cadmium telluride (CdTe) solar cells*: Known for their cost-effectiveness and ease of manufacturing, CdTe cells offer moderate efficiency but contain cadmium, a toxic material.
- *Copper indium gallium selenide solar cells*: These cells demonstrate high absorption efficiency and flexibility, making them suitable for portable applications.

C-) Dye-Sensitized Solar Cells

DSSCs can be categorized into different types depending on the electrode materials and the dyes employed. DSSCs operate on distinct principles and materials unlike traditional silicon or thin-film solar cells. Conventional solar cells rely on the photovoltaic effect in semiconductor materials like silicon to convert light into electricity. In contrast, DSSCs utilize a light-sensitive dye to release electrons, generating electricity through a process analogous to photosynthesis. This innovative technology offers a low-cost alternative to traditional solar cells, with promising applications across various fields. The dye should absorb light over a wide range of wavelengths, encompassing both the visible spectrum and near-infrared regions.

Additionally, it should exhibit chemical stability and resistance to degradation from exposure to ultraviolet radiation. After absorbing light, the dye rapidly releases electrons and efficiently retrieves lost electrons from the electrolyte. The dyes used in DSSCs can be either organic or inorganic compounds, such as ruthenium-based or organic dyes. The positive electrode is commonly composed of semiconductor materials such as TiO_2, known for its high porosity and ability to facilitate efficient electron transport. The electrolyte, usually consisting of iodide/triiodide (I^-/I_3^-), facilitates charge transfer between the two electrodes. The negative electrode generally comprises materials like platinum or carbon, which collect electrons and complete the electric circuit. Transparent glass serves as the substrate supporting both electrodes while allowing light to reach the dye.

The working principle of DSSCs relies on the generation of electrons from dye molecules upon light exposure, followed by their transfer to the conduction band of the semiconductor (Figure 6.1(a)). The dye is subsequently regenerated by accepting electrons from the electrolyte, typically consisting of an iodide/triiodide (I^-/I_3^-) redox couple. The photogenerated electrons travel through the transparent conductive substrate and a semiconductor layer, ultimately reaching the counter electrode *via* an external circuit, thereby completing the electrical pathway. Under illumination, the device functions in light-to-electricity conversion mode, with its stability maintained through continuous electron regeneration. However, the overall efficiency is limited by recombination processes, including electron back-transfer from the semiconductor to the oxidized dye or electrolyte before contributing to the external load [9–11].

DSSCs employ dyes to absorb sunlight and convert it into electrical energy, serving as intermediaries between solar radiation and nanomaterials such as TiO_2. These dyes are categorized as organic or inorganic.

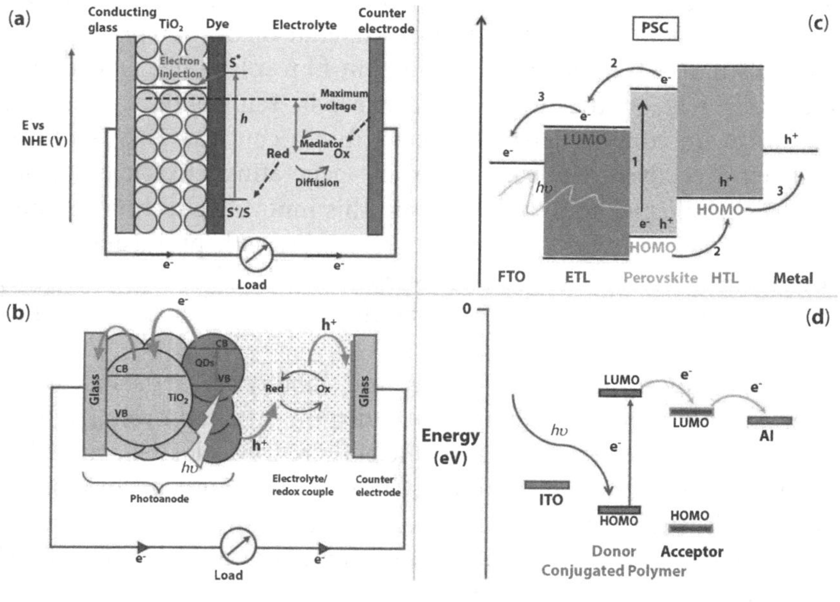

Figure 6.1 Schematic representation of the working mechanisms of various solar cell technologies: (a) DSSCs, highlighting photon absorption by the dye and subsequent electron transfer through the TiO_2 layer; (b) QDSSCs, illustrating quantum dot excitation and electron injection into the semiconductor layer; (c) PeSCs, depicting light absorption by the perovskite layer and charge separation at the interfaces; and (d) OSCs, showing the role of donor–acceptor materials in generating charge carriers.

Organic dyes, often containing conjugated double bonds, enable broad-spectrum light absorption. Certain organic polymers, such as carotenoids and ruthenium-based complexes, are widely used for their flexibility and efficiency. Inorganic dyes, particularly ruthenium-based compounds like N3 and N719, offer superior absorption characteristics and electron injection properties, making them common in DSSC research. By integrating innovative dye materials with nanostructured semiconductor architectures, DSSCs provide a cost-effective and versatile alternative to conventional photovoltaic systems [12–14].

DSSCs stand out for their ability to be fabricated on flexible and transparent substrates, enabling integration into windows, building facades, and foldable electronic devices. Their efficient performance under low-light and indirect illumination conditions makes them suitable for indoor and shaded environments. However, DSSCs face efficiency challenges, with conversion rates typically ranging from 7% to 11%, lower than traditional silicon-based solar cells. Environmental factors such as humidity and temperature fluctuations can also degrade materials over time, impacting long-term stability. Another limitation is using liquid electrolytes, which are prone to leakage and evaporation, potentially shortening the cell's lifespan. To overcome these challenges, researchers are investigating alternative electrolyte materials, such as solid-state and gel-based systems, to improve the stability and durability of the devices. Despite these challenges, DSSCs have promising applications in energy-generating building materials such as semi-transparent solar windows and rooftops, which allow light transmission while producing electricity. They are also being integrated into portable solar chargers for mobile phones and small electronic devices, expanding their potential in sustainable energy solutions [15, 16].

D-) Quantum Dot-Sensitized Solar Cells

These cells share a structural similarity with DSSCs but utilize quantum dots—nanometer-sized semiconductor particles—instead of dyes to convert solar energy (Figure 6.1b). The energy levels of quantum dots can be tuned by adjusting their size, enabling the absorption of different light wavelengths. Upon photon absorption, electrons transition from the quantum dots to a semiconductor material, typically TiO_2, and are subsequently transferred to an external circuit to generate electricity [17].

E-) Perovskite Solar Cells

PeSCs are an emerging photovoltaic technology that leverages materials with a perovskite crystal structure, known for their high light absorption and efficient charge transport. These cells typically incorporate lead-halide perovskites, such as methylammonium lead iodide ($MAPbI_3$) (Figure 6.1c). With efficiencies currently ranging between 15% and 25%, PeSCs present a compelling alternative to conventional silicon-based solar cells, thanks to their wide-spectrum light absorption and cost-effective manufacturing processes [18]. However, stability and long-term durability challenges remain, necessitating further research and development. PeSCs consist of multiple functional layers, each playing a distinct role in the photovoltaic process:

- *Transparent conductive electrode:* Typically made of indium tin oxide (ITO) or fluorine-doped tin oxide (FTO), this layer offers a combination of electrical conductivity and optical transparency, enabling light to pass through while facilitating charge collection.
- *Electron transport layer (ETL):* Responsible for directing photogenerated electrons from the perovskite layer to the external circuit, ETLs are commonly made from TiO_2 or organic nanocomposites such as phenyl-C61-butyric acid methyl ester.
- *Perovskite layer:* Serving as the active light-absorbing medium, this layer is composed of compounds like $MAPbI_3$, which can be fabricated using vapor deposition or solution-processing techniques.
- *Hole transport layer (HTL):* Facilitates the movement of holes (electron vacancies) from the perovskite layer to the counter electrode, enhancing charge separation and collection. Common HTL materials include Spiro-OMeTAD and PEDOT: PSS.
- *Metal back electrode:* Functions as the final charge-collecting layer, typically made from gold or silver, to ensure efficient charge extraction.

When the perovskite layer absorbs light, electrons are excited from their ground state to a higher energy level, initiating the charge generation process. These electrons are transferred to the ETL and move through the external circuit toward the metal back electrode, generating an electric current. Simultaneously, holes migrate toward the transparent conductive electrode *via* the HTL. The opposing movement of electrons and holes results in charge separation, producing a usable electrical output. Although silicon-based solar cells achieve high efficiency, their production is expensive and technologically complex. Thin-film solar technologies, including PeSCs, present a viable alternative due to their cost-effectiveness, lightweight design, and potential for integration with energy storage systems such as SCs. The primary goal in hybrid solar energy systems is to combine photovoltaic and electrochemical storage functionalities within a single integrated device.

F-) Organic Solar Cells

OSCs employ carbon-based materials to convert light into electricity. Their structure consists of a transparent conductive electrode, typically ITO, an active layer made of conjugated polymers or small organic molecules that absorb light and facilitate charge separation, and a negative electrode, usually composed of a metal such as aluminum (Figure 6.1d) [19, 20]. The active layer in OSCs enables efficient light absorption and charge transport. At the same time, choosing organic materials allows for flexibility, lightweight construction, and potential for large-scale production using cost-effective printing techniques. However, challenges such as lower efficiency and stability compared to inorganic solar cells continue to be active research areas.

6.2.2 Supercapacitors

SCs are highly promising energy storage devices, garnering significant research interest due to their superior characteristics compared to conventional capacitors and batteries. They offer high capacitance, enhanced power density, and notable energy density, along with rapid charging capabilities and extended operational lifetimes. Based on their energy storage mechanisms, SCs are classified into three categories: electrostatic double-layer capacitors (EDLCs), pseudocapacitors, and hybrid capacitors. In EDLCs, energy storage occurs through charge adsorption at the electrode/electrolyte interface (Figure 6.2a). The capacitance depends on

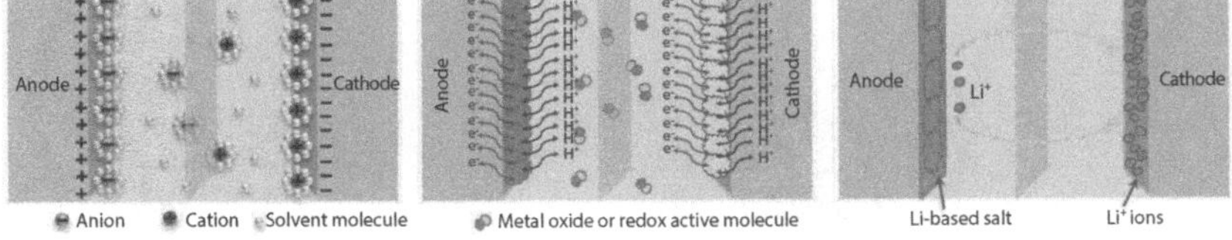

Figure 6.2 Mechanisms of different capacitor types: (a) EDLCs, (b) pseudocapacitors, and (c) hybrid capacitors [6].

the electrode surface area and the separation between the electrode and the electrolyte. Pseudocapacitors, by contrast, store energy via fast surface redox reactions (Figure 6.2b), a mechanism similar to that of batteries but with significantly faster charge/discharge rates. Hybrid capacitors integrate the properties of both EDLCs and pseudocapacitors, combining double-layer charge storage and redox reactions to enhance performance (Figure 6.2c). These advantageous properties position SCs as excellent candidates for integration with solar cells, facilitating the development of advanced PSCs hybrid systems.

6.2.3 Photo-Supercapacitors

PSCs integrate solar cells with SCs, allowing self-charging and maximizing clean energy collection. Various types of solar cells, such as silicon, DSSCs, OSCs, and PeSCs, can be integrated with SCs to form a unified energy storage system [7, 12, 21].

The photovoltaic component of PSCs consists of solar cells that absorb sunlight and convert it into electrical energy, which is then stored in an SC [22]. The SC retains energy between its electrodes and releases it when connected to a load. PSCs represent an innovative integration of solar cells and SCs into a single device, enabling shared functionality. These systems can

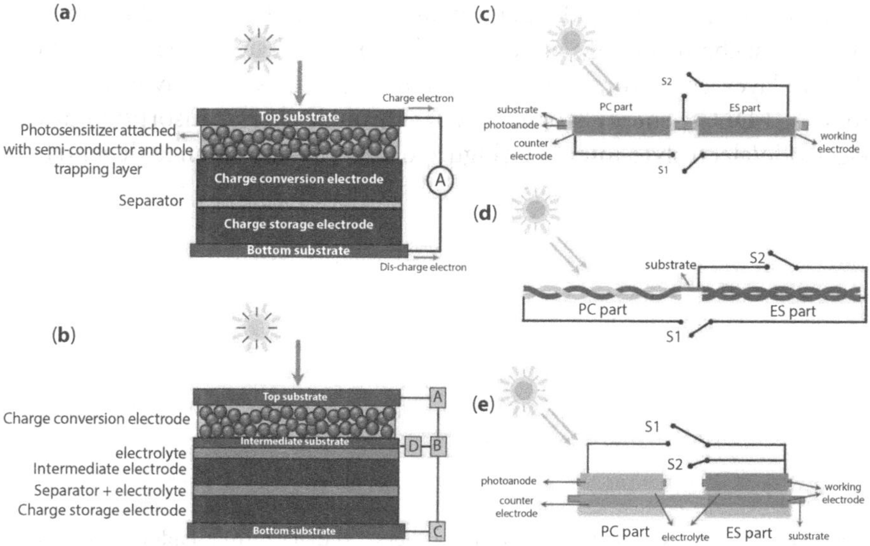

Figure 6.3 Design structures of various types of PSCs: (a) two- and (b) three-electrode planar PSCs, (c) twisted, (d) coaxial, and (e) parallel fiber PSCs.

be configured in either a three-electrode or two-electrode design (Figure 6.3a, b), with electrodes arranged in a flat or fibrous structure (Figure 6.3c, d, e). In a three-electrode configuration, an additional counter electrode facilitates charge storage and redox reactions, improving efficiency by providing a dedicated surface for electron transfer while maintaining separation between the solar cell and SCs. In contrast, two-electrode systems offer a more straightforward design but may reduce efficiency due to increased resistance. The structural flexibility of PSCs allows for tailored configurations to optimize performance for specific applications. Ongoing research focuses on enhancing material properties and structural designs to improve energy conversion and storage efficiency, further advancing sustainable energy solutions [23, 24].

6.2.4 Efficiency and Factors Affecting Performance

The overall efficiency of PSCs can be determined either by multiplying the individual efficiencies of the solar cells and SCs or by calculating the ratio of the light energy stored during illumination to the total energy stored throughout the charging process:

$$\eta_{PV} = \eta_s \cdot \eta_{sc} \tag{6.1}$$

where η_{PV} is the photovoltaic conversion efficiency (PCE), η_{sc} is the efficiency of the solar cell, and η_s denotes the energy storage efficiency of the SC.

The PCE of a solar cell under illumination can be calculated using the following equation:

$$\eta_{PC} = \frac{V_{oc} J_{sc} FF}{P_{in}} \tag{6.2}$$

V_{oc}, J_{sc}, FF, and P_{in} refer to the open-circuit voltage, short-circuit current density, fill factor (FF), and incident light power density (100 mW cm^{-2}), respectively.

The FF in PSCs represents the efficiency with which light energy is converted into electrical energy stored within the device. It is calculated using

$$FF = \frac{P_{\max}}{V_{oc} \cdot J_{sc}} \tag{6.3}$$

where P_{max} is the maximum power output of the solar cell, V_{oc} is the open-circuit voltage, and J_{sc} is the short-circuit current density.

The capacitance of an SC is crucial in defining the overall efficiency of the PSC system. It can be evaluated through techniques such as cyclic voltammetry (CV), galvanostatic charge–discharge (GCD) measurements, or electrochemical impedance spectroscopy.

The output power during the charging of an SC (E_{output}), power density (P_A, W cm^{-2}), and energy density (E_A, Wh cm^{-2}) can be evaluated using the following equations:

$$E_{output} = \frac{1}{2}CV^2 \tag{6.4}$$

$$P_A = \frac{E_A}{3600t} \tag{6.5}$$

$$E_A = \frac{0.5C_AV^2}{3600} \tag{6.6}$$

The overall efficiency of an integrated PSC unit can be obtained using the following equations:

$$\eta_{overall} = \frac{E_{output}}{E_{light}A_{solar}} \tag{6.7}$$

$$E_{light} = P_{in}t \tag{6.8}$$

Here, A_{solar} denotes the area of the solar cell, while E_{light} represents the energy absorbed by the solar cell during the charging process. The overall efficiency is determined by dividing the light energy stored during photoluminescence by the total energy accumulated during photo-recharging, as calculated using the previously derived equations.

Several factors can affect the overall efficiency of PSCs:

- *Light energy source*: Variability in the light source can cause fluctuations in charging time, which impacts the amount of energy the SC can store in each cycle.

- *Efficiency of solar cells:* Even under optimal conditions, the efficiency of solar cells is less than 100%, meaning some energy is inevitably lost during conversion.
- *Adjustment of open-circuit voltage:* Adjusting the open-circuit voltage of integrated solar cells can take 2 to 3 minutes. Consequently, the charging duration with light is crucial in determining overall efficiency.

The PSCs unit is typically characterized by low voltage, which may sometimes be insufficient to power small electronic devices. To address this, connecting multiple solar cells in series increases the voltage, while connecting SCs in parallel improves energy storage capacity, providing an effective solution for high voltage requirements [22, 25, 26].

6.2.5 Two-Electrode PSCs

In 2004, Miyasaka *et al.* developed the first PSCs capable of efficiently storing light energy as electrical energy [27]. This device comprises light and counter electrodes separated by an electrolyte. The light electrode is coated with a light-sensitive dye and connected to a glass substrate, functioning as a solar cell. Conversely, its opposite surface interfaces with the counter electrode, forming an SC. Nanomaterials within the light electrode facilitate electron capture from photon interactions with the dye. Following charge separation, the electrons migrate through the light electrode to the counter electrode, thereby charging the SC and establishing a double-layer charge.

6.2.6 Three-Electrode PSCs

Three-electrode PSCs have been proposed to overcome the high internal resistance and slow discharge rates observed in two-electrode PSCs. Miyasaka and Murakami introduced a third electrode to enable rapid discharge through oxidation and reduction processes [27]. This additional electrode serves dual purposes: the solar cell electrode on the photoelectrode side and the SC electrode on the counter electrode side. Two types of electrolytes can be employed: an oxidation–reduction electrolyte on the photoelectrode side and a suitable electrolyte (aqueous, organic, or solid) on the energy storage side. The photoelectrode and counter electrode are connected during the charging process, while the additional electrode remains electrically isolated. When light illuminates the photoelectrode, photo-excited electrons are transferred to the counter electrode, activating the solar cell. Once the charging process is complete or the light source is turned off, the counter electrode connects to the additional electrode, disconnecting from the photoelectrode. This transition enables the SC to enter its charge storage phase, during which a double-layer charge is formed. In this phase, the redox-active electrolyte interacts with the photoelectrode, facilitating the transfer of electrons to maintain the stored charge.

6.2.7 Classification of Planar/Uniaxial PSCs

6.2.7.1 Planar/Uniaxial PSCs Based on DSSC

Planar PSC systems are energy storage devices that combine the functions of SCs and photovoltaic energy collection within a single structure. The term "planar" refers to the internal arrangement of the device, where photovoltaic active materials—responsible for converting light into electrical energy—and energy storage components (typically electrodes) are positioned within the same plane or flat surface [25]. This integrated design optimizes the interaction between energy collection and storage functions. It allows sunlight to be directly converted into electrical energy by photovoltaic materials, which are then stored in the SC, enabling the device to perform dual functions: harvesting and storing energy for future use. The planar configuration enhances efficiency by maximizing the capture and conversion of solar energy into electricity while minimizing energy losses typically associated with external connections in conventional systems where solar cells and storage units function independently (Figure 6.4).

In planar designs of PSC systems, semiconductor materials are typically employed in the photovoltaic section to absorb light and convert it into electrical energy. These materials can include advanced photovoltaic materials such as perovskites, semiconductor compounds, organic photovoltaic cells, and nanomaterials. On the SC side, the energy storage component consists of electrodes made from materials with high surface area and excellent electrical conductivity, such as graphene, MXene, CNTs, or transition metal oxides. In this design, the electrodes are arranged side by side in a flat configuration rather than stacked in a sandwich-like structure, reducing internal resistance and enhancing the device's rapid charge and discharge capabilities. Such PSCs have numerous potential applications, including portable electronic devices that charge *via* solar energy, devices requiring rapid energy storage, and wearable and flexible electronics that can be produced on flexible substrates. Additionally, they are suitable for small-scale self-powered sensors and energy-harvesting systems. However, planar PSCs are still in the early development and research stages, with limited reports on the topic.

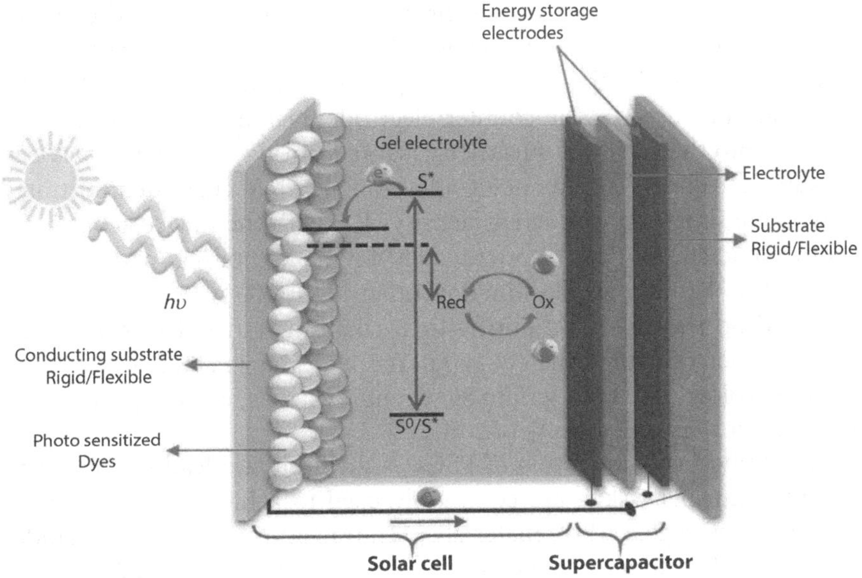

Figure 6.4 Schematic representation of PSC-based DSSC.

Miyasaka has made several attempts to integrate DSSCs with SCs [27]. However, a significant issue arises: electrons released due to photon absorption can return through the photoelectrode, potentially affecting the discharge process in the energy storage section. The liquid electrolyte on the photovoltaic side increases the likelihood of self-discharge or short-circuiting in the energy storage section (Figure 6.4) [28, 29]. The counter electrode in DSSCs is essential for reducing tri-iodide shuttle types by accepting electrons from the external circuit. While platinum is commonly used for this function, its high cost and wear susceptibility have led to the adoption of more affordable and eco-friendly alternatives, such as polymer and carbon-based materials.

FTO glass substrates are favored for their good conductivity and low cost. They are suitable as transparent light substrates in the photovoltaic section and effective conductors in the energy storage section. However, FTO's disadvantages include fragility and lack of flexibility. FTO has been successfully used in DSSCs, while ITO and substrates like polyethylene naphthalate (PEN) or polyethylene terephthalate (PET) offer better flexibility and performance. ITO-PET, for instance, combines good conductivity with appropriate flexibility. A primary limitation of polymer-based flexible DSSCs is their low operating temperature (below 150°C), while glass-based DSSCs can operate at higher temperatures (around 450–500 °C), facilitating better material adhesion to the electrode [30].

Chen *et al.* synthesized a three-electrode PSC system [31]. The photoelectrode was composed of TiO_2 and N719-type photosensitive dye on an ITO substrate, while the energy storage electrode was made of PEDOT, with platinum used as the counter electrode. Nanomaterials were compressed onto the substrate surface to ensure a strong connection between the substrate and TiO_2, enhancing adhesion. The J–V curve demonstrated a favorable short-circuit current density (J_{sc}), with the device achieving a photovoltaic efficiency of approximately 4.37%, a J_{sc} of 8.4 mA cm^{-2}, and an FF of 0.71 under an illumination intensity of 100 mW cm^{-2}. A slight voltage drop was observed during light charging, attributed to an internal resistance of approximately 160 μΩ. The device reached a maximum charging voltage of 0.68 V within 80 seconds, with a specific capacitance of 0.52 F g^{-1} at a current density of 2 mA cm^{-2}.

In the photovoltaic section of PSCs, ruthenium and lead iodide methylammonium perovskite are commonly used to achieve high PCE [32]. For energy storage, carbon materials [33], transition metal chalcogenides [25, 34], and conductive polymers [35] are utilized. Recent advances in nanomaterial synthesis have facilitated the use of nanoparticles in the photovoltaic section, enhancing performance. These developments allow for the tailored design of nanomaterials, which can be optimized for various applications, including photovoltaic systems [33]. Research continues to focus on reducing synthesis costs and improving the efficiency of DSSCs, making this a critical area for future work. A significant challenge with these sensitive solar cells is their reliance on ruthenium-based dyes, which offer broad absorption spectra but are both toxic and expensive. As an alternative, organic dyes and quantum dots, which provide broad absorption spectra and good conductivity, are being explored [36, 37].

CNTs are considered exceptional electrode materials due to their remarkable mechanical strength and excellent electrical conductivity. In DSSCs, CNTs have been employed as counter electrodes and catalytic agents, which are similar to platinum electrodes, owing to their high PCE. Yang *et al.* developed a three-electrode PSC system based on DSSCs, incorporating MWCNTs as inter-electrodes and energy storage electrodes [8]. The active electrode in the photovoltaic section consisted of TiO_2 integrated with dye. The electrolyte in the storage section was a gel electrolyte, while the electrolyte in the photovoltaic section was an oxidizing electrolyte. The MWCNTs were synthesized on a Si substrate using the CVD method, achieving an outstanding electrical conductivity of 104 S cm^{-1}. The PSC demonstrated a PCE of 6.1%, with an open-circuit voltage (V_{oc}) of 0.75 V, a short-circuit current density (J_{sc}) of 13.41 mA cm^{-2}, and an FF of 0.61. The specific capacity of the MWCNTs was measured at 48 F g^{-1}, and the device demonstrated good stability with 95% capacity retention after 100 cycles. To achieve better results, MWCNTs were synthesized with polyaniline (PANI). MWCNT-PANI was utilized in the energy storage section, and electrochemical measurements indicated peaks in the CV curve due to PANI's oxidation and reduction processes. MWCNT-PANI exhibited a capacity of 83 F g^{-1}, significantly higher than MWCNT alone. The vertically aligned MWCNTs enable the production of integrated, flexible units, enhancing their application potential.

Sangeetha *et al.* developed PSC systems using SnO_2 as a photosensitive dye-based photovoltaic electrode. Graphene-coated FTO substrates were employed for the energy storage electrode [35]. The electrolyte, responsible for electron conductivity, consisted of a mixture of polyvinyl alcohol (PVA) and various concentrations of potassium iodide. The SC demonstrated a rapid charge and discharge profile after exposure to sunlight for 6.4 seconds. The device exhibited satisfactory performance, achieving a specific capacitance of 20 F g^{-1}, an FF of 0.42, and a PCE of 0.56%.

6.2.7.2 Flexible Single-Layer Photo-Supercapacitors Based on Quantum Dot Solar

PSC systems can utilize semiconductor materials commonly used in the photovoltaic section and nanomaterials such as quantum dots [25]. These quantum dots can potentially replace traditional semiconductors and dyes in photovoltaic electrodes. Similar to quantum dots, inorganic nanomaterials offer excellent photocatalytic sensitivity and do not require photosensitive dyes (Figure 6.5). The operating principle of QD-based PSCs is analogous to that of DSSC-based PSCs, with the primary distinction in light-sensitive material. In QD-based PSCs, inorganic quantum dots are employed as the light-absorbing component. This configuration allows for a higher theoretical capacity and the potential to generate more than one electron and hole per photon, providing a distinct advantage over traditional DSSC technology.

In 2015, Narayanan *et al.* developed PSC systems based on light-sensitive CdS QDs [38]. The photovoltaic electrode had a TiO_2/CdS (QD)/Au structure. The inter-electrode was designed with Ag/MWCNT, while the counter electrode in the photovoltaic conversion section was structured as MWCNT/FTO/glass. The electrolyte in the photovoltaic section was a mixture of Na_2S and polyacrylamide, whereas the electrolyte in the energy storage section was a gel consisting of $LiCF_3SO_3$, PC, and PMMA. To render the glass conductive, it was coated with a silver paste. The PSC demonstrated satisfactory performance, achieving a light conversion efficiency of 3.4%,

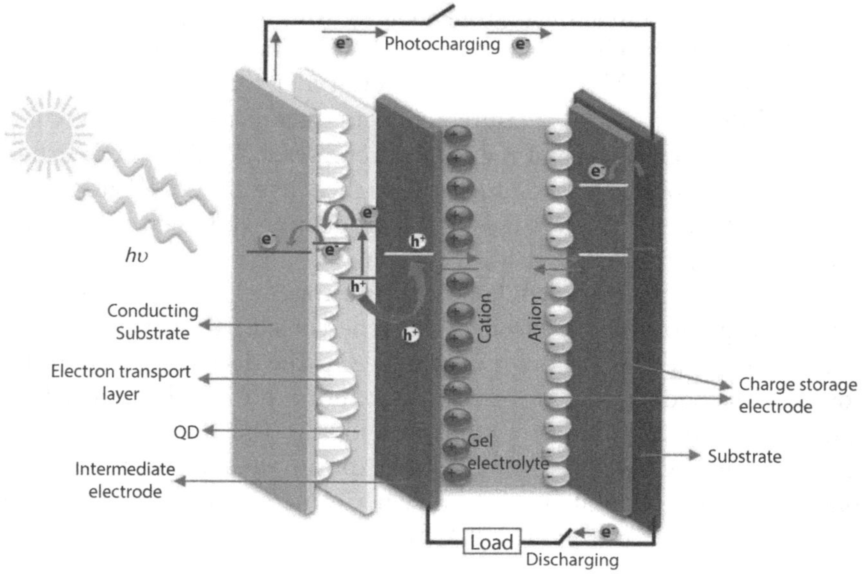

Figure 6.5 Schematic diagram of a QDSSC PSC.

J_{sc} of 10.64 mA cm^{-2}, and V_{oc} of 735 mV. The energy storage performance was promising, with a specific capacitance of 150 F g^{-1}. However, the device demonstrated relatively low stability, retaining only 68% of its initial capacity after 25 cycles.

Various semiconductor materials can serve as alternatives to TiO_2 in photovoltaic electrodes, with zinc oxide (ZnO) standing out due to its excellent conductivity, ease of crystallization, and versatile properties. Wide-bandgap semiconductor materials such as CdS, CdSe, PbS, Ag_2S, and PbSe have recently been investigated as light-sensitive materials, replacing traditional pigments. Among these materials, Ag_2S has shown promising potential for use in PSC systems. Cortes *et al.* developed a PSC using a ZnO/Ag_2S/PEDOT combination as the photovoltaic electrode, where the surface of ZnO/Ag_2S quantum dots (QDs) was coated with ZnS to minimize interfacial effects [39]. In this configuration, PEDOT was utilized as the energy storage electrode, while PVP/[HEMIm][BF_4] served as the electrolyte. Conductive polymers such as P_3HT, PEDOT, and PEDOT:PSS are known for their excellent hole transport capabilities and favorable energy storage properties. The device achieved a specific capacitance of 0.667 μF/cm^2, an energy storage efficiency of 6.38%, and an overall efficiency of 8.25 ± 1.04%.

In another study, Curtis *et al.* developed a QDSSC-based PSC. They synthesized a photoelectrode consisting of ZnO nanorods modified with Ag_2S

quantum dots, used PEDOT material for charge storage as the counter electrode, and employed PVP/[HEMIm][BF_4] as the electrolyte. When light strikes the Ag_2S quantum dots, these dots become excited, allowing electrons to move through the ZnO nanorods, exit the external circuit, and accumulate negative charge in the counter electrode material. The ZnO nanorods become positively charged when facing the counter electrode, forming a double-layer charge. The electrolyte between the negatively charged counter electrode and the positively charged ZnO nanorods contains negative BF_4^- and positive $HEMIm^+$ ions. These ions migrate toward the oppositely charged electrodes, contributing to the formation of the double-layer charge. During the photoelectric conversion, the SC charges and discharges occur without light. Notably, it has been observed that light sensitivity can be achieved directly through QDs without the need for traditional dyes [25].

6.2.7.3 *Flexible Perovskite-Based Solar Cell Photo-Supercapacitor*

Perovskite-based solar cells are distinguished by their straightforward manufacturing process, high photovoltaic efficiency, and effective light absorption. Organic halide PeSCs possess a narrow bandgap ranging from 1.55 to 3.2 eV [25]. These cells are noted for their low charge carrier diffusion length at the micrometer scale, which is conducive to achieving optimal photovoltaic efficiency. Perovskite cells demonstrate superior photovoltaic efficiency compared to DSSCs, QDSSCs, and OSCs. A schematic representation of perovskite-based photovoltaic solar cells is illustrated in Figure 6.6.

In 2015, Xu *et al.* integrated a PeSC with an SC [40]. The photovoltaic electrode was made from $CH_3NH_3PbI_3$ with an n-i-p structure, while the energy storage electrode was a symmetric SC composed of MWCNTs. SCs are used for energy storage and to stabilize fluctuations in energy sources, which can be influenced by varying weather conditions. Three devices were synthesized using TiO_2 as the base material in the photovoltaic electrodes. All devices exhibited distinct characteristics, with a high overall efficiency of 10%.

Xu *et al.* developed a fully porous perovskite-based PSC. The PEDOT electrode was an interlayer for the SC and the solar cell. The photovoltaic electrode was synthesized using TiO_2, ZrO_2, and perovskite materials on FTO glass. The perovskite layer ($CH_3NH_3PbI_3$) was deposited using a direct immersion method with the raw materials. The interlayer was deposited onto the substrate *via* electropolymerization within an appropriate voltage range, utilizing a $LiClO_4$ electrolyte between the porous PEDOT layers. Following GCD under low-light conditions, the PSC displayed a high voltage of 0.7 V

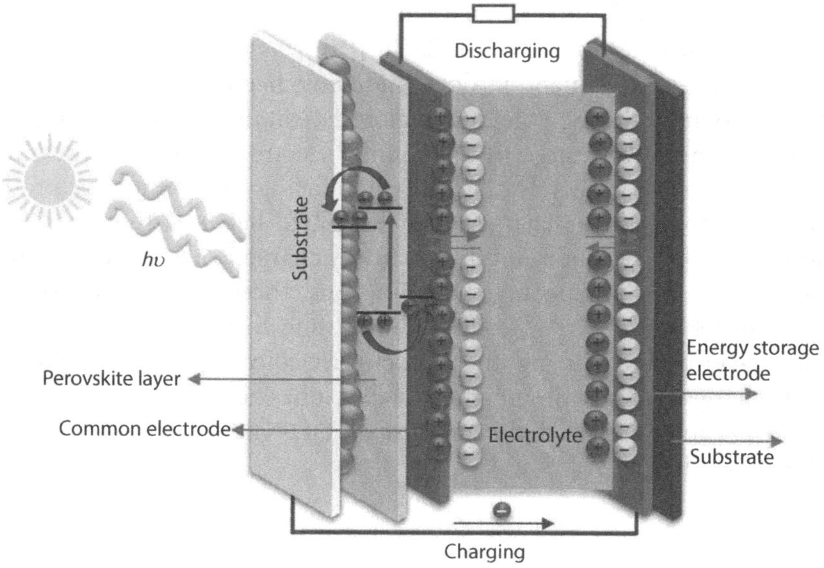

Figure 6.6 Schematic diagram of perovskite-based solar cell PSCs.

within 7 seconds. The areal capacitance was measured at 12.8 μF.cm^{-2}. The device demonstrated an energy density of 7.83 × 10^{7} Wh.cm^{-2} and a power density of 7.4 μW.cm^{-2}, achieving impressive efficiencies of 73.7% in energy density and 4.7% in overall efficiency [41].

6.2.8 2D Materials for Photo-Supercapacitors

Innovative configurations have led to significant advancements, achieving notable improvements in charge storage efficiency under illumination. These developments highlight the potential of PSCs as effective devices for solar energy conversion and storage, paving the way for future research and practical applications in renewable energy technologies. Currently, PSC development is in the research and development phase, driven by advancements in nanomaterials and integrated devices. Inorganic semiconductors perform well in the photovoltaic section, while 2D nanomaterials such as graphene and MXene excel in energy storage. Liu *et al.* developed an integrated unit on a Au substrate. The photovoltaic electrode consisted of a PEDOT:PSS/Si nanowire mesh, while the energy storage electrode was a symmetric SC containing laser-processed graphene (LSG) (Figure 6.7) [41].

Graphene is widely regarded as a significant material for energy storage in SCs due to its high electrochemical performance and excellent

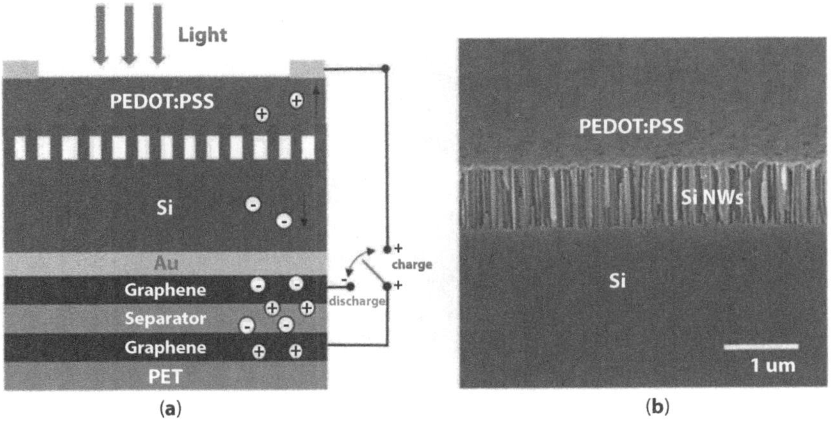

Figure 6.7 A schematic illustrating the structure of the self-charging power pack. (b) A cross-sectional scanning electron microscopy image showing a silicon nanowire array coated with a PEDOT:PSS film [41].

mechanical properties. Du *et al.* synthesized a PSC that integrates a PeSC and graphene for energy conversion and storage [42]. The photovoltaic section is structured as ITO/PEDOT: PSS/$CH_3NH_3PbI_{3x}Cl_x$/$PC_{61}BM$/Al, while the energy storage section consists of a graphene-layered symmetric SC. A single SC exhibited a remarkable specific capacitance of 245 F g^{-1}, retaining 83% of its initial capacity after 10,000 charge–discharge cycles. The PSC demonstrated excellent performance, achieving a high current density of 22.50 mA cm^{-2}, an open-circuit voltage of 0.9 V, and an FF of 69%, resulting in a PCE of 14.13%. The charged SC discharged at 0.75 V, below the open-circuit voltage. These results highlight the device's high efficiency and suitability for energy conversion and storage applications.

A DSSC has been designed by integrating a compressed and porous TiO_2 film as the anode with rGO/PPy electrodes in a symmetric SC configuration, forming a PSC (Figure 6.8) [43]. The electrochemically deposited PPy/rGO material on both surfaces serves as a naturally dual-functional inter-electrode, allowing for the regeneration of the electrolyte while simultaneously functioning as the SC electrode. The DSSC and isolated SC characteristics prior to integration have been evaluated. The energy conversion efficiency of the DSSC was measured at 2.4%, while the specific capacitance of the SC was recorded at 308.1 F g^{-1}. The performance of the integrated PSC was tested under low light conditions of 100 mW cm^{-2}. An extended lifespan was achieved by employing a PPy/rGO-type electrode at the cell/SC interface, lasting up to 50 charge/discharge cycles. The PSC

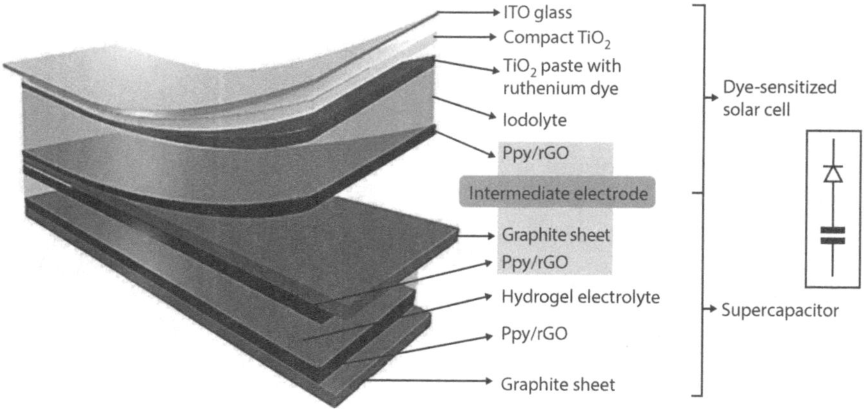

Figure 6.8 The configuration of the three-electrode PSC system consists of a DSSC and a symmetric SC based on PPy/rGO. A corresponding circuit diagram is provided in the inset [43].

exhibited a specific capacity of 124.7 F g^{-1} and maintained a capacity retention rate of 70.9% after 50 consecutive charge/discharge cycles.

A flexible, 3D porous graphene foam-based supercapacitor (GFSC) has been developed for energy storage applications [44]. The GFSC features an innovative layered structure comprising highly conductive electrodes of graphene, Ag conductive epoxy, and graphene foam, forming an electrochemical double layer that delivers exceptional electrochemical and supercapacitive performance. At a current density of 0.67 mA cm^{-2}, the GFSC achieves an areal capacitance of 38 mF cm^{-2}, approximately three times higher than previously reported values for flexible carbon-based SCs. The device also demonstrates superior energy and power densities of 3.4 μWh cm^{-2} and 0.27 mW cm^{-2}, respectively, outperforming traditional carbon-based SCs. The GFSC exhibits remarkable stability under static and dynamic bending conditions, retaining 68% of its initial capacitance after 25,000 charge–discharge cycles. Its lightweight design, cost-effective manufacturing, and lack of self-heating make the GFSC an attractive alternative for a wide range of power density applications, from nW cm^{-2} to mW cm^{-2}. In this context, the GFSC has been successfully integrated with a flexible photovoltaic cell to create a flexible self-charging power pack. This system has effectively powered a continuously powered CuO nanorod-based chemically resistant pH sensor.

Tuc Altaf *et al.* demonstrated that PSCs can be enhanced through the integration of nanomaterials, specifically a combination of 2D graphene oxide (GO) and one-dimensional zinc oxide (ZnO) [45]. In their study,

PSCs were synthesized using various ratios of GO to ZnO, focusing on optimizing the composite properties and device performance. The electrolyte consisted of a gel mixture of PVA/LiCl, with FTO used in the photovoltaic section and carbon paper in the energy storage section. Photoelectrochemical measurements indicated that the best performance was achieved with a GO/ZnO ratio of 1:5. After exposure to UV light, the PSC exhibited an excellent capacitance value of 6612 mF g^{-1}, as calculated from CV analysis (scan rate: 10 mV s^{-1}). The PSC demonstrated remarkable stability, maintaining capacity retention rates of 99.6% and 100% after 30,000 cycles. Additionally, under UV light, the PSC showed an operating voltage of 2.5 V, achieving an energy density of 6.3 Wh kg^{-1} and a power density of 625 W kg^{-1}. As a practical application, the GO/ZnO-based PSC successfully powered a digital clock for over one hour and forty minutes when charged for 50 and 130 seconds at current densities of 0.2 A g^{-1} under UV and AM1.5 sunlight, respectively.

In another study, Tuc Altaf *et al.* explored two-electrode PSC systems incorporating 3D ZnO nanoflakes combined with rGO nanocomposites to enable *in situ* solar energy conversion and storage, as illustrated in Figure 6.9 [46]. The study examined three compositions with varying rGO weight percentages ranging from 8% to 32%. It was found that while increasing the rGO content enhanced energy density, excessive amounts resulted in a loss of light sensitivity. The optimal performance was achieved with electrodes containing 16% rGO, which maintained full functionality over 40,000 cycles. This configuration demonstrated an areal capacitance of 40 mF cm^{-2} and an energy density of 22 μWh cm^{-2}, representing a 170% increase compared to the values measured under dark conditions.

Tuc Altaf *et al.* studied innovative composites of graphitic carbon nitride (g-C_3N_4) and ZnO nanowires employed as photoactive electrode materials in all-solid-state two-electrode PSC systems [47]. The study utilized a lithiated Nafion® membrane as the electrolyte and separator, significantly enhancing performance compared to standard filter paper separators. Under UV illumination at a working potential of 1.5 V and a current density of 26.7 mA g^{-1}, the PSC with the Nafion® membrane exhibited an energy density five times higher than that of traditional systems. The g-C_3N_4/ZnO NW PSC also demonstrated exceptional cycling stability, retaining 90.2% of its capacitance and achieving a Coulombic efficiency of 99.9% over 25,000 charge/discharge cycles. UV illumination increased the energy density by 21.5 times, reaching 11 Wh kg^{-1} at the same voltage and current density. These findings underscore the potential of PSCs for efficient solar energy utilization and storage in future applications.

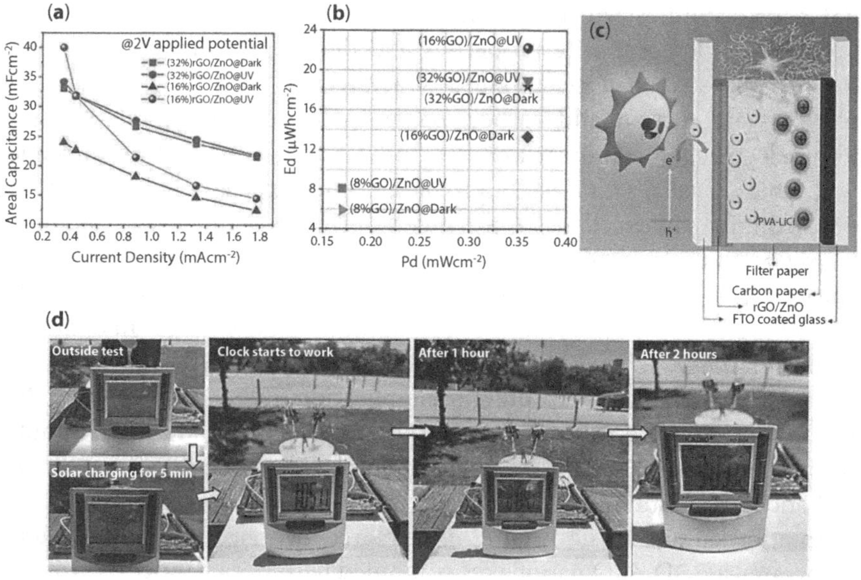

Figure 6.9 (a) Comparative areal capacitance values are plotted as a function of the applied current density within a 2-V potential range. (b) Ragone plot illustrates the relationship between energy and power densities. (c) Schematic representation of the photo charge–discharge process. (d) Photographs captured during the outdoor testing of the device [46].

Recent advancements in PSCs have focused on integrating materials such as zinc oxide (ZnO) and MXene-based compounds to enhance performance metrics, including energy density and capacitance. The architecture of these two-electrode systems typically consists of a photoanode paired with a charge storage counter electrode. Colak *et al.* synthesized ZnO/MXene (based on $Ti_3C_2T_x$) hybrid electrodes [48]. The PSC was configured in three device architectures: a ZnO-based PSC, a ZnO/MXene-based PSC, and a ZnO/MXene/MXene-based PSC. Adding MXene to the SCs significantly enhanced the total capacity, approximately quadrupling it due to oxidation and reduction surface reactions, thus forming a double-layer electrical structure (Figure 6.10). Under UV irradiation at 2.5 V, the ZnO/MXene//MXene-PSC achieved the highest energy density of 57.3 μWh cm^{-2} and a power density of 11.1 mW cm^{-2}. The area capacitance calculated from CV analysis for this device was 151.5 mF cm^{-2}. Additionally, the ZnO/MXene//MXene-based PSC exhibited remarkable stability, maintaining a 100% capacity retention rate over 10,000 cycles during GCD tests. The study demonstrated that using ZnO nanoflake electrodes in conjunction with $Ti_3C_2T_x$ MXene paper significantly increased total capacitance

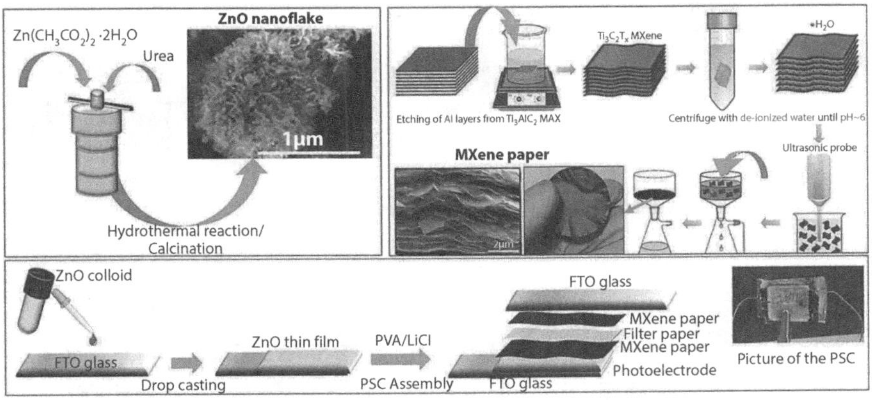

Figure 6.10 A schematic illustration depicting the material synthesis process and the step-by-step fabrication of the device [48].

compared to traditional designs, owing to enhanced surface redox reactions and electric double-layer formation.

MXenes have gained significant attention as advanced materials due to their exceptional photovoltaic, electrical, chemical, and mechanical properties. They are highly suitable for diverse applications such as sensors, photovoltaic devices, and energy storage systems. Qin *et al.* developed a flexible and semi-transparent PSC based on MXene (Figure 6.11) [49]. The photovoltaic section is mounted on a PET substrate, with the initial layer consisting of MXene coated with PEI/P_3HT: ICBA/PEDOT: PSS. The energy storage section comprises a series of symmetric structures made of semi-transparent $Ti_3C_2T_x$/ionogel/$Ti_3C_2T_x$ placed on a PDMS base. This PSC demonstrates excellent performance, with a PCE of 2.5%, an open-circuit voltage (V_{oc}) of 0.81 V, a short-circuit current density (J_{sc}) of 6.89 mA cm^{-2}, and an FF of 0.45. The MXene-based transparent SC exhibits a high volumetric capacitance of 502 F cm^{-3} and remarkable stability. The energy storage efficiency is reported at 88%, while the overall device efficiency is 2.2%. These findings offer valuable insights into the fabrication of integrated devices. The MXene-based PSC offers excellent performance, good flexibility, and reasonable cost, facilitating the synthesis and exploration of integrated devices by combining nanomaterials and devices.

Recent advancements in synthesizing electrode materials for energy storage devices, such as batteries, SCs, and solar cells, have led to continued research into how to harness the capabilities of these devices and integrate them into a unified system. In this context, substantial advancements have been achieved in developing PSCs that integrate photo-energy conversion

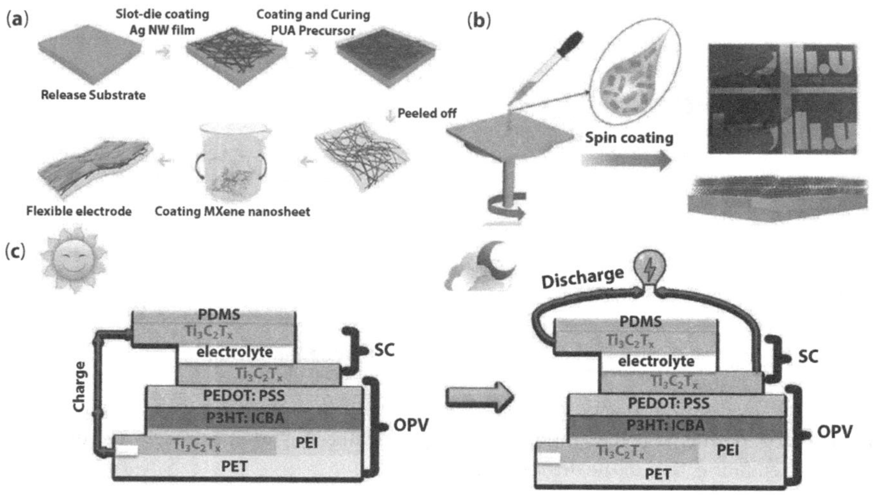

Figure 6.11 (a) The fabrication process of the MXene-based flexible electrode involves slot-die coating of ag NW films, curing a PUA precursor, and coating MXene nanosheets. (b) Spin coating is used to prepare the transparent flexible electrode, as shown alongside the final product images. (c) The device integrates OPVs with a SC, enabling solar energy storage and discharge [49].

and storage capabilities. This chapter explores the fundamental operating principles of PSCs, the electrode materials involved, and methods for their integration. Research on PSCs remains in its early stages, requiring further advancements to overcome challenges in achieving seamless system integration.

Material compatibility: A critical factor in improving PSC efficiency is carefully selecting compatible materials. The efficiency of converting light energy into electrical and chemical energy remains limited, often due to energy losses during charging and discharging caused by high internal resistance or material incompatibility between the solar cells and SCs. Appropriate material selection can prevent the premature reversion of generated electrons to their ground states through oxidation–reduction reactions. Material deposition methods and adhesives should also be chosen to avoid increasing the device's resistance. Developing high-efficiency, thermally stable materials, such as perovskites for solar cells, could enhance light conversion efficiency. Moreover, integrating perovskite cells with 2D nanomaterials like MXenes and graphene can improve light absorption and energy storage capacity. Advanced electrode designs using highly conductive hybrid materials (such as conductive polymers and metal oxides) can optimize electrode/electrolyte interface reactions.

Stability concerns: Materials used in PeSCs or DSSCs may degrade chemically when exposed to moisture, oxygen, or UV light, leading to a decline in electrochemical performance over time. To address this, protective coatings, insulators, and environmentally resilient electrolytes need to be developed to extend the lifespan of PSCs.

Electrolyte selection: Aqueous electrolytes, commonly used in SCs, offer high efficiency but are limited by a narrow voltage window. Organic electrolytes provide a broader voltage range but present environmental risks and usability challenges. Aqueous electrolytes can also negatively affect the performance of integrated devices due to leakage and evaporation. Solid electrolytes present a promising alternative, though challenges remain in achieving complete electrode surface coverage. Gel electrolytes have shown potential when combined with electrode materials, offering high conductivity and excellent filling properties.

Utilization of PeSCs: PeSCs demonstrate high light conversion efficiency but have not yet been widely utilized in PSCs. Future perovskite-based PSCs are expected to offer high performance, unlocking significant potential for energy conversion and storage.

The potential of flexible and micro-supercapacitors: Micro-supercapacitors provide an effective energy storage solution for modern, flexible electronics and wearable devices. Their integration with solar cells could significantly enhance the development of modern electronic devices and smart systems. Despite the challenges faced by PSCs, their potential in future energy technologies is substantial. Addressing these issues can lead to significant advancements in the performance and application of PSCs.

References

1. Hosenuzzaman, M., Rahim, N.A., Selvaraj, J., Hasanuzzaman, M., Malek, A.B.M.A., Nahar, A., Global prospects, progress, policies, and environmental impact of solar photovoltaic power generation. *Renew. Sustain. Energy Rev.*, 41, 284–297, 2015. https://doi.org/10.1016/J.RSER.2014.08.046.
2. Kannan, N. and Vakeesan, D., Solar energy for future world: - A review. *Renew. Sustain. Energy Rev.*, 62, 1092–1105, 2016. https://doi.org/10.1016/J.RSER.2016.05.022.
3. Battaglia, C., Cuevas, A., De Wolf, S., High-efficiency crystalline silicon solar cells: status and perspectives. *Energy Environ. Sci.*, 9, 1552–1576, 2016. https://doi.org/10.1039/C5EE03380B.
4. Wee, G., Salim, T., Lam, Y.M., Mhaisalkar, S.G., Srinivasan, M., Printable photo-supercapacitor using single-walled carbon nanotubes. *Energy Environ. Sci.*, 4, 413–416, 2011. https://doi.org/10.1039/C0EE00296H.

5. Fraas, L.M., Low-cost solar electric power. *Low-Cost Sol. Electr. Power*, 9783319075303, 1–181, 2014. https://doi.org/10.1007/978-3-319-07530-3/COVER.
6. Muzaffar, A., Ahamed, M.B., Deshmukh, K., Thirumalai, J., A review on recent advances in hybrid supercapacitors: Design, fabrication and applications. *Renew. Sustain. Energy Rev.*, 101, 123–145, 2019. https://doi.org/10.1016/J.RSER.2018.10.026.
7. Chirauri, S.K., Dehury, A.K., Chaudhary, Y.S., Photosupercapacitors: A perspective of planar and flexible dual functioning devices, Wiley Interdiscip. *Rev. Energy Environ.*, 9, e377, 2020. https://doi.org/10.1002/WENE.377.
8. Yang, Z., Li, L., Luo, Y., He, R., Qiu, L., Lin, H., Peng, H., An integrated device for both photoelectric conversion and energy storage based on free-standing and aligned carbon nanotube film. *J. Mater. Chem. A*, 1, 954–958, 2012. https://doi.org/10.1039/C2TA00113F.
9. Rong, Y., Hu, Y., Mei, A., Tan, H., Saidaminov, M.I., Il Seok, S., McGehee, M.D., Sargent, E.H., Han, H., Challenges for commercializing perovskite solar cells. *Sci.*, 361, 6408, eaat8235, 2018. https://doi.org/10.1126/science.aat8235.
10. Li, Z., Klein, T.R., Kim, D.H., Yang, M., Berry, J.J., Van Hest, M.F.A.M., Zhu, K., Scalable fabrication of perovskite solar cells. *Nat. Rev. Mater.*, 3, 1–20, 2018. https://doi.org/10.1038/natrevmats.2018.17.
11. Suresh Kumar, N. and Chandra Babu Naidu, K., A review on perovskite solar cells (PSCs), materials and applications. *J. Mater.*, 7, 940–956, 2021. https://doi.org/10.1016/j.jmat.2021.04.002.
12. Vadivel, R., Hariganesh, S., Ramyakrishna P. and Boddula, S., Photo-Supercapacitor. *Mater. Res. Found.*, 61, 223–232, 2019. https://doi.org/10.21741/9781644900499-9.
13. Altaf, C.T., Coskun, O., Kumtepe, A., Rostas, A.M., Iatsunskyi, I., Coy, E., Erdem, E., Sankir, M., Sankir, N.D., Photo-supercapacitors based on nanoscaled ZnO. *Sci. Rep.*, 12, 1–15, 2022. https://doi.org/10.1038/s41598-022-15180-z.
14. Devadiga, D., Selvakumar, M., Shetty, P., Santosh, M.S., Recent progress in dye sensitized solar cell materials and photo-supercapacitors: A review. *J. Power Sources*, 493, 229698, 2021. https://doi.org/10.1016/J.JPOWSOUR.2021.229698.
15. Gong, J., Sumathy, K., Qiao, Q., Zhou, Z., Review on dye-sensitized solar cells (DSSCs): Advanced techniques and research trends. *Renew. Sustain. Energy Rev.*, 68, 234–246, 2017. https://doi.org/10.1016/J.RSER.2016.09.097.
16. Gong, J., Liang, J., Sumathy, K., Review on dye-sensitized solar cells (DSSCs): Fundamental concepts and novel materials. *Renew. Sustain. Energy Rev.*, 16, 5848–5860, 2012. https://doi.org/10.1016/J.RSER.2012.04.044.
17. Lee, Y.L. and Lo, Y.S., Highly Efficient Quantum-Dot-Sensitized Solar Cell Based on Co-Sensitization of CdS/CdSe. *Adv. Funct. Mater.*, 19, 604–609, 2009. https://doi.org/10.1002/ADFM.200800940.

18. Chen, P., Xiao, Y., Li, S., Jia, X., Luo, D., Zhang, W., Snaith, H.J., Gong, Q., Zhu, R., The Promise and Challenges of Inverted Perovskite Solar Cells. *Chem. Rev.*, 124, 10623–10700, 2024. https://doi.org/10.1021/acs.chemrev.4c00073.
19. Li, D., Lian, Q., Du, T., Ma, R., Liu, H., Liang, Q., Han, Y., Mi, G., Peng, O., Zhang, G., Peng, W., Xu, B., Lu, X., Liu, K., Yin, J., Ren, Z., Li, G., Cheng, C., Co-adsorbed self-assembled monolayer enables high-performance perovskite and organic solar cells. *Nat. Commun.*, 15, 2024. https://doi.org/10.1038/s41467-024-51760-5.
20. Tang, H., Bai, Y., Zhao, H., Qin, X., Hu, Z., Zhou, C., Huang, F., Cao, Y., Interface Engineering for Highly Efficient Organic Solar Cells. *Adv. Mater.*, 36, 1–45, 2024. https://doi.org/10.1002/adma.202212236.
21. Meng, H., Pang, S., Cui, G., Photo-Supercapacitors Based on Third-Generation Solar Cells. *ChemSusChem.*, 12, 3431–3447, 2019. https://doi.org/10.1002/CSSC.201900398.
22. Sun, Y. and Yan, X., Recent Advances in Dual-Functional Devices Integrating Solar Cells and Supercapacitors. *Sol. RRL*, 1, 2017. https://doi.org/10.1002/solr.201700002.
23. Rahman, M., A Comprehensive Review on Perovskite Solar Cells Integrated Photo-supercapacitors and Perovskites-Based Electrochemical Supercapacitors. *Chem. Rec.*, 1–24, 2024. https://doi.org/10.1002/tcr.202300183.
24. Ng, C.H., Lim, H.N., Hayase, S., Harrison, I., Pandikumar, A., Huang, N.M., Potential active materials for photo-supercapacitor: A review. *J. Power Sources*, 296, 169–185, 2015. https://doi.org/10.1016/j.jpowsour.2015.07.006.
25. K., N. and Rout, C.S., Photo-powered integrated supercapacitors: a review on recent developments, challenges and future perspectives. *J. Mater. Chem. A*, 9, 8248–8278, 2021. https://doi.org/10.1039/D1TA00444A.
26. Yang, M., Wang, D., Ling, Y., Guo, X., Chen, W., Emerging Advanced Photo-Rechargeable Batteries. *Adv. Funct. Mater.*, 34, 2410398, 2024. https://doi.org/10.1002/ADFM.202410398.
27. Miyasaka, T. and Murakami, T.N., The photocapacitor: An efficient self-charging capacitor for direct storage of solar energy. *Appl. Phys. Lett.*, 85, 3932–3934, 2004. https://doi.org/10.1063/1.1810630.
28. Freitag, M., Teuscher, J., Saygili, Y., Zhang, X., Giordano, F., Liska, P., Hua, J., Zakeeruddin, S.M., Moser, J.E., Grätzel, M., Hagfeldt, A., Dye-sensitized solar cells for efficient power generation under ambient lighting. *Nat. Photonics*, 11, 372–378, 2017. https://doi.org/10.1038/nphoton.2017.60.
29. Ananthakumar, S., Balaji, D., Ram Kumar, J., Moorthy Babu, S., *Role of co-sensitization in dye-sensitized and quantum dot-sensitized solar cells, SN Appl. Sci.*, 1, 186, 2019, https://doi.org/10.1007/s42452-018-0054-3.
30. Park, N.G., Kim, K.M., Kang, M.G., Ryu, K.S., Chang, S.H., Shin, Y.J., Chemical sintering of nanoparticles: A methodology for low-temperature fabrication of dye-sensitized TiO2 films. *Adv. Mater.*, 17, 2349–2353, 2005. https://doi.org/10.1002/adma.200500288.

31. Chen, H.W., Hsu, C.Y., Chen, J.G., Lee, K.M., Wang, C.C., Huang, K.C., Ho, K.C., Plastic dye-sensitized photo-supercapacitor using electrophoretic deposition and compression methods. *J. Power Sources*, 195, 6225–6231, 2010. https://doi.org/10.1016/j.jpowsour.2010.01.009.
32. Liu, Q., Ju, M.G., Zeng, X.C., Unconventional perovskite-to-perovskite tandem cell designed by stacking with large-gap phosphonium-based analogs. *Mater. Today Energy*, 42, 101556, 2024. https://doi.org/10.1016/J.MTENER.2024.101556.
33. Liu, T., Lee, B., Lee, M.J., Park, J., Chen, Z., Noda, S., Lee, S.W., Improved capacity of redox-active functional carbon cathodes by dimension reduction for hybrid supercapacitors. *J. Mater. Chem. A*, 6, 3367–3375, 2018. https://doi.org/10.1039/C7TA10881H.
34. Sree Raj, K.A., Shajahan, A.S., Chakraborty, B., Rout, C.S., Two-Dimensional Layered Metallic VSe2/SWCNTs/rGO Based Ternary Hybrid Materials for High Performance Energy Storage Applications. *Chem.– A Eur. J.*, 26, 6662–6669, 2020. https://doi.org/10.1002/CHEM.202000243.
35. Sangeetha, D.N., Hegde, N., Poojari, V., Devadiga, D., Sudhakar, Y.N., Santosh, M.S., Selvakumar, M., Conductivity/Electrochemical Study of Polyvinyl pyrrolidone-Poly(vinyl alcohol)/I3– Thin Film Electrolyte for Integrated Dye-Sensitized Solar Cells and Supercapacitors. *J. Electron. Mater.*, 49, 6325–6335, 2020. https://doi.org/10.1007/s11664-020-08432-z.
36. Khan, M.Z.H., Al-Mamun, M.R., Halder, P.K., Aziz, M.A., Performance improvement of modified dye-sensitized solar cells. *Renew. Sustain. Energy Rev.*, 71, 602–617, 2017. https://doi.org/10.1016/j.rser.2016.12.087.
37. Cole, J.M., Pepe, G., Al Bahri, O.K., Cooper, C.B., Cosensitization in Dye-Sensitized Solar Cells. *Chem. Rev.*, 119, 7279–7327, 2019. https://doi.org/10.1021/acs.chemrev.8b00632.
38. Narayanan, R., Kumar, P.N., Deepa, M., Srivastava, A.K., Combining Energy Conversion and Storage: A Solar Powered Supercapacitor. *Electrochim. Acta*, 178, 113–126, 2015. https://doi.org/10.1016/j.electacta.2015.07.121.
39. Solís-Cortés, D., Navarrete-Astorga, E., Schrebler, R., Peinado-Pérez, J.J., Martín, F., Ramos-Barrado, J.R., Dalchiele, E.A., A solid-state integrated photo-supercapacitor based on ZnO nanorod arrays decorated with Ag2S quantum dots as the photoanode and a PEDOT charge storage counter-electrode. *RSC Adv.*, 10, 5712–5721, 2020. https://doi.org/10.1039/c9ra10635a.
40. Xu, X., Li, S., Zhang, H., Shen, Y., Zakeeruddin, S.M., Graetzel, M., Cheng, Y.B., Wang, M., A power pack based on organometallic perovskite solar cell and supercapacitor. *ACS Nano*, 9, 1782–1787, 2015. https://doi.org/10.1021/nn506651m.
41. Liu, H., Li, M., Kaner, R.B., Chen, S., Pei, Q., Monolithically Integrated Self-Charging Power Pack Consisting of a Silicon Nanowire Array/Conductive Polymer Hybrid Solar Cell and a Laser-Scribed Graphene Supercapacitor. *ACS Appl. Mater. Interfaces*, 10, 15609–15615, 2018. https://doi.org/10.1021/acsami.8b00014.

42. Du, P., Hu, X., Yi, C., Liu, H.C., Liu, P., Zhang, H.L., Gong, X., Self-powered electronics by integration of flexible solid-state graphene-based supercapacitors with high performance perovskite hybrid solar cells. *Adv. Funct. Mater.*, 25, 2420–2427, 2015. https://doi.org/10.1002/adfm.201500335.
43. Lau, S.C., Lim, H.N., Ravoof, T.B.S.A., Yaacob, M.H., Grant, D.M., MacKenzie, R.C.I., Harrison, I., Huang, N.M., A three-electrode integrated photo-supercapacitor utilizing graphene-based intermediate bifunctional electrode. *Electrochim. Acta*, 238, 178–184, 2017. https://doi.org/10.1016/j.electacta.2017.04.003.
44. Manjakkal, L., Núñez, C.G., Dang, W., Dahiya, R., Flexible self-charging supercapacitor based on graphene-Ag-3D graphene foam electrodes. *Nano Energy*, 51, 604–612, 2018. https://doi.org/10.1016/j.nanoen.2018.06.072.
45. Altaf, C.T., Colak, T.O., Rostas, A.M., Mihet, M., Lazar, M.D., Iatsunskyi, I., Coy, E., Yildirim, I.D., Misirlioglu, F.B., Erdem, E., Sankir, M., Sankir, N.D., GO/ZnO-based all-solid-state photo-supercapacitors: Effect of GO:ZnO ratio on composite properties and device performance. *J. Energy Storage*, 68, 107694, 2023. https://doi.org/10.1016/j.est.2023.107694.
46. Altaf, C.T., Colak, T.O., Rostas, A.M., Socaci, C., Lazar, M.D., Tudoran, L.B., Aleinawi, M.H., Misirlioglu, F.B., Yildirim, I.D., Erdem, E., Sankir, N.D., Sankir, M., Zinc oxide nanoflake/reduced graphene oxide nanocomposite-based dual-acting electrodes for solar-assisted supercapacitor applications. *Energy Adv.*, 3, 1965–1976, 2024. https://doi.org/10.1039/d4ya00253a.
47. Altaf, C.T., Colak, T.O., Lufrano, F., Unal, G.S., Sankir, N.D., Sankir, M., Graphitic carbon nitride/zinc oxide composite electrodes for all-solid-state photo-supercapacitor with ion exchange membrane separator. *J. Energy Storage*, 55, 105784, 2022. https://doi.org/10.1016/j.est.2022.105784.
48. Colak, T.O., Altaf, C.T., Yesilbag, F.N.T., Yesilbag, Y.O., Yildirim, I.D., Erdem, E., Misirlioglu, F.B., Sankir, N.D., Sankir, M., Photo-supercapacitors based on zinc oxide/MXene paper dual acting electrodes. *J. Energy Storage*, 86, 111274, 2024. https://doi.org/10.1016/j.est.2024.111274.
49. Qin, L., Jiang, J., Tao, Q., Wang, C., Persson, I., Fahlman, M., Persson, P.O.A., Hou, L., Rosen, J., Zhang, F., A flexible semitransparent photovoltaic supercapacitor based on water-processed MXene electrodes. *J. Mater. Chem. A*, 8, 5467–5475, 2020. https://doi.org/10.1039/d0ta00687d.

Part 3

SOLAR-ASISTED INTEGRATED SYSTEMS

7

Unlocking the Potential of Sustainable Energy: Exploring the Role of Supercapacitors in Enhancing Energy Storage Efficiency of Photovoltaic Systems

R.H.M.D. Premasiri, P.L.A.K. Piyumal, A.L.A.K. Ranaweera* and S.R.D. Kalingamudali

Electronic Design and Innovation Center (EDIC), Department of Physics and Electronics, Faculty of Science, University of Kelaniya, Kelaniya, Sri Lanka

Abstract

Transitioning to sustainable energy systems is vital for addressing environmental challenges and reducing reliance on finite fossil fuels. Photovoltaic (PV) systems are emerging as a key solution, but fluctuating solar irradiance affects their energy output and efficiency. This chapter explores the integration of supercapacitors (SCs) into PV systems to enhance energy storage, system reliability, and sustainability. SCs excel with high-power density, rapid charge–discharge capabilities, and durability, effectively complementing batteries by managing energy fluctuations and stabilizing systems. SCs improve power quality and ensure consistent energy output even during low solar irradiance, meeting short-term energy storage needs and delivering high electric power on demand. Beyond PV systems, SCs find applications in regenerative braking systems in electric vehicles, uninterruptible power supplies, and Internet of Things devices. Advanced techniques such as novel methods in Maximum Power Point Tracking, Supercapacitor Assisted Low Dropout regulators, and Supercapacitor Assisted Loss Management further optimize energy management and performance. This analysis emphasizes the transformative potential of SCs in advancing sustainable energy systems, particularly in PV energy storage. In conclusion, this chapter illustrates the critical role of SCs in unlocking the potential of sustainable energy by enhancing energy storage efficiency within PV systems.

**Corresponding author*: arunaran@kln.ac.lk

Nurdan Demirci Sankir and Mehmet Sankir (eds.) Solar Capacitors and Batteries, (177–242) © 2025 Scrivener Publishing LLC

***Keywords*:** Energy storage, photovoltaic systems, renewable energy integration, supercapacitors, sustainable energy

7.1 Introduction

7.1.1 Overview of Sustainable Energy Systems

7.1.1.1 Importance of Transitioning to Sustainable Energy

Rapidly increasing environmental, economic, and social constraints severely affecting the global community have made it urgent to transition from conventional fossil fuels to renewable energy systems. For over a century, fossil fuels have driven industrialization but are also the largest contributor to climate change, as primarily they emit large amounts of carbon. When coal, oil, and natural gas are burned, they emit an extreme amount of carbon dioxide (CO_2) and other greenhouse gases that cause significant global warming, increasing the frequency of extreme weather occurrences and leading to rising sea levels. Additionally, the environmental damage associated with fossil fuel extraction, such as oil spills and habitat destruction, exacerbates ecological imbalance.

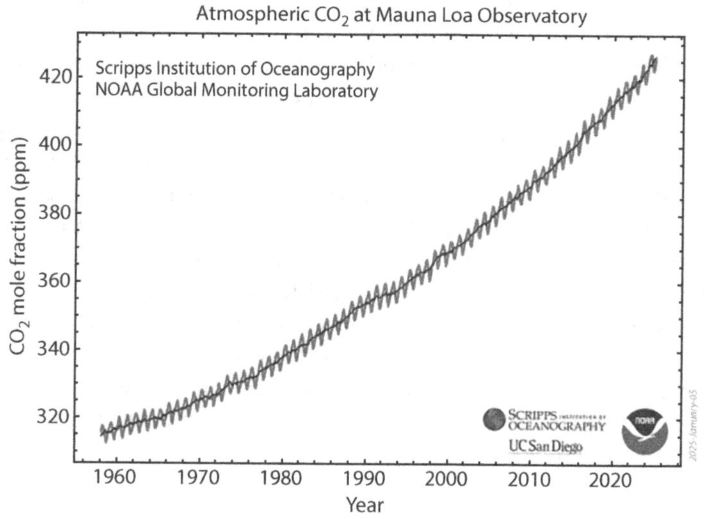

Figure 7.1 Atmospheric CO_2 at Mauna Loa Observatory [1].

The National Oceanic and Atmospheric Administration Global Monitoring Lab reports that the global average atmospheric CO_2 level has reached a new record high of 419.3 parts per million (ppm) in 2023. The rise of 2.8 ppm from 2022 signifies the 12th consecutive year of an increase in atmospheric CO_2 exceeding 2 ppm. At Mauna Loa Observatory in Hawaii, where contemporary CO_2 measurements commenced in 1958, the annual average CO_2 concentration for 2023 was 421.08 ppm as shown in Figure 7.1 [1]. These data highlight the considerable and persistent influence of non-renewable energy sources on atmospheric CO_2 levels, emphasizing the critical necessity for a shift toward more sustainable energy alternatives.

Worldwide, the ratio of fossil fuels in primary energy use remains significant. According to information from "Our World in Data," in 2020, Figure 7.2 shows the proportion of global primary energy consumption derived from fossil fuels was roughly 76.88% (with oil accounting for about 28.88%, coal for around 25.06%, and natural gas for approximately 22.94%). Renewable sources such as hydro, wind, and solar have experienced significant growth in recent years, reaching a level of 11%. The continued predominance of fossil fuels highlights the critical need for this shift, given the increasingly unsustainable environmental and health impacts associated with their use [3].

Renewable energy sources such as solar, wind, hydropower, and geothermal offer a clean, abundant, and sustainable supply of energy. These sources produce minimum to no greenhouse gas emissions during their

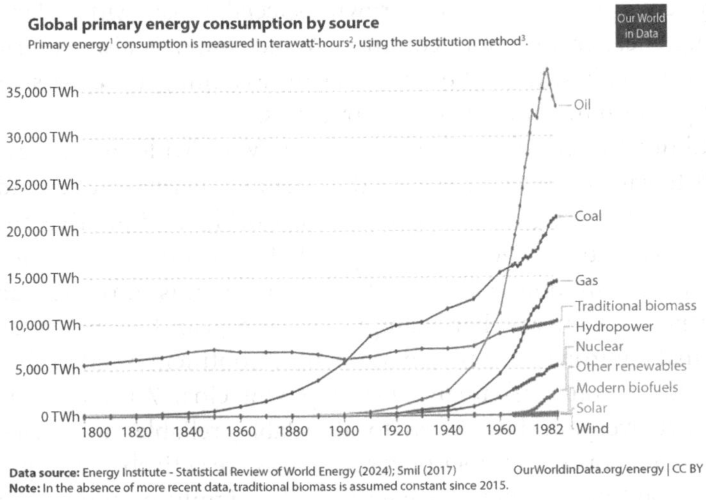

Figure 7.2 Global primary energy consumption by source [2].

operation, making them essential in the global effort to reduce carbon footprints and meet international climate goals, including the aims outlined in the Paris Agreement [4]. Renewable energy does not just offer environmental benefits; it also plays a major role in boosting energy security. In contrast to fossil fuels, which tend to be spatially limited and susceptible to political volatility, renewable resources are more uniformly distributed and can be locally harnessed. This reduces the need to rely on imported fossil fuels, helping countries achieve greater energy independence without being dependent on other countries.

Economically, the transition to sustainable energy can drive innovation, create new job opportunities, and support the growth of green industries. Investing in renewable energy technologies not only lowers long-term energy costs but also shields economies from the volatility of fossil fuel markets. The move toward sustainable energy brings important social benefits, particularly in improving public health by cutting down on air and water pollution, which are major causes of respiratory and cardiovascular diseases. It also promotes social equity by delivering affordable, reliable energy to underserved and remote communities, supporting more inclusive and equitable development.

7.1.1.2 Current Trends and Global Initiatives

The world's energy landscape is quickly shifting toward more sustainable and renewable sources. More countries are adopting renewable energy technologies with the rapid advancements in solar, wind, hydro, and biomass systems. Solar power has grown especially fast, with solar panels becoming cheaper and more common. Wind power is also experiencing growth, especially with offshore wind farms assuming a progressively significant position in national energy strategies.

In addition to technological advancements, worldwide efforts are driving the shift toward more environmentally friendly energy sources. Adopted in 2015, the Paris Agreement is a crucial worldwide initiative aimed at addressing climate changes by limiting global temperature rises to less than 2°C. Countries are now setting bold goals to cut emissions and boost the use of renewable energy. For example, the European Union's Green Deal aims to make Europe the first climate-neutral continent by 2050. Similarly, the United Nations' Sustainable Development Goal 7 (SDG 7) aims to ensure that everyone has access to affordable, reliable, and sustainable energy. An increasing number of corporations are actively participating in the transition toward renewable energy, with a significant number making a firm commitment to only use renewable sources through initiatives such as RE100 [5]. This effort unites enterprises that are dedicated to achieving clean energy goals. Concurrently, there is a significant surge in investments in green technologies, driven by the recognition of the economic advantages of sustainable energy by both governments and private sectors. In conclusion, recent advancements are advancing global progress towards a sustainable future, emphasizing the crucial significance of clean energy in addressing environmental challenges and constructing a more robust and resilient economy.

7.1.2 The Role of PV Systems in Sustainable Energy

7.1.2.1 Basics of PV Technology

Photovoltaic (PV) technology, commonly known as solar power, is a technique that transforms sunlight directly into electricity using semiconductor materials that utilize the PV effect. This entity has emerged as a significant contributor to renewable energy systems due to its ability to generate clean, sustainable, and dependable energy. With the rising need for sustainable energy options, PV technology has become essential in the worldwide transition to renewable energy sources. The sun produces an enormous quantity of energy approximately 3.8×10^{23} kW each second, yet only a fraction of that energy, about 1.8×10^{14} kW, successfully arrives at Earth, located at around 150 million kilometers distance.

Understanding the amount of solar energy intercepted by the Earth is crucial for recognizing the possibilities of solar power as a sustainable energy alternative. The energy that reaches the Earth's surface is vast in scale and holds significant potential for generating power. For example, if merely 0.1% of this captured solar energy were to be transformed with an efficiency of 10%, it would exceed the global generating capacity, which is approximately 3,000 GW. The total annual solar radiation that reaches the Earth is estimated to exceed 7,500 times the global primary energy consumption, which stands at approximately 450 exajoules (EJ) per year [6].

Solar cells are inherently semiconductor devices, so they may be produced using identical processing and manufacturing methods as other semiconductor devices. Nevertheless, rigorous standards for cleanliness and quality control in semiconductor production are essential. The cells are typically constructed using silicon wafers, which can be either monocrystalline or polycrystalline. The process begins with the formation of p-n junctions on the surface of the silicon wafer, which is essential for creating an electric field. The front layer is designed to maximize the absorption of incident radiation, while the back layer typically covers the entire rear area of the cell.

The process of generating electricity in solar cells involves several steps. When sunlight strikes the cell's surface, it passes through the antireflective coating and reaches the p-n junction, formed at the interface between n-type and p-type semiconductors. Photons with energy greater than the semiconductor bandgap are absorbed, exciting electrons to a higher energy level and creating electron–hole pairs. The built-in electric field at the p-n junction, resulting from the spatial charge difference between the n-type and p-type materials, drives electrons toward the n-type region and holes toward the p-type region, preventing recombination and enabling charge flow. Electrons diffuse to the n-region, and holes move to the p-region, maintaining separation across the junction and facilitating electricity generation. This separation creates a potential difference, generating voltage, while the flow of separated charges produces the current necessary for power output. Figure 7.3 shows the basic construction of a solar cell consisting of the most important parts mentioned above.

Photovoltaic cells are consolidated into panels, and several panels can be linked together to construct modules. In general, the performance of a solar cell is defined by its current–voltage (I–V) properties, which is shown in Figure 7.4, which indicates the magnitude of electric current produced at varying degrees of irradiation. In reference to Figure 7.4, I and U represent the current and voltage, respectively. I_{sc}, I_{mpp}, U_{oc}, U_{mpp}, and P_{mpp} stand for short circuit current, current at maximum power point, open circuit voltage, voltage at maximum power point, and power at maximum power point, respectively. When a solar panel operates under ideal conditions with

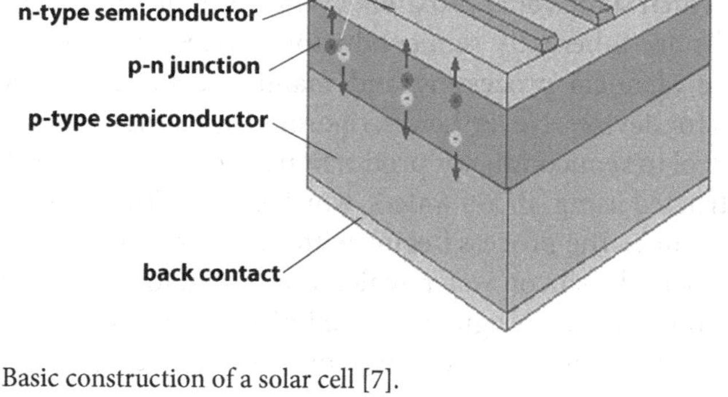

Figure 7.3 Basic construction of a solar cell [7].

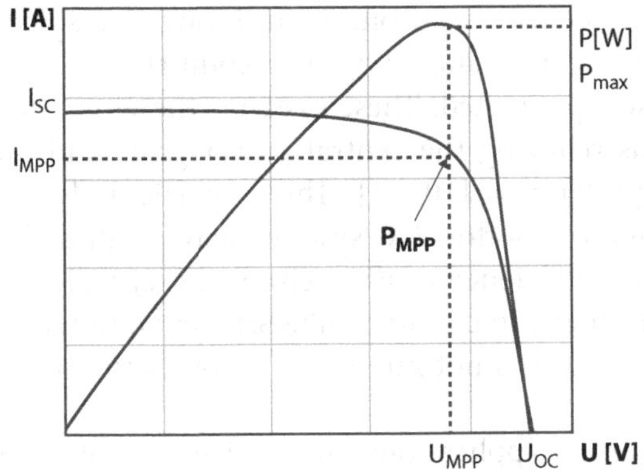

Figure 7.4 Representation of current–voltage characteristics and the power of a solar cell [7].

minimum resistance, it is able to produce a significant amount of current. This is because the panel is composed of numerous individual cells, which are connected in parallel to each other. As a result, the current produced by each cell is aggregated to create a combined total current. The total voltage output of the panel is also affected by the configuration of the cells and modules. Specifically, when they are linked in a series configuration, the voltage output increases. This is because the voltage of each cell is added together to create a higher total voltage. It is worth noticing that solar cells exclusively produce electricity in response to light exposure. They do not possess the capability to store energy for future usage, unlike energy storage devices. Consequently, if the solar panel is exposed to light, it will continue to generate electricity [7].

7.1.2.2 The Potential of PV Systems to Meet Future Energy Demands

The extensive potential of PV systems to fulfil future energy requirements is becoming more apparent due to technological developments, escalating climate concerns, and the investigation of novel uses. Global projections indicate that by 2050, PV systems have the capacity to produce a maximum of 175,111 TWh per year [8]. However, this potential may fluctuate because of climate change. The integration of supercapacitors (SCs) into PV systems represents a substantial advancement in energy storage and system dependability, which will be further discussed in the following sections. This integration not only improves the effectiveness of PV systems but also allows them to more effectively adjust to variations in energy demand, therefore guaranteeing a more reliable energy provision [9].

Rooftop PV systems are becoming an innovative solution for providing power in urban areas, including such countries as Latvia and Sweden, which are densely populated. These systems, therefore, distributed power generation, thus relieving the central power grids and directly addressing local energy demand [10, 11]. This is particularly evident from the enhanced uptake of rooftop PV systems across cities that have proven the versatility and efficiency of PV technology regardless of geographical and climate variations. In countries like Sri Lanka, India, Bangladesh, and Pakistan, rooftop solar is not rare when it comes to household electricity generation.

Apart from urban applications, innovative uses of PV technology are finding applications in transport systems. For instance, the Netherlands is considering the feasibility of installing PV systems along the highway, a move that will greatly shape the future of green mobility. These PV systems are intended for electricity consumption of EV charging infrastructure, which, in turn, contributes to the evolution of efficient and eco-friendly transportation and lowered emissions of greenhouse gases [12]. Such applications are some inspiring samples that prove the relevance and importance of PV technology for achieving sustainable development targets around the world.

Therefore, the PV systems provide ample evidence for their applicability across the extended spectrum, from the rooftop of the city building to the transport network; this indeed highlights the significance of these systems in meeting the world's future energy requirements. While society struggles with climate change and in the search for renewable energy sources, PV systems, especially when integrated with other systems such as the SCs, will go a long way in defining the future energy regime. Some of these will be discussed in the subsequent topics of this chapter as follows.

7.1.3 Challenges in PV Systems

7.1.3.1 Fluctuating Irradiance

Several issues affecting the operation of the PV systems of particular interest are the issue of irradiance variability, which greatly affects the system's performance and durability. The first one is the fluctuating nature of irradiance, which is a result of clouds, shading, and atmospheric conditions, to mention but a few. These fluctuations make it hard to maintain the stable output of the PV system and thus call for more sophisticated control mechanisms to address this issue [13]. This inconsistency in the fluctuations in irradiance also results to a high loss of power especially in the micro grid systems in off-grid or isolated areas. *Nowcasting System* that forecasts irradiance variability on a short-term basis is being piloted to improve the performance of these systems [14]. As for short-term fluctuation in irradiance that lasts for seconds, this can put a lot of stress in power converters and other parts, thus requiring efficient designs that can easily handle such fluctuations [15]. Last but not the least, soiling, shading, and thermal variations further compound the issues of efficiency in PV systems; in the form of cleaning, cooling, or managing the effect of shading tactics, extensive strategies are needed [16]. However, because of irradiance levels, cells generally tend to lower outputs of energy; hence, their performance diminishes.

Conversely, high irradiance can enhance performance but may also increase the risk of overheating, especially in shaded cells. Figure 7.5 represents the variation of PV system efficiency vs the irradiance and temperature where it concluded the reduction in efficiency with the change of irradiance [17]. The literature indicates that shaded cells can reach temperatures significantly higher than non-shaded cells, which can degrade their performance and potentially shorten their lifespan [18].

In their present state, PV systems can offer a lot in meeting the world's energy requirements; however, they suffer from fluctuations and instability

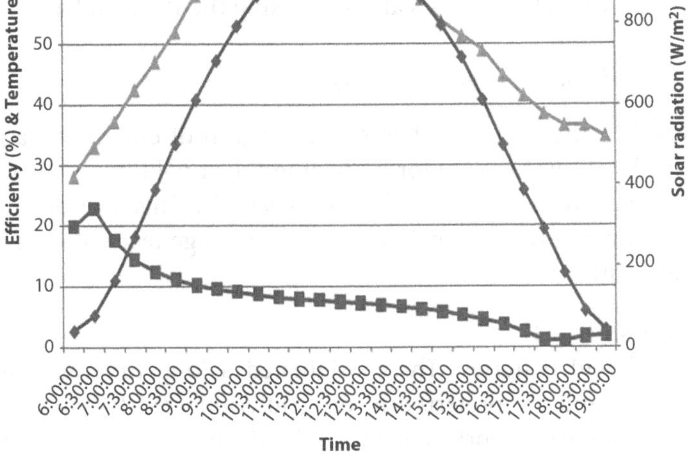

Figure 7.5 Variation of PV system efficiency vs the irradiance and temperature [17].

issues in systems like the DC to AC conversion and voltage control. In a similar way, when it is used for DC applications, this instability caused by fluctuation irradiance makes fluctuations in voltages. Such voltage deviations can lower system efficiency and, in certain worst-case scenarios, harm essential system components of the appliance as well as the PV system. This instability presents a great problem in the expected reliability of as well as durability of PV installations [19].

Apart from increasing the stability, the incorporation of efficient energy storage systems is valuable toward enhancing the reliability of PV systems due to the fluctuations in solar irradiation. The battery and SC are the main forms of energy storage that continue to support the availability of power from the PV systems in case of a low solar yield [20]. SC Li-ion batteries-based hybrid energy storage systems (HESSs), which provide the feature of maximum power transfer rate with the capability of extended charge and discharge cycles, have great potential to improve the performance of charge and discharge cycles. Besides prolonging the cycle time of the battery, this also increases the efficiency of the entire PV system [20]. Also, when choosing the storage type, it is a necessity to study energy storage efficiency, which refers to the effectiveness with which an energy storage system can store and later release energy. It is typically expressed as a percentage and indicates how much of the energy input can be retrieved as useful energy output.

However, hybrid AC power systems in conjunction with energy management tactics to be discussed in subsequent subtopics give a strong solution in the off-grid power system where energy storage has stronger significance in PV systems. Combined with other storage technologies, these systems can properly coordinate the energy demand and supply, therefore decreasing the operational costs and increasing the PV installations.

7.1.3.2 Available Storage Solutions

As discussed in the previous chapter, integration of energy storage solutions with PV systems is vital for managing the variability of solar energy generation and ensuring a reliable power supply. This section discusses in detail the overview of the primary energy storage technologies used in conjunction with PV systems.

Batteries [21, 22]

- Lead-acid batteries: These originated a long time ago, are cheap, and are preferred in off-grid connected PV systems. But they have short cycle life and reduced energy density in comparison with the modern types of storage systems. Figure 7.6 (a) shows the general structure of a Lead-Acid battery.

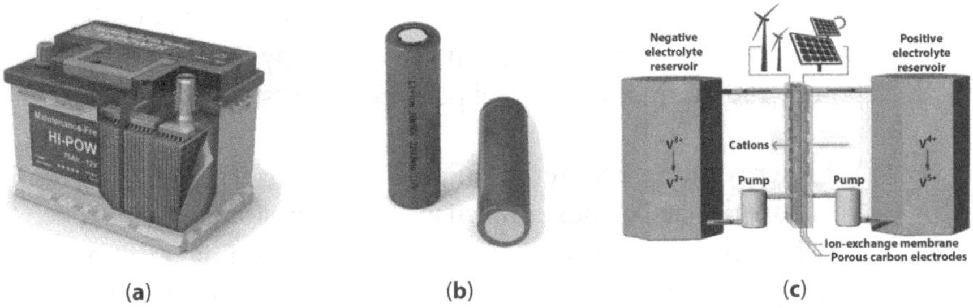

Figure 7.6 Representation of different types of batteries in the market. (a) Lead-acid battery (Source: https://jmhpower.com/). (b) Commercially available li-ion battery 18650. (c) Structure of a flow battery (Source: C&EN – American Chemical Society).

- Lithium-ion batteries: These are characterized by a high energy density, a longer cycle length, and a shorter charging time; hence, they are adopted in most residential and commercial PV systems. They can be applied for the applications that require low energy consumption. Figure 7.6 (b) shows a common general purpose Li-Ion battery (18659) available in the market.
- Flow batteries: Some of the types of flow batteries include the vanadium redox flow batteries that hold energy in a liquid electrolyte form. They provide scalability of power density and very long cycle life, thus suitable for large energy storage applications. Both can be charged and discharged at the same time allowing for the control of energy. Figure 7.6 (c) refers to the structure of a flow battery.

Supercapacitors (SCs)

- SCs are the advanced versions of capacitors that offer charging and discharging at faster rates. They are well suited to operations where high power is needed for short time periods. SCs can allow the control of voltage deviations and increase the response rate of the PV systems at the time of maximum load [23]. The image in Figure 7.7 (a) shows a commercially available Series XL SC with 3000F. More details about SCs will be presented in the subsequent chapters.

Hybrid Energy Storage System (HESS)

- A HESS integrates two or more storage technologies with power delivering devices to achieve better system characteristics. The principal advantages of employing HESS are reduction of stress and temperatures affecting battery life through decreasing the load and power request on the battery, enhancement of efficiency of energy extraction from the battery by less request of power, and increase in the total power of the system [24].

Flywheel Energy Storage Systems (FESS)

- These are the latest kinds of mechanical energy storage systems, which store energy in the form of flywheel spinning at ultra-high speeds. Such systems are steadily gaining appreciation for features that include high-power density, efficiency, and eco-friendliness in applications where fast response and cycling are typical, including grid support and storage, renewable energy interfacing, and data centers [25]. Figure 7.7 (b) illustrates a hybrid energy storage application, where various energy sources like a PV array, wind turbine, and storage devices such as SCs and batteries work together. Figure 7.7 (c) depicts the structure of a Flywheel Energy Storage System (FESS).

Compressed Air Energy Storage (CAES)

- These systems utilize compression of air into underground energetic cavities or chambers that can be tanks. In the same way, when electricity is required, the compressed air is reheated and, in the process of expansion, a turbine is used to produce electricity. Even though CAES systems are better for applications on a large scale, they can be used in combination with PV systems [27]. Figure 7.8 (a) shows a schemetic diagram of a CAES.

Pumped Hydro Storage (PHS)

- This is also a well-proven solution that applies to two water storage facilities located at different levels. In the case of PV

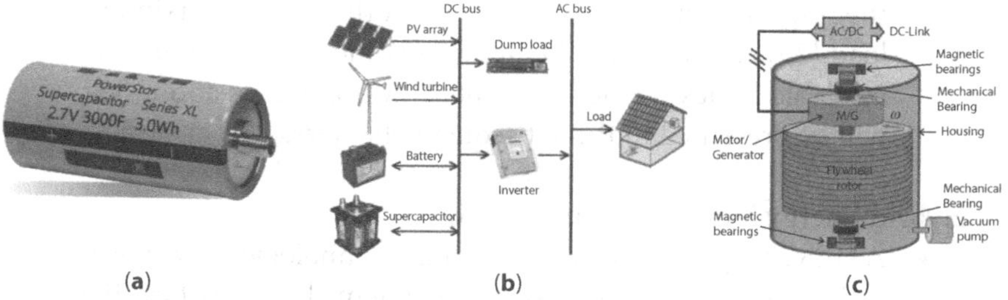

Figure 7.7 Graphical views of (a) commercially available Series XL SC, (b) an example of hybrid energy storage application, and (c) structure of FESS [26].

system excess generation of energy, water is taken to the upper reservoir for storage. This water is used to produce energy when needed with the stored water being released to generate electricity. This solution is ideal for large-scale energy storage; however, it requires certain geographical requisites [28]. Figure 7.8 (b) illustrates a PHS system, with water being pumped from a lower reservoir to an upper reservoir, where it is later released to generate electricity through a turbine.

Thermal Energy Storage (TES)

- TES incorporates thermal energy in excess generated by PV systems for use later in heating applications or in generation

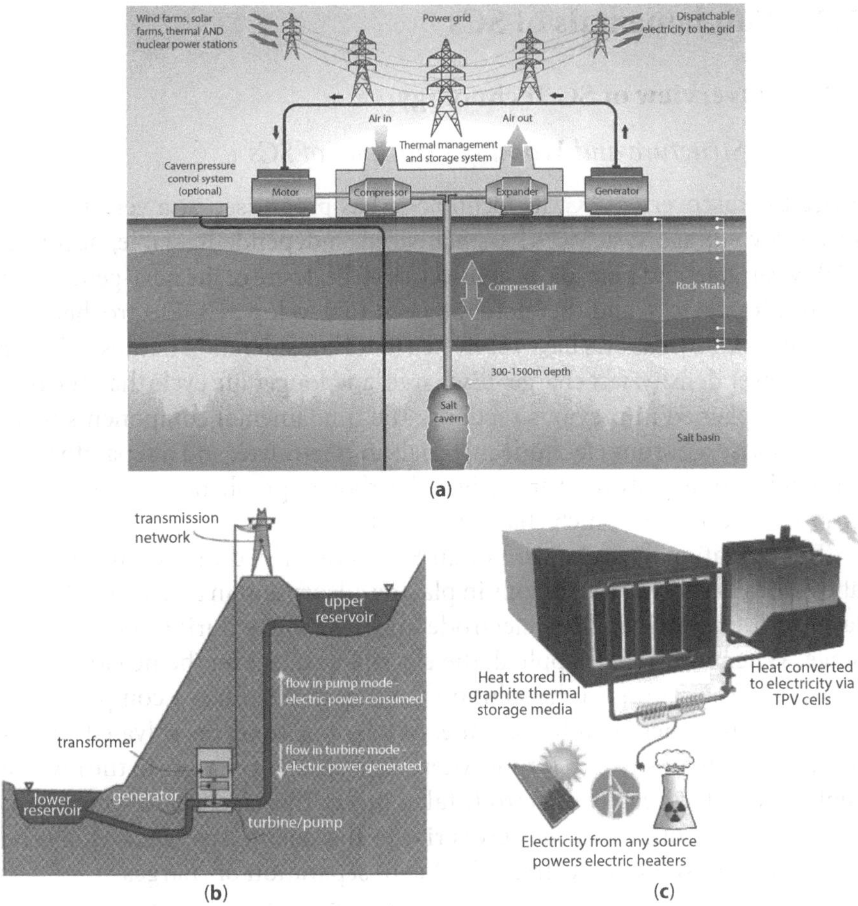

Figure 7.8 Schematic diagram of (a) a CAES [27], (b) a PHS [30], and (c) TES [31].

of electricity. Some of the most frequently used are molten salt storage as well as phase change materials, which enable good control of energy and load [29]. Figure 7.8 (c) depicts a TES system, where heat is stored in graphite thermal media and converted to electricity.

Grid Energy Storage Solutions

- Certain PV systems are connected to grid-level storage systems where excess energy can be fed back to the grid. Large-scale battery storage and other distributed power management applications on the electricity grid can be beneficial in managing the dilemmas of supply and demand of renewable energy sources.

7.2 Fundamentals of SCs

7.2.1 Overview of SC Technology

7.2.1.1 Structure and Working Principles of SCs

Since the discovery of SC, its technology has progressed at a very fast rate in energy storage systems. The small-sized, independent, stable, accurate, highly variable, and portable power source is the desire of the next-generation electronics devices, and therefore the need to develop SCs. Electrochemical capacitors known as SCs have received a lot of consideration because of their high-power density, fast charge–discharge, and longer life cycle than the battery as explained in previous sections. The fundamental components of an SC comprise a porous electrode material, an electrolyte, and a separator; the electrodes commonly used are activated carbon or graphene oxide that has a large surface area on which charge is stored.

The operational mechanism of an SC, as described in the image given above, lies in the inability of ions in placing adsorption in an electric double layer at the interface of the electrode and electrolyte during the charging stage. When a voltage is applied, the electrons collect on the negative electrode, and the cations in the solution are attracted to form a compact layer known as the Inner Helmholtz Plane. At the same time, the solvated anions are attracted to positive electrode where deposition of charges in the inverse replica of that at negative electrode takes place. This electrical arrangement of charges on both electrodes gives rise to the double layers, in which the energy is deposited by electric field due to separation of charges.

As an SC charges, an applied voltage on the capacitor affirms electrons to the negative electrode and thus gives it the negative charge. This attracts

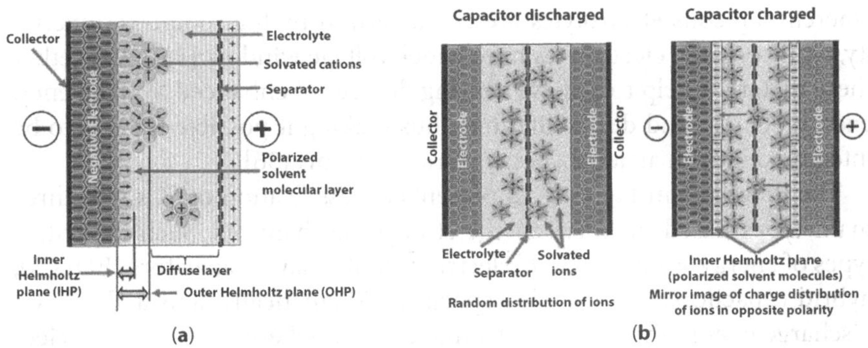

Figure 7.9 Formation of EDLC capacitor: (a) conceptual electrochemical capacitor and (b) arrangement of anions and cations at charged and discharged states [49].

positively charged cations from the electrolyte, which moves toward the negatively charged electrode and forms a layered structure known as the Inner Helmholtz Plane. At the same time, negatively charged anions can be attracted to the positive electrode and thereby also create a double layer. This brings about the storage of energy at the dying electric field between such separated charges. During the discharging process, the energy is liberated when the circuit is joined, and the electrons flow back from the negative terminal and reach the positive terminal. As the charges accumulated at the electrodes equilibrate, the coulombic force of attraction reduces and the ions in the electrolyte are reorganized thus discharging the SC. (See Figure 7.9 for the electric double-layer formation and ion arrangement during charging and discharging.)

7.2.1.2 Classification and Materials Used in SC Construction

The materials that have been used to develop SC include carbon-based materials, metal oxides, conducting polymers, and electrolytes. Carbon-based materials like activated carbon, graphene, and carbon nanotubes (CNTs) are very useful materials due to their large surface area as well as high electrical conductivity that enables high charge storage through formation of electric double layers [33]. Pseudocapacitive materials are incorporated into the electrode structure to improve the overall capacitance; these include MnO_2, RuO_2, and NiO, and these compounds facilitate fast redox reactions in the electrode [34]. Other materials that are being employed are the conducting polymers such as polyaniline (PANI) and polypyrrole (PPy) due to their pseudo capacitance behavior, flexibility, and charge storage *via* doping–dedoping mechanisms [35]. The electrolyte between the electrodes can be either aqueous, organic, or solid state wherein aqueous electrolytes are characterized by high ionic conductivity, while organic electrolytes have wider voltage window [35]. Altogether these materials help to achieve varying degrees of enhanced SC efficiency, energy density, and device lifetime, thus making it possible for SCs to be integrated into an array of energy storage systems [36].

The flowchart in Figure 7.10 presents a classification of SCs into three main types based on their charge storage mechanisms. Such capacitive types of energy storage include electrical double-layer capacitors (EDLCs), hybrid capacitors, and pseudo capacitors. Magnetically activated EDLCs discharge energy through the Helmholtz layer, where ions from the electrolyte accumulate near the electrode surface. To maximize charge storage, they use high-surface-area materials such as activated carbons, CNTs, graphene, carbon aerogels, carbon foam, and carbon-derived carbons. These advanced materials enhance energy density, conductivity, and performance, making EDLCs highly efficient for energy storage applications. Pseudo capacitors are based on electrochemical double-layer formation using materials such as metal oxides and conducting polymers through Faradaic processes. Finally, a hybrid capacitor makes use of both electrostatic as well as electrochemical storage technologies, and subcategories include asymmetrical pseudo/EDLCs, composite types, and rechargeable battery-like capacitors. This classification provides example of how SC is designed and shows that they are technologically versatile enough to be used in a variety of applications. A comparison of commercially available single-cell supercapacitors, highlighting their voltage ratings, capacitance, energy density, and ESR values, is provided in Table 7.1. This comparison

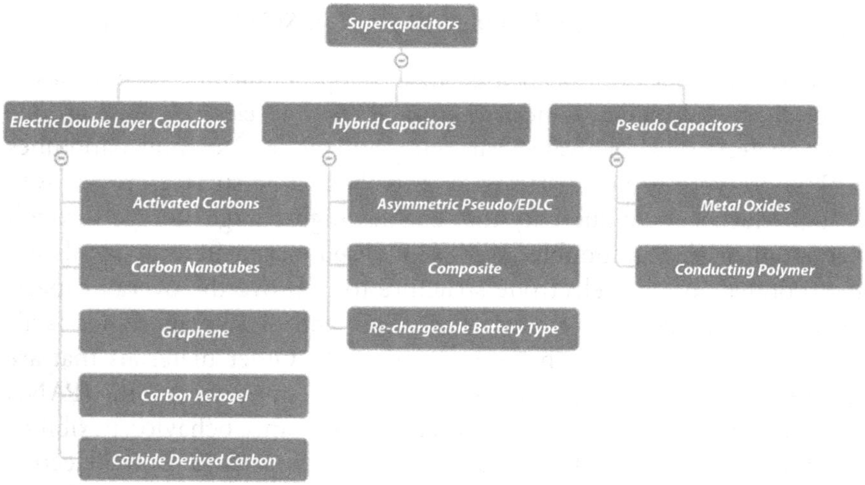

Figure 7.10 Classification of supercapacitors.

Table 7.1 Comparison of commercially available single-cell SCs [37].

Parameter	EDLCs	Hybrid SCs	CAPA batteries
Energy density (WhL^{-1})	5–8	10–14	50–120
Power density (WL^{-1})	8,000	2,500–4,000	1,600–3,200
Cycle life	1,000,000	40,000–50,000	15,000–20,000
Rated voltage (V)	2.7	2.7	2.8
Capacitance (F)	1–3,000	200–7,500	1,000–70,000

helps in selecting the most suitable SC for regenerative braking applications based on performance and efficiency metrics.

7.2.2 Comparison with Traditional Capacitors and Batteries

7.2.2.1 Differences in Energy and Power Density

The following Ragone plot in Figure 7.11 depicts the comparative evaluation of several energy storage technologies in terms of power densities, which is expressed in watts per kilogram and the energy densities that are in watt-hours per kilogram. Each technology is represented by a different color, highlighting their unique characteristics: Capacitors (green): Placed at the bottom right of the graph where power density is high while the energy

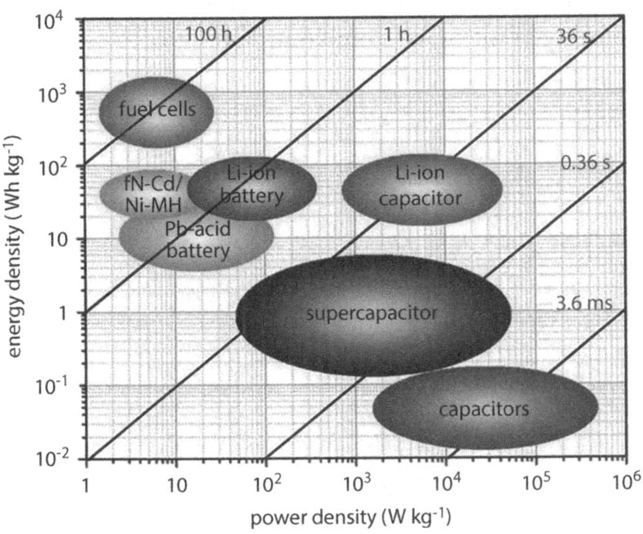

Figure 7.11 Ragone plot of different energy storage devices [38].

density is low. This shows versatility in delivering energy, and this makes them ideal for applications in areas whereby energy is needed in short bursts, for instance in electronics and power control. SCs (blue): Middle of the range of power and energy density that provides a highly effective product in the form of SCs as it allows fast energy generation and high usage capability.

They let out energy rapidly and can also accumulate more energy than conventional capacitors, though they have a higher energy density than capacitors but less energy density than batteries. This versatility makes them suitable for applications that involve large power ratings and moderate amounts of energy density as is the case when using them in automobiles for regenerative braking. Batteries (brown): Placed on the left side of the graph where energy density is high, and power density is comparatively less than the capacitors. They are long-lasting indicating that they can be used for long-term energy storage systems like electric cars and renewable energy storage. However, they have a slow rate of energy release, and consequently, they are not very suitable in applications that demand a quick power output. Fuel cells (teal): Alike batteries but apparently placed even closer to high energy density, fuel cells also exhibit the feature of high energy density with lower power density. It lasts for longer durations and hence is useful in applications that require steady power delivery including the static cogeneration of electricity and several automobiles, although they release energy at a slower rate than capacitors.

The charging and discharging characteristics of SCs have been plotted in Figure 7.12 and compared with rechargeable batteries. In these cases, the charging and discharging time is very less compared with the batteries as the power density of the SCs is comparatively high.

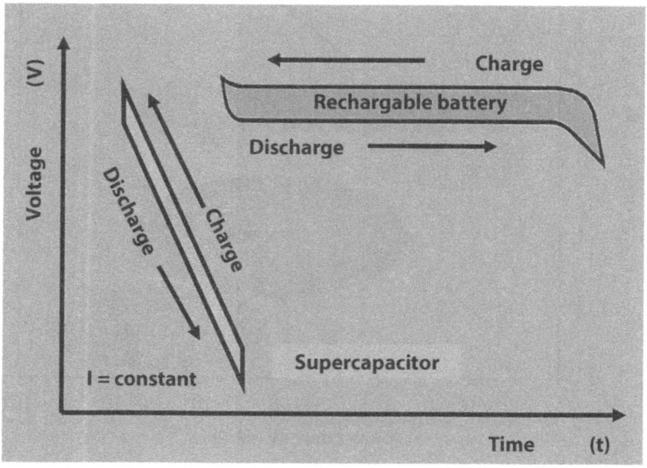

Figure 7.12 Different charge /discharge behavior of SCs and rechargeable batteries [39].

7.2.2.2 Lifecycle and Durability Comparisons with Batteries

SCs or batteries' lifespan differs by the product type because of different techniques and functions. Compared to the other battery types, SCs usually have a greater number of charge/discharge cycles. It can operate with the capacity of more than millions of cycles, ideal for applications that call for the quick charge and discharge of energy. On the other hand, lead-acid batteries that have been in use for more than a century suffer from a life cycle that stands at 500–1500 cycles depending on its design and utilization characteristics. According to literature, despite the higher durability in comparison with modern lead-acid batteries, they are less effective due to a lower life cycle and specific energy density compared to the emerging battery technologies such as lithium-ion batteries [40].

7.2.3 Advantages of SCs

7.2.3.1 High-Power Density and Rapid Charging

They have high-power density mainly because they can store energy in the form of electrostatic charges and not chemically. Therefore, they can absorb and discharge energy rapidly, meaning they take less time to be charged or discharged than batteries, and therefore are suitable for applications that require high energy in short time intervals or surge energy. This fast-charging mode also helps in cutting down on the time taken to charge in applications like energy storage, which is the reason more and more organizations are opting for it in sectors like electric vehicles (EVs), regenerative brake systems, and emergency power sources. Also, it is their high and instantaneous power factor that is desirable in applications where the energy is required in small periods. Though they have lower energy density than batteries, high-power density along with fast charging capability of the SCs makes them the core component of HESSs where both power and energy are required. This will be discussed further in subsequent sections of this chapter.

7.2.3.2 Longevity and Low Maintenance Requirements

Compared to a normal battery, SCs have a very long life cycle and very low self-discharge rates, making them valuable for many applications. In contrast with batteries that degrade chemically during the use and charging and discharging cycles, SCs retain energy in the form of the electrostatic attraction between the ions and the conducting plates mentioned above and, therefore, do not degrade as much throughout their usage. In the same respect, SCs possess a much longer operational cycle as compared to the system where they are incorporated into. Besides, SCs require minimal maintenance because they are not affected by issues like the memory effect or deep discharge effect, which commonly impact batteries. Due to the low maintenance that they require, along with their ability to last for a long time, SCs are especially suited for usage either in areas that are difficult to access or if maintenance is expensive. These factors cumulatively greatly diminish the total cost of ownership of SCs especially when it is compared with the long-term reliability of energy storage devices. Further details concerning this will be discussed in the next subsequent sections.

7.2.3.3 Environmental Benefits Compared to Chemical Batteries

Compared to chemical batteries, SCs have several environmentally friendly features based on the material they are made of, the process of production, and their end-of-life disposal. In contrast to batteries that involve the use of chemical reactions involving toxic and relatively rare elements including lead, cadmium, and lithium, SCs primarily involve materials like activated carbon in their operations, which are relatively cheap, are eco-friendly, and can be recycled easily. The fabrication of SCs also, as a rule, implies the use of less toxic reagents and makes a smaller number of pollutants. Moreover, as has been previously stated, SCs have a much longer cycle; the frequency at which they will require replacement is significantly lower, and hence, the amount of waste produced in the long run is considerably less. That is why, in addition to their ability to last considerably longer than batteries and be inherently less toxic, they impose lower environmental burden during their life cycle. Also, at the end of charge/discharge cycle, SCs require a smaller number of efforts for disposal and recycling as they do not offer critical quantity of hazardous materials, which are highly injurious to environment and health. These factors explain why SCs are increasing in their demand especially in areas where sustainability is a key concern [41].

7.3 Integration of SCs in PV Systems

7.3.1 Addressing Instability in PV Systems

7.3.1.1 How SCs Mitigate the Effects of Fluctuating Irradiance

The incorporation of SC into PV systems has a positive impact on their stability and production, as it diminishes the impact of power variation, which resulted from the change of irradiance, for instance owing to the presence of cloud or shading. SCs with high-power density as compared to conventional capacitors serve the purpose of storing energy and stabilizing the output. Due to their ability to store energy in a very short time and release it when the generation is low, SCs assist in maintaining the voltage level of the DC bus and thus increase the efficiency of the system. It is evident from various studies that in the absence of SCs, the voltage of the PV system can fluctuate and conjecture the stability of the PV system, which is made in milliseconds by integrating SCs [42].

Apart from stability enhancement of a system, SCs allow PV systems to meet or provide power quality for sudden loads or power surges. This capability is more essential in those systems having an irregular or stochastic power consumption profile, or in cases where power requirements vary frequently, such as in industrial or grid-integrated systems. SCs also play the role of one of the ways through which management of energy for renewable energy system can be achieved because they can respond to swift changes in weather situations without having to incorporate other mechanical facilities as well as complex control systems. Therefore, the current variation in the power electronic converters shown in the study may cause ±15% DC bus voltage oscillation if not regulated by the SC or the grid. On the other hand, either support reduces the variation in the DC bus voltage to ±8% during the dynamic irradiance conditions and ±2.5% at dynamic load conditions. The study also emphasizes the importance of having an SC or grid support in the power management system of a PV system, more especially in controlling the volatility of the voltage on the DC bus [43].

Where both SC and batteries are integrated into the hybrid system, SCs lower the loads on the battery system, increase battery durability, and enhance the system ability to adapt to changes in irradiance. One of the advantages for using a hybrid combination of SCs and batteries is it is capable of short-term and long-term energy storage where SCs cover the short-term energy rate variation and batteries will take care of the energy supply for the longer term. Among these, one of the studies shows that the integration of SCs is pivotal to minimizing stresses on batteries and, therefore, enhancing their lifetime in hybrid systems. It particularly analyzes various operating parameters that include RMS current gain in the battery, loss reduction, total effective energy efficiency, and the power reduction of surge load eliminated in the battery. This study identifies the fact that when the capacitors are connected in parallel and when the filter constant has been adjusted, it does, in fact, give the system a boost and lower energy losses increasing the overall system performance. Furthermore, the study also compares these improvements in terms of cost with respect to SCs, which reveals that improvements in performance and increase in the power capability of SCs justify additional

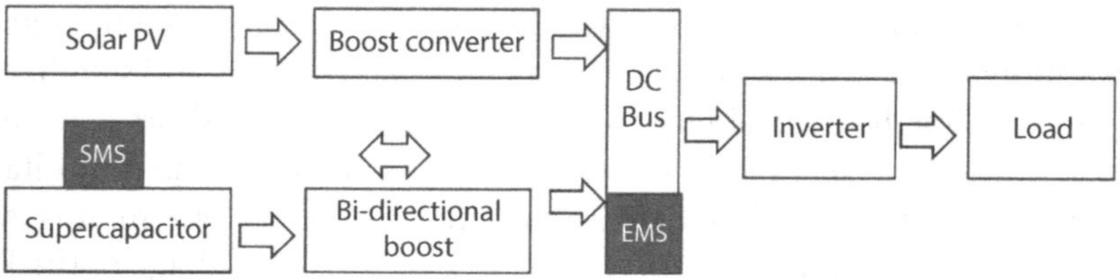

Figure 7.13 PV and SC system integration [42].

investment making SCs a cost-effective substitute for increasing the reliability and life cycle of PV energy storage systems [44]. Furthermore, hybrid PV systems with the incorporation of SCs and inverters have been demonstrated to deliver constant power for fluctuating irradiance, and thus indispensable for off-grid and remote systems. The study introduces a control scheme for efficient power flow management between generation and load sides, stabilizing voltage under varying conditions [45]. Figure 7.13 illustrates the integration of a solar PV system with a SC. A boost converter regulates PV output to a DC bus, which powers the load *via* an inverter. The SC connects through a bi-directional boost converter, with Supercapacitor Management System (SMS) and Energy Management System (EMS) managing power flow and voltage stability.

The literature also outlined that SCs are capable of operating in different temperatures and other environmental conditions. Therefore, they are essential in critical applications like telecommunication, military operations, disaster avoidance systems, and any other event that requires continuity of power supply. In the first stage of the study, a boost converter was employed with a Perturb and Observe Maximum Power Point Tracking (MPPT) algorithm. The results of the study showed that, without energy storage, the system struggled to maintain a constant current during a drop in irradiance from 1000 to 600 W/m^2. However, with the SC in place, the current injected into the grid remained stable throughout the transient, indicating the effectiveness of the energy buffer [46].

7.3.1.2 Role in Stabilizing Voltage and Improving Power Quality

Irradiance varies causing variations in power generation, which creates voltage fluctuations that may bring about instabilities in the system apart from declining power quality. These difficulties have been tackled by SCs, since they are used in the roles of energy buffers, which are charged with the energy in the instances of a high irradiance level and discharged to the circuits in the instances of a low irradiance level. This dynamic response capability helps to mitigate volatile variations in the voltage level and facilitates a steadier output power, serving to improve the stability and reliability of the PV system. Several case studies have shown how SCs played a significant and vital function in regulating voltage and enhancing power quality irrespective of its fluctuating irradiance. For an example, Saripalli *et al.* [47] performed the simulations of a system that implemented SCs, and fuzzy logic controller for MPPT improved output voltage ripple, which contributes to the system stability during maximum irradiation level variation. Through operating at the DC bus voltage, SCs maintained continuous power distribution with reduced issues such as voltage drops and instability in the power distribution systems. This highlights the need for SCs in the advancement of PV systems' performance, particularly when the energy flow is managed through intelligent control systems such as the fuzzy logic. Also, several studies have shown that MPPT with SCs implemented in the system is capable, and it increases the overall efficiency of the system [48, 49].

In another case study by Cabrane *et al.* [50], it was observed that integrating SCs with batteries in the form of HESS benefited from voltage stabilization as well as reduced cycle stress on batteries, hence enhancing their durability. Furthermore, it made it clear that in the face of transient condition, the system kept the DC bus voltage constant at 400 V showing that the SCs were capable of handling voltage variations and sudden load variations with ease. This ability of maintaining stability in voltage is highly critical in the PV systems especially because the energy generation is clearly erratic, and SCs play the role of providing the required burst of energy when there is a change in the level of irradiance. Additionally, the benefits of SCs for battery stress make the system cost-effective since it increases the usage of batteries before they are replaced.

Furthermore, in high-concentration photovoltaic (HCPV) systems, where precise solar tracking is crucial, Huang and Tsai [51] illustrated that SCs offset tracking errors, enhancing both the power quality and stability of the output. Even with swiftly changing weather, the SCs in the HCPV system were able to rapidly self-regulate differing energy input, stabilizing power output and ensuring continuous, premium energy distribution. This case highlights the crucial role of SCs in sustaining voltage levels and in removing inefficiencies linked to mechanical or environmental inabilities in the system.

The study of González *et al.* [52] analyzed the implementation of SCs in PV networks of low voltage, finding that SCs significantly boost power quality at the point of common coupling (PCC). By leveraging energy compensation techniques, notably ramp control and SCs, power fluctuations were substantially lowered, allowing for a smoother connection of renewable energy into the grid. Moreover, the deployment of an SC-based energy storage system can administer at the very least 5.76 daily charge/discharge cycles, improving the system's capability to manage changes in energy. Eventually, the proposed system could decrease the rate of change of PV energy injected into the PCC by approximately 6.66 times, proving the potential value of SCs in enhancing the stability of PV energy output.

In summary, applying SCs to solar PV systems is key for voltage stabilization and the improvement of power quality, particularly in situations that experience fluctuating irradiance. Their efficient and rapid charging capacity for energy storage and release help prevent fluctuations in voltage, leading to better energy delivery and enhanced system resilience. Not only do SCs improve power quality, but they also contribute to the longevity and economic benefit of energy storage technologies like batteries, which makes them an essential part of contemporary PV applications whether in grid-connected or standalone systems. The studies discussed highlight the indispensable role of SCs in ensuring the reliable operation of PV systems, especially under variable environmental conditions.

7.3.2 Role of SCs in Energy Storage

7.3.2.1 Short-Term Vs. Long-Term Energy Storage Needs in PV Systems

Energy storage in the PV system is basically categorized in two types, short-term energy storage and long-term energy storage, where the distinction between these two important facets plays a very crucial role in determining the capacity of the systems and the overall stability. Now that the dynamics of transient irradiance have been identified as a key factor, it is apparent that short-term energy storage is necessary to offset these effects. SCs are particularly effective in functioning for short-term fluctuations because they can charge and discharge in the shortest time possible, meeting these fluctuations instantly. This guarantees that voltage stability shall be ensured, and power quality is kept intact in short duration of time when there is fluctuation in output produced by the solar panel. Other related research like Saripalli *et al.*'s [47] work shows that incorporating the SCs into the PV system minimizes power swings and voltage fluctuations during the short-term disturbances.

However, for consistency in supply and demand over several hours, days, or even weeks, long duration storage is important especially at night or during cloudy and hence reduced solar generation. As they provide much higher energy density, batteries are more useful for storage to utilize the excess of the storable solar energy produced during the time of day with abundance of sunlight. Long-term storage systems make sure that power is always available as a continuous supply, thus making the system more reliable. Cabrane *et al.* [50] have also revealed that integrating SC and battery systems optimizes both forms of energy requirements. This approach does not only increase the overall efficiency of the system, as well as the longevity of the batteries, minimizing the amount of time they are subjected to deep discharge cycles, while SCs take care of the high-frequency, short-duration energy fluctuations.

The combination of both short-term and long-term storage solutions makes it possible to have the most efficient use and management of energy in PV systems. SCs respond to real-time, high-power requirements, whereas batteries, on the other hand, provide long-term, sustainable storage fulfilling a synergy that enhances the functionality of the PV system.

7.3.2.2 How SCs Complement Traditional Batteries

Based on the studies, SCs and batteries integrate in PV energy storage systems to give a HESS that offers short-term and long-term storage solutions. SCs help to balance the transient irradiance changes, which causes fluctuations in power with their high-power density. Batteries are more ideal for the storage of energy as they can supply energy over long intervals of time when solar generation is not present. Both play an effective role in the energy management in PV applications and make certain circuit configurations.

SCs are used together with batteries to partly alleviate the batteries during routine short-time high current demands. This reduces the pressure put on batteries to avoid reaching a deep discharge cycle, which is healthy for batteries. For instance, Aouini *et al.* [53] found that the proposed hybrid SC–battery system in the case of residential PV applications averted the concentration of current demand on batteries, thus enhancing the batteries' lifespan by shifting the peak load to SCs. In a similar study, Loor-Cevallos *et al.* [54] described the role of SC in isolated rural PV systems by providing reserve energy to control sudden energy spikes so that batteries can store energy during periods of high irradiation to compensate for longer periods of low irradiation thus enhancing the reliability of the overall system.

It is also necessary to point out that the interdependence of SCs and batteries enhances energy efficacy. Harinkhede and Vaidya [55] stated that incorporating the SCs with batteries in PV systems leads to improvements in the sharing of power as well as enhances the energy storage system to manage fluctuations in loads. Mbendane *et al.* [42] also extended the understanding, which stated that SCs provide steady voltage during transient status by providing a more efficient and energy-smart operation instead of batteries.

7.3.3 Enhancing System Reliability and Efficiency

7.3.3.1 Case Studies of Reliability Improvements with SC Integration

As discussed in early sections of the chapter, SCs boost the operation and performance of PV systems by controlling surge currents and maintaining voltage levels in the face of irregular solar exposure. Research has shown how the incorporation of SCs can avoid system anomalies, enhance battery cycles, and reduce energy oscillations.

Aouini *et al.* [53], in their study of residential PV systems, further explained that adding SCs to common energy storage systems with hybrid batteries proved more reliable because of the efficiency of energy delivery. The interdependency between the SCs and batteries decreased the amount of stress on the battery systems and increased their life cycle. This improvement in battery life directly transfers into reduced system failures and higher reliability of home usage of solar energy systems; hence, SCs integrated with hybrid systems are fit for residential applications of solar power. It can be concluded that the discussed approach is most effective in scenarios in which reliability in terms of the constant availability of energy and the lifespan of storage components are significant.

In their field studies on the performance of isolated, standalone PV systems in rural areas, Loor-Cevallos *et al.* [54] offered further data on the effectiveness of SCs. In their study involving distant areas that are not connected to the grid or where the connection is intermittent, the authors noted that SCs acted as stabilizers of energy storage systems. These include that, in these isolated PV systems, energy can vary greatly because of the sudden changes in the environment such as when the clouds appear or there is a sudden rise or fall in the level of solar irradiation. SCs were able to mitigate these short-term oscillations and guarantee a constant and uninterrupted energy output to the users in off-grid areas. This aided in reducing the chances of power blackouts, especially in remote areas, which are a big issue since dependable power is necessary for basic operations. According to Loor-Cevallos *et al.* [54], the performance of SCs made it possible to stabilize the supply of energy while at the same time reducing the stress or wear on the battery systems, a factor that is crucial in rural areas of the world where it could be difficult to get the replacement parts needed for maintaining energy storage systems.

Concerning the aspects of the model, the effectiveness of the HESSs that include SCs was shown by Harinkhede and Vaidya [55] during the comprehensive simulation-based study. Their study demonstrated that SCs sustain voltage while alleviating stress on batteries and mitigating system failures linked to high-frequency energy demands, in contrast to traditional storage systems. This study decreased the functional stress on the batteries and therefore improved the functional life of the batteries, thus improving the general capacity of the energy storage system. The study also showed how to avoid failures of the system and regulating voltage fluctuations; SCs allowed for the constant delivery of power, especially when the input power within the PV system is fluctuating because of changes in weather conditions.

Moreover, Mbendane *et al.* [42] extended the assertion of the importance of SCs in renewable energy systems and confirmed that SCs are more effective in regulating voltage fluctuations during transient conditions to prevent energy losses while preserving the reliability of power output. Their study was aimed at investigating the ability of SCs to respond to sudden changes in the energy supply, especially in renewable energy systems such as the PV system, where fluctuations in the intensity of the irradiation lead to drastic changes in the energy production. The study also focused on the rate of response of the SCs to such variations to maintain the necessary voltage levels in some key points in the system, for example, the DC bus. In this way, SCs mitigated voltage fluctuations that are typical for renewable energy sources and provided power quality and other system components from detrimental effects.

In summary, these studies support the necessity of SCs in managing voltage fluctuations, improving power quality, and contributing to increased reliability of energy storage systems in PV applications.

7.3.3.2 Quantitative Benefits in Terms of System Efficiency and Uptime

Through integrating SCs into PV systems, certain enhancements in the system performance and availability can be observed due to the discussed effects of moderating short-term power management and load sharing with other types of energy storage like batteries. Numerical data from various research works demonstrate how SCs act effectively in improving energy distribution in PV systems and the overall efficiency of those systems.

System Efficiency Improvements

In a study conducted by Aouini *et al.* [53], the authors found that the energy efficiency of the residential PV system increases by 12% with the integration of SCs and batteries. This efficiency was credited to the high-rate capability of SCs in managing transient power, thus enabling the battery system to store energy in the long term. This made the deep discharge cycles in batteries to be less frequent, hence enhancing the operational efficiency. A study carried out by Chang *et al.* [56] shows that SCs improve energy efficiency by up to 10% when combined with electrolytic capacitors for high-current, short-duration applications. This increase in efficiency is important in low-voltage ride-through and grid code compliance applications such as distributed generation systems. Also, on a similar note, Kularatna *et al.* [57] revealed that SCs can act as lossless voltage-dropping components across power conversion stages, hence enhancing the overall efficiency of electronic devices.

Increased System Uptime

Loor-Cevallos *et al.* [54], who analyzed isolated PV systems in rural areas, presented that the inclusion of SCs enhances system uptime by 15%. SCs maintained the voltage and regulated minor disruptions resulting from changes in environmental conditions and thus provided a consistent power supply irrespective of the fluctuating conditions from solar panels. This was of the essence in off-grid locations that require a constant and reliable power supply during the execution of various operations. A study by Parvini *et al.* [58] stated that the SCs provide $1x10^6$ charge–discharge cycles, far much exceeding conventional batteries, where the maximum is around $5x10^3$ cycles. Also, the study reveals that SCs work effectively at low temperatures, such as near freezing at –40°C, or high temperatures, such as 60°C, and the energy dissipation is relatively low in both these conditions.

Reduction in Energy Losses

Harinkhede and Vaidya [55] have also done simulations to demonstrate that SCs have a role in cutting down energy losses in PV systems by 10%. SCs successfully catered for the high-frequency, short-term power demands without suffering from voltage drop or inefficiency that comes with poor power quality. The findings of the study revealed that the application of SCs enabled enhancement of the quality of power fed to the grid while lowering the overall power loss magnitude. In addition, the study done by Piyumal *et al.* [59] also offers a novel approach for maximum energy extraction from PV systems employing SCs. This paper proposes the use of dual storage systems that can help in the reinforcement of energy storage while allowing the use of the stored energy behind the loads from SCs, hence improving the PV energy management system (EMS) in standalone applications.

Battery Lifespan Extension
Mbendane *et al.* [38, 41] stated that the incorporation of SCs in a PV system increased battery durability by 20%. Through smoothening of rapid energy variations, deep cycle discharge of the battery was minimized by SCs, thereby reducing degradation of batteries and enhancing the lifespan of the energy storage system.

Improved Power Output Stability
Another study performed by Saripalli *et al.* [47] demonstrated that there has been an enhancement in power output balance by 25% of PV systems when the system was integrated with SCs. The quick response of SCs to waveform variations concerning irradiance levels made it possible to maintain the output voltage within the required operating range, thereby contributing to clearer power output and minimal disruption of the power grid.

7.4 Advanced Techniques in SC–PV Integration

The integration of SC-assisted techniques has become one of the key enablers of energy storage and management as these techniques make use of the high-power density and short charging and discharging time of the SCs, thereby increasing the system efficiency and stability. These techniques are particularly important in countering the intermittent issues of renewable resources, including solar and wind, by energy storage and regulation. By incorporating SCs with other standard power parts like batteries, regulators, and converters, methods like Supercapacitor Assisted Loss Management (SCALOM) and Supercapacitor Assisted Low Dropout Regulator (SCALDO) facilitate improved energy conversion, lower power loss, and effective power supply. Besides, this integration also enables a more efficient, lasting, and adaptive utilization of energy system performance and, therefore, enhancing the stability of industrial and renewable energy installations.

7.4.1 SCALOM Technique

The SCALOM is an innovative energy management technique created to enhance the efficiency of PV systems particularly at the moments when their output is low or varies significantly. The main aim of SCALOM is to ensure system stability in energy delivery, properly control the energy, and increase the dependable probability of power delivery with the SCs through rapid charge and discharge.

This technique is most efficient in terms of stabilizing and enabling the enhancement of the efficiency of PV systems. Some of these features include low output regulation in which SCs can store extra power during the day and use them to generate electricity during the night. While it stably operates at a relatively low voltage without power losses, SCALOM demonstrates great efficiency in the regularization of rapidly occurring fluctuations. Furthermore, the efficiency of energy harvesting increases by using the SCALOM since all the accumulated energy during very short intervals of high irradiance is to be stored in the SC and utilized in those systems that do not incorporate the use of such technology. There, it also accompanies the twine energy storage integration, which comprises the SCs besides the batteries where the SCs discharge a short-term power supply normally, and the batteries are prevented from deep discharging, leading to an increased battery lifespan. These features make SCALOM a great tool for increasing the PV systems' effectiveness, reliability, and durability.

The study also put forward SCALOM, which is said to be an optimum solution to improve the efficiency of the solar inverter. SCALOM also manages and mitigates static losses, high-frequency switching losses, and transformer losses. This approach reduces voltage stress and RMS current in inverter components for improvement in operational performance. By comparing with the reference design of the study, test outcome demonstrates a 30% increase in efficiency or capacity enhancement to accept higher voltage input and operate more freely to prove that SCALOM produces less loss and enhances the functionalities of the inverters [60].

7.4.1.1 Advantages of SCALOM in PV Systems

Enhanced system performance: By considering low output conditions and more optimum use of the energy storage system more effectively, SCALOM can enhance the efficiency of PV systems by 10–15%, as analyzed by researchers Harinkhede and Vaidya [55].

Enhanced uptime: In particular, SCALOM increases system availability exponentially in the event of limited energy generation. This helps in the persistence of energy supply, especially where grid support is not available as is the case with some standalone PV systems. It has also been established in the literature that systems utilizing SCALOM work 20% more time [53].

Faster response times: The short charge and discharge time of SCs enables SCALOM to enhance the system efficiency, meeting the fluctuating

Table 7.2 Summary of statistical results [38, 43, 47–49].

Metric	Improvement with SCALOM
Reduction in voltage fluctuations	30% reduction
Increase in system uptime	20% increase
Improvement in energy harvesting	10–15% improvement
Extension in battery lifespan	15–20% longer battery life
Reduction in energy losses	10% reduction

power demand within milliseconds, thus providing a consistent power supply to the system.

These statistical results (Table 7.2) showcase the significant impact SCALOM has on system reliability, especially in managing the inherent variability of solar power. By integrating SCs to handle short-term power fluctuations, SCALOM ensures that PV systems maintain a stable, reliable power output, even under challenging solar conditions.

7.4.2 SCALDO Technique

SCALDO regulators are the latest in the category of power electronics and take the technology of voltage regulation systems to the next level. Incorporating SCs in the common low-dropout (LDO) regulator design, SCALDO systems eliminate loss-associated power while sustaining steady voltage with less oscillation. Subasinghage *et al.* [61] also proved the feasibility of employing SCALDO regulators for high-efficiency DC–DC conversion with efficiency enhancement of up to 88% over conventional LDO regulators. This improvement is primarily attributed to the energy recovery method utilized from the structure of SCs, which acts as near lossless voltage droppers in the system thus reducing energy loss during voltage regulation.

While SCALDO regulators contribute to the reduction of energy consumption and improve efficiency, they also provide better stability and performance in dynamic load conditions. A preliminary analysis of SCALDO by Subasinghage *et al.* analyzed the ripple control in the system and proved that the integration of SCs reduces switching ripple and the output voltage is smooth in transient conditions [61]. The use of SCs in the energy recovery mechanism ensures that the system charges and discharges the capacitors quickly enough to counter any load demand changes, hence maintaining output voltage stability without affecting performance.

The subsequent work in developing SCALDO was done by Jayasekara *et al.* [62], where different ways of optimizing were presented, with a focus on the use and control of SCs depending on the supply voltage. The authors proved that through the regulation of the number of capacitors depending on the input voltage, it is possible to achieve the desired efficiency of SCALDO systems, which by the results of the theoretical and experimental ranged from 81.5% to 89%, and in different varying conditions. This level of optimization is important in the use of hardware, especially where power management is of vital importance.

Another important feature that remains essential in SCALDO systems is its stability. Have further confirmed the stability of SCALDO regulators through empirical analysis, which has established that the incorporation of SCs with LDO regulators unleashed no stability problems. The study reassured the fact that SCALDO can hold the stable voltage regulation for a system regardless of the variation of inputs to guarantee its lifetime and performance in real-world applications [63].

These statistical results (Table 7.3) demonstrate the substantial effect SCALDO has on enhancing system efficiency, particularly in reducing energy losses in low dropout regulators. By incorporating SCs to manage voltage regulation, SCALDO significantly minimizes power dissipation and heat generation, ensuring that power systems operate more efficiently and reliably, even under varying load conditions. This leads to improved energy conservation and extended component lifespan in applications where precise voltage control is critical. Figure 7.14 shows the SCALDO regulator's two phases: (a) Charging Phase: The input supplies power to both the LDO and charges the SC. (b) Discharging Phase: The SC powers the LDO, reducing input load and improving efficiency.

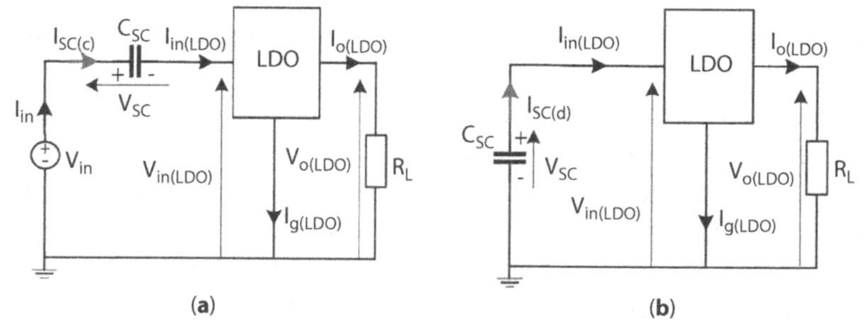

Figure 7.14 The fundamental circuit configuration of the SCALDO technique: (a) SC charging phase and (b) SC discharging phase [63].

Table 7.3 Summary of statistical results [61–65].

Metric	Improvement with SCALDO
Efficiency improvement	81.5% to 88% efficiency range
Reduction in output ripple	Significant reduction in ripple
Extension in battery lifespan	15–20% longer battery life
Reduction in energy losses	Up to 20% reduction in losses
Improved system stability	No instability in voltage regulation

7.4.3 Design Considerations for Parallel Integration

There are certain design aspects that need to be considered when connecting the SCs in parallel with the PV systems for higher efficiency and system performance. The first crucial consideration relates to the selection of the SCs of the right size, especially regarding the intermittence of solar power. This calls for the adjustment of the capacitance of SCs to the variation in expected solar irradiance and vice versa, a way that they can absorb and release energy without being loaded. For example, Juying *et al.* [66] have shown that the efficiency of photo-SCs depends on the matching between capacitance and charging time depending on changes in the external conditions, including irradiance and temperature. Furthermore, the authors believe that SCs should be connected in parallel with a bidirectional DC–DC converter to enhance the regulation of the energy flow of the PV modules to support the efficient charge–discharge cycle. Figure 7.15 shows the integration of SC with solar energy. The first part illustrates a photo-integrated SC with energy storage electrodes and photo-sensitized dyes. The second part presents a parallel

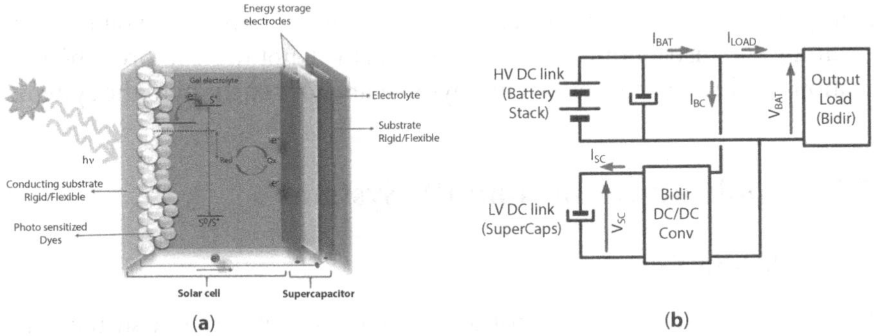

Figure 7.15 (a) Schematic diagram of a photo-integrated SC [67]. (b) Parallel configuration for interfacing SCs with primary energy source (battery stack) [68].

configuration for interfacing the SC with a battery stack, enabling efficient power flow between the sources and the load.

Thermal management is another critical design factor that must be taken into consideration. Due to the high possibility of thermal stress in SCs when charged and discharged at high rates, there is always the need for an efficient cooling system or a passive cooling system. Finally, EMS should be incorporated into power conversion to enable proper distribution of power between SCs and batteries. According to the work of Zineb *et al.* [69], the incorporation of SCs into PV systems provides relief to batteries to increase their lifespan as well as improve the reliability of the overall system.

7.4.4 Efficiency Improvements through Energy Harvesting and Waste Reduction

SCs connected in parallel with PV systems enhance energy-harvesting effectiveness and minimize wasted energy. By integrating SCs, the excess energy that is generated, especially at solar noon, is quickly stored, and hence more of the energy that is generated is captured in the system. For instance, Li *et al.* [70] discussed in their study how the efficiency of power management systems can be increased by 25% using SCs as energy storage devices, proving charging the SC before delivering power to the load. Markus *et al.* [71], in their study, also found that photo-SC systems based on SCs were able to enhance energy conversion efficiency, as the energy from periods of high irradiance was stored and discharged during low solar output conditions, thus eliminating energy wastage. Furthermore, the use of SCs decreases the usage of batteries for short-term energy variations, lowering energy losses connected with cyclic battery charging. Harinkhede and Vaidya [55] also showed that if a battery–SC system is incorporated, the overall efficiency can be boosted by 15%, and losses occurring during peak load periods can be minimized. SCs are useful in controlling voltage and retaining and distributing energy in a way that will not waste it through fluctuations or through conversion in ways that are not beneficial to the system.

7.5 Applications Beyond PV Systems

7.5.1 SCs in EVs

EVs are innovating the transportation sector by delivering a sustainable option to conventional fuel vehicles. Using electricity stored in batteries, EVs serve to lower greenhouse gas emissions, improve energy efficiency,

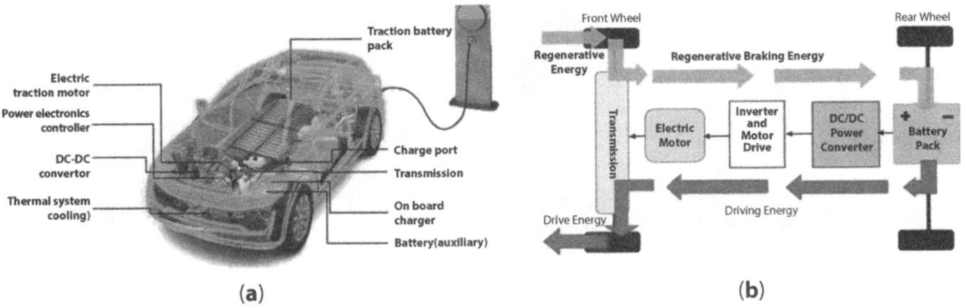

Figure 7.16 (a) General EV with main components [72]. (b) Regenerating and driving energy flows for electrified vehicles [73].

and cut back on the overall carbon footprint. The regenerative braking system stands as one of the remarkable features of EVs, a tech that improves energy efficiency by reconverting kinetic energy, normally wasted during braking, back into electrical energy. As illustrated in Figure 7.16, this captured energy is stored in the vehicle's battery, which boosts driving range and overall efficiency and makes EVs an intelligent option for sustainably mobile travel.

7.5.1.1 Role in Regenerative Braking Systems

A regenerative braking system is an important feature in EVs when using SCs, in which energy is stored that would otherwise be dissipated as heat during the braking process. Their characteristic of charge and discharge cycles makes them suitable for this application because they can quickly capture the high power produced whenever the vehicle is braking and deliver an equal amount of power whenever it is required to boost the acceleration of the vehicle or supply power to other facilities. It also enhances the generality of the vehicle and extends the operational life of traditional batteries because they are minimally subjected to confident charge–discharge cycles.

In the work of Teasdale *et al.* [74], the authors explained the necessity of using SCs in combination with batteries in the schemes of regenerative braking to relieve the battery during the energy humps. Figure 7.17(a) shows the experimental setup of the regenerative braking system used by Teasdale. In their research, they found out that the SCs effectively captured the braking energy, and therefore the battery's lifespan was enhanced. Another clear advantage of SCs is tolerance to charge and discharge cycling power, which makes them suitable to be used in traffic conditions that require frequent energy exchange, otherwise known as the process whereby energy is stored and released in short bursts with very high-power ratings [74].

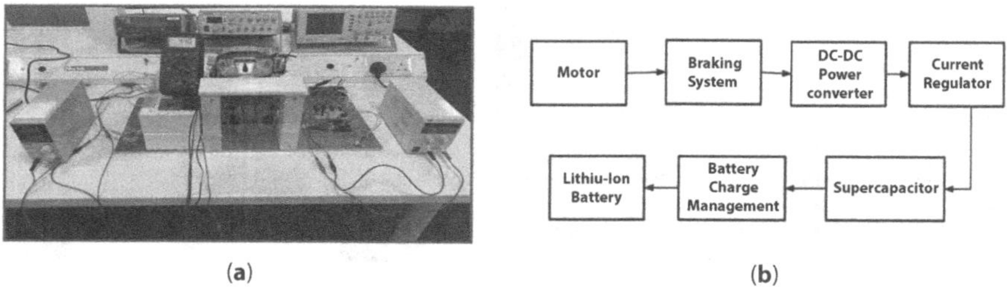

Figure 7.17 (a) Regenerative braking system: construction overview [74]. (b) Block diagram of the overall structure of the system [75].

In another study, Mtengi *et al.* [75] provided the conceptual design of a regenerative braking system, which makes use of SCs as energy reservoir. They enhanced the power flow of the various components of the system, including the SC, battery, and motor, through which the braking energy was stored and reused more efficiently. The SCs stored the high-power energy bursts during braking, thereby minimizing the energy dissipation and increasing the overall energy density of the car by 20%.

In their study, Milan and Dragana [76] examined the hydrogen fuel cell vehicles' regenerative braking that they estimated could not be managed by fuel cells alone but required SCs to store the braking energy. The literature proved that the integration of regenerative braking system through SCs enhanced the power factor of the car by 15–20% enabling the car to capture more energy during the braking mode [76]. Electrification of vehicles has been studied in detail by Mishra and Pattnaik [77], who proposed a hybrid EV model using battery and SC for regenerative brake. According to their simulation results, this hybrid system could increase energy recovery by 25% in the electrical network compared to the systems that were using only batteries. Last of all, Kale *et al.* [78] described the application of SCs in increasing braking effectiveness and decreasing energy dissipation. Their findings also indicated that the use of regenerative braking system using SCs meant for supercharging enhances more braking efficiency and reduces power consumption by 30% at urban environments that require regular brakes.

7.5.1.2 Enhancing Energy Efficiency and Reducing Reliance on Batteries

SCs have attracted much attention in EVs because of the potential of improving power quality and mitigating the use of batteries. This gives the battery system lower pressures and, hence, a longer cycle and overall efficiency of the system. As illustrated in Figure 7.18, another study by Mariasiu and Kelemen

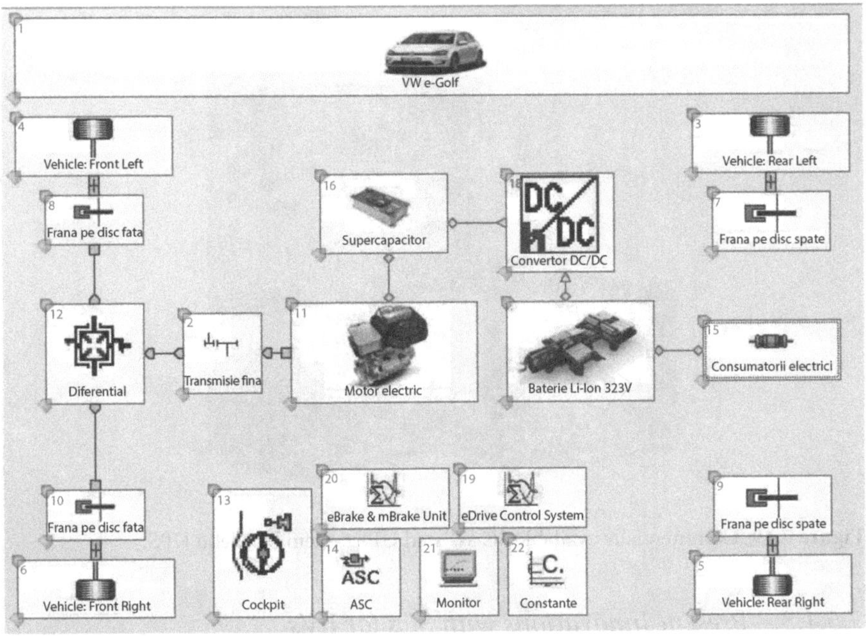

Figure 7.18 Virtual model of e-Golf EV without SC [79].

[79] shows that the integration of a hybrid power storage in the E-Golf is likely to result in decreased energy consumption by 2.95% per 100 km when operating under the Worldwide harmonized Light Vehicles Test Procedure (WLTP). Analyses of the impact of incorporating these devices into vehicle designs also revealed that SCs led to the reduction in the energy spikes that are drawn off the battery and disparate deep discharge cycles that degrade batteries.

Moreover, Bharani *et al.* [80] pointed out that the use of hybrid systems with SC has the benefits in terms of energy where these systems could be charged or discharged at a high rate without a large loss. Substantial benefits of SCs were obtained, and they boosted the efficiency of electric power management systems by up to 20% and reduced the load on batteries. In the same regard, Michael *et al.* [75] established that energy storage in conjunction with battery systems could enhance recharging from regenerative brakes by 25%, cutting on energy loss during brake operations and increasing the overall energy efficiency of EVs. This way, the vehicle was designed to perform better when required while at the same time maintaining battery capacity. As shown in Figure 7.19, commercially available SC-based systems, such as the Riello UPS, demonstrate how SCs are effectively applied in real-world power storage solutions.

Figure 7.19 Commercially available SC assisted UPS system by Riello UPS.

7.5.1.3 Present Innovations with SCs for EVs

New trends in SCs for EVs are concerned with energy density enhancement, miniaturization of the energy storage systems, and cyclic stability, which was achieved by integrating SCs. Such a combination enables energy restitution during regenerative braking and enhances acceleration in EVs [75]. Moreover, combining it with a new generation of materials such as graphene and CNTs in designing the SCs has brought enhancements in the energy density of the device as well as the rates of charging. Mouli [81] covers the incorporation of nanostructured materials such as reduced graphene oxide and enhances the SC energy density by up to 95% for 5000 cycles, making SCs more appropriate for EVs. These materials improve SC functionality by maximizing the conductivity of the tissues along with an increased surface area for energy storage. SCs are also being integrated into higher level power control in EVs for refinement of power output and alleviation of battery stresses under high power demands. This leads into increases in the efficiency of the overall system and longevity of the battery, which is important for the range extension of EVs [82].

7.5.1.4 Future Trends in EV SC Technology

Among future trends for SC technology for EVs, the progression of solid-state SCs, which plan to get rid of liquid electrolyte materials, is important.

Some of these SCs' advantages include a potential for a higher energy density and safety in part due to their ability to be designed into considerably smaller architectures; essentially, it makes them perfect for use in EVs where space is limited. Paravannoor [83] has discussed in one of his studies that the implementation of solid-state SCs can significantly revolutionize the industry through faster charging as well as better integration with the structure of EVs. Another possible future development relates to the use of structural SCs where energy storage is an integral part of the car structure. In their article, Greenhalgh *et al.* [84] consider possibilities of employing body panels of vehicles as energy storage and note that it may bring several optimizations, among which a decrease in the weight of EVs up to 30% and an improvement in their efficiency. These multi-purpose materials serve as the car's structural support and battery component, which has the potential of increasing the driving efficiency of future EVs. Moreover, the incorporation of artificial intelligence (AI) in the EMS will further contribute to the control of power flow between the SCs and batteries in the EVs. AI algorithms will help to make real-time changes so that energy is conserved for better performance and less energy consumption.

7.5.2 SCs in Uninterruptible Power Supply (UPS) Systems

7.5.2.1 Importance of Power Density and Rapid Discharge Capabilities

SCs are now integral parts of the UPS systems because of their very high-power densities and their ability to discharge very quickly. In contrast with batteries that may take a certain amount of time to discharge energy, SCs can discharge large amount of energy in a very short time and are thus well suited in situations where power needs to be delivered immediately and without delay, for example, in situations of power failure or system breakdowns. Since SCs have high-power density, they are capable of charging and supplying short term high-power pulses needed to sustain the operation of critical loads [85].

This technology can potentially provide a power density as high as 10 kW/kg, thus making SC more efficient UPS systems where instantaneous back-up power is required for critical loads [86]. This high-power density allows UPS systems to transmit power with better results and thus enhance the reliability of the UPS system. Moreover, due to their high cycling capability, the SCs can offer extended and self-sustained service to UPS configurations.

7.5.2.2 Use Cases in Critical Power Applications

SCs are being applied in more critical power applications where system reliability is fundamental. Some of these applications are hospitals, data centers, and industrial facilities where a power interruption no matter how small could cause disastrous effects. In such situations, SCs incorporated in UPS systems allow for the best energy storage system due to its capability of delivering power in microseconds whenever there has been a transfer to the backup generator or batteries. As seen in Figure 7.20, SCs are also finding innovative uses in wearable electronics, highlighting their versatility beyond traditional UPS setups. For example, they keep the life-supporting devices running in case of power loss in the hospitals or help to keep the servers operating during the downtime in data centers [87].

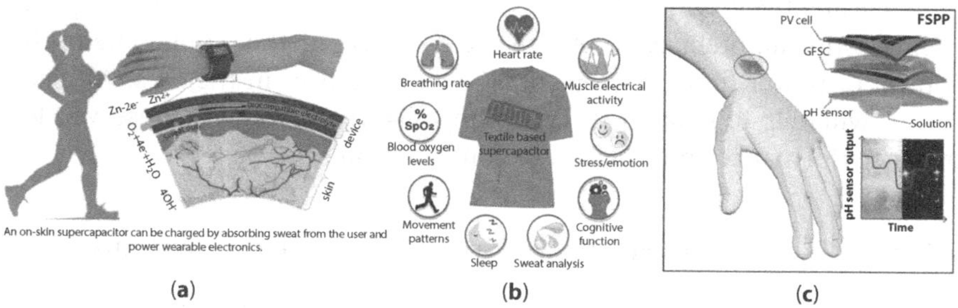

Figure 7.20 (a) An on-skin SC can be charged by absorbing sweat from the user and power wearable electronics [90]. (b) Textile-based flexible SCs for powering up wearable devices to monitor physiological parameters [91]. (c) Flexible self-charging SC based on graphene–Ag–3D graphene foam electrodes [92].

In addition, Mouli [81] shows that UPS systems that incorporated SC as the storage device have 50% of the power consumption of battery-only UPS systems in industrial settings. The decrease in energy consumption, combined with a high frequency of response of SCs, makes it possible to maintain the power supply of critical facilities without interruptions. The combination of SCs with conventional battery-backup UPS systems has also been demonstrated to increase the battery lifespan by taking some of the high-power loads as well as preventing frequent deep discharge cycles.

7.5.3 SCs in Internet of Things (IoT) Devices

7.5.3.1 Power Management for Smart Sensors and Wearable Devices

Power management is one of the significant challenges facing IoT devices, specifically smart sensors and wearable technologies, because the sources of energy are periodically available, and the devices are expected to have a long lifetime. SCs are quite advantageous in handling power for such applications primarily due to their high-power density, short term charge and discharge times, and long lifespan. SCs can effectively harvest energy from several sources including solar panels, vibrations, or human body movements and then release a boost of energy to the sensors or devices when required in short duration intervals. This means that smart sensors and wearables can work for several days or weeks without charging the battery or replacing it, especially for sites that are out of reach. For example, Sarkar *et al.* [88] proved that SCs that shunt with piezoelectric materials can capture and store energy from mechanical motion, therefore providing energy for wearable devices with minimal external charging. Some of the existing types of wearable energy storage systems were discussed by Ganeshbabu and Selvan [89] in which how SCs could supply the energy bursts associated with smart textiles and other health-monitoring wearables was presented. SCs in such devices help to provide a consistent power supply to support constant monitoring as well as data transfer in healthcare and fitness usages.

Likewise, the high cycle life of the SCs also means less frequency of replacement, hence improving the sustainability of the IoT systems. In million cycle tests, Zhu *et al.* [93] also proved that SCs have long cycle life and can work for a long time without much capacity fading, which make them suitable in smart sensor and wearable technology devices where battery maintenance is not feasible.

Moreover, SCs are ideal to be used in IoT system's main power management because energy is stored and used at high rates. In the power supply of IoT systems under PV energy, SCs as mentioned by Mouli [81] enable fast charging, hence ensuring that the energy generated in the few hours of sunlight is not wasted. Finally, in smart sensors, SCs are also used in conjunction with batteries, where batteries are loaded for long-term energy storage, and SCs are loaded for high-power delivery. Naeem *et al.* [94] revealed that the integration of the above hybrid systems enhances power efficiency while minimizing the size and cost of power systems for IoT devices.

7.5.3.2 Advantages of Fast Charging and Long Cycle Life in IoT Applications

SCs bring a lot of benefits to IoT devices because of the fast-charging ability and high cycle durability of the million times charge–discharge. They are well suited for operating smart sensors, wearable devices, and any other IoT systems that demand uninterrupted power supply with low upkeep. A study done by Wang *et al.* [95] on the advanced charge circuit for indoor PV energy was demonstrated as enhancing the capability of SCs in decreasing the activation time for IoT devices from 19.8 hours to less than 2 seconds. This fast-charging capability is important in that IoT devices like smart sensors should always be ready to work especially where energy is available in intermittent supply. That makes it suitable in energy-harvesting systems because of the ultrafast energy storage and release ability of SCs. The other benefit is the long life cycle of SCs. Asl *et al.* [96] have explained that solid-state SCs are long-lasting devices that can carry millions of charge and discharge cycles with high efficiency that makes them more suitable to connect IoT devices in remote areas.

Abraham *et al.* [97] state that the fast charge nature of SCs enables them to charge IoT devices efficiently, adding to the energy storage aspect of the batteries, which makes IoT devices operate optimally even in a high-power consumption scenario where the battery degrades much faster than expected. Literature also revealed that utilizing SCs in parallel with Li-ion storage for charging, devices have the benefits of quick charging. SCs could provide better energy density and very high-power densities that are essential in ensuring a continuous energy feed in the faster-responding IoT devices. Moreover, Abraham *et al.* [97], in their study on the SCs for energy storage devices, pointed out that SCs have long life of as much as 100% capacitance right after 10000 charge–discharge cycles and therefore would efficiently support IoT devices that require reliable energy in the long run.

7.5.3.3 Integration Challenges and Potential Solutions

Integrating SCs into IoT devices faces a range of difficulties, mainly emanating from the requirement to balance size, energy density, and how well they work with existing technologies. A significant difficulty is the size restriction since IoT devices generally need to be small and light, limiting the area for energy storage options. Even though SCs are efficient in power density, they have historically had a lower energy density, holding less energy per unit volume in comparison to batteries. This feature may act as a barrier in processes that need prolonged operation without needing to recharge. Addressing this need, modern developments in material science have concentrated on inventing new electrode materials, such as graphene and CNTs, that improve the energy density of SCs without increasing their dimensions [98].

Still another integration difficulty is in the compatibility regarding pre-existing circuit designs. Because of the efficiency in charge and discharge cycles in SCs, they require power management systems, especially when paired with battery energy storage systems. Emerging as a viable answer is the combination of batteries' high energy density with the fast charge/discharge cycles of SCs in the form of HESS. These systems can use the strengths of both technologies, delivering a compromise between energy density and power capability. In addition, sophisticated energy management circuits are under development to affirm the stable operation of hybrid systems, which allows for efficient energy usage in IoT devices [99].

The cost is an important element too. Although the development of SC technology is accelerating, the costs of materials and manufacturing processes remain high. Investigations are underway into cost-effective solutions such as bio-based materials or more straightforward manufacturing processes that allow for reductions in costs, all while supporting performance quality. This initiative intends to make SCs easier to access for widespread deployment in IoT situations, where the cost factor is vital [100].

7.6 Challenges and Limitations

7.6.1 Technical Challenges in SC Integration

The addition of SCs to a variety of systems, mainly in renewable energy and storage technologies, creates multiple technical challenges. These problems consist of physical and electrical integration, energy management, and problems concerning thermal regulation and cost. In PV systems, to deal with energy intermittency, for example, one needs solutions for charge control, capacitor matching, and handling thermal effects, as discussed by Ibrahim *et al.* [101]. HESS combining SCs and batteries can enhance energy delivery and battery lifespan, but issues such as voltage fluctuations and stability present significant hurdles, as highlighted by Abdelhalim *et al.* [102]. In the field of wearable and implantable electronics, SCs must be both flexible and multifunctional to match the changing energy demands of these devices, creating difficulties in the development of materials that both perform well and are physically flexible [103]. The use of SCs in conjunction with renewable energy technologies like solar cells encounters further challenges, including bulkiness and high costs of manufacture, which discourage their applicability for major deployments [104]. Addressing these technical barriers is fundamental for the broad adoption of SCs in energy systems, from individual equipment up to renewable energy infrastructure.

7.6.1.1 Issues Related to Scalability and Cost

Integrating SCs into multiple energy storage systems encounters the important challenges of scalability and cost. The key problem is to balance the advancement of energy storage capacity alongside cost-effective and scalable manufacturing approaches. Jiang *et al.* [105] discuss how the improvement of both the voltage output and capacitance in fabric-based SCs can be realized through original electrode designs and screen-printing techniques that offer both scalability and flexibility. On the other hand, reaching this scalability with affordable costs requires further development of materials and processes. Likewise, Sundriyal and Bhattacharya [106] indicate that even though inkjet printing is a hopeful technology for making flexible, solid-state SCs, there are challenges in the scaling of this process for production at a large volume and a low cost. Issues with the cost of materials such as graphene oxide along with the challenges of device design, notably in reaching high-performance metrics like energy density and flexibility, still restrict extensive use. Additionally, Yoon [93] points out that the high costs of advanced wafer-level packaging used in miniaturized products continue to be a challenge not withstanding technological progress because this process requires substantial capital investment for production at large.

7.6.1.2 Integration Challenges with Existing PV Infrastructure

There are many difficulties integrating SCs to existing PV systems, largely due to both electrical and physical integration issues. SCs' high-power density makes them perfect for storing solar energy that is intermittent; however, using them in PV systems, especially at the module or cell stage, brings about challenges in charge control, capacitor matching, and thermal management. The work of Ibrahim *et al.* [101] points out that SCs can indeed reduce the variations seen in PV output, but this integration requires sophisticated power electronics. Also, the incorporation of SCs, especially graphene varieties, introduces apprehensions about dimensioning, thermal management, and the overall lifetime of the system when it faces changing outdoor conditions. Rosier *et al.* [107] put forward the concept of integrating ramp-rate control with hybrid SC–battery systems into grid-tied PV systems. This method improves efficiency and also leads to a lower number of power converters, which can make system design simpler, but could lead to new technical hurdles related to controlling power fluctuations and maintaining enduring stability. Nordin *et al.* [108] additionally examine how systems integrating solar cells with SCs may present solutions for small and wearable devices, yet scaling this approach for extensive PV infrastructure is an important technical obstacle.

7.6.2 Economic Considerations

7.6.2.1 Cost–Benefit Analysis of SC Integration

The advantages of SCs include ability to deliver rapid charge/discharge cycles and handle short-term power transients in systems with varying energy sources likes solar and wind. For example, Guven *et al.* [109] state that incorporating SCs with HESS has been found to improve the general cost of the system; the annual system cost drops to $603,537.85 with competitive Levelized Cost of Electricity (LCOE) of $0.238 per kWh. The initial costs of SCs exceed those of conventional energy storage systems, as noted by Cabrane *et al.* [110] in their study. This includes expenses associated with large-scale production of SCs and the installation of control systems to address the disparity between the high-power output and low-energy density of SCs compared to batteries.

Innovative designs like the direct integration schemes presented by Vilathgamuwa and Jayasinghe [111] also point in bringing down the costs of interfacing the converters as low as possible. Although the economic advantages of SC integration are evident for system robustness and lifetime, the substantial initial costs provide a significant obstacle, requiring additional progress in manufacturing and cost-reduction methodologies.

7.6.2.2 Potential for Cost Reduction through Technological Advancements

Improvements in technology have played an important part in considerably lowering the costs associated with SCs, mainly *via* sustainable material developments, hybrid systems, and better electric and physical designs. The use of nanocellulose-based materials appears prominently in the literature, as shown by Choudhary *et al.* [112]. Extracted from renewable sources, nanocellulose supplies a financially viable and eco-friendly choice for SC electrodes. As a result of its exceptional structural qualities, nanocellulose provides superior electrochemical performance with a large surface area and excellent conductivity, making it a top material for large energy storage applications. This sustainable method helps to bring production costs down while also being in harmony with worldwide initiatives to reduce environmental pollution.

Most recently, work has been carried out on increasing the operating temperature range of SCs, which is essential for greater efficiency and lower service costs in harsher environmental conditions. Jiang *et al.* [113] highlight how optimization of electrolytes and electrode designs can broaden the operational temperature window in which these SCs function. With robust SCs that operate within broad temperature ranges, industries can decrease repair and replacement costs, which, in return, enhances the cost efficiency of energy storage solutions.

Also, developments in SC design created for targeted uses, particularly light transportation systems, are helping to reduce costs. Iannuzzi and Lauria [114] suggest a design that is deliberately designed to better manage peak demand in the light transportation sector, including electric buses and trams. SCs aid in boosting the energy efficiency of these systems, thereby helping to decrease operational expenditures and attracting transportation industries seeking to diminish fuel consumption and maintenance charges.

Finally, these technological advancements in green materials and hybrid systems, temperature optimization, and application-specific designs are crucial in bringing down total costs for manufacturing and implementation. By prioritizing the environment, boosting energy density, and increasing operational efficiency, these innovations present SCs as a more viable option for everything from massive-scale energy storage to everyday transport systems.

7.6.3 Environmental Impact and Sustainability Concerns

7.6.3.1 Lifecycle Analysis of SCs

Exploring the environmental perception of SC lifecycle assessment identifies and examines how these SC devices can affect the sustainability of energy storage systems from extraction to end-of-life disposal. One of the most important research topics is the impact of SCs on the battery systems of EVs, enhancing the battery cycle life, power density, and overall vehicle performance. Vukajlovic *et al.* [115] show how the use of SC improves performance in conjunction with Li-ion batteries in EVs and minimizes environmental implications by increasing battery lifetime. This leads to low battery replacement and the utilization of new batteries, and hence a small impact is left on the environment.

Also, the degradation behavior of SCs is an important concern with life cycle assessment since its performance over time influences its environmental and economic impact. De Bernardinis *et al.* [116] describe the influence of the cycling behavior and ageing processes on the storage components and particularly on applications in transport. It is important to know these factors if the lifespan of SCs will be sustained to be useful in long-term renewable energy storage systems. One of the studies by Kowal *et al.* [117] looked at the problem of self-discharge in SCs and how it decreases the efficiency of the product and causes energy losses over time. Their effort in this area of research on the reduction of self-discharge resulting from design enhancements is useful for lengthening the life of SCs as well as minimizing the damage that they cause to the environment.

7.6.3.2 Disposal and Recycling Challenges

Environmental risks from improper disposal and complex material waste are the most problematic issues in SC disposal and recycling. One of the emerging solutions is transforming plastic waste into carbon nanomaterials for SC electrodes, which provides a dual benefit of waste management and creating sustainable energy storage solutions. Silva *et al.* [118] discuss how carbon nanostructures produced from the pyrolysis of plastic waste can be used to replace materials manufacturing SCs. It is a promising solution to the increasing problem of plastic waste and its recycling in the framework of a circular economy *via* SCs.

A novel application of SCs involves reusing the activated carbons from depleted SCs to remove harmful ions, including silver and chromium, from water. Wu *et al.* [119] demonstrate that SC carbon derived from recycled electrode materials can serve as effective chemical adsorbents for water treatment, hence enhancing the utility of waste beyond its initial life cycle. This not only resolves the challenge of successful disposal of SCs but also enhances water purification, positioning it as one of the most effective recycling methods.

7.6.3.3 Environmental Benefits Compared to Alternative Storage Solutions

Due to their long cycle life, fast charging and discharging rates, and built composition, they have better environmental properties than other energy storage solutions, such as batteries. Unlike batteries that contain harmful heavy metals, SCs use carbon-based electrodes and other sustainable materials to increase sustainability as the decades roll by. The study by Burcea *et al.* [120] concludes that using freely available, affordable materials, such as activated charcoal in SCs, can lower environmental impact and the demand for harmful specialized materials.

A study by El Maati *et al.* [121] states that SCs, notably when merged with sustainable nanomaterials such as Sm_2O_3/rGO composites, present clean energy storage possibilities that involve less resource consumption than conventional batteries.

Moreover, Kumar *et al.* [122] indicate that incorporating SCs into renewable energy systems induces an improvement in both efficiency and sustainability for managing the fluctuations in solar and wind energy. The capability of SCs to stabilize power output means they are an important asset in the reduction of reliance on less sustainable storage systems, including lithium-ion batteries, that feature shorter lifespans and greater disposal considerations.

7.7 Future Prospects and Technological Advancements

7.7.1 Innovations in SC Technology

7.7.1.1 Emerging Materials and Fabrication Techniques

With the advent of new materials and fabrication techniques, SCs have undergone a great evolution. Carbon-based nanomaterials like graphene and CNTs are compelling candidates because they combine high electrical conductivity, large surface area, and excellent charge storage properties. These properties make energy storage easier, and so they directly affect the SC performance as well. Notably, carbon nanomaterials are being used increasingly in miniaturized energy storage device design and development, as recent trends have demonstrated. These devices are compact in size and hence can be an excellent application for portable and wearable technologies, which presents itself as a trendy go-to modifier for energy storage [123]. The transition metal sulfides (TMSs), such as binary composites, have been extensively studied for their good electrochemical performance. There is a great promise in these materials for high energy and power densities, which are essential for enhancing the overall performance of SCs [124]. Recently, less expensive and environmentally more sustainable zinc-based materials have triggered a notable part of interest since they are stable in water. These composites enhance charge–discharge efficiency, increase energy density, and provide better overall functionality for SCs [125]. Furthermore, progress in the production of flexible SCs constructed from textiles, increasingly using covalent organic frameworks (COFs), metal-organic frameworks (MOFs), and a new class of graphene MXenes, has opened important avenues in the field of wearable technology.

Table 7.4 Summary of emerging materials and fabrication techniques.

Material/technique	Key benefits	Application focus
Carbon-based nanomaterials (graphene, CNTs)	Excellent conductivity, large surface area, efficient charge storage	Miniaturized energy storage for portable/ wearable tech
Transition metal sulfides (TMSs, binary composites)	High energy and power densities, impressive electrochemical performance	Hybrid SCs
Zinc-based materials	Cost-effective, environmentally friendly, stable	Energy storage systems
Composite materials	Enhanced charge–discharge efficiency, increased energy density	SCs with improved overall performance
Flexible, textile-based SCs (COFs, MOFs, MXenes)	Integration into wearable devices, flexibility, superior energy storage	Wearable devices

These materials make it possible for SCs to seamlessly integrate into wearable devices, delivering flexibility along with improved energy storage capacity [126]. Table 7.4 provides a summary of the emerging materials and fabrication techniques discussed in this section.

7.7.1.2 Advances in Energy Density and Charge–Discharge Efficiency

Current improvements in SC technology are enhancing energy density and improving charge–discharge efficiency. An important innovation is the fabrication of hybrid nanocomposites, including copper ferrite/carbon sphere ($CuFe_2O_4$/C-sphere NC), which yield outstanding energy densities up to 40.9 Wh/kg and maximum power densities of 11,248 W/kg while ensuring great cyclic stability throughout 5,000 cycles [127]. The next creative method centers on potassium-ion hybrid capacitors (PIHCs). These capacitors greatly boost energy storage by implementing heterojunction interfaces with better ion adsorption and electron transfer, dealing with the common low-energy-density difficulties encountered in major applications [128]. Also, the application of green synthesis approaches, including the use of Albizia amara (*Mimosoideae Boivin*) plant extracts to create $NdMn_2O_5$ nanocomposites, has accomplished energy densities of 42 Wh/kg and extraordinary cyclic stability, encouraging environmentally safe and affordable solutions for powerful SCs [129]. The development of jute-derived activated carbon composites, with sulfur doping, has shown marked improvements in specific capacitance and energy density, reaching 32 Wh/kg at a power density of 500 W/kg, giving bio-based materials a notable place in SC technology [130]. In conclusion, existing research on versatile and elevated-temperature SCs has rolled out novel electrode materials that sustain exceptional energy and power densities, which improves versatility for industrial purposes in future energy storage systems.

7.7.2 Potential Developments in PV Systems and Energy Storage

7.7.2.1 Integration with Other Renewable Energy Sources

Adopting a system that combines PV technology with wind, biomass, and energy storage capabilities such as SCs or batteries offers a fundamental way to meet the inconsistency of renewable energy and assure a continuous power supply. The combination of wind and PV energy works well because wind power is generally more plentiful at night and in times of diminished sunlight, in contrast to PV systems, which produce their greatest energy during the day. Hybridization leads to a smoothing of energy production and results in a more even energy delivery. The combination of Wind Energy Conversion Systems with PV systems controlled by advanced mechanisms, such as genetic algorithm supported artificial neural networks, increases grid stability and manages both active and reactive power flows. This system design improves performance throughout the entire system and offers improved compensation for the variability in energy output [131].

As well as the combination of PV and wind, energy storage technologies such as SCs and batteries are also important for the effective management of the fluctuating output from renewable energy sources. As a result of their potent claim in power density and fast charge–discharge cycles, SCs assist greatly in maintaining power systems by delivering brief energy surges during peak loads or when voltages fall. When in combination with PV systems, SCs provide the ability to save excess energy produced at solar peak times for later use during periods of reduced sunlight. Furthermore, when it is integrated with wind power, it will do the same, resulting in an enhanced overall efficiency of the energy system. The fast response to fluctuations makes SCs the ideal storage alternative for systems that need immediate power equilibrium [132].

On the other hand, batteries, particularly lithium-ion batteries, excel in longer-duration energy storage systems. When SCs are combined within AC/DC grids, energy storage systems attached to bidirectional converters facilitate seamless energy flow from PV systems to the grid supporting voltage regulation and maintaining the stability of power. A multi-port autonomous reconfigurable solar PV system (MARS) that links PV systems to battery storage has demonstrated a dramatic enhancement of grid stability, seeking to assure consistent energy supply even in unpredictable weather conditions [133].

In urban scenarios, combining decentralized PV systems with either SCs or battery storage has allowed cities to lessen their dependence on centralized power grids. This design contributes to a reduction in greenhouse gas emissions and boosts energy efficiency by turning rooftops into spaces for power generation. These combined systems are reasonably priced, require very little upkeep, and supply a reliable means for tackling energy consumption peaks. These types of situations greatly gain from SCs that efficiently adapt to sudden changes in power demand. In contrast, batteries provide sustained energy storage and contribute to a balanced strategy for the management of energy in urban settings [134]. Also, the unification of PV systems with renewable energy technologies, including geothermal or biomass, in conjunction with energy storage solutions, supports the establishment of a more stable energy supply. Being advantaged by the reliable base-load quality of biomass and geothermal energy, hybrid systems offer a stable power supply, which PV systems supplement with extra energy throughout the day. Combining SCs with batteries enhances these systems by effectively capturing energy and maintaining it for use when required, which enables a smooth supply of power considering variations in renewable energy supply.

The Grey Wolf Optimized Adaptive Neuro-Fuzzy Inference System is an important part of managing the energy flow between PV systems, other renewable resources, and energy storage systems. Supportive algorithms enable the stabilization of voltage levels, the effective distribution of energy, and increased dynamic response for integrated renewable energy systems, maximizing the use of SCs and batteries [127].

Ultimately, incorporating PV systems with other renewable energy generations and energy storage technologies gives a complete answer to the difficulties considered by renewable energy variability. The systems not only provide improved energy efficiency but also enhance the reliability,

Table 7.5 Summary of advances in energy density and charge–discharge efficiency.

Advancement	Key achievements
Copper ferrite/carbon sphere	Energy density up to 40.9 Wh/kg, power density of 11,248 W/kg, cyclic stability over 5000 cycles
Potassium-ion hybrid capacitors (PIHCs)	Improved energy storage, enhanced ion adsorption and electron transfer
Nanocomposites ($NdMn_2O_5$)	Energy density of 42 Wh/kg, environmentally friendly and cost-effective
Jute-derived activated carbon composites with sulfur doping	Specific capacitance improvements, energy density of 32 Wh/kg at 500 W/kg power density
Flexible, high-temperature-resistant SCs (novel electrode materials)	Maintains high energy and power densities, adaptable for industrial applications

sustainability, and resilience of the energy future we are headed toward. Table 7.5 presents a summary of recent advances in energy density and charge–discharge efficiency.

7.7.2.2 Future Trends in Distributed Energy Storage

Distributed energy storage (DES) is about to change dramatically due to technological progress in materials, energy storage systems, and their pairing with renewable energy sources. The trend of utilizing advanced materials, especially MXenes and nanocomposites, displays great promise in improving the capability of SCs, facilitating higher energy densities and quicker charge–discharge cycles. Partially oxidized MXenes deliver better electrochemical performance, which puts them in a crucial position for future energy storage methods [122].

In addition, the assimilation of renewable energy systems with DES is showing an increased degree of sophistication. These systems achieve energy demand fulfilment while decreasing environmental impacts, thanks to the integration of uncertainty-based techno-economic assessments and life-cycle analyses for optimization. With zero and net-zero energy buildings of the future, this technique is important for having reliable and flexible energy storage systems to address varying energy demands [135]. These inventions are propelling DES ahead, making it both cheaper and more scalable. Also, their ability to be biocompatible permits merging with wearable and biomedical devices, which enlarges their application.

As the technology for 5G and the IoT progresses, it will affect the role of DES. Advanced energy storage technologies working alongside 5G communication networks will enable smart grids to carry out real-time performance monitoring and tuning of their energy storage solutions. This interaction allows improved management of distributed energy resources, reducing carbon emissions and increasing the durability of energy infrastructure. The merger of solid-state batteries and advanced flow batteries with 5G connectivity creates a scalable solution to meet future energy challenges in both microgrids and decentralized energy networks [136]. Innovations in the technologies of solid-state and flow batteries are significantly hopeful, offering greater energy densities, better longevity, and improved safety. As they evolve, these technologies will probably turn into core elements of DES systems, particularly in large grid lighting systems [134]. In addition to increasing SC efficiency, these techniques and materials also support more sustainable energy storage systems that are important for distributed grids.

7.7.3 The Role of SCs in Future Energy Systems

7.7.3.1 Potential for SCs in Grid-Level Energy Storage

Due to their fast charge–discharge capabilities and considerable operational longevity, SCs show considerable promise for grid-level energy storage. Unlike common batteries, SCs' skill in swiftly delivering and absorbing large amounts of power at once makes them suitable for roles such as frequency regulation and voltage stabilization. As an example, hybrid SCs, including those designed using copper ferrite/carbon nanocomposites, have shown high-energy densities (till 40.9 Wh/kg) and outstanding cyclic stability, thus making them perfect for grid applications requiring prompt energy buffering [127].

In addition, SCs play an important part in supporting renewable energy sources, particularly solar and wind, which are unpredictable by nature. The rapid, transient power handling capability of SCs ensures smoother operation of grids that are largely powered by renewable energy, improving grid reliability and lessening the requirement for fossil fuel backup systems. The latest improvements in nanocomposites, particularly $ZnFe_2O_4$/$NiFe_2O_4$/CeO_2, have improved SC performance and made them more efficient for grid-level energy storage applications [137]. It is particularly effective for managing short-term power fluctuations and stabilizing electrical grids.

7.7.3.2 Contribution to Smart Grids and Microgrids

SCs have become an important component of smart grids and microgrids related to the stability and efficiency of the grid. The dynamic energy management and real-time monitoring of smart grids depend heavily on the crucial function of SCs in coping with fluctuations in power supply and demand. They stand out in applications aimed at rapid energy delivery, including frequency regulation and peak shaving, thus amplifying the dependability of energy systems that feature renewable energy sources, such as solar and wind. The rapid charge–discharge performance of SCs allows them to lessen sudden events, permitting the grid to respond to fluctuations without compromising stability.

In microgrids that add renewable energy, SCs contribute to low-inertia systems by mitigating the common instability found in such systems. The typical approach is to combine SCs with batteries in a HESS that uses SCs to stabilize short-term energy fluctuations and has batteries supplying extended energy storage. Not only does this hybrid method better the microgrid's response speed under power surges, but it also improves the battery's lifespan by lowering its operating stress during transient events [138]. Research shows, for example, that adding SC banks connected to DC microgrids provides stability to voltage and helps protect critical parts from current surges, leading to a more resilient and efficient system [57].

In addition, SCs have served a role in hybrid AC/DC microgrids by supporting virtual inertia control and improving grid stability during transitions between grid-connected and isolated modes. This method is notably valuable in making certain that both AC and DC sub-grids remain stable regardless of shifting power conditions [139].

7.8 Conclusion

The integration of SCs into PV systems represents a transformative step toward achieving sustainable energy solutions. By addressing the inherent challenges of fluctuating solar irradiance and dynamic energy demands, SCs offer a robust mechanism for stabilizing energy output, enhancing power quality, and ensuring reliability. Their high-power density, rapid charge–discharge capabilities, and long cycle life make them indispensable for modern energy storage needs, effectively complementing traditional batteries.

This chapter investigated the utilization of SCs in alleviating instability in solar systems, optimizing short-term energy storage needs, and improving overall system efficiency. Advanced methodologies like SCALDO, SCALOM, and innovative MPPT methods demonstrate that additional performance optimization continues to be achieved in PV systems. SCs contribute to energy efficiency in various applications, including EVs, UPS systems, and IoT devices.

In the near future, SC technology will require significant development, the discovery of novel materials, and cost-effective designs for broader applications. The future advancement of renewable energy systems is promising due to their potential use in grid-scale energy storage, smart grids, and distributed energy systems. This chapter underscores the essential role of SCs in facilitating the shift toward sustainable energy systems, hence fostering a more robust and environmentally sustainable energy future.

Acknowledgments

The authors would like to acknowledge the financial assistance provided by the University of Kelaniya, Sri Lanka, under Grant No. RP/03/02/05/01/2023 for project titled "Development of Supercapacitor – Battery Hybrid PV System for Efficient Solar Energy Harnessing" and Grant No. RP/03/02/05/02/2023 for project titled "Design and Development of Supercapacitor Assisted DC Microgrid for Solar PV-Based LED Lighting System for the Long Run of SMEs".

References

1. National Oceanic and Atmospheric Administration (NOAA), *NOAA Global Monitoring Laboratory - Atmospheric CO2 Levels*, Accessed: Aug. 26, 2024. [Online]. Available: https://gml.noaa.gov/ccgg/trends/.
2. Smil, Energy Institute - Statistical Review of World Energy 2024, in: *Our World in Data*.
3. Ritchie, H. and Roser, M., Our World in Data - Global Primary Energy Consumption, in: *Energy. Our World in Data*, Accessed: Aug. 26, 2024. [Online]. Available: https://ourworldindata.org/energy.
4. United Nations Framework Convention on Climate Change (UNFCCC), *The Paris Agreement. UNFCCC*, Accessed: Aug. 26, 2024. [Online]. Available: https://unfccc.int/process-and-meetings/the-paris-agreement.
5. The Climate Group & CDP. RE100: *Companies Committed to 100% Renewable Power*. Retrieved from https://www.there100.org.
6. World Energy Council. (2013). World energy resources: Solar 2013. Retrieved from https://www.worldenergy.org.

7. Dobrzański, L.A., Szczęsna, M., Szindler, M., Drygała, A., Electrical properties mono- and polycrystalline silicon solar cells. *J. Achiev. Mater. Manuf. Eng.*, 59, 2, 67–74, 2013.
8. Saxena, A., Brown, C., Arneth, A., Rounsevell, M., Modelling the global photovoltaic potential on land and its sensitivity to climate change. *Environ. Res. Lett.*, 18, 10, 104017, Oct. 2023. doi: 10.1088/1748-9326/acf86f.
9. Karabacak, K., Towards sustainable solar energy solutions: Harnessing supercapacitors in PV systems. *Energy Storage Convers.*, 2, 1, 436, Apr. 2024. doi: 10.59400/esc.v2i1.436.
10. Lebedeva, K., Borodinecs, A., Odineca, T., The potential of multiapartment rooftop PV systems as citizen' energy communities in Latvia. *Front. Energy Res.*, 11, Dec. 2023. doi: 10.3389/fenrg.2023.1297297.
11. Ruan, T., Topel, M., Wang, W., Laumert, B., Potential of grid-connected decentralized rooftop PV systems in Sweden. *Heliyon*, 9, e16871, 1–12 https://doi.org/10.1016/j.heliyon.2023.e16871.
12. Peerlings, J., Reinders, A., Catita, C., Brito, M.C., The photovoltaic potential for electric vehicle charging along highways: A Dutch case study. *Prog. Photovoltaics Res. Appl.*, 32, 4, 244–252, Apr. 2024. doi: 10.1002/pip.3759.
13. Alami, A.H., *et al.*, Management of potential challenges of PV technology proliferation. *Sustain. Energy Technol. Assess.*, 51, 101942, Jun. 2022. doi: 10.1016/j.seta.2021.101942.
14. R., S., *et al.*, Solar Irradiance Nowcasting System Trial and Evaluation for Islanded Microgrid Control Purposes. *Energies (Basel)*, 15, 17, 6100, Aug. 2022. doi: 10.3390/en15176100.
15. Yao, Y., Ertugrul, N., Kani, A.P., Investigation of Short-Term Intermittency in Solar Irradiance and Its Impacts on PV Converter Systems, in: *2022 32nd Australasian Universities Power Engineering Conference (AUPEC)*, IEEE, pp. 1–6, Sep. 2022, doi: 10.1109/AUPEC58309.2022.10215771.
16. Tsafarakis, O., *Chasing The Sun: A Journey in Monitoring Solar Panels – Data Science Approaches for Analysis of PV Performance*, Doctoral dissertation, Utrecht University, Utrecht, the Netherlands, 2023. https://doi.org/10.33540/1855.
17. Tsoutsos, T., Savvakis, N., Kanakis, I., Papadakis, G., Kyriakarakos, G., Schwartz, B., Temperature Effect on PV Performance: Experimental Results from a 2.18 kWp Thin Film PV System, in: *26th European Photovoltaic Solar Energy Conference and Exhibition*. Hamburg, Germany, 2011, September.
18. Smara, Z. and Aissat, A., Investigation of the Relationship between Shaded Cell Temperature and the Operating Point of PV Systems. *Environ. Res. Eng. Manage.*, 80, 2, 133–143, Jul. 2024. doi: 10.5755/j01.erem.80.2.35188.
19. Arezki, S. and Boudour, M., Solutions to the instability problem of the DC input voltages of neutral point clamping multilevel inverter in photovoltaic chain connected to the network, in: *2nd International Symposium on Environment Friendly Energies and Applications*, EFEA 2012, pp. 13–18, 2012, doi: 10.1109/EFEA.2012.6294088.

20. Arunan, R., Janardhanan, V., Sundaram, N.M., Kumar, M.M., Improving the Efficiency of Photovoltaic Systems through Hybrid Energy Storage Management for DC Applications. *IOP Conf. Ser. Earth Environ. Sci.*, 1375, 1, 012014, Jul. 2024. doi: 10.1088/1755-1315/1375/1/012014.
21. Bukhari, S.M.A.S., Maqsood, J., Baig, M.Q., Ashraf, S., Khan, T.A., Comparison of Characteristics-Lead Acid, Nickel Based, Lead Crystal and Lithium Based Batteries, in: *Proceedings - UKSim-AMSS 17th International Conference on Computer Modelling and Simulation, UKSim 2015*, pp. 444–450, Institute of Electrical and Electronics Engineers Inc, Sep. 2016, . doi: 10.1109/UKSim.2015.69.
22. Xu, Q., Comparison and optimization of different types of lithium batteries for new energy vehicles, in: *Proceedings of the Eighth International Conference on Energy System, Electricity, and Power (ESEP 2023)*, Y. Zhang, M. Yao, Z. Xu (Eds.), vol. 13159, Paper 131591X. SPIE – The International Society for Optics and Photonics, 2024, May. https://doi.org/10.1117/12.3024451.
23. Ameta, R.K., Mehetre, S.S., Inwati, G.K., Behere, S.S., General introduction about electrochemistry and supercapacitors, in: *Smart Supercapacitors*, pp. 3–16, Elsevier, 2023, doi: 10.1016/B978-0-323-90530-5.00020-4.
24. Ibanez, F. and Ibanez, F., Hybrid Energy Storage Systems in Electric Vehicle Applications, in: *Electric Vehicles - Design, Modelling and Simulation*, Sep. 2023, doi: 10.5772/INTECHOPEN.113000.
25. Xu, K., Guo, Y., Lei, G., Zhu, J., A Review of Flywheel Energy Storage System Technologies. *Energies*, 16, 6462, Sep. 2023. doi: 10.3390/EN16186462.
26. Amiryar, M.E. and Pullen, K.R., Analysis of standby losses and charging cycles in flywheel energy storage systems. *Energies*, 13, 17, 4441, 1–22, https://doi.org/10.3390/en13174441.
27. Ferrell, M., *How Compressed Air Batteries are FINALLY Here.* Undecided with Matt Ferrell, 2024, August 27 Retrieved from https://undecidedmf.com/how-compressed-air-batteries-are-finally-here/.
28. Koritarov, V., Kwon, J., Ploussard, Q., Balducci, P., *A review of technology innovations for pumped storage hydropower*, HydroWIRES Technical Report, Argonne National Laboratory, U.S. Department of Energy, 2022, April. https://doi.org/10.2172/1867238.
29. Cabeza, L.F., Martorell, I., Miró, L., Fernández, A.I., Barreneche, C., Introduction to thermal energy storage (TES) systems, in: *Advances in Thermal Energy Storage Systems: Methods and Applications*, pp. 1–28, Elsevier Inc, 2015, doi: 10.1533/9781782420965.1.
30. Viadero Jr., R., Sinkler, R., Wine, K., *Western IllInoIs UnIversIty-QUad CItIes 3300 rIv er drIv e MolIn e, I l*, 2016.
31. Massachusetts Institute of Technology, ASE Group. (n.d.). Thermal Energy Grid Storage (TEGS) concept. Retrieved from https://ase.mit.edu/projects/thermal-energy-grid-storage-tegs/.
32. Piyumal, P.L.A.K., Ranaweera, A.L.A.K., Kalingamudali, S.R.D., Kularatna, N., Novel Approach for Harnessing Maximum Energy from PV Systems using Supercapacitors, in: *2018 IEEE Region 10 Humanitarian*

Technology Conference (R10-HTC), pp. 1–6, IEEE, Dec. 2018, doi: 10.1109/R10-HTC.2018.8629801.
33. Ramkumar, R., *et al.*, MnO/Mn_2O_3 Aerogels as Effective Materials for Supercapacitor Applications. *Energies*, 17, 10, 2258, 1–13, https://doi.org/10.3390/en17102258.
34. Jayakumar, M. and V. P., S., Recent Developments in the Field of Supercapacitor Materials, in: *Multidimensional Nanomaterials for Supercapacitors: Next Generation Energy Storage*, pp. 284–302, Bentham Science Publishers, 2024, doi: 10.2174/9789815223408124010017.
35. Yang, Y., The construction of an asymmetric hybrid supercapacitor with 2D materials MXene and perovskite. *Appl. Comput. Eng.*, 59, 1, 131–138, May 2024. doi: 10.54254/2755-2721/59/20240781.
36. Zhu, S., Sun, Y., Li, K., Dang, Y., Guan, X., Construction of structural supercapacitor with high energy density and mechanical strength based on dual-carbon electrodes and polyacrylamide-Portland cement-Na2SO4 electrolyte. *J. Power Sources*, 597, 234150, Mar. 2024. doi: 10.1016/J.JPOWSOUR.2024.234150.
37. *ESD-SCAP Catalogue*, SAMWHA Capacitor Co., Ltd, 2018.
38. Jagadale, A., Zhou, X., Xiong, R., Dubal, D.P., Xu, J., Yang, S., Elsevier B.V, May 01, 2019, doi: 10.1016/j.ensm.2019.02.031.
39. Elcap • CC0, *Charge-Discharge-Supercap-vs-Battery*, https://en.m.wikipedia.org/wiki/File:Charge-Discharge-Supercap-vs-Battery.png.
40. Zaidan, M.R., Hasan, G.T., Al-Obaidi, M.A., Comparative Study Between a Battery and Super-capacitor of an Electrical Energy Storage System for a Traditional Vehicle. *Gazi Univ. J. Sci.*, 35, 4, 1405–1415, Dec. 2022. doi: 10.35378/GUJS.969972.
41. Gopinath, K.P., Vo, D.V.N., Gnana Prakash, D., Adithya Joseph, A., Viswanathan, S., Arun, J., Environmental applications of carbon-based materials: a review. *Environ. Chem. Lett.*, 19, 1, 557–582, Feb. 2021. doi: 10.1007/S10311-020-01084-9.
42. Mbendane, B., Abe, B., Richards, C., Analysis of Supercapacitors in Renewable Energy Systems for Managing Power Fluctuations. *Int. J. Energy Res.*, 2024, 4575761, 2024. https://doi.org/10.1155/2024/4575761.
43. Soni, Y.P. and Fernandez, E., Transients Analysis of Photovoltaic System with Supercapacitor Participation in Power Management in Dynamic Scenarios, in: *2023 11th National Power Electronics Conference, NPEC 2023*, 2023, doi: 10.1109/NPEC57805.2023.10384983.
44. Cabrane, Z., Ouassaid, M., Maaroufi, M., Elsevier Ltd, Dec. 07, 2016, doi: 10.1016/j.ijhydene.2016.06.141.
45. Zakzouk, N.E. and Lotfi, R.A., Power Flow Control of a Hybrid Battery/Supercapacitor Standalone PV System under Irradiance and Load Variations; Power Flow Control of a Hybrid Battery/Supercapacitor Standalone PV System under Irradiance and Load Variations. *2020 10th International*

Conference on Power and Energy Systems (ICPES), 2020, doi: 10.1109/ICPES51309.2020.9349655/20/$31.00.

46. Palma, L., Analysis of supercapacitor connection to PV power conditioning systems for improoved performance, in: *5th International Conference on Clean Electrical Power: Renewable Energy Resources Impact, ICCEP 2015*, pp. 198–203, Aug. 2015, doi: 10.1109/ICCEP.2015.7177623.
47. Saripalli, B.P., Singh, G., Singh, S., Supercapacitors based energy storage system for mitigating solar photovoltaic output power fluctuations. *World J. Eng.*, 20, 5, 815–834, Aug. 2023. doi: 10.1108/WJE-08-2021-0468/FULL/XML.
48. Premasiri, M.D., Piyumal, K., Ranaweera, A.K., Kalingamudali, S.R., Investigation of MPPT in photovoltaic systems with step-down DC-DC converter and supercapacitor energy storage, in: *Proceedings of the 5th International Conference on Applied and Pure Sciences (ICAPS)*, University of Kelaniya, Sri Lanka, 2024, October.
49. Piyumal, K., Ranaweera, A., Kalingamudali, S., Kularatna, N., Maximum Power Point Tracking of an Off-Grid Photovoltaic System Consisting of a Series Connected Supercapacitor with a Step-Down Converter. *ECS Trans.*, 107, 1, 16233–16250, Apr. 2022. doi: 10.1149/10701.16233ecst.
50. Cabrane, Z., Kim, J., Yoo, K., Ouassaid, M., HESS-based photovoltaic/batteries/supercapacitors: Energy management strategy and DC bus voltage stabilization. *Sol. Energy*, 216, 551–563, Mar. 2021. doi: 10.1016/j.solener.2021.01.048.
51. Huang, Y.P. and Tsai, P.F., Improving the Output Power Stability of a High Concentration Photovoltaic System with Supercapacitors: A Preliminary Evaluation. *Math. Probl. Eng.*, 2015, 1, 127949, Jan. 2015. doi: 10.1155/2015/127949.
52. González, L.G., Chacon, R., Delgado, B., Benavides, D., Espinoza, J., Study of Energy Compensation Techniques in Photovoltaic Solar Systems with the Use of Supercapacitors in Low-Voltage Networks. *Energies*, 13, 15, 3755, Jul. 2020. doi: 10.3390/EN13153755.
53. Aouini, R., Djeddou, Y., Saadi, Y., Mesbahi, T., Solar Energy Management System with Hybrid Battery/Supercapacitor Storage for Residential Applications, in: *2023 9th International Conference on Control, Decision and Information Technologies (CoDIT)*, IEEE, pp. 2342–2347, Jul. 2023, doi: 10.1109/CoDIT58514.2023.10284115.
54. Loor-Cevallos, M.E., Rodríguez-Gámez, M., Medranda-Cobeña, G.I., Loor-Vélez, C.A., Lituma-Ramírez, E.D., Energy Storage Using Supercapacitors in a Photovoltaic Installation for a House in the Rural Area. *J. Eng. Res.*, 4, 8, 2–11, Mar. 2024. doi: 10.22533/AT.ED.317482406038.
55. Harinkhede, S. and Vaidya, P., Photovoltaic with Battery and Supercapacitor Energy Storage System for Better Performance Devices and Modelling. *Lect. Notes Electr. Eng.*, 1109, 151–162, 2024. doi: 10.1007/978-981-99-8289-9_12.
56. Chang, C., *et al.*, in: *Proc. APEC 2013: Twenty-Eighth Annual IEEE Applied Power Electronics Conf. and Exposition,* Long Beach, CA, USA, Mar. 17–21, 2013.

57. Kularatna, N., Subasinghage, K., Gunawardane, K., Jayananda, D., Ariyarathna, T., Supercapacitor-Assisted Techniques and Supercapacitor-Assisted Loss Management Concept: New Design Approaches to Change the Roadmap of Power Conversion Systems. *Electronics*, 10, 14, 1697, Jul. 2021. doi: 10.3390/ELECTRONICS10141697.
58. Parvini, Y., Siegel, J.B., Stefanopoulou, A.G., Vahidi, A., Supercapacitor Electrical and Thermal Modeling, Identification, and Validation for a Wide Range of Temperature and Power Applications. *IEEE Trans. Ind. Electron.*, 63, 3, 1574–1585, Mar. 2016. doi: 10.1109/TIE.2015.2494868.
59. Piyumal, P.L.A.K., Ranaweera, A.L.A.K., Kalingamudali, S.R.D., Kularatna, N., Novel approach for harnessing maximum energy from PV systems using supercapacitors, in: *IEEE Region 10 Humanitarian Technology Conference, R10-HTC*, vol. 2018-December, Jul. 2018, doi: 10.1109/R10-HTC.2018.8629801.
60. Naligama, C.A., Kularatna, N., Steyn-Ross, A., Extending Supercapacitor Assisted Loss Management Approach for Efficiency Improvements in Solar Inverters, in: *5th IEEE International Conference on DC Microgrids, ICDCM 2023*, 2023, doi: 10.1109/ICDCM54452.2023.10433603.
61. Subasinghage, K., Gunawardane, K., Kularatna, N., Supercapacitor-Assisted Low-Dropout Regulator Technique for Low Output Ripple DC–DC Conversion. *IEEE J. Emerging Sel. Top. Ind. Electron.*, 3, 4, 1028–1037, Feb. 2022. doi: 10.1109/JESTIE.2022.3149708.
62. Jayasekara, L.D.P.S., Piyumal, P.L.A.K., Ranaweera, A.L.A.K., Kalingamudali, S.R.D., Efficiency Optimization of Supercapacitor Assisted Low Dropout (SCALDO) Regulators, in: *IEEE Region 10 Humanitarian Technology Conference, R10-HTC*, vol. 2020-December, Dec. 2020, doi: 10.1109/R10-HTC49770.2020.9356952.
63. Subasinghage, K., Gunawardane, K., Kularatna, N., Stability analysis and experimental validation of the supercapacitor-assisted low-dropout regulator. *IET Power Electron.*, 13, 15, 3213–3225, Nov. 2020. doi: 10.1049/IET-PEL.2019.1601/CITE/REFWORKS.
64. Subasinghage, K. and Weranga, K., *Advanced Analysis of Supercapacitor-Assisted Low-Dropout (SCALDO) Regulator*, Tuwhera Open Access Repository, Auckland University of Technology, New Zealand, 2019. http://hdl.handle.net/10292/12559.
65. Subasinghage, K., Gunawardane, K., Kularatna, N., A Preliminary Study on the Switching Ripple of Supercapacitor-Assisted Low-Dropout Regulator, in: *2021 10th International Conference on Information and Automation for Sustainability, ICIAfS 2021*, pp. 530–535, Aug. 2021, doi: 10.1109/ICIAFS52090.2021.9606003.
66. Xiao, J., Fu, T., Chen, J., Su, S., Wang, J., Chen, J., Optimum design of a supercapacitor charged by a photovoltaic module. *J. Renew. Sustain. Energy*, 13, 5, 1ENG, Sep. 2021. doi: 10.1063/5.0063202.

67. Abeeb Olalekan, M., Adedokun, O., Taiwo Bello, I., Ayinde Kareem, M., Kwong Yam, F., Recent Advances in Photo-supercapacitor: A Mini Review. *Adv. Mater. Sci. Technol.*, 4, 2, 31–49, 2022. doi: 10.37155/2717-526x-0402-2.
68. Garcia, J., Garcia, P., Giulii Capponi, F., De Donato, G., Analysis, Modeling, and Control of Half-Bridge Current-Source Converter for Energy Management of Supercapacitor Modules in Traction Applications. *Energies (Basel)*, 11, 9, 2239, Aug. 2018. doi: 10.3390/en11092239.
69. Mohamed, E. and Youssef, Z., *Proceedings of 2015 IEEE International Renewable and Sustainable Energy Conference: (IRSEC'15)*, IEEE, 2015.
70. Li, S., Zhao, Z., Li, B., Wei, T., Jiang, H., Yan, Z., Supercapacitors accumulating energy harvesting from stacked sediment microbial fuel cells and boosting input power for power management systems. *Int. J. Hydrogen Energy*, 47, 19, 10689–10700, Mar. 2022. doi: 10.1016/J.IJHYDENE.2021.11.081.
71. Diantoro, M., Al Ittikhad, A., Albadi'Ah, I.V., Latifah, E., Taufiq, A., Meevasana, W., Integrated Array of Dye-Sensitized Solar Cells and Supercapacitor for Photo-Supercapacitor Multifunctional Harvesting Device. *J. Phys. Conf. Ser.*, 2243, 1, 012052, Jun. 2022. doi: 10.1088/1742-6596/2243/1/012052.
72. Acar, E., Jain, N., Ramu, P., Hwang, C., Lee, I., Springer Science and Business Media Deutschland GmbH, Mar. 01, 2024, doi: 10.1007/s00158-024-03737-7.
73. Balali, Y. and Stegen, S., Review of energy storage systems for vehicles based on technology, environmental impacts, and costs. *Renew. Sustain. Energy Rev.*, 135, 110185, Jan. 2021. doi: 10.1016/J.RSER.2020.110185.
74. Teasdale, A., Ishaku, L., Amaechi, C.V., Adelusi, I., Abdelazim, A., A Study on an Energy-Regenerative Braking Model Using Supercapacitors and DC Motors. *World Electr. Veh. J.*, 15, 7, 326, Jul. 2024. doi: 10.3390/WEVJ15070326.
75. Mtengi, B., Prabaharan, S.R.S., Murtala Zungeru, A., Garba Ambafi, J., Michael, S.M., Design of Regenerative Braking System and Energy Storage with Supercapacitors as Energy Buffers. *Int. J. Electr. Comput. Eng. Syst.*, 15, 4, 321–333, Mar. 2024. doi: 10.32985/ijeces.15.4.3.
76. Prokin, M. and Prokin, D., Regenerative braking for fuel cell powertrain, in: Proceedings of the 13th Mediterranean Conference on Embedded Computing (MECO), pp. 1–4, Budva, Montenegro, 2024, July. https://doi.org/10.1109/MECO62516.2024.10577868.
77. Mishra, S. and Pattnaik, S., A Battery-Ultracapacitor Based Electric Vehicle with Regenarative Braking System, in: *5th International Conference on Energy, Power, and Environment: Towards Flexible Green Energy Technologies, ICEPE 2023*, 2023, doi: 10.1109/ICEPE57949.2023.10201541.
78. Kale, A., Gajbhiye, A., Khubalkar, S., Nangrani, S.P., Analysis of Regenerative Braking System in Electric Vehicles, in: *2023 IEEE Renewable Energy and Sustainable E-Mobility Conference, RESEM 2023*, 2023, doi: 10.1109/RESEM57584.2023.10236088.

79. Mariasiu, F. and Kelemen, E.A., Analysis of the Energy Efficiency of a Hybrid Energy Storage System for an Electric Vehicle. *Batteries*, 9, 8, Aug. 2023. doi: 10.3390/batteries9080419.
80. Bharani Prakash, T., Naveen, P., Sri Vishnujah, A., Subhanivethaa, R., Varsha, M.T., Effective Energy Storage for EV Using Battery Supercapacitor, in: *2023 9th International Conference on Advanced Computing and Communication Systems, ICACCS 2023*, pp. 613–617, 2023, doi: 10.1109/ICACCS57279.2023.10112898.
81. Mouli, C.L.R., Supercapacitors as Future Energy Storage Devices for Electric Vehicles, in: *2022 IEEE Delhi Section Conference, DELCON 2022*, 2022, doi: 10.1109/DELCON54057.2022.9753349.
82. Berrueta, A., Ursua, A., Martin, I.S., Eftekhari, A., Sanchis, P., Supercapacitors: Electrical Characteristics, Modeling, Applications, and Future Trends. *IEEE Access*, 7, 50869–50896, 2019. doi: 10.1109/ACCESS.2019.2908558.
83. Paravannoor, A., Market trends, innovations and challenges, in: *Supercapacitors and Their Applications: Fundamentals, Current Trends and Future Perspectives*. A. Paravannoor and B.K.V. (Eds.), pp. 181–200, Taylor & Francis, 2023, April. https://doi.org/10.1201/9781003258384-12 radiohata.com+6.
84. Greenhalgh, E.S., Nguyen, S., Valkova, M., Shirshova, N., Shaffer, M.S.P., Kucernak, A.R.J., A critical review of structural supercapacitors and outlook on future research challenges. *Compos. Sci. Technol.*, 235, 109968, Apr. 2023. doi: 10.1016/J.COMPSCITECH.2023.109968.
85. Sumangala, T.P., Sreekanth, M.S., Rahaman, A., Applications of Supercapacitors. *Springer Ser. Mater. Sci.*, 313, 367–393, 2021. doi: 10.1007/978-3-030-68364-1_11.
86. Alexander, L., Choi, S., Siegel, J., Thompson, L., *2016 Ndia Ground Vehicle Systems Engineering High Power Supercapacitors For Hybrid Energy Storage Systems For Soldiers And Vehicles.*
87. Gekakis, N., Nadeau, A., Hassanalieragh, M., Modeling of Supercapacitors as an Energy Buffer for Cyber-Physical Systems, in: *Cyber-Physical Systems*, pp. 190–209, Chapman and Hall/CRC, 2015, doi: 10.1201/b19206-12.
88. Sarkar, A., Student, B.T., Saha, A., Podder, S., Piezoelectric-Driven Graphene-Based Self-Chargeable Supercapacitor For Wearable Devices, in: *Futuristic Trends in Chemical*, Material Sciences & Nano Technology, vol. 3, Book 2, pp. 98–114. IIP Series. https://doi.org/10.58532/V3BECM2CH7.
89. Ganeshbabu, M. and Kalai Selvan, R., Wearable supercapacitor, in: *Handbook of Energy Materials*. R.K. Gupta (Ed.), pp. 1–45, Springer Nature, Singapore, 2023. https://doi.org/10.1007/978-981-16-4480-1_53-1.
90. Luo, Z., Wang, Y., Kou, B., Liu, C., Zhang, W., Chen, L., 'Sweat-chargeable' on-skin supercapacitors for practical wearable energy applications. *Energy Storage Mater.*, 38, 9–16, Jun. 2021. doi: 10.1016/J.ENSM.2021.02.046.
91. Islam, M.R., Afroj, S., Novoselov, K.S., Karim, N., *Smart Electronic Textile-Based Wearable Supercapacitors*, John Wiley and Sons Inc, Nov. 03, 2022, doi: 10.1002/advs.202203856.

92. Manjakkal, L., Núñez, C.G., Dang, W., Dahiya, R., Flexible self-charging supercapacitor based on graphene-Ag-3D graphene foam electrodes. *Nano Energy*, 51, 604–612, Sep. 2018. doi: 10.1016/J.NANOEN.2018.06.072.
93. Zhu, J., Han, S., Wu, Q., Zhang, J., *et al.*, All-in-one configured flexible supercapacitor for wide-temperature operation and integrated application. *ACS Appl. Energy Mater.*, 6, 8, 2023. https://doi.org/10.1021/acsaem.2c03800.
94. Naeem, S., Patil, A. V., Shinde, U. P., Ali, A., Memon, K., and Bavluwala, M., *Applications and aspects of supercapacitor in IoT devices as a sustainable energy source.*
95. Wang, C.H., Huang, K.H., Wu, C.Y., Enhanced charge circuitry for indoor photovoltaic energy harvesting with fast activation and high efficiency. *IET Power Electron.*, 16, 13, 2228–2237, Oct. 2023. doi: 10.1049/PEL2.12542.
96. Asl, M.S., *et al.*, Flexible all-solid-state supercapacitors with high capacitance, long cycle life, and wide operational potential window: Recent progress and future perspectives. *J. Energy Storage*, 50, 104223, Jun. 2022. doi: 10.1016/J.EST.2022.104223.
97. Abraham, T.R., Sunil, K., Shah, M., Ashok, N., Thomas, S., Energy storage devices: Batteries and supercapacitors, in: *Nanobiohybrids for Advanced Wastewater Treatment and Energy Recovery*, pp. 61–84, IWA Publishing, 2023, doi: 10.2166/9781789063592_0061.
98. Simon, P. and Gogotsi, Y., Materials for electrochemical capacitors. *Nat. Mater.*, 7, 11, 845–854, Nov. 2008. doi: 10.1038/nmat2297.
99. Pandolfo, A.G. and Hollenkamp, A.F., Carbon properties and their role in supercapacitors. *J. Power Sources*, 157, 1, 11–27, Jun. 2006. doi: 10.1016/J.JPOWSOUR.2006.02.065.
100. *The Electrochemical Society Interface*, p. 53, Spring, 2008.
101. Ibrahim, T., Stroe, D., Kerekes, T., Sera, D., Spataru, S., An overview of supercapacitors for integrated PV-energy storage panels, in: *Proceedings - 2021 IEEE 19th International Power Electronics and Motion Control Conference, PEMC 2021*, Institute of Electrical and Electronics Engineers Inc., pp. 828–835, Apr. 2021, doi: 10.1109/PEMC48073.2021.9432540.
102. Abdelhalim, T., Kouider, L., Abdelkader, R., Hartani, M.A., Integration of Photovoltaic, Lithium Battery, and Supercapacitor Technologies for Enhanced DC Voltage in Standalone Microgrids, in: *2nd International Conference on Energy Transition and Security, ICETS 2023*, 2023, doi: 10.1109/ICETS60996.2023.10410817.
103. Sheng, H., Ma, Y., Zhang, H., Yuan, J., Li, F., Li, W., Xie, E., Lan, W., Integration of supercapacitors with sensors and energy-harvesting devices: A review. *Adv. Mater. Technol.*, 9, 21, 2301796, 2024. https://doi.org/10.1002/admt.202301796.
104. Plebankiewicz, I., Bogdanowicz, K.A., Iwan, A., Photo-Rechargeable Electric Energy Storage Systems Based on Silicon Solar Cells and Supercapacitor-Engineering Concept. *Energies*, 13, 15, 3867, Jul. 2020. doi: 10.3390/EN13153867.

105. Jiang, L., Meng, F., Hu, J., Yan, X., Scalable self-integrated all-in-one fabric-based supercapacitors with simultaneous enhancement of output voltage and capacitance. *J. Power Sources*, 556, 232512, Feb. 2023. doi: 10.1016/J.JPOWSOUR.2022.232512.
106. Sundriyal, P. and Bhattacharya, S., Scalable Micro-fabrication of Flexible, Solid-State, Inexpensive, and High-Performance Planar Micro-supercapacitors through Inkjet Printing. *ACS Appl. Energy Mater.*, 2, 3, 1876–1890, Mar. 2019. doi: 10.1021/acsaem.8b02006.
107. Rosier, A., Mabo, P., Chauvin, M., Burgun, A., An ontology-based annotation of cardiac implantable electronic devices to detect therapy changes in a national registry. *IEEE J. Biomed Health Inform.*, 19, 3, 971–978, May 2015. doi: 10.1109/JBHI.2014.2338741.
108. Nordin, N.A., *et al.*, *Integrating photovoltaic (Pv) solar cells and supercapacitors for sustainable energy devices: A review*, MDPI, Nov. 01, 2021, doi: 10.3390/en14217211.
109. Guven, A. F., Abdelaziz, A. Y., Samy, M. M., and Barakat, S., Optimizing energy dynamics: A comprehensive analysis of hybrid energy storage systems integrating battery banks and supercapacitors. *Energy Convers. Manag.*, 312, 118560, Jul. 2024, doi: 10.1016/j.enconman.2024.118560.
110. Cabrane, Z., Ouassaid, M., Maaroufi, M., Performance analysis of supercapacitor integration in photovoltaic energy storage system, in: *Proceedings of 2015 IEEE International Renewable and Sustainable Energy Conference, IRSEC 2015*, Apr. 2016, doi: 10.1109/IRSEC.2015.7454983.
111. Vilathgamuwa, D.M. and Jayasinghe, S.D.G., A three-level flying-capacitor rectifier based supercapacitor direct integration scheme for distributed generation systems, in: *IEEE International Symposium on Industrial Electronics*, pp. 1566–1571, 2012, doi: 10.1109/ISIE.2012.6237325.
112. Choudhary, N., *et al.*, Recent advancements in nanocellulose-based supercapacitors for energy storage devices: A review. *Carbohydr. Polym. Technol. Appl.*, 7, 100416, 2024. https://doi.org/10.1016/J.CARPTA.2023.100416.
113. Jiang, X., *et al.*, Beyond conventional limits: Advancements and insights in broadening operating temperature ranges of supercapacitors. *Energy Storage Mater.*, 70, 103462, Jun. 2024. doi: 10.1016/J.ENSM.2024.103462.
114. Iannuzzi, D. and Lauria, D., A new supercapacitor design methodology for light transportation systems saving, in: *Energy Management Systems*. Iannuzzi, D. and Lauria, D. (Eds.), pp. 183–198, InTech, 2011. https://doi.org/10.5772/17876.
115. Vukajlović, N., Milićević, D., Dumnić, B., Popadić, B., Comparative analysis of the supercapacitor influence on lithium battery cycle life in electric vehicle energy storage. *J. Energy Storage*, 31, 101603, Oct. 2020. doi: 10.1016/J.EST.2020.101603.
116. De Bernardinis, A., *et al.*, Supercapacitors aging study: Models theoretical analysis and attempt to physical correlation related to applied cycles for transport application, in: *2018 International Conference on Applied and*

Theoretical Electricity, ICATE 2018 - Proceedings, Nov. 2018, doi: 10.1109/ICATE.2018.8551485.
117. Kowal, J., *et al.*, Detailed analysis of the self-discharge of supercapacitors. *J. Power Sources*, 196, 1, 573–579, Jan. 2011. doi: 10.1016/J.JPOWSOUR.2009.12.028.
118. da Silva, E.P., *et al.*, Sustainable energy and waste management: How to transform plastic waste into carbon nanostructures for electrochemical supercapacitors. *Waste Manage. (Oxford)*, 171, 71–85, Nov. 2023. doi: 10.1016/J.WASMAN.2023.08.028.
119. Wu, F., Zhao, T., Yao, Y., Jiang, T., Wang, B., Wang, M., Recycling supercapacitor activated carbons for adsorption of silver (I) and chromium (VI) ions from aqueous solutions. *Chemosphere*, 238, 124638, 2020. https://doi.org/10.1016/j.chemosphere.2019.124638.
120. Burcea, I.M., Negroiu, R.C., Mihailescu, B., Svasta, P., Moise, M.V., Ionescu, C., Realization of Cost-Effective Supercapacitors Utilizing Substitutable Materials, in: *Proceedings of the International Spring Seminar on Electronics Technology*, 2024, doi: 10.1109/ISSE61612.2024.10604057.
121. El Maati, L.A., *et al.*, Facile synthesis of environmental friendly Sm2O3/rGO electrode as an alternative source for energy storage devices. *Diam. Relat. Mater.*, 142, 110769, Feb. 2024. doi: 10.1016/J.DIAMOND.2023.110769.
122. Kumar, Y.A., *et al.*, Supercharging the future: MOF-2D MXenes supercapacitors for sustainable energy storage. *J. Energy Storage*, 80, 110303, Mar. 2024. doi: 10.1016/J.EST.2023.110303.
123. John, B. and Amreen, K., Emerging trends of miniaturized carbon nanomaterial based energy storage microdevices. *Univers. J. Carbon Res.*, 2, 1, 2024. http://dx.doi.org/10.37256/ujcr.2120244941.
124. Ali, Z., Iqbal, M.Z., Hegazy, H.H., Recent advancements in redox-active transition metal sulfides as battery-grade electrode materials for hybrid supercapacitors. *J. Energy Storage*, 73, 108857, Dec. 2023. doi: 10.1016/J.EST.2023.108857.
125. Adedoja, O.S., Sadiku, E.R., Hamam, Y., An Overview of the Emerging Technologies and Composite Materials for Supercapacitors in Energy Storage Applications. *Polymers*, 15, 10, 2272, May 2023. doi: 10.3390/POLYM15102272.
126. Mehetre, S.S., Ameta, R.K., Behere, S.S., Inwati, G.K., Supercapacitors—new developments. *Smart Supercapacitors Fundam. Struct. Appl.*, 39–64, Jan. 2023. doi: 10.1016/B978-0-323-90530-5.00021-6.
127. Rushmittha, J.J., Radhika, S., Alrashidi, K.A., Maheshwaran, G., Dhinesh, S., Sambasivam, S., Hybridization of CuFe2O4 by carbon microspheres with improved charge storage characteristics for high energy density solid-state hybrid supercapacitor. *Mater. Today Sustain.*, 28, 100950, Dec. 2024. doi: 10.1016/J.MTSUST.2024.100950.
128. Chao, S., *et al.*, Hexagon MXene based heterojunction interface with high ion adsorption and electron transfer efficiency for soft-package potassium-ions

aqueous supercapacitors. *J. Colloid Interface Sci.*, 673, 922–933, Nov. 2024. doi: 10.1016/J.JCIS.2024.06.115.

129. Balaji, V., *et al.*, Unleashing nature's energy: Enchanted Albizia amara leaf extract empowers supercapacitors with NdMn2O5 nanocomposite. *Mater. Lett.*, 370, 136862, Sep. 2024. doi: 10.1016/j.matlet.2024.136862.
130. Shah, S.S., Aziz, M.A., Oyama, M., Advancing Electrochemical Energy Storage: The Synergistic Impact of Sulfur Nano-Confinement in Hierarchically Porous Jute-Derived Activated Carbon for Enhanced Supercapacitor Performance. *ECS Meet. Abstr.*, MA2024-01, 1, 11, Aug. 2024. doi: 10.1149/MA2024-01111MTGABS.
131. Jessy Mol, E. and Mary Linda, M., Integration of Wind and PV Systems Using Genetic-Assisted Artificial Neural Network. *Intell. Autom. Soft Comput.*, 35, 2, 1471–1489, 2023. doi: 10.32604/iasc.2023.024027.
132. Muthamizhan, T., Aijaz, M., Prakash, G., Rathnavel, P., Sivakumar, A., A Multi-Port Autonomous Reconfigurable Solar PV System for Renewable Energy Integration in Hybrid AC/DC System, in: *Proceedings of the 9th International Conference on Electrical Energy Systems, ICEES 2023*, pp. 687–693, 2023, doi: 10.1109/ICEES57979.2023.10110073.
133. Vlad, C. and Lungu, C.V., Considerations regarding PV systems. *Ann. "Dunarea de Jos" Univ. Galati Fascicle II Math. Phys. Theor. Mech.*, 45, 2, 126–132, Dec. 2022. doi: 10.35219/ann-ugal-math-phys-mec.2022.2.14.
134. Khan, H.A., *et al.*, A comprehensive review on supercapacitors: Their promise to flexibility, high temperature, materials, design, and challenges. *Energy*, 295, 131043, May 2024. doi: 10.1016/J.ENERGY.2024.131043.
135. Le, S.T., Nguyen, T.N., Bui, D.K., Ngo, T.D., Techno-economic and life cycle analysis of renewable energy storage systems in buildings: The effect of uncertainty. *Energy*, 307, 132644, 2024, October. https://doi.org/10.1016/j.energy.2024.132644.
136. Adenekan, O.A., Ezeigweneme, C., Chukwurah, E.G., Driving innovation in energy and telecommunications: next-generation energy storage and 5G technology for enhanced connectivity and energy solutions. *Int. J. Manage. Entrepreneurship Res.*, 6, 5, 1581–1597, May 2024. doi: 10.51594/IJMER.V6I5.1124.
137. Manohar, A., *et al.*, Ternary nanocomposites of ZnFe2O4/NiFe2O4/CeO2: Investigating electrochemical energy storage and biocompatible properties. *J. Environ. Chem. Eng.*, 12, 5, 113337, Oct. 2024. doi: 10.1016/J.JECE.2024.113337.
138. Perez, F., Hoffmann, B.V., Dos S Pereira, G.M., Küster, K.K., Durce, C.C., The use of Supercapacitors in Microgrids as Hybrid Energy Storage Systems - A review, in: *2019 IEEE PES Conference on Innovative Smart Grid Technologies, ISGT Latin America 2019*, Sep. 2019, doi: 10.1109/ISGT-LA.2019.8895484.
139. Jithin, T., Rajeev, T., Jithin, S., Inertia Control of Hybrid AC/DC Microgrid Using Supercapacitors, in: *ICPC2T 2022 - 2nd International Conference on Power, Control and Computing Technologies, Proceedings*, 2022, doi: 10.1109/ICPC2T53885.2022.9776860.

8

A Combination of Energy Conversion and Storage: A Solar-Driven Supercapacitor

Mohammed Arkham Belgami and Chandra Sekhar Rout*

Centre for Nano and Material Sciences, JAIN (Deemed-to-be University), Jain Global Campus, Bangalore, Karnataka, India

Abstract
This chapter pertains to the extensive research on harvesting the radiant energy, which posed a challenge to storing the same energy. This led to an open door to construct and manipulate devices that can convert as well as store energy. Solar-driven supercapacitors (SDSCs)/photosupercapacitors serve best for this purpose; they can be a promising alternative for the upcoming power sources and suit extremely well in the high and green energy demanding generations to come. A high multitude of integrated energy conversion and storage devices have been already developed, among which solar cells combined with supercapacitors having an appreciable solar conversion and storage properties are well known. Thereby this chapter will enlighten us with the different facets of SDSCs, the different integration forms of photovoltaics and supercapacitors, and efficiency calculations of SDSCs.

Keywords: Photosupercapacitors, dye sensitized solar cells, perovskite solar cells, organic polymer-based solar cells, electrical double-layer capacitance, pseudocapacitance

8.1 Introduction

Growing interest in a variety of clean energy solutions has been prompted in recent years by factors like rising global temperatures, increasing environmental challenges, and depleting fossil fuel sources. Particularly potential

**Corresponding author*: r.chandrasekhar@jainuniversity.ac.in; csrout@gmail.com

Nurdan Demirci Sankir and Mehmet Sankir (eds.) Solar Capacitors and Batteries, (243–266) © 2025 Scrivener Publishing LLC

options include hydrogen, solar, wind, geothermal, tidal power, and many more. Out of which, solar energy has drawn interest from all around the world pertaining to its abundant supply, widespread distribution, environmentally benign and sustainable adaptation, as well as its affordability [1, 2]. Accordingly, researchers in academia and industry faced a bigger task in figuring out how to use solar energy effectively and efficiently.

Since the finding of the solar cell in Bell Laboratory in 1954 [3], researchers have invested significant effort into developing advanced solar cell technologies that possess desirable features such as smaller size, cost efficiency, straightforward fabrication, lightweight construction, flexibility, and enhanced performance in low-light environments [4]. The total reliance of solar cells on sunshine for their correct operation is widely recognized. However, because sunlight is sporadic, there is increasing stability concern regarding the functioning of solar cells. It has become clear that the right solar energy storage technology is required to meet this problem. Along with the advanced technologies of electrochemistry including batteries based on lead acid and Li-ion, fuel cells, flow batteries, and supercapacitors (SCs), a diversity of solar energy storage systems have been investigated over time. When a photo-battery is exposed to light, it experiences a voltage surge, but as soon as the illumination stops, the cell voltage quickly drops to zero, which is the fundamental problem of solar-based battery systems. The cell voltage no longer instantly goes to zero when coupled in tandem with an SC, thereby overcoming this problem. SCs provide uninterrupted power transmission as a result, increasing the dependability and reliability of device power output [5]. SCs and solar cells were initially externally coupled to create a hybrid model with an energy conversion and storage mechanism [6, 7]. Nevertheless, reliance on integrated systems *via* external connections comes with a number of difficulties. The energy storage efficiency of the SC component tends to be reduced by the usage of external connecting circuits [8]. This decrease happens as a result of the external cables' intrinsic internal resistance, which uses some of the power and lowers the energy storage efficiency. Additionally, conventional energy storage and conversion systems are sometimes large and heavy, making them unsuitable for supplying the needs of portable and wearable electronic gadgets.

To answer these problems, a unique energy storage and photoelectric conversion device with dual functionality was developed. They are known as photosupercapacitors (PSCs) or solar-driven SCs (SDSCs), and they stand out as incredibly promising and effective energy conversion and storage technologies [9–11]. They are attractive because of their fast charging and discharging rates, high power density, extended life cycles, environmental friendliness, and nearly maintenance-free nature. The chapter goes into the intriguing realm of PSCs, illuminating its important developments and prospects for the future. The review of PSCs in Section 8.2 emphasizes their function as cutting-edge energy storage technologies. The focus of Section 8.2.1 shifts to several of these extraordinary gadgets' key parts. The adaptable dye-sensitized solar cell (DSSC), the promising organic photovoltaic-based solar cell, and the astonishing perovskite-based solar cells are the three main categories of solar cells. The electrical double-layer-based SC and the pseudocapacitance-based SC are both included in Section 8.2.1.2's exploration of the world of SCs.

The third part of the essay explores alternative integration strategies for PSCs. In Section 8.3.1, the combination of PSCs with DSSCs is highlighted, revealing potential synergy. The focus changes to organic photovoltaic-based PSCs in Section 8.3.2, which provides details on their special benefits. The combination involving perovskite-based solar cells with SCs is explored in Section 8.3.3, emphasizing their extraordinary possibilities.

The difficult science of gauging the performance of PSCs is explained in Section 8.4 as it delves into the area of efficiency computation. Finally, Section 8.5 addresses the difficulties and presents a convincing picture of the opportunities that lie ahead for these cutting-edge technologies. As the essay progresses, it becomes clear that PSCs are extremely promising and are positioned to be a key component of the next cohort of sustainable devices for energy storage.

8.2 What are Photosupercapacitors

Because of its remarkable energy density, lithium-ion batteries (LIBs) have become incredibly popular solution for energy storage. They do, however, have a number of shortcomings. These include their relatively high cost, the requirement for improved safety precautions, and worries about the scarce supply of lithium, which is frequently produced by mining techniques associated with abusive labor practices and significant environmental impact [12]. These batteries' lifespan is also shortened by the limited number of charge and discharge cycles they can sustain, making frequent replacements necessary.

In situations where quick, high-performance, and dependable energy storage technologies are needed to meet these difficulties, SCs are the go-to choice [13]. SCs can charge or discharge within seconds or minutes, enabling them to deliver substantial power bursts for short durations when required. SCs, on the other hand, provide a sustainable charging alternative

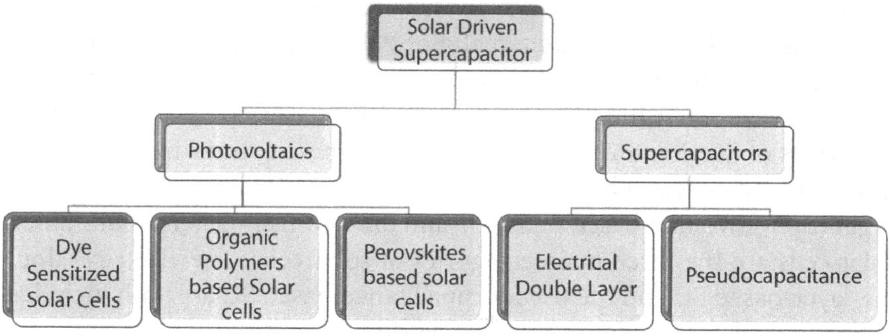

Figure 8.1 Representation of different aspects related to photosupercapacitor.

when combined with reasonably priced, solution-processed photovoltaic or solar cells. A PSC, which has the potential to revolutionize the field of self-sustaining electronics, is the result of this creative combination. PSCs are capable of operating on their own, obtaining their energy from as basic a source as indoor illumination or sunshine. This innovation is crucial in increasing the use of renewable energy solutions in a variety of applications [14, 15] (Figure 8.1).

8.2.1 Major Components of PSCs

8.2.1.1 Solar Cell

Photovoltaic effect refers to the process that allows solar cells to transform light directly to electrical energy. This phenomenon occurs when semiconductor materials are exposed to light. Specifically, when photons possessing energy equal to or exceeding the material's bandgap (E_g) are absorbed, they excite electrons, promoting them to higher energy states. This excitation results in the creation of electron–hole pairs, which then are separated within the semiconductor. The positive terminal efficiently gathers the positively charged holes, generating an electric current from the incoming light, while the liberated electrons are captured by an external circuit as electrical energy.

Silicon-based solar cell, which predominantly uses crystalline silicon and has outstanding efficiency of about 26.7%, is the most well-known and well-established solar cell technology. For a very long time, this technology controlled the market for commercial solar modules [16–18]. It does, however, have several disadvantages, such as stiffness, significant weight, and high production costs as a result of the heavy usage of pricey raw materials.

Amorphous silicon has been used as a substitute, despite having a lower efficiency of 11.9%, mostly because it is more affordable than its crystalline equivalent [16]. These first-generation photovoltaic technologies on silicon are classified as such.

GaAs, CdTe, and other metal–metalloid combinations are found in the second generation of solar devices. Unfortunately, these materials are less trustworthy for extensive photovoltaic usage because of their high cost and restricted availability [19, 20].

The third-generation solar technologies, such as DSSCs, organic material-based photovoltaics (OPVs), perovskite solar cells (PvSC), and others, are currently leading advancements in the field. These cutting-edge technologies are currently the subject of intensive study and use in the solar energy sector. Third-generation solar cells are becoming more and more popular because of their exceptional compatibility with a variety of flexible substrates, affordability, and simplicity of integration into houses and buildings, even if previous generations attained better conversion efficiencies [4, 21].

8.2.1.1.1 Dye-Sensitized Solar Cells

DSSCs have appeared to be a popular and cost-effective substitute to silicon-based solar cells, which are often expensive and rigid [22]. They are a leader among third-generation photovoltaics, which provide an appealing remedy for the significant shortcomings of their first- and second-generation equivalents [23, 24]. Because of their basic production procedures, abundant supply of safe components, and cost-effectiveness, they can accommodate stiff or flexible materials such as lightweight plastics, metal foils, wires, or fibers. As depicted in Figure 8.2(a), an electrolyte, a dye substance, a photoanode, and a counter electrode are the components of a DSSC device. All these parts influence the device's operation in one way or another, highlighting how the kind of photoanode material employed has a significant impact on the current density and circuit voltage produced by photogeneration, etc.

Materials with similar conduction band properties and enhanced carrier mobility, such as ZnO and TiO_2, have garnered significant scientific interest. In DSSCs, the fundamental process involves using light to initiate the photogeneration of electrons within the dye sensitizer. These energized electrons then enter the semiconductor's conduction band. The I/I_3^- redox system is used to simultaneously reduce the dye and receive electrons from the electrolyte. The photogenerated electron sets out toward the conducting substrate, where it travels until it arrives at the counter electrode, where it leaves

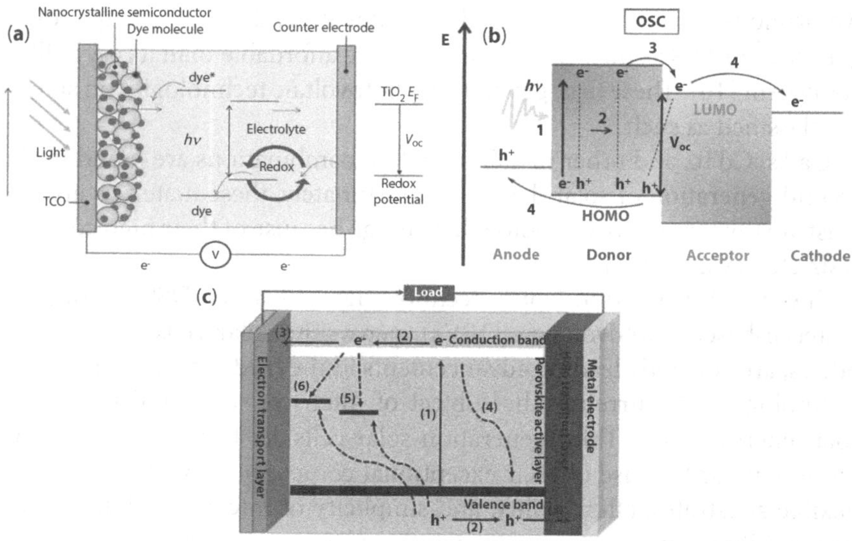

Figure 8.2 Operational processes of (a) DSSC (Reprinted with permission from [24].) (b) organic (Reprinted with permission from [27].) (c) Perovskite photovoltaics.

the device through an externally linked circuit [14, 25, 26]. These solar cells are versatile and powerful due to their light weight and thin design, functioning well even under diffuse and low-intensity lighting circumstances.

8.2.1.1.2 Organic Photovoltaics (OPVs)

OPVs, which utilize semiconductors composed of carbon-based materials—particularly polymers in conjugated form—within the functional layers, are transforming the generation of renewable energy [27]. When exposed to light, these organic semiconductor films have a special property that causes them to produce excitons with a short diffusion length. The electron donor, typically an organic semiconductor, and the electron acceptor semiconductors are essential components in the development of OPVs. The effectiveness of this process is enhanced by the high electron affinities and ionization potentials of both the donor and acceptor materials. When solar energy is absorbed, an electron from the highest occupied molecular orbital (HOMO) of the donor is excited to its lowest unoccupied molecular orbital (LUMO). If the bandgap of the acceptor material is lower than that of the donor, the electron is then shifted to the LUMO of the acceptor. The difference of energy between the HOMO and the LUMO of the donor and the acceptor, respectively, results in a higher voltage output. OPVs are particularly attractive due to their accessibility, non-toxic nature, and cost-effective manufacturing processes, which can be conducted at low temperatures. OPVs also have the benefit of being printed on broad surfaces and flexible substrates, which opens up a world of opportunity for their widespread use in renewable energy applications [28–30] (Figure 8.2(b)).

8.2.1.1.3 Perovskite Solar Cells

PvSCs have generated significant interest in both academic and commercial sectors due to their outstanding properties and the ease with which they can be fabricated using solution processing methods. These perovskite materials, particularly those with specific light-absorbing characteristics, exhibit a remarkable range of attributes, including a high extinction coefficient, a tunable bandgap that varies with the composition of the perovskite, a low rate of carrier recombination, and long carrier diffusion lengths that can exceed one micron [31–33]. The structure of PvSC consists of essential components such as metal contacts, an electron transport layer (ETL), and a hole transport material (HTM), as illustrated in Figure 8.2(c). The ETL plays a crucial role by collecting light and generating charge carriers when the material absorbs light. The HTM intervenes to collect and transport the hole carriers as the ensuing electron–hole pairs are successfully separated. Researchers have looked into mesoscopic, planar, and inverted architectural designs for PvSC during the past ten years [34].

In addition, perovskite tandem devices used with silicon-based solar cells are becoming more and more popular. Due to its ability to outperform solo silicon cells in terms of efficiency, this tandem technique is generating a lot of attention [35]. The combination of silicon-based and perovskite technologies lays up the possibility of even more effective and long-lasting solar energy conversion.

8.2.1.2 Supercapacitor

The energy storage component is crucial in the context of PSCs. Capacitors and SCs have become viable alternatives to rechargeable batteries, which have historically dominated the energy storage market in electronic gadgets. Although the charge storage process of SCs is similar to that of normal capacitors, their energy storage capability is truly outstanding, having the ability to store almost 105 times more energy than conventional capacitors [36]. They do, however, have lower energy densities than batteries, which is a crucial distinction to make.

The primary advantage of SCs lies in their rapid charging/discharging capability and excellent cyclability. These unique attributes make them particularly well suited for applications within the field of PSCs, where the capacity for rapid energy storage and release is critically important.

Because of the electric field-induced migration of electrolytic ions, charge storage in SCs mostly occurs at the interface between the electrode and the electrolyte [37]. The key to achieving high specific capacitance in SCs lies in their electrode materials, which feature expansive surface areas, nano-scale dielectric distances, and additional faradaic contributions induced by pseudocapacitance [38].

Depending on the mechanism used in the energy storage process, SCs are classified as either electric double-layer capacitors (EDLCs) or pseudocapacitors [39].

8.2.1.2.1 Electric Double-Layer Capacitors

EDLCs stand out among the wide variety of SCs offered on the market as one of the most popular and in-depth investigated versions. The distinctive behavior of the EDLC is caused by the formation of an electric double layer (EDL) at the interface between the electrolyte and the conductive and electrochemically active electrode material. Static electricity can be stored using this amazing method without the need for electron movement between the electrode and the electrolyte [40] (Figure 8.3(a)). When electric polarization takes place at the interface between the electrode and the electrolyte, the creation of the EDL is visible [41]. Notably, neither the anode nor the cathode electrodes are affected by the charge, which is only held inside this double layer. The lack of any faradaic reactions inside the charge storage mechanism is a distinguishing feature of EDLCs. Due to their large surface area and strong electrical conductivity, carbon-based materials, such as graphene, activated carbon, carbon nanotubes (CNTs), and other carbonaceous nanomaterials, display EDL behavior [42, 43]. Two electrodes are separated by ion-permeable membranes within an electrolyte in the creation of SCs using electrodes attached to metallic current collectors [44]. It is important to note that the charge storage capability of EDLCs is somewhat constrained since double-layer production relies primarily on the electrode surface rather than utilizing the full electrode bulk.

Additionally, studies have examined how pore size affects how well EDLCs store energy. Studies have shown that electrode materials with ultra-micropores may store more charge. This finding may be explained by the presence of pseudofaradaic contributions in the development of double layers, where the pore size nearly matches the ionic size [45].

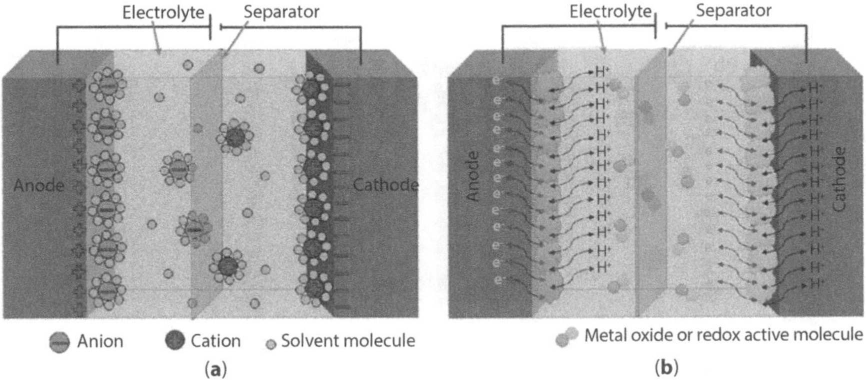

Figure 8.3 Depiction of the working mechanism of (a) EDLC-based supercapacitor and (b) pseudocapacitance-based supercapacitor. Reprinted with permission from [38].

8.2.1.2.2 Pseudocapacitors

Pseudocapacitors, which have a charge storage mechanism derived from faradaic processes, constitute a separate class of energy storage technologies. Rapid electron transfer activities take place at or close to the electrode surface of these pseudocapacitors when the electrode material interacts with the electrolyte. The distinctive electrochemical characteristics of pseudocapacitors are quasi-rectangular cyclic voltammetry (CV) and quasi-triangular galvanostatic charge–discharge (GCD) curves, which are caused by the faradaic nature of their energy storage mechanism. In terms of electrochemical behavior, pseudocapacitors fall somewhere in the middle, between solid-state diffusion processes and the static EDL. The discovery of ruthenium oxide (RuO_2), which had symmetric and reversible CV behavior [46], led to the first study of pseudocapacitance in 1971. This behavior was later elucidated as reversible redox reactions in 1990 [47] (Figure 8.3(b)).

Conway divided the types of faradaic charge storage in pseudocapacitors into three groups in 1999 [48, 49]:

1. Adsorption pseudocapacitance, also known as underpotential deposition, is the process of depositing a monolayer of adsorbed ions while simultaneously transferring charge (Figure 8.4(a)).
2. Redox pseudocapacitance, in which ions interact with the electrode material in faradaic processes (Figure 8.4(b)).
3. Intercalation pseudocapacitance, which results from ions intercalating into a host material's surface and subsurface

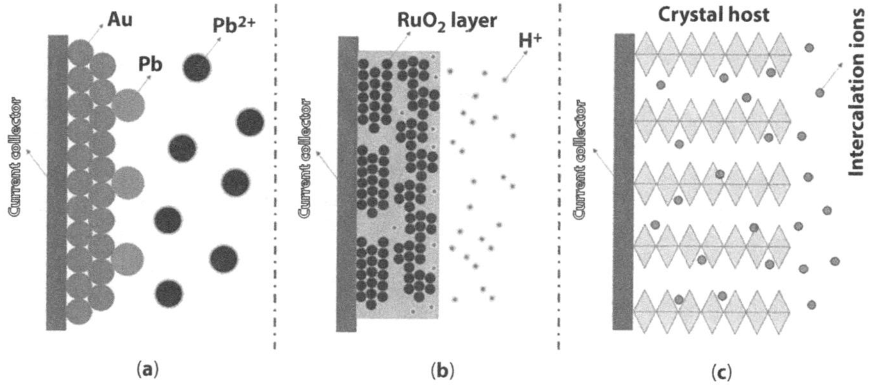

Figure 8.4 Mechanism of charge storage in (a) underpotential, (b) redox, and (c) intercalation type of pseudocapacitance. Reprinted with permission from [48].

layers. In contrast to battery behavior, this mechanism involves charge-transfer processes but does not result in crystallographic phase shifts (Figure 8.4(c)).

When compared to EDL materials, pseudocapacitive materials often have larger theoretical capacitance, making them interesting options for high-energy and high-power devices. Their limited conductivity and poor cycle stability are obstacles to their general use, demanding more research and development efforts to get beyond these drawbacks.

8.3 Different Integration Methods of PSCs

8.3.1 PSCs Involving DSSC-Based Charging Unit

As early as in 2004, Miyasaka and Murakami were able to integrate DSSC device with an SC system based on two electrodes and generate a discharge capacity of 75 mC/cm^2 and a voltage of 0.45 V [50]. The structural design of PSCs consists of essential components, including a photoelectrode optimized for effective sunlight absorption, a counter electrode, and a redox-free liquid electrolyte. The counter electrode and the photoelectrode both exhibit a heterojunction featuring a porous activated carbon layer, which enhances their charge storage capabilities. Notably, photogenerated electrons as well as holes get stored as dual-layer charges near the interface between the activated carbon electrode and the electrolyte. A significant challenge in this two-electrode configuration arises during the discharge process, where there is a risk of electrons retracing their path from the energy storage electrode back to the photoelectrode due to Schottky barriers present in the semiconductor associated with the photoelectrode. This backtracking can severely hinder the discharge process.

Moreover, a novel kind of bifunctional internal electrode has been incorporated amid the photoelectrode and the counter electrode. This feature allows for redox electron transfer on one side while storing charge on the other, thereby improving the overall performance of the system, with charge-state voltage reaching up to 0.8 V [51]. Since then, a wide variety of combinations have been created based on the three-electrode sandwich structure. Platinum has long been used as the counter electrode in DSSCs, but due to its high cost and corrosion vulnerability, there has been a shift toward less expensive and more ecologically friendly alternatives, like materials based on carbon and polymers. Conductive polymers have garnered attention as potential electrode materials for SCs due to their high specific capacitance, excellent electrical conductivity, affordability, dominating charge density, and ease of manufacture.

The mechanical characteristics of DSSC-based PvSC are another crucial component. Because of their inexpensive production methods, conventional DSSCs are usually constructed on glass substrates that are hard and brittle. While glass substrates have contributed to the widespread adoption of DSSCs due to cost efficiency, their lack of flexibility has led to increased interest in flexible DSSCs made from fluorine-doped tin oxide (FTO) plastic materials (Figure 8.5). These alternatives offer significant flexibility, simplify mass production processes, and reduce weight. As an alternative for glass substrates, ITO-polyethylene naphthalate or even polyethylene terephthalate has been in use to create TiO_2-coated photoanodes, allowing for greater shape flexibility. However, it is crucial to recognize a significant limitation of polymer-based flexible DSSCs: they function optimally at lower temperatures, generally below 150°C. Operating above this temperature can lead to degradation. In contrast, glass-based DSSCs can operate effectively at much higher temperatures, typically between 450 and 500°C. Such an elevated annealing heat improves the bonding of photoanode to the glassy substrate, ensuring reliable performance across a broader temperature range.

In the making of plastic-based DSSCs, the traditional high-temperature sintering process, typically conducted at temperatures ranging from 450 to 550°C, is inapplicable. Consequently, alternative techniques such as chemical sintering, the use of binder additives, and mechanical compression methods become invaluable for enhancing particle connectivity in these systems. Among these approaches, the mechanical compression method

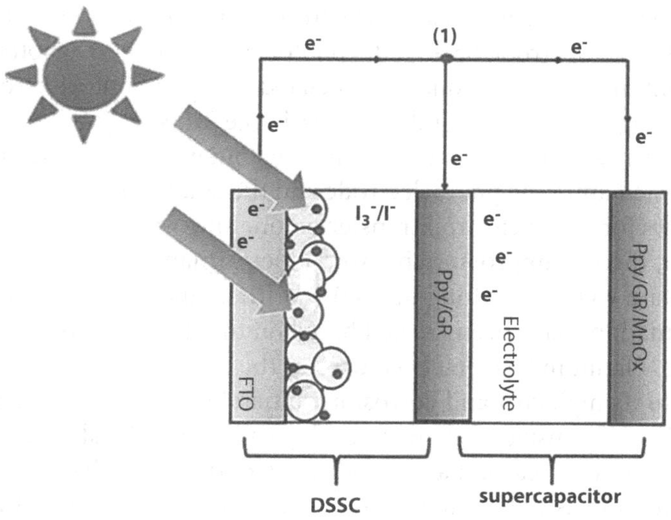

Figure 8.5 Schematic representation of DSSC-based photosupercapacitor. Reprinted with permission from [53].

stands out as the most efficient means of preparing the TiO_2 film. It offers distinct advantages in terms of ease of process control and considerably shorter processing times [52–54].

In the evolution of PSCs, the selection for SC electrode materials led the researchers to opt for a flexible and conducting polymer film known as PEDOT (poly(3,4-ethylenedioxythiophene)). This choice was driven by several appealing attributes, including its commendable conductivity, swift switching capability, low toxicity profile, mechanical flexibility, and impressive specific capacitance, which can reach values approaching 5 F cm^{-2}. These qualities collectively position PEDOT as a highly promising candidate for use in SC applications [55].

An exciting development in the realm of photoelectrochemical solar energy conversion devices involves integration of a DSSC that utilizes TiO_2 coupled with dye along with redox flow batteries (RFBs) containing 2,2,6,6-tetramethyl-1-piperidinyloxy/1,4-benzoquinone [56]. This innovative device captures solar energy to charge itself without requiring an external bias and can discharge similarly to traditional RFBs over multiple cycles while maintaining an energy storage density of 3.5 mAh/g. One of the standout features of RFBs is their extensive list of advantages, including prolonged lifespans, cost-effectiveness, and versatile design, all of which render them highly appealing for large-scale electrical energy storage applications. These systems rely on two soluble redox couples to facilitate their energy storage processes [56].

8.3.2 PSCs Involving OPV-Based Charging Unit

OPV devices that utilize conducting polymers have emerged as a promising technology, receiving significant attention in research over the last twenty years. These polymers, which are employed for photosensitization, integrate both donor and acceptor materials, leading to commendable photo-conversion efficiencies (Figure 8.6) [11]. A notable trend in contemporary research is when a solar cell device is combined with an SC to realize bifunctional capabilities. Several interesting studies have been published on this topic.

One major challenge faced by DSSC-based PSCs is the utilization of liquid redox electrolyte medium, which can hinder the assembly of elastic devices because of the risk related to electrolyte leakage. In contrast, organic PSCs do not rely on liquid-based electrolytes, making them ideal candidates for flexible electronics. This characteristic opens exciting opportunities for creating versatile and flexible electronic devices. Srinivasan *et al.* developed a printable system that combines a solar device of bulk heterojunction made from poly(3-hexylthiophene) and phenyl-C61-butyric acid methyl ester with an SC [8]. The resulting device demonstrated a 43% reduction in internal resistance compared to

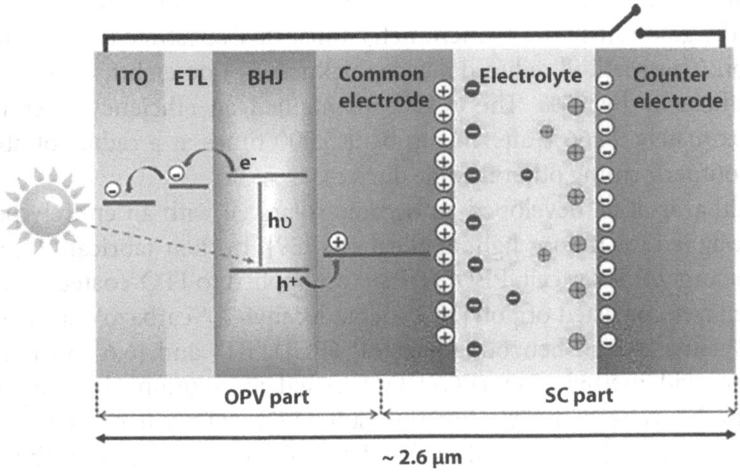

Figure 8.6 Schematic illustration of organic photovoltaic-based photosupercapacitor. Reprinted with permission from [11].

systems using wire connections. A CNT coated with Al layer aided as a counter electrode, while a variant of CNT/PVA-H_3PO_4 network was used for charge storage, yielding a capacity of 28 Fg^{-1}. The device was lightweight and thin, with electrolyte leakage addressed through patterned molding on the electrode.

A stable, flexible, and fiber form of PSC was fabricated by Zhitao *et al.* [57]. For the photovoltaic component, nanotubes of titania were perpendicularly grown on the surface of a titanium wire *via* electrochemical anodization, tailed by the application of poly(3-hexylthiophene):phenyl-C61-butyric acid methyl ester (P3HT:PCBM) and poly(3,4-ethylenedioxythiophene):poly(styrene sulfonate) (PEDOT:PSS) layers. Multi-walled carbon nanotube (MWCNT) sheets were wrapped around the improved titanium wire in order to complete the photovoltaic section. As an SC part, a poly(vinyl alcohol) (PVA)/H_3PO_4 gel electrolyte was coated onto the titania nanotube-modified titanium wire, with sheets of MWCNT attached to finalize the device.

Additionally, another integrated device measuring less than 50 micrometers in thickness was created using an OPV device built on a 1.3-micrometer-thick transparent polyimide substrate [58]. A treated ZnO layer served as an ETL, while the bulk heterojunction made from poly[4,8-bis(5-(2-ethylhexyl)thiophen-2-yl)benzo[1,2-b;4,5-b′]dithiophene-2,6-diyl-alt-(4-octyl-3-fluorothieno[3,4-b]thiophene)-2-carboxylate-2-6-diyl] (PBDTTT-OFT) polymer and [6,6]-phenyl-C71-butyric acid methyl ester (PC71BM) formed the photoactive layer. A 30-nanometer-thick film of gold was deposited as a connecting electrode, and PEDOT:PSS/CNT complex films were applied onto the gold films on a 1-micrometer-thick PET substrate to create the SC section. This method reduced device thickness by one-eighth while improving efficiency by 15%. The device maintained an efficiency retention of approximately 94.66% after being bent 5,000 times at a radius of about 2 mm, outperforming other flexible devices.

Balthazar *et al.* developed an organic solar cell with an entirely printed SC designed for indoor light applications [59]. In their fabrication process for the organic solar cell, PEDOT:PSS was applied to ITO-coated glass. The active layer consisted of poly[N-9′-heptadecanyl-2,7-carbazole-alt-5,5-(4,7-di-2-thienyl-2′,1′,3′-benzothiadiazole)] (PCDTBT) and [6,6]-phenyl-C71-butyric acid methyl ester (PC71BM) mixed in solution (1:3.7 in ortho-dichlorobenzene with 5% dimethyl sulfoxide), followed by coating with diluted polyethylenimine ethoxylated (PEIE) to complete the photoconversion component of the device. The SC was printed using a mixture of N-methyl-2-pyrrolidone and powdered carbon as an electrode alongside a gel polymer electrolyte [59].

8.3.3 PSCs Involving Perovskites-Based Charging Unit

Halide perovskites, known for their exceptional energy conversion properties, have recently expanded their horizons to encompass energy storage applications as well. A schematic illustration of the device has been depicted in Figure 8.7. This exciting development highlights the multifunctional potential of halide perovskites, where a single material can excel in both solar energy conversion and energy storage.

A notable example of this versatility comes from the work of Michael De Volder and his team [60] who introduced a multifunctional 2D halide perovskite, specifically $(C_6H_9C_2H_4NH_3)_2PbI_4$. This perovskite material effectively serves dual roles, functioning as both a light harvester and a catalyst for photocharging. To enable efficient transport of photogenerated electrons from the perovskite layer to the current collector (copper), materials like phenyl-C61-butyric acid methyl ester (PCBM) and reduced form of graphene oxide (rGO) are utilized. Since the LUMO of $(C_6H_9C_2H_4NH_3)_2PbI_4$ (-3.0 eV) is higher than that of PCBM (-3.8 eV) and copper (-4.6 eV), this choice is predicated on the materials' advantageous energy characteristics. Positively charged photogenerated holes are produced during the photocharging process, and these holes are expected to either interact with lithium iodide (LiI) to produce Li^+ ions or expel Li^+ ions from the perovskite matrix. Concurrently, the photoelectrons accumulate on the counter electrode, facilitating the reduction of Li^+ ions in the electrolyte. This coordinated interaction completes the photocharging cycle of the battery, underscoring the versatile potential of halide perovskites in energy conversion and storage applications.

Xu *et al.* undertook an innovative project by developing an entirely mesoporous perovskite-built PSC [61]. In their remarkable form of integration, a PEDOT-carbon electrode served as an essential intermediary junction electrode, fulfilling two functions within the SC, which was symmetrically assembled by two electrodes composed of MWCNTs, polypyrrole nanofibers, and bacterial cellulose membrane. Concurrently, this electrode was integral to the solar cell component, featuring a photoanode constructed from a mesoporous n-i-p structure of $CH_3NH_3PbI_3$. The fabrication process involved creating the photoanode on FTO glass using a combination of ZrO_2, TiO_2, and perovskite materials. The layer of perovskite ($CH_3NH_3PbI_3$) was applied using a direct dipping method with precursors. Through electropolymerization carried out within a suitable voltage range, the intermediate electrode was meticulously deposited onto the substrate. In order to maximize performance, the $LiClO_4$ electrolyte was notably positioned between the ClO_4^- PEDOT completely

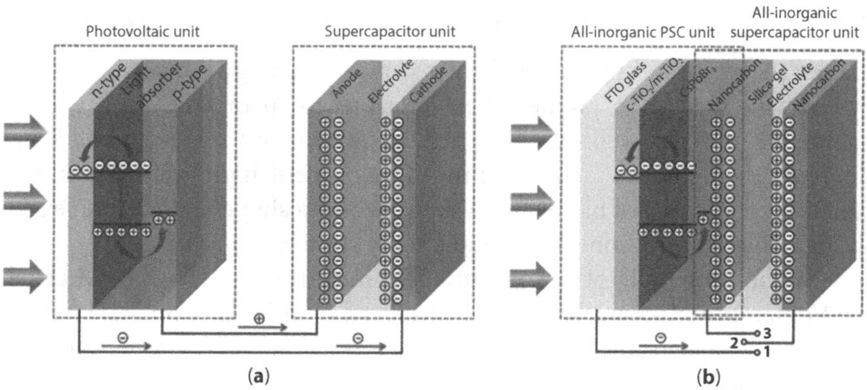

Figure 8.7 Schematic illustration of perovskite-based photosupercapacitor. Reprinted with permission from [35].

mesoporous layers. Through the procedures of photocharging and discharging, the functionality of this integrated device was assessed under light illumination. Surprisingly, the SC equivalent showed a considerable amount of energy storage capability by reaching a very high voltage of 0.7 V in just 7 seconds.

By creating an integrated asymmetric SC that blends in seamlessly with inorganic perovskite-based solar cell systems based on cesium lead halide, Huey *et al.* made a groundbreaking contribution [62]. A distinctive feature of this integration is its lack of external wire connections, resulting in a more efficient and streamlined system. The electrode for photoconversion was meticulously constructed with a layer of mesoporous titania as the semiconducting material, then $CsPbBr_3$-xI_x was used as light-harvesting component, and for the purpose of hole transport, spiro-OMeTAD was used. An asymmetric module with a PEDOT:PSS-coated RZCo layer for the positive electrode and PyR for the negative electrode was skillfully combined with this well-designed photoanode. To enhance performance, they adjusted the ratios of Br to I in the $CsPbBr_3$-xI_x photoanode. This modification proved crucial since higher iodine concentrations were linked to decreased stability. Impressively, the solar cell within this integrated module exhibited remarkable durability, retaining 70% of its power conversion efficiency (PCE) after a week under highly humid environments. Even after 24 hours in similar challenging environments, the device maintained an impressive 33% retention in PCE [62].

8.4 Efficiency of PSCs

The domain of solar cell devices encompasses various types, including silicon solar cells, DSSCs, OPVs, and PvSCs. These diverse technologies have found successful integration with SCs, giving rise to the concept of self-charged SCs.

In this integrated PSC module, the solar cell component is responsible for capturing solar energy and quickly converting it into electrical energy, which is a crucial aspect of the system's functionality. Conversely, the energy storage section of the module serves as a reservoir for the electrical energy produced by the solar cells. When connected to an external load, this initiates the discharge process, allowing for power distribution. To evaluate the overall efficiency of this integrated system, a straightforward calculation method is used. This method involves multiplying the productivities of the individual solar cell and SC components. This approach provides valuable insights into the system's performance and its effectiveness in capturing, storing, and distributing energy efficiently.

$$\eta_{TE} = \eta_S \eta_{SC}$$

The following equation is used to calculate the efficiency of photoconversion of any photovoltaic system on the exposure of light.

$$\eta_{PC} = \frac{V_{OC} J_{SC} FF}{P_{in}}$$

Here V_{OC}, J_{SC}, FF, and P_{in} represent the open-circuit voltage, short-circuit current, fill factor, and incident light power density, respectively [14].

For an integrated PC module, the overall efficiency is evaluated by keeping the capacitance in consideration. To assess the specific capacitance of an SC, one of three mathematical equations is utilized, with each equation designed for a particular electrochemical technique. Below, we explore the mathematical formulas used to determine specific capacitance through CV, GCD curves, and electrochemical impedance spectroscopy.

$$C = \frac{1}{\frac{dv}{dt}}$$

$$C = \frac{I(t_2 - t_1)}{v_2 - v_1}$$

$$C = \frac{-1}{2\pi f Im(z)}$$

where I represents the current, t represents the time, dv/dt represents the potential scan rate in the CV, and frequency is represented by f.

Power density (P_A), energy density (E_A), and output energy (E_{out}) of an SC during the process of charging can be assessed by the use of the following equation.

$$E_{out} = \frac{1}{2}CV^2$$

$$P_A = \frac{E_A}{3600t}$$

$$E = \frac{0.5C_A V^2}{3600}$$

Overall efficiency of the PSC system is evaluated using the following equations.

$$\eta_{overall} = \frac{E_{out}}{E_{light} A_{solar}}$$

$$E_{light} = P_{in} t$$

Here, E_{light} denotes the energy the solar cell receives throughout the charging process, and A_{solar} denotes the solar cell's area [14]. The performance of various PSC's have been listed in Table 8.1.

Table 8.1 Summarizing the performance of various PSC cited in this chapter.

Sl. no.	Photovoltaic	Supercapacitor	Specific capacitance	Photocharge voltage	Overall efficiency	Reference
01	DSSC	Activated carbon (EDLC)	0.69 F/cm^2	>0.45 V	32 %	[50]
02	DSSC	PProDOT-Et2	6.5 F/cm^2	0.75 V	0.6 %	[51]
03	DSSC	PEDOT	5 F/cm^2	--	--	[55]
04	OP	CNT	28 Fg^{-1}	0.6 V	3.39 %	[8]
05	OP	MWCNT	--	--	1.01 %	[57]
06	OP	PEDOT:PSS/CNT coated on Au	20.13 mF/cm^{-2}	--	6 %	[58]
07	OP	NMP/carbon powder	130 mF/cm^{-2}	--	2.9 %	[59]
08	PvSC	PCBM / rGO	100 mAh/g	3.05 V	--	[60]
09	PvSC	MWCNT/Polypyrrole	8.5 mF/cm^2	0.7 V	6.37 %	[61]
10	PvSC	PEDOT:PSS-RZCo/ PyR	150 mF/cm^2	--	--	[62]

8.5 Challenges and Future Perspectives

Since solar cell-based integrated devices are still in their infancy, further research is required to ensure their continuous development and practical applications. Future developments in PSC technology present both many potentials and difficulties. Future studies should examine crucial technical factors in order to fully realize the likely potential of PSCs and encourage their broad application. Improving our knowledge of the chemistry and materials science that underpin these devices is crucial.

The creation of sophisticated materials for the photovoltaic and SC components of PSCs may be a key area for advancement. New materials with special qualities that could significantly improve the efficiency and sustainability of PSC devices include metal halide perovskites, organic semiconductors, and two-dimensional materials.

Addressing the challenges associated with liquid electrolytes, which are susceptible to leakage and packing issues, is another critical focus area. Research should prioritize the creation of high-quality solid-state electrolytes that can overcome these limitations.

Additionally, there is an urgent need for studies aimed at enhancing the operational efficiency and cycling characteristics of these devices. While new methods have been suggested, achieving the reliability necessary for real-time applications and commercial viability remains a significant hurdle. SDSCs present a blend of challenges and opportunities, with the potential to transform energy systems. Therefore, substantial research efforts are essential for advancing these innovative integrated devices.

References

1. Lewis, N.S., Introduction: Solar Energy Conversion. *Chem. Rev.*, 115, 23, 12631–12632, 2015.
2. Kannan, N. and Vakeesan, D., Solar energy for future world: - A review. *Renew. Sustain. Energy Rev.*, 62, 1092–1105, 2016.
3. Chapin, D.M., Fuller, C.S., Pearson, G.L., A New Silicon p-n Junction Photocell for Converting Solar Radiation into Electrical Power. *J. Appl. Phys.*, 25, 5, 676–677, 1954.
4. Fakharuddin, A., Jose, R., Brown, T.M., Fabregat-Santiago, F., Bisquert, J., A perspective on the production of dye-sensitized solar modules. *Energy Environ. Sci.*, 7, 12, 3952–3981, 2014.

5. Yang, Z., Li, L., Luo, Y., He, R., Qiu, L., Lin, H., Peng, H., An integrated device for both photoelectric conversion and energy storage based on free-standing and aligned carbon nanotube film. *J. Mater. Chem. A*, 1, 3, 954–958, 2012.
6. Meng, Q., Wu, H., Meng, Y., Xie, K., Wei, Z., Guo, Z., High-Performance All-Carbon Yarn Micro-Supercapacitor for an Integrated Energy System, *Adv. Mater.*, 26, 24, 4100–4106, 2014.
7. Shabangoli, Y., Rahmanifar, M.S., El-Kady, M.F., Noori, A., Mousavi, M.F., Kaner, R.B., An Integrated Electrochemical Device Based on Earth-Abundant Metals for Both Energy Storage and Conversion. *Energy Storage Mater.*, 11, 282–93, March 1, 2018.
8. Wee, G., Salim, T., Lam, Y.M., Mhaisalkar, S.G., Srinivasan, M., Printable photo-supercapacitor using single-walled carbon nanotubes. *Energy Environ. Sci.*, 4, 2, 413–416, 2011.
9. Muzaffar, A., Ahamed, M.B., Deshmukh, K., Thirumalai, J., A review on recent advances in hybrid supercapacitors: Design, fabrication and applications. *Renew. Sustain. Energy Rev.*, 101, 123–145, 2019.
10. Hassan, Q., Jaszczur, M., Abdulateef, A.M., Abdulateef, J., Hasan, A., Mohamad, A., An analysis of photovoltaic/supercapacitor energy system for improving self-consumption and self-sufficiency. *Energy Rep.*, 8, 680–695, 2022.
11. Nguyen, D.C.T., Shin, J., Kim, S.-K., Lee, S.-H., Solar-Powered Supercapacitors Integrated with a Shared Electrode. *ACS Appl. Energy Mater.*, 4, 12, 14014–14021, 2021.
12. Kim, T., Song, W., Son, D.-Y., Ono, L.K., Qi, Y., Lithium-ion batteries: outlook on present, future, and hybridized technologies. *J. Mater. Chem. A*, 7, 7, 2942–2964, 2019.
13. Shao, Y., El-Kady, M.F., Sun, J., Li, Y., Zhang, Q., Zhu, M., Wang, H., Dunn, B., Kaner, R.B., Design and Mechanisms of Asymmetric Supercapacitors. *Chem. Rev.*, 118, 18, 9233–9280, 2018.
14. K, N. and Rout, C.S., Photo-powered integrated supercapacitors: a review on recent developments, challenges and future perspectives. *J. Mater. Chem. A*, 9, 13, 8248–8278, 2021.
15. Michaels, H., Rinderle, M., Freitag, R., Benesperi, I., Edvinsson, T., Socher, R., Gagliardi, A., Freitag, M., Dye-sensitized solar cells under ambient light powering machine learning: towards autonomous smart sensors for the internet of things. *Chem. Sci.*, 11, 11, 2895–2906, 2020.
16. Yoshikawa, K., Kawasaki, H., Yoshida, W., Irie, T., Konishi, K., Nakano, K., Uto, T., Adachi, D., Kanematsu, M., Uzu, H., Yamamoto, K., Silicon heterojunction solar cell with interdigitated back contacts for a photoconversion efficiency over 26%. *Nat. Energy*, 2, 5, 1–8, 2017.
17. Gaudiana, R., Third-Generation Photovoltaic Technology – The Potential for Low-Cost Solar Energy Conversion. *J. Phys. Chem. Lett.*, 1, 7, 1288–1289, 2010.
18. Olivares-Amaya, R., Amador-Bedolla, C., Hachmann, J., Atahan-Evrenk, S., Sánchez-Carrera, R.S., Vogt, L., Aspuru-Guzik, A., Accelerated computational

discovery of high-performance materials for organic photovoltaics by means of cheminformatics. *Energy Environ. Sci.*, 4, 12, 4849–4861, 2011.
19. Papež, N., Dallaev, R., Ţălu, S., Kaštyl, J., Overview of the Current State of Gallium Arsenide-Based Solar Cells. *Materials*, 14, 11, 3075, 2021.
20. Romeo, A. and Artegiani, E., CdTe-Based Thin Film Solar Cells: Past, Present and Future. *Energies*, 14, 6, 1684, 2021.
21. Saga, T., Advances in crystalline silicon solar cell technology for industrial mass production. *NPG Asia Mater.*, 2, 3, 96–102, 2010.
22. Chen, H.-W., Hsu, C.-Y., Chen, J.-G., Lee, K.-M., Wang, C.-C., Huang, K.-C., Ho, K.-C., Plastic dye-sensitized photo-supercapacitor using electrophoretic deposition and compression methods. *J. Power Sources*, 195, 18, 6225–6231, 2010.
23. Murakami, T.N., Kawashima, N., Miyasaka, T., A high-voltage dye-sensitized photocapacitor of a three-electrode system. *Chem. Commun.*, 26, 3346–3348, 2005.
24. Cole, J.M., Pepe, G., Al Bahri, O.K., Cooper, C.B., Cosensitization in Dye-Sensitized Solar Cells. *Chem. Rev.*, 119, 12, 7279–7327, 2019.
25. Arjunan, T.V. and Senthil, T.S., Review: Dye sensitised solar cells. *Mater. Technol.*, 28, 1–2, 9–14, 2013.
26. Zhang, S., Yang, X., Numata, Y., Han, L., Highly efficient dye-sensitized solar cells: progress and future challenges. *Energy Environ. Sci.*, 6, 5, 1443–1464, 2013.
27. Marinova, N., Valero, S., Delgado, J.L., Organic and perovskite solar cells: Working principles, materials and interfaces. *J. Colloid Interface Sci.*, 488, 373–389, 2017.
28. Riede, M., Spoltore, D., Leo, K., Organic Solar Cells—The Path to Commercial Success. *Adv. Energy Mater.*, 11, 1, 2002653, 2021.
29. Di Carlo Rasi, D. and Janssen, R.A.J., Advances in Solution-Processed Multijunction Organic Solar Cells. *Adv. Mater.*, 31, 10, 1806499, 2019.
30. Tong, Y., Xiao, Z., Du, X., Zuo, C., Li, Y., Lv, M., Yuan, Y., Yi, C., Hao, F., Hua, Y., Lei, T., Lin, Q., Sun, K., Zhao, D., Duan, C., Shao, X., Li, W., Yip, H.-L., Xiao, Z., Zhang, B., Bian, Q., Cheng, Y., Liu, S., Cheng, M., Jin, Z., Yang, S., Ding, L., Progress of the key materials for organic solar cells. *Sci. China Chem.*, 63, 6, 758–765, 2020.
31. Ansari, M.I.H., Qurashi, A., Nazeeruddin, M.K., Frontiers, opportunities, and challenges in perovskite solar cells: A critical review. *J. Photochem. Photobiol. C Photochem. Rev.*, 35, 1–24, 2018.
32. Kim, J.Y., Lee, J.-W., Jung, H.S., Shin, H., Park, N.-G., High-Efficiency Perovskite Solar Cells. *Chem. Rev.*, 120, 15, 7867–7918, 2020.
33. Correa-Baena, J.-P., Saliba, M., Buonassisi, T., Grätzel, M., Abate, A., Tress, W., Hagfeldt, A., Promises and challenges of perovskite solar cells. *Science*, 358, 6364, 739–744, 2017.
34. Liang, J., Zhu, G., Wang, C., Zhao, P., Wang, Y., Hu, Y., Ma, L., Tie, Z., Liu, J., Jin, Z., An all-inorganic perovskite solar capacitor for efficient and stable spontaneous photocharging. *Nano Energy*, 52, 239–245, 2018.

35. Yan, L.L., Han, C., Shi, B., Zhao, Y., Zhang, X.D., A review on the crystalline silicon bottom cell for monolithic perovskite/silicon tandem solar cells. *Mater. Today Nano*, 7, 100045, 2019.
36. Jiang, Y. and Liu, J., Definitions of Pseudocapacitive Materials: A Brief Review. *Energy Environ. Mater.*, 2, 1, 30–37, 2019.
37. Simon, P. and Gogotsi, Y., Materials for electrochemical capacitors. *Nat. Mater.*, 7, 11, 845–854, 2008.
38. Wang, Y., Song, Y., Xia, Y., Electrochemical capacitors: mechanism, materials, systems, characterization and applications. *Chem. Soc. Rev.*, 45, 21, 5925–5950, 2016.
39. Chen, X., Paul, R., Dai, L., Carbon-based supercapacitors for efficient energy storage. *Natl. Sci. Rev.*, 4, 3, 453–489, 2017.
40. Henderson, D., Recent progress in the theory of the electric double layer. *Prog. Surf. Sci.*, 13, 3, 197–224, 1983.
41. Schmickler, W., Electrochemical Theory | Double Layer, in: *Encyclopedia of Electrochemical Power Sources*, J. Garche, (Ed.), pp. 8–13, Elsevier, Amsterdam, 2009.
42. Simon, P. and Gogotsi, Y., Charge storage mechanism in nanoporous carbons and its consequence for electrical double layer capacitors. *Philos. Trans. R. Soc. Math. Phys. Eng. Sci.*, 368, 1923, 3457–3467, 2010.
43. Sharma, P. and Bhatti, T.S., A review on electrochemical double-layer capacitors. *Energy Convers. Manag.*, 51, 12, 2901–2912, 2010.
44. Kim, B.K., Sy, S., Yu, A., Zhang, J., Electrochemical Supercapacitors for Energy Storage and Conversion, in: *Handbook of Clean Energy Systems*, pp. 1–25, John Wiley & Sons, Ltd, 2015.
45. Largeot, C., Portet, C., Chmiola, J., Taberna, P.-L., Gogotsi, Y., Simon, P., Relation between the Ion Size and Pore Size for an Electric Double-Layer Capacitor. *J. Am. Chem. Soc.*, 130, 9, 2730–2731, 2008.
46. Trasatti, S. and Buzzanca, G., Ruthenium dioxide: A new interesting electrode material. Solid state structure and electrochemical behaviour. *J. Electroanal. Chem. Interfacial Electrochem.*, 29, 2, A1–A5, 1971.
47. Ardizzone, S., Fregonara, G., Trasatti, S., "Inner" and "outer" active surface of RuO2 electrodes. *Electrochim. Acta*, 35, 1, 263–267, 1990.
48. Conway, B.E., *Electrochemical Supercapacitors: Scientific Fundamentals and Technological Applications*, Springer Science & Business Media, 2013.
49. Liu, Y., Jiang, S.P., Shao, Z., Intercalation pseudocapacitance in electrochemical energy storage: recent advances in fundamental understanding and materials development. *Mater. Today Adv.*, 7, 100072, 2020.
50. Miyasaka, T. and Murakami, T.N., The photocapacitor: An efficient self-charging capacitor for direct storage of solar energy. *Appl. Phys. Lett.*, 85, 17, 3932–3934, 2004.
51. Hsu, C.-Y., Chen, H.-W., Lee, K.-M., Hu, C.-W., Ho, K.-C., A dye-sensitized photo-supercapacitor based on PProDOT-Et2 thick films. *J. Power Sources*, 195, 18, 6232–6238, 2010.

52. Lindström, H., Holmberg, A., Magnusson, E., Malmqvist, L., Hagfeldt, A., A new method to make dye-sensitized nanocrystalline solar cells at room temperature. *J. Photochem. Photobiol. Chem.*, 145, 1, 107–112, 2001.
53. Boschloo, G., Lindström, H., Magnusson, E., Holmberg, A., Hagfeldt, A., Optimization of dye-sensitized solar cells prepared by compression method. *J. Photochem. Photobiol. Chem.*, 148, 1, 11–15, 2002.
54. Ng, C.H., Lim, H.N., Hayase, S., Harrison, I., Pandikumar, A., Huang, N.M., Potential active materials for photo-supercapacitor: A review. *J. Power Sources*, 296, 169–185, 2015.
55. Snook, G.A., Peng, C., Fray, D.J., Chen, G.Z., Achieving high electrode specific capacitance with materials of low mass specific capacitance: Potentiostatically grown thick micro-nanoporous PEDOT films. *Electrochem. Commun.*, 9, 1, 83–88, 2007.
56. Zhang, W., Liu, P., Zhang, X., A solar rechargeable flow battery based on electroactive organic redox couples. *Mater. Sci. Eng.*, 452, 032068, 2018.
57. Zhang, Z., Chen, X., Chen, P., Guan, G., Qiu, L., Lin, H., Yang, Z., Bai, W., Luo, Y., Peng, H., Integrated Polymer Solar Cell and Electrochemical Supercapacitor in a Flexible and Stable Fiber Format. *Adv. Mater.*, 26, 3, 466–470, 2014.
58. Liu, R., Takakuwa, M., Li, A., Inoue, D., Hashizume, D., Yu, K., Umezu, S., Fukuda, K., Someya, T., An Efficient Ultra-Flexible Photo-Charging System Integrating Organic Photovoltaics and Supercapacitors. *Adv. Energy Mater.*, 10, 20, 2000523, 2020.
59. Lechêne, B.P., Cowell, M., Pierre, A., Evans, J.W., Wright, P.K., Arias, A.C., Organic solar cells and fully printed super-capacitors optimized for indoor light energy harvesting. *Nano Energy*, 26, 631–640, 2016.
60. Ahmad, S., George, C., Beesley, D.J., Baumberg, J.J., De Volder, M., Photo-Rechargeable Organo-Halide Perovskite Batteries. *Nano Lett.*, 18, 3, 1856–1862, 2018.
61. Xu, J., Ku, Z., Zhang, Y., Chao, D., Fan, H.J., Integrated Photo-Supercapacitor Based on PEDOT Modified Printable Perovskite Solar Cell. *Adv. Mater. Technol.*, 1, 5, 1600074, 2016.
62. Ng, C.H., Lim, H.N., Hayase, S., Zainal, Z., Shafie, S., Lee, H.W., Huang, N.M., Cesium Lead Halide Inorganic-Based Perovskite-Sensitized Solar Cell for Photo-Supercapacitor Application under High Humidity Condition. *ACS Appl. Energy Mater.*, 1, 2, 692–699, 2018.

9

Exploring the Potential of a Battery-Assisted Solar Cooking System

Mohammed Hmich[1], Bilal Zoukarh[1], Sara Chadli[1], Rachid Malek[1], Olivier Deblecker[2], Khalil Kassmi[1,3*] and Najib Bachiri[3]

[1]*Faculty of Science, Mohamed First University, LETSER Laboratory, Oujda, Morocco*
[2]*University of Mons, Faculty of Engineering-Electrical Power Engineering Unit, Mons, Belgium*
[3]*Human and Environment Association, Berkane, Morocco*

Abstract
This chapter presents research on the operation and evaluation of an innovative photovoltaic (PV) solar cooker that incorporates battery storage for enhanced efficiency, reliability, and autonomy. The system is designed to optimize energy utilization by integrating a heating plate (thermal resistor) powered by a 48 V/250 Ah battery bank, which is charged *via* 600-Wp PV panels. The charging process is managed by a BOOST-based DC/DC converter and a digital control circuit. Experiments conducted both during the day and at night, based on user needs, confirm that the stored electrical energy ensures continuous cooking, independent of sunlight fluctuations. The power conversion system operates with an efficiency exceeding 95%, while the heating element reaches 400°C within 30 minutes, with a temperature increase rate ranging from 12 to 33°C per minute. Boiling one liter of water (100°C) requires approximately 45 minutes, corresponding to a heating rate of 3.2°C per minute. The analysis of experimental data highlights that the battery system successfully delivers the necessary electrical energy for heating and daily cooking, providing an operational autonomy of up to 14 days. The overall results validate the feasibility of using a PV-powered battery system for cooking applications, catering to the energy needs of rural and urban households alike.

Keywords: Photovoltaic energy, battery storage, electrical power, solar-powered cooker, energy management, power conversion, DC/DC power regulator, electronic control unit

**Corresponding author*: khkassmi@yahoo.fr; k.kassmi@ump.ac.ma

Nurdan Demirci Sankir and Mehmet Sankir (eds.) Solar Capacitors and Batteries, (267–288) © 2025 Scrivener Publishing LLC

9.1 Introduction

For a long time, cooking has predominantly depended on energy sources like coal, firewood, and natural gas, whether for household use or in the food service industry [1, 2]. According to the International Energy Agency, global coal consumption hit a record high in 2022, rising by 3.3% and bringing total demand to 8.3 billion tonnes [3]. A study carried out in Kisangani reveals that coal accounts for 81.4% of energy consumption, while firewood represents 31% [4]. As a response to the challenges brought about by climate change and the rise in greenhouse gas emissions [5, 6], the transition to renewable and environmentally friendly energy sources [7], such as solar power, has become crucial. Photovoltaic-powered solar cookers present a viable alternative [8, 9], enabling cooking without relying on fossil fuels, firewood, or coal, thereby contributing to forest conservation and improved air quality. However, despite their benefits in terms of efficiency and continuous operation, modern solar cookers still face notable limitations: Their efficiency and reliability do not always align with user expectations, and their operation remains inconsistent due to insufficient monitoring. Furthermore, the solar cooking systems found in the literature—including parabolic cookers, Fresnel reflector cookers [10–12], box cookers [13–16], and photovoltaic-powered models [17–19]—often lack reliability and have yet to see widespread household adoption. To address these limitations, integrating battery-based energy storage and enhancing the digital control of solar cookers are essential steps toward improving their performance, efficiency, and autonomy, particularly in the absence of sunlight.

Battery energy storage systems (BESS) are available in a range of technologies, each with its own specific characteristics, adapted to different needs:

- Lithium-ion batteries are widely used in solar energy systems due to their numerous advantages. They are known for their high energy density, which allows a significant amount of energy to be stored in a compact form, and their impressive energy efficiency (85–95%), reducing energy loss during charging and discharging. Additionally, lithium-ion batteries have a long lifespan (typically 10 to 15 years or more, with around 4,000 to 6,000 cycles), making them ideal for both residential and commercial solar applications [20, 21]. These batteries are also favored for their quick response time and their ability to function across a broad range of temperatures. Their characteristics make them an excellent choice for storing energy generated by photovoltaic panels during daylight hours and delivering it when sunlight is limited or absent [22, 25].

- Sodium–sulfur (NaS) batteries, on the other hand, operate at high temperatures and are suitable for large-scale, long-term storage. They have a longer service life of around 4,500 cycles, but a slower response [20, 26].
- While lead-acid batteries are more affordable, they provide less energy storage capacity and have a reduced lifespan, typically ranging from 1,000 to 1,500 cycles, making them more suitable for backup purposes or small off-grid solar setups. In contrast, redox flow battery systems utilize liquid electrolytes stored in separate tanks, which allows for scalable energy storage. These systems are well suited for long-term applications, offering a service life of over 20 years and virtually unlimited cycles, though their efficiency is lower (65–75%) [27–30].
- Finally, lithium-iron-phosphate ($LiFePO_4$) batteries, a variant of lithium-ion batteries, are distinguished by their high thermal and chemical stability, offering enhanced safety and longer service life (2,000 to 4,000 cycles). They are commonly used for applications requiring enhanced safety [31–33].

This overview of batteries on the market shows that each type of BESS offers specific advantages in terms of energy storage requirements, charge capacity and service life. This diversity and performance of batteries enables us to meet the challenges of integrating renewable energies. In the context of our application, the performance and characteristics of lead-acid batteries are best suited to storing the electricity needed for cooking in the absence of sunlight. The batteries are charged on sunny days and supply the electrical energy needed for cooking all year round (day and night).

In addition, solar cookers suffer from regulation and optimization problems, with poorly controlled performance and insufficient monitoring. On current solar cookers, these regulation and control systems are practically non-existent [17, 19]. The integration of digital technologies would enable real-time monitoring of electrical quantities such as voltages, currents, power, and duty cycles of DC/DC converters, identifying recurring malfunctions and proposing appropriate solutions. This approach aims to optimize the design of cookers and facilitate their use in both rural and urban households.

To incorporate electrical energy storage in batteries into innovative solar cookers that are adaptable and meet the needs of domestic users, we are conducting research within the framework of the European WBI 3.3 and LEAP-RE 'SoCoNexGen' projects. The goal is to provide rural and urban users with hot-plate-style solar cookers, which combine energy generated by photovoltaic panels during sunlight hours utilizing energy stored in solar-powered batteries. The aim is to offer a reliable, accessible technology for households in regions with high solar potential year-round (e.g., Africa).

This chapter presents the initial results from the innovative solar cooker, where the electrical energy needed for cooking is supplied by lead-acid solar batteries, charged through photovoltaic panels (600 Wp) *via* a microcontroller-based Maximum Power Point Tracking (MPPT) solar charge controller. The MPPT method used is Perturbation and Observation (P&O), which optimizes the power extracted from the PV panels. Special focus is given to the electronic control system, which controls the energy supplied by the batteries to power the heating plate, made up of a thermal resistor. By comparing our results with those found in existing literature, we demonstrate the viability of operating the solar cooker by storing photovoltaic energy in batteries.

9.2 Innovative Cooker Structure

9.2.1 Specifications

The proposed prototype cooker, a griddle-type model (Figure 9.1), allows cooking using solar batteries on days with limited sunlight and at night. The cooker operates through the integration of a power board (Block 1) and a digital electrical board (Block 2). The digital board features a microcontroller (Raspberry Pi), which manages all the cooker's functions *via* the photovoltaic panels and batteries. These functions include the acquisition and display of electrical data, fault detection, and the manual or automatic operation of the cooker. Based on our fieldwork, we present the prototype cooker (Figure 9.1), which is designed to meet users' daily cooking energy needs of approximately 2 kWh/d. The proposed cooker must therefore meet the following specifications:

- Produce daily electrical energy for cooking (day and night) using batteries (48 V/250 Ah),

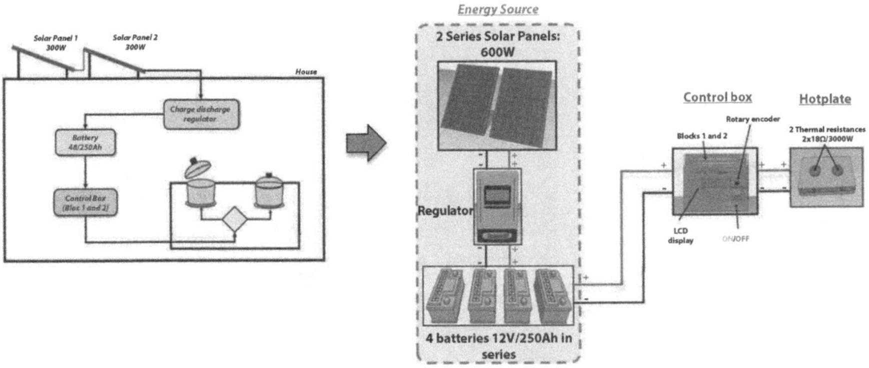

Figure 9.1 Schematic diagram of the prototype cooker to be designed and produced in the WBI and LEAP-RE projects.

- Two photovoltaic panels (600 Wp) charge the batteries, producing between 3.5 and 4.2 kWh per day,
- Cooking on nights and days with little sunshine, depending on user requirements, at 200–450 W for 5 hours (equivalent to 1–2.25 kWh/d of energy).
- The batteries' full charge (12 kWh/d) ensures 5 to 12 days' cooking autonomy, without sunshine.

9.2.2 System Schematic

Figure 9.2 shows the schematic diagram of Cooker 2 and the structure of the photovoltaic heating system (Blocks 1 and 2) developed as part of this study. The key components of this cooker include:

- Power source: This system provides energy to the cooker and is made up of 48 V/250 Ah solar batteries (four 12 V/250 Ah batteries connected in series). These batteries are charged by two 600 Wp photovoltaic panels on sunny days. They store between 4 and 5 kWh per day, ensuring autonomy for 5 to 7 days. They also guarantee the continuous operation of critical components (e.g., Raspberry Pi board, power switches).
- Heating element (length: 50 cm; width: 32 cm; height: 10 cm): It includes two 18 Ω thermal resistors. One of these resistors is powered by the solar batteries, allowing cooking even during cloudy weather or at night.

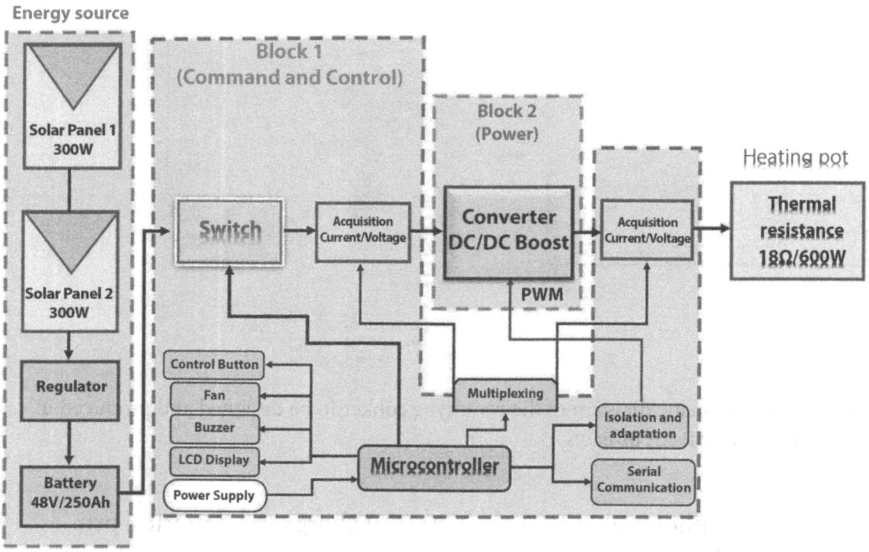

Figure 9.2 Block diagram of photovoltaic cooker and system structure (Blocks: 1 and 2).

- Power supply unit (Block 1): This unit consists of a boost DC/DC converter functioning at 20 kHz, delivering a power output of 1 kW. It is powered by the solar batteries during periods of low sunlight or at night.
- Control system (Block 2): This system includes both traditional and modern electronic circuits, supplied by the batteries *via* a voltage regulation circuit. The microcontroller manages the entire operation of the cooker, overseeing and coordinating all its functions.
- A computer (optional) for which an interface (application) is set up to store, process, and display the various electrical quantities acquired. This application can also be used to monitor operation locally and remotely (*via* the Internet). It can also be used to remotely control certain functions (system ON/OFF). This is achieved by activating/deactivating the various power relays, enabling the cooker to be switched on/off.

Drawing on field studies conducted with several rural and urban households, we developed the cooker shown in Figure 9.1 to accomplish the following functions:

- Function 1: Ensure cooker operation using solar batteries during periods of low sunlight or nighttime. This is achieved through an external control button (Button 2) equipped with a potentiometer (Potentiometer 2), which regulates the heating of the thermal resistor. The microcontroller controls this process by activating the switch and generating a PWM signal operating at 20 kHz, with an adjustable duty cycle between 0.1 and 0.45.
- Function 2: Acquire and display electrical parameters of the converters (voltages, current, and efficiency) on an LCD (Liquid Crystal Display) screen (2004 model used for the solar cooker), with both local and remote monitoring capabilities. A digital multiplexer is used to collect these values.
- Function 3: Detect potential malfunctions such as overvoltage, overcurrent, or disconnection of the thermal resistors from the DC/DC converter outputs. In case of an anomaly, the microcontroller deactivates the switch to isolate the power sources of the two converters.

9.3 Cooker Design and Operation

9.3.1 Solar Cooker Test with Battery Storage

To ensure effective monitoring and validation of the proposed cooker, it is essential to assess weather conditions (illumination levels and surrounding temperature), the characteristics (voltages, currents, and power) of the batteries and heating plate, as well as the thermal characteristics of the cooker (temperature of the heating element). To achieve this, we developed a weather station and a comprehensive computer-controlled measurement system, programmed in Python, to track meteorological conditions and the electrical and thermal performance of the cooker. The next sections give a concise summary of the structure and operation of each component of this station and measurement system.

9.3.1.1 Weather Station

As part of our work, the station is designed to collect data on ambient temperature and illuminance from the weather station installed in the laboratory, ensuring validation of the cooker's performance across different seasons of the year. Figure 9.3 shows the synoptic diagrams and equipment of the meteorological

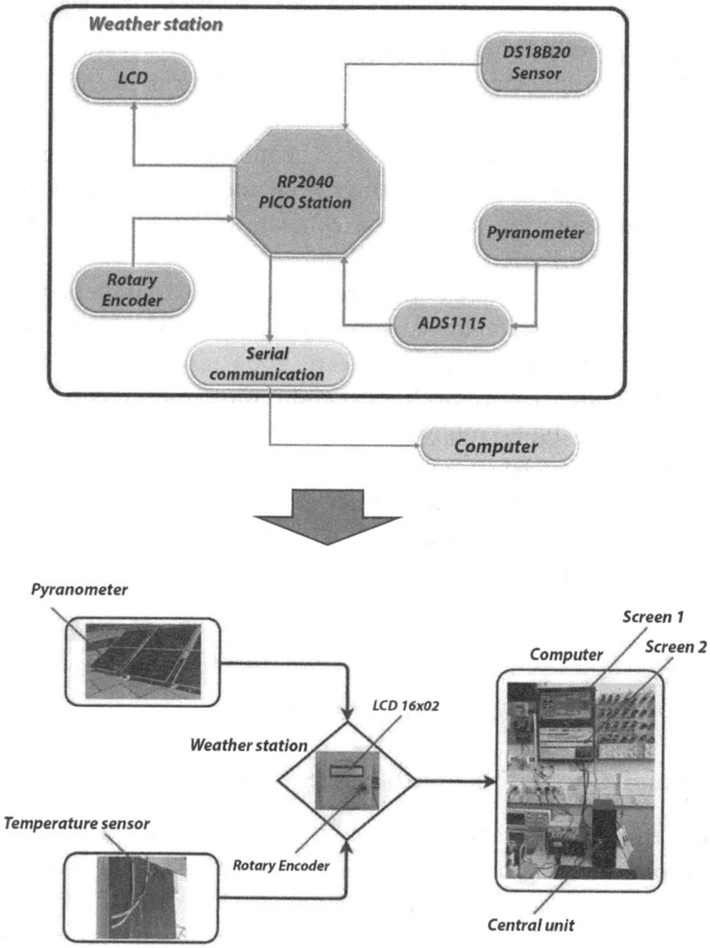

Figure 9.3 Synoptic diagram and meteorological station equipment.

station and the system for acquiring, controlling, and displaying meteorological quantities from the meteorological station. The station consists of:

- A pyranometer to measure light intensity.
- A digital sensor (DS18B20) to measure ambient temperature.
- An electronic module incorporating a 'Raspberry Pi' microcontroller to perform the following tasks:
  - Acquisition of ambient temperature *via* digital sensor.
  - Acquisition of illumination intensity, using the ADS1115 pyranometer and analog-to-digital converter.

- ➢ Transmission of measured values from the microcontroller to the local computer *via* FT232 level converter.
- ➢ LCD display of acquired values (Liquid Crystal Display, 1602 model).
- ➢ Use a rotary encoder to scroll and select, on the LCD, the type of quantities to be displayed: illuminance intensity, ambient temperature, or both simultaneously.

- A computer that receives data from the weather station box for storage, display, and subsequent analysis.

9.3.1.2 Measurement Bench

Figure 9.4 illustrates the experimental setup used to monitor both the operation of the cooker and the weather station. This comprehensive test bench has been installed as part of the research projects led by the MEER team. The setup includes:

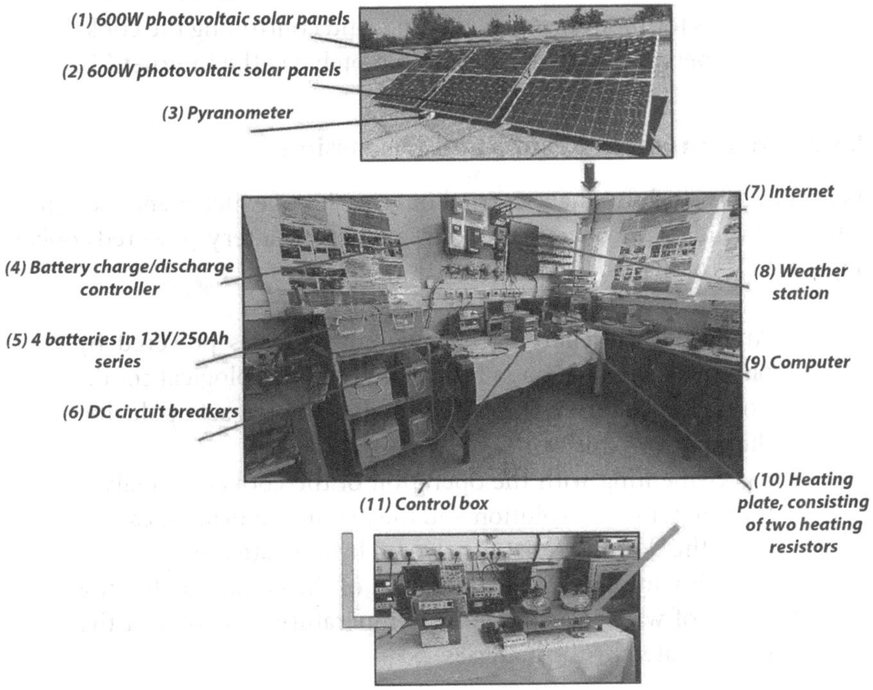

Figure 9.4 Photo of the complete measurement bench set up at the LETSER laboratory.

- Two 600-Wp photovoltaic panels (2) connected in series, supplying energy to the batteries (5) through a charging and discharging regulator (4).
- A charge/discharge controller (4) responsible for managing the charging process of the solar batteries (5) *via* the photovoltaic panels.
- Four solar batteries (5), each rated at 12 V/250 Ah, charged by the 600-Wp photovoltaic panels (2) through the charge/discharge controller (4).
- An energy management and control unit (11), powered by the batteries (5). It controls the heating of the hotplate (10) through a converter and an electronic circuit that oversees the entire cooker system.
- A heating plate (10) equipped with a heating resistor, rated at 2 kW, capable of reaching a temperature of 1000°C.
- A meteorological station (8), including a pyranometer (3) and a thermal sensor, to measure ambient temperature and solar irradiance (Figure 9.3).
- A computer setup with dual screens (9), used for programming the control board (Raspberry Pi microcontroller), as well as for monitoring, regulating, and controlling the cooker's operation both locally and remotely *via* the Internet (7).

9.3.2 Measurement Results and Discussion

As part of this study, we have carried out a series of experiments designed to test and validate the operation of our innovative battery-powered cooker (Figure 9.1). To do this, we propose:

- Analyzing the charging behavior of batteries powered by photovoltaic panels, as a function of the metrological conditions deduced from the meteorological station set up during the course of this work.
- Experimenting with the operation of the cooker to analyze the temperature evolution and the maximum heating capacity of the thermal resistor powered by the batteries.
- Experimenting with the operation of the cooker by boiling 1 liter of water to analyze the temperature evolution of the water heated by the batteries.

9.3.2.1 *Battery Charging by Photovoltaic Panels*

We monitored the charging process of the 48 V/250 Ah batteries by the 600-Wp photovoltaic panels using an MPPT solar charge controller (Figure 9.4). From this controller, we extracted the electrical parameters of both the panels and batteries, including voltages, currents, power, and efficiencies. From the meteorological station's data acquisition system (Figure 9.4), we extracted the weather conditions (illumination intensity and ambient temperature). The results obtained (Figure 9.5) show the following:

- The day's measurements were characterized by cloud passages, resulting in sudden variations in illuminance intensity, without affecting the ambient temperature value. Illuminance peaked at around 1000 W/m^2 at midday. However, ambient temperature varies between 25°C and 35°C.
- The voltages of the photovoltaic panels and batteries vary very little throughout the day. They are around 60 and 48 V, respectively. Fluctuations in PV panel voltage are due to passing clouds. Throughout the day, battery voltage is lower than that of the photovoltaic panels.
- Photovoltaic panel currents and battery charge currents vary according to variations in irradiance. If irradiance increases (decreases), currents increase (decrease). Photovoltaic panel and battery currents reach 9 and 10 A, respectively. Throughout the day, the battery current is higher than that of the photovoltaic panels.
- The energy provided by the photovoltaic panels and the battery charging rate increase with rising illuminance intensity and decrease as it declines. They reach a peak of approximately 500 W around midday.
- The regulator's efficiency is highly satisfactory, at over 95%.

All the results obtained demonstrate that the controller's power block (DC/DC converter) effectively operates as a buck converter. The voltage (current) from the photovoltaic panels is higher (lower) than that of the batteries, with an efficiency exceeding 95%. During the day, the batteries charge at a rate of approximately 64 Ah, which represents 26% of the total battery capacity (250 Ah). From this, we can conclude that, under similar meteorological conditions, a full battery charge would require 31 hours (or 4 days) of charging.

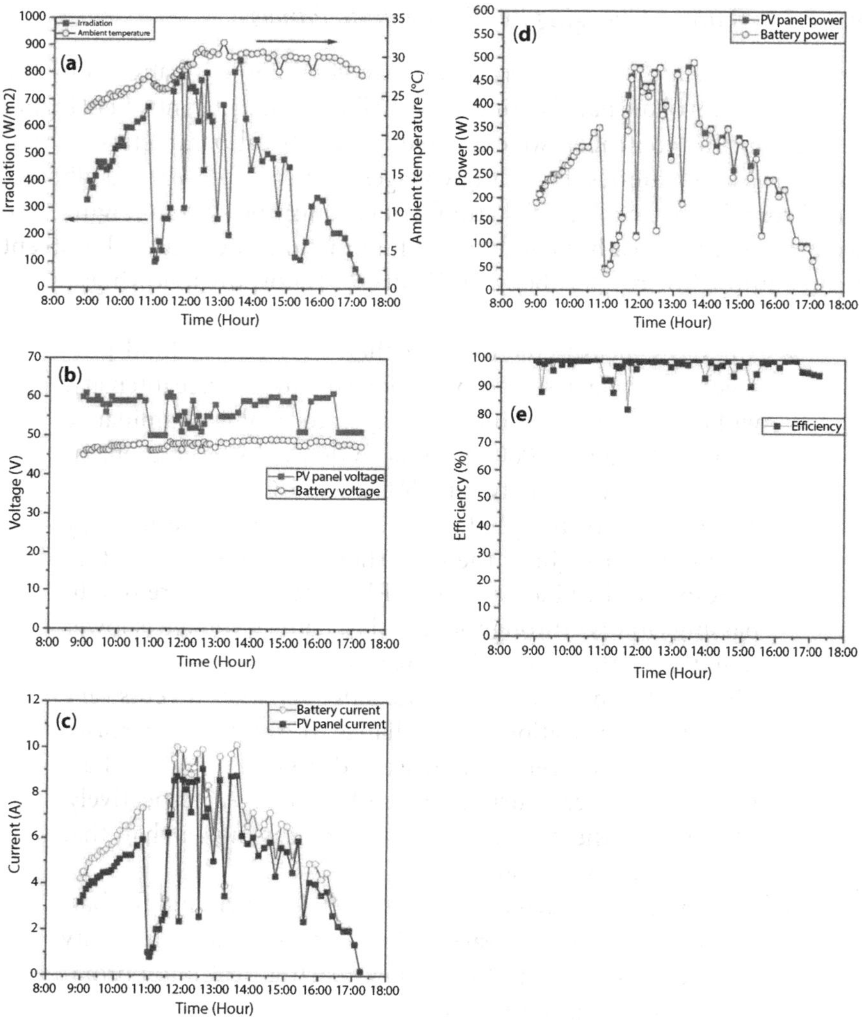

Figure 9.5 Meteorological parameters of the station and electrical parameters of the battery charge controller (Figure 9.4): (a) illuminance and ambient temperature, (b) voltages of photovoltaic panels and batteries, (c) photovoltaic panel and battery currents, (d) power of photovoltaic panels and batteries, (e) controller efficiency.

9.3.2.2 *Solar Vacuum Cooker*

To evaluate the temperature evolution of the thermal resistor, we tested the vacuum cooker for 2 hours, operating at a power of approximately 400 W. The electrical values of the batteries and the thermal resistance (voltages, currents, power) are shown in Figure 9.6. From these measurements, we

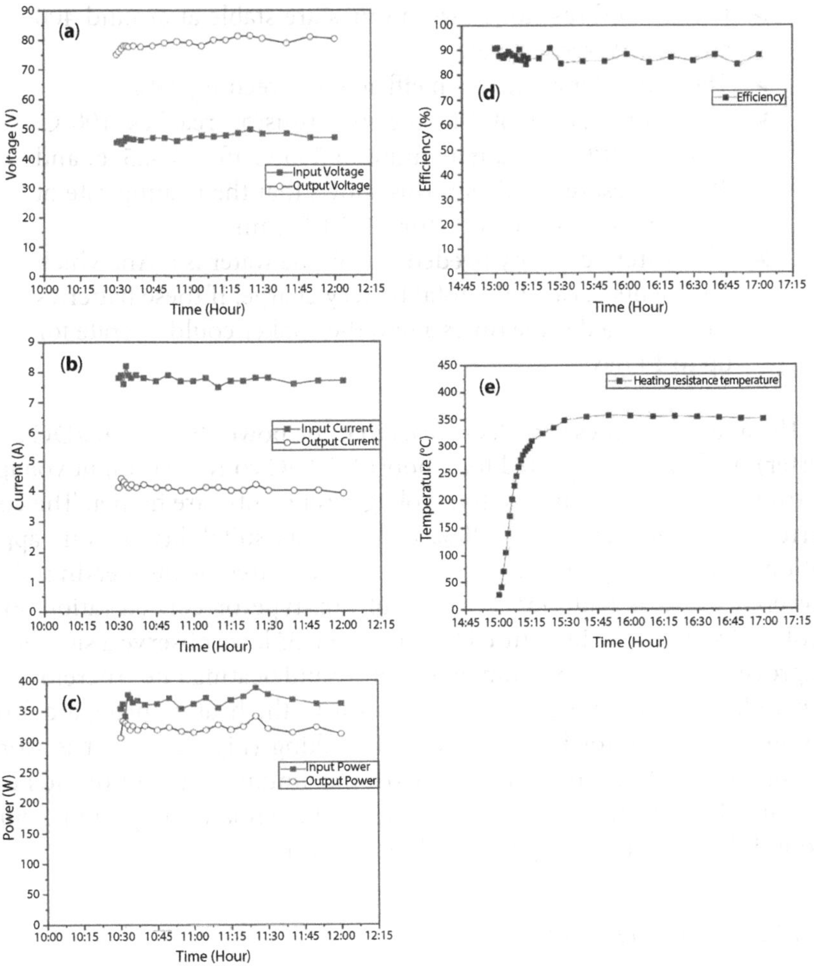

Figure 9.6 Typical cooker operation with vacuum batteries. (a) the voltages at the input and output of the converter, (b) the currents at the input and output of the converter, (c) the powers at the input and output of the converter, (d) power conversion efficiency, (e) temperature of the heating element.

deduced and plotted on the same Figure 9.6 the efficiency of the Block 1 converter (Figure 9.2). From these measurements we conclude the following:

- Battery and resistor voltages stabilize at around 48 and 78 V, respectively,
- Battery and resistor currents stabilize at around 8 and 4 A, respectively,

- Battery and resistor power ratings are stable at around 400 and 320 W, respectively,
- The converter achieves an efficiency exceeding 88%,
- The temperature of the thermal resistor reaches 100°C, 150°C, 200°C, and a maximum of 350°C after 3, 4.5, 6, and 30 minutes, respectively. This shows that the heating rate of the resistor is 0.2–0.56°C (or 12–33°C/min).
- The battery capacity needed to boil the water is 6 Ah, which represents 2.4% of the total battery charge. If these batteries were charged three times a day, the cooker could operate for up to 14 days.

The analysis of these results reveals that the power block (DC/DC converter) performs as expected for a boost DC/DC converter: input voltages (currents) are lower, while output voltages (currents) are higher. The converter's efficiency, exceeding 88%, confirms its suitability for our application. When comparing the heating temperatures of the heating plate, powered by solar batteries, to those of box-type or concentration solar cookers found in the literature [11, 14, 15, 34, 35], we observe a significant improvement in the maximum temperature and heating rate. Therefore, we conclude that both blocks 1 and 2, along with the heating plate, are functioning effectively for battery-powered cooking (Figure 9.2). It is worth noting that, as stated in Section 9.2.3, the maximum temperature and heating rate of the heating element could be further enhanced by adding rings around the element to trap heat and reduce heat loss.

9.3.2.3 Water Heating

To assess the evolution of the thermal resistance heating and cooking temperatures, we assessed the cooker shown in Figure 9.4 by boiling one liter of water with a power consumption of approximately 400 W. The electrical parameters of the batteries and thermal resistance (voltages, currents, power), along with the temperatures of the heating element and the boiling water, are presented in Figure 9.7. From these measurements, we deduced and plotted on the same Figure 9.7 the efficiency of the Block 1 converter (Figure 9.2). From these measurements we conclude the following:

- Battery and resistor voltages stabilize at around 48 and 77 V, respectively,
- Battery and resistor currents stabilize at around 8 and 4 A, respectively,

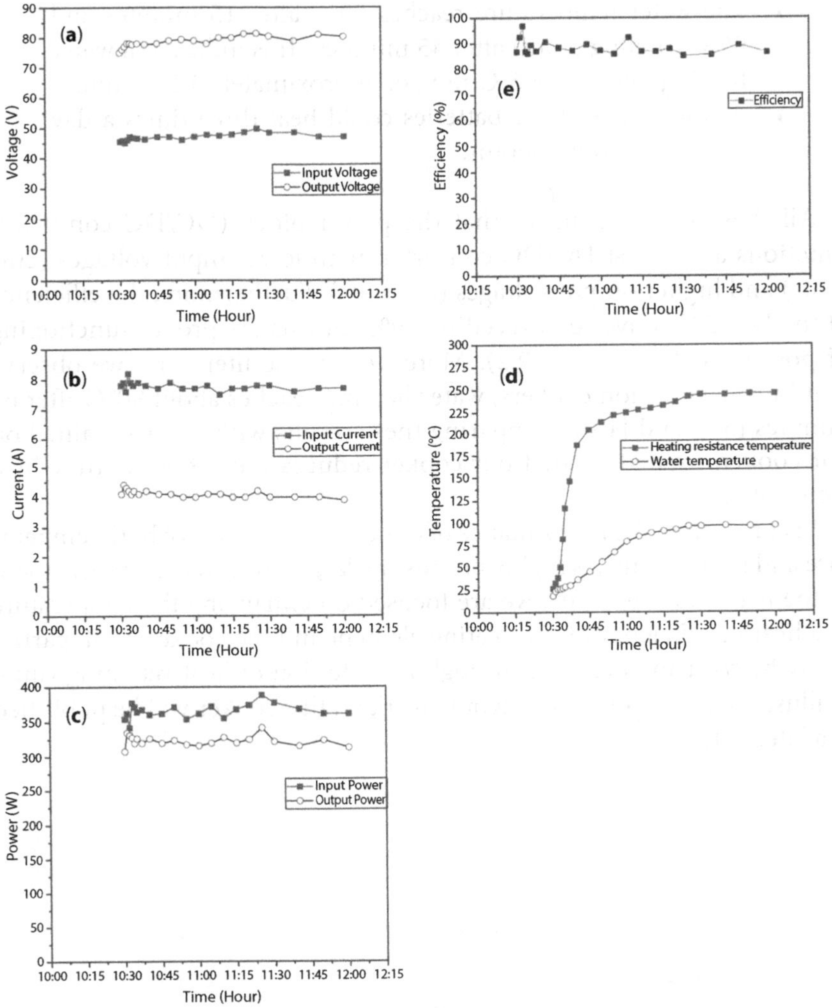

Figure 9.7 Heating 1 liter of water with the cooker shown in Figure 9.4. (a) the voltages at the input and output of the converter, (b) the currents at the input and output of the converter, (c) the powers at the input and output of the converter, (d) power conversion efficiency, (e) temperature of the heating element.

- Battery and resistor power ratings are stable at around 350 and 320 W, respectively,
- The DC/DC converter achieves an efficiency greater than 89%,
- After 45 minutes of heating, the thermal resistor's temperature peaks at 255°C,

- The water temperature reaches 60°C after 15 minutes and 95°C (boiling point) after 45 minutes. This indicates a water heating rate of 4–4.6°C/min, or approximately 3.2°C/min.
- A full charge of the batteries could heat three times a day with a 14-day autonomy.

All these results confirm that the power block (DC/DC converter) functions as a boost DC/DC converter, with lower input voltages (currents) and higher output voltages (currents). Furthermore, the efficiency of the DC/DC converter, exceeding 89%, affirms the proper functioning of power block 1 (Figure 9.2). Moreover, in the literature, we observe that in concentration cookers, water heating reaches about 90°C after 67 minutes [36], and [13]. Comparing these results with those obtained on our cooker, we deduce that our cooker reduces water boiling time by a factor of 3.

In conclusion, the performance of our cooker, in terms of both temperature and heating time, is highly promising, largely due to the energy stored in the batteries. Currently, we are focused on enhancing the temperature and heating efficiency of the heating element and the cooker itself, particularly by reducing heat loss through the addition of heat-retaining rings, as illustrated in Figure 9.8. Ongoing work on this cooker will be published at a later date.

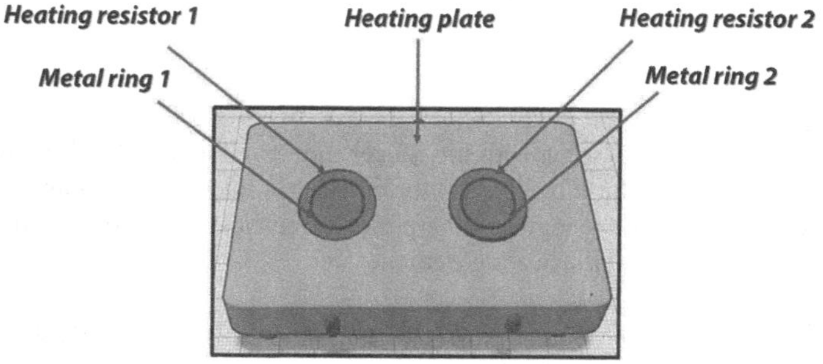

Figure 9.8 Hotplate shape with heat-retaining rings.

9.4 Conclusion

This chapter outlines the design, construction, and operation of a novel cooking device, consisting of a hotplate with two thermal resistors, fueled by solar energy stored in batteries. The cooker is tailored to address the needs of users in both rural and urban settings. Experimentation with this 400-W cooker has shown us the following:

- Charging behavior of batteries powered by 600-Wp photovoltaic panels, as a function of weather conditions. We observed an average charging current of 8 A, allowing around 64 Ah of charge to be accumulated in the batteries over 1 day, representing 25.6% of their total capacity. This allowed us to conclude that 3–4 days of similar sunshine would be needed to fully charge the batteries.
- During heating, battery voltage and current are around 48 V and 4 A, respectively.
- The proposed DC/DC converter achieves an efficiency greater than 89%.
- When boiling 1 liter of water, the temperature increases by 4–5°C/min, reaching 95°C after 45 minutes. Meanwhile, the hotplate temperature reaches a maximum of 350°C following 30 minutes of operation, with a temperature increase of 12–33°C per minute.

A comparison of these results with conventional solar cookers, whether concentrated or canned, shows an improvement in heating temperature and a threefold reduction in the time required to boil water (a factor of 3). These performances show that the solar battery cooker proposed in this work is an effective solar cooking solution, tailored to address the challenges brought about by climate change and energy crises. Additionally, the integration of the digital techniques we have proposed (control, supervision, optimization of operation) strengthens the impact and scope of this initiative, contributing to the wider use of renewable energies in the daily lives of users, in particular for cooking using photovoltaic electrical energy stored in solar batteries.

In the future, this work will be continued to improve water temperature and heating time, and hence cooking time, by reducing power losses through the use of rings placed around the hotplate's heating resistors.

Acknowledgments

This research is supported through the following projects:

- Project 3.3, under the Morocco–Wallonia–Brussels Cooperation Program (2023-2027), Wallonia-Brussels International (Belgium).
- Project LEAP-RE (2022-2025): SoCoNexGen—Solar Indoor Cooking Systems of the Next Generation, Europe–Africa Research and Innovation Call on Renewable Energy.
- Project 'ARICA23_703' (2024-2026): 'A Smart Model for Sustainable Development to Ensure Energy and Water Security', Federation of Arab Scientific Research Councils, The Arab Alliances for Scientific Research and Innovation ARICA Initiative.

References

1. Marín, A.I., Abdul Malak, D., Bastrup-Birk, A., Chirici, G., Barbati, A., Kleeschulte, S., Mapping forest condition in Europe: Methodological developments in support to forest biodiversity assessments. *Ecol. Indic.*, 128, 107839, Sept. 2021. doi: 10.1016/j.ecolind.2021.107839.
2. Yolcan, O.O., World energy outlook and state of renewable energy: 10-Year evaluation. *Innov. Green Dev.*, 2, 4, 100070, Déc. 2023. doi: 10.1016/j.igd.2023.100070.
3. International Energy Agency (IEA). Global coal demand set to remain at record levels in 2023 – News. Accessed June 20, 2025. [Online]. Available at: https://www.iea.org/news/global-coal-demand-set-to-remain-at-record-levels-in-2023.
4. Imani, G. and Moore-Delate, E., *Study report on wood-energy consumption and cooking equipment in the city of Kisangani*. Center for International Forestry Research (CIFOR), Bogor, Indonesia, 2021. [Online]. Available at: https://www.cifor-icraf.org/publications/pdf_files/Reports/Imani-GML-Report.pdf.
5. Li, B., *et al.*, Study of greenhouse gas emissions from smoldering coal fires: Estimation considering the indirect greenhouse effect of precursors. *J. Clean. Prod.*, 443, 141113, Mars 2024. doi: 10.1016/j.jclepro.2024.141113.
6. Pang, X., *et al.*, Study of spatiotemporal variation and annual emission of CH4 in Shaoxing Yangtze River Delta, China, Using a portable CH4 detector on the UAV. *J. Environ. Sci. (China)*, 151, 140–149, 2025. doi: 10.1016/j.jes.2024.03.045.

7. Molla, S., Farrok, O., Alam, M.J., Electrical energy and the environment: Prospects and upcoming challenges of the World's top leading countries. *Renew. Sustain. Energy Rev.*, 191, 114177, Mars 2024. doi: 10.1016/j.rser.2023.114177.
8. Simon Prabu, A., Chithambaram, V., Anto Bennet, M., Shanmugan, S., Janarthanan, B., Performance of solar cooker with evacuated tubes and photovoltaic panels with phase change materials. *Energy Effic.*, 16, 7, 1–14, Oct. 2023. doi: 10.1007/s12053-023-10151-3.
9. Bigelow, A.W., Tabatchnick, J., Hughes, C., Testing solar cookers for cooking efficiency. *Sol. Energy Adv.*, 4, 100053, Janv. 2024. doi: 10.1016/j.seja.2024.100053.
10. Mawire, A., Lentswe, K., Owusu, P., Performance of two solar cooking storage pots using parabolic dish solar concentrators during solar and storage cooking periods with different heating loads. *Results Eng.*, 13, 100336, Mars 2022. doi: 10.1016/j.rineng.2022.100336.
11. Sagade, A.A., Mawire, A., Palma-Behnke, R., Sagade, N.A., Low-cost solar concentrating collector-receiver system as an effective enabler for clean cooking and heating in urban areas. *Sol. Energy*, 263, 111813, Oct. 2023. doi: 10.1016/j.solener.2023.111813.
12. Mekonnen, B.A., Liyew, K.W., Tigabu, M.T., Solar cooking in Ethiopia: Experimental testing and performance evaluation of SK14 solar cooker. *Case Stud. Therm. Eng.*, 22, 100766, Déc. 2020. doi: 10.1016/j.csite.2020.100766.
13. Verma, S., Banerjee, S., Das, R., A fully analytical model of a box solar cooker with sensible thermal storage. *Sol. Energy*, 233, 531–542, Févr. 2022. doi: 10.1016/j.solener.2021.12.035.
14. Vengadesan, E. and Senthil, R., Experimental investigation of the thermal performance of a box type solar cooker using a finned cooking vessel. *Renew. Energy*, 171, 431–446, Juin 2021. doi: 10.1016/j.renene.2021.02.130.
15. Getnet, M.Y., Gunjo, D.G., Sinha, D.K., Experimental investigation of thermal storage integrated indirect solar cooker with and without reflectors. *Results Eng.*, 18, 101022, Juin 2023. doi: 10.1016/j.rineng.2023.101022.
16. Masum Ahmed, S.M., Al-Amin, M.R., Ahammed, S., Ahmed, F., Saleque, A.M., Rahman, M.A., Design, construction and testing of parabolic solar cooker for rural households and refugee camp. *Sol. Energy*, 205, 230240, 15 July 2020.
17. Lecuona-Neumann, A., Nogueira-Goriba, J.I., Famiglietti, A., del C. Rodríguez-Hidalgo, M., Boubour, J., Solar Photovoltaic Cooker with No Electronics or Battery. *Energies*, 17, 5, Art. no 5, Janv. 2024. doi: 10.3390/en17051192.
18. Opoku, R., *et al.*, Unlocking the potential of solar PV electric cooking in households in sub-Saharan Africa – The case of pressurized solar electric cooker (PSEC). *Sci. Afr.*, 17, e01328, Sept. 2022. doi: 10.1016/j.sciaf.2022.e01328.
19. Altouni, A., Gorjian, S., Banakar, A., Development and performance evaluation of a photovoltaic-powered induction cooker (PV-IC): An approach for

promoting clean production in rural areas. *Clean. Eng. Technol.*, 6, 100373, Févr. 2022. doi: 10.1016/j.clet.2021.100373.

20. An In-Depth Look at Grid-Scale Energy Storage Systems, in: *Alsym Energy*, 7 Octobre 2024, https://www.alsym.com/blog/an-in-depth-look-at-grid-scale-energy-storage-systems/.
21. *Powerwall | Tesla France*, Tesla, 7 Octobre 2024, [En ligne]. Disponible sur: https://www.tesla.com/fr_fr/powerwall.
22. Huang, R., *et al.*, A novel framework for low-temperature fast charging of lithium-ion batteries without lithium plating. *Chem. Eng. J.*, 497, 154729, Oct. 2024. doi: 10.1016/j.cej.2024.154729.
23. Zhang, X., Han, Y., Zhang, W., A Review of Factors Affecting the Lifespan of Lithium-ion Battery and its Health Estimation Methods. *Trans. Electr. Electron. Mater.*, 22, 5, 567–574, Oct. 2021. doi: 10.1007/s42341-021-00357-6.
24. Lai, X., *et al.*, Comprehensive assessment of carbon emissions and environmental impacts of sodium-ion batteries and lithium-ion batteries at the manufacturing stage. *J. Clean. Prod.*, 423, 138674, Oct. 2023. doi: 10.1016/j.jclepro.2023.138674.
25. Galuppini, G., Berliner, M.D., Lian, H., Zhuang, D., Bazant, M.Z., Braatz, R.D., Efficient computation of safe, fast charging protocols for multiphase lithium-ion batteries: A lithium iron phosphate case study. *J. Power Sources*, 580, 233272, Oct. 2023. doi: 10.1016/j.jpowsour.2023.233272.
26. Yasser, D., Haj Mohamad, A., Auger, F., Réduction des fluctuations de la puissance d'un parc éolien par système de stockage d'énergie avec supercondensateurs, in: *Symposium de Génie Electrique*, Nancy, France: Université de Lorraine [UL], Juill. 2018, 7 Octobre 2024, [En ligne]. Disponible sur: https://hal.science/hal-02983370.
27. *Centre énergétique ESSTM*, ESS, Inc, 7 Octobre 2024, https://essinc.com/energy-center/.
28. Hao, H., *et al.*, A Review of the Positive Electrode Additives in Lead-Acid Batteries. *Int. J. Electrochem. Sci.*, 13, 3, 2329–2340, Mars 2018. doi: 10.20964/2018.03.70.
29. Imadan, A. I., Semassou, G. C., Tossa, A. K., Saley, H. A., Étude de faisabilité technico-économique d'installation d'une filière de régénération de batteries stationnaires solaires au plomb acide au Bénin. *Afrique Sci.*, 24, 2, 122–131, 2024. Disponible sur : http://www.afriquescience.net.
30. Vangapally, N., *et al.*, Lead-acid batteries and lead–carbon hybrid systems: A review. *J. Power Sources*, 579, 233312, Sept. 2023. doi: 10.1016/j.jpowsour.2023.233312.
31. Toh, W.D., Xu, B., Jia, J., Chin, C.S., Chiew, J., Gao, Z., Lithium Iron Phosphate (LiFePO4) Battery Power System for Deepwater Emergency Operation. *Energy Procedia*, 143, 348–353, Déc. 2017. doi: 10.1016/j.egypro.2017.12.695.
32. Gao, Y., Pan, Z., Sun, J., Liu, Z., Wang, J., High-Energy Batteries: Beyond Lithium-Ion and Their Long Road to Commercialisation. *Nano Micro Lett.*, 14, 1, 94, Avr. 2022. doi: 10.1007/s40820-022-00844-2.

33. Li, H., Xing, S., Liu, Y., Li, F., Guo, H., Kuang, G., Recovery of Lithium, Iron, and Phosphorus from Spent LiFePO4 Batteries Using Stoichiometric Sulfuric Acid Leaching System. *ACS Sustain. Chem. Eng.*, 5, 9, 8017–8024, Sept. 2017. doi: 10.1021/acssuschemeng.7b01594.
34. Gorjian, A., Rahmati, E., Gorjian, S., Anand, A., Jathar, L.D., A comprehensive study of research and development in concentrating solar cookers (CSCs): Design considerations, recent advancements, and economics. *Sol. Energy*, 245, 80–107, Oct. 2022. doi: 10.1016/j.solener.2022.08.066.
35. Rekha, S.M.S., *et al.*, Efficient heat batteries for performance boosting in solar thermal cooking module. *J. Clean. Prod.*, 324, 129223, Nov. 2021. doi: 10.1016/j.jclepro.2021.129223.
36. Lentswe, K., Mawire, A., Owusu, P., Shobo, A., A review of parabolic solar cookers with thermal energy storage. *Heliyon*, 7, 10, e08226, Oct. 2021. doi: 10.1016/j.heliyon.2021.e08226.

Part 4

PHOTOELECTROCHEMICAL BATTERIES AND PEROVSKITE-BASED PHOTO SUPERCAPACITORS

10

Solar Flow Batteries

Tuluhan Olcayto Colak[1], Emine Karagoz[1], Ecenaz Yaman[1], Mehmet Kurt[1], Cigdem Tuc Altaf[2], Nurdan Demirci Sankir[1,2*] and Mehmet Sankir[1,2†]

[1]*Micro and Nanotechnology Graduate Program, TOBB University of Economics and Technology, Ankara, Turkey*
[2]*Department of Materials Science and Nanotechnology Engineering, TOBB University of Economics and Technology, Ankara, Turkey*

Abstract
Incentives in the utilization of renewable energy have driven advances in numerous technologies to enable the simultaneous conversion and storage of intermittent solar energy. This is accomplished by capturing solar radiation and converting or storing it as chemical energy. By utilizing redox couples for accumulation and releasing solar energy as electricity, solar flow battery (SFB) systems, which are produced by photoelectrochemical cells or monolithically integrated photovoltaics to redox flow batteries, are advantageous solutions to circumvent the intermittent nature of solar energy. Throughout this chapter, SFBs and the components that make them up are broken down in chronological order based on the improvements in materials' performance and configurations. In this respect, photo-active electrodes, device configurations, counter electrodes, electrolytes, and separators have been summarized. The final section of this chapter covers the methods for evaluating the performance of SFBs and the electrochemical characterization of these batteries.

***Keywords*:** Solar energy, energy storage, flow batteries, redox couples

**Corresponding author:* nsankir@etu.edu.tr
†*Corresponding author:* msankir@etu.edu.tr

Nurdan Demirci Sankir and Mehmet Sankir (eds.) Solar Capacitors and Batteries, (291–350) © 2025 Scrivener Publishing LLC

10.1 Introduction

Flow batteries are fully rechargeable electrical energy storage devices in which the electrolyte passes as a flowing medium from one or more electrochemical cells. They offer superior features with their long cycle life, low self-discharge rates, and advanced safety features and can be manufactured with low-cost materials such as thermoplastic and carbon. Besides, they are modular systems ranging from watts to megawatts, with operating times lasting for days [1]. Redox flow battery (RFB) systems consist of two main parts where the electrochemical reactions occur (the battery stack), and the electrolytes are stored (the external tanks). The battery stack typically contains two sets of electrodes and bipolar current collector plates. In addition, a membrane is placed between the two electrodes, which prevents the electrolytes on both sides from mixing and acts as a charge-carrying conductor [2]. A distinctive factor of flow batteries is that the separated power and energy components can be scaled individually. In a flow battery, power is a function of the electrode area within the cell, and capacity is a function of the amount of electrolyte and the concentration of active ions [1].

The energy conversions in RFBs rely on the reversible electrochemical reactions (10.1) of two redox couples (A^{2+}/A^{+} and C^{3+}/C^{2+}), which are generally dissolved in electrolyte solutions [2]. The simplified redox mechanism for RFB is given below. In addition, redox reactions for hybrid flow batteries (zinc–polyiodide), Vanadium redox, and iron–chromium flow batteries are tabulated in Table 10.1.

$$A^{2+} + C^{2+} \underset{Discharge}{\overset{Charge}{\rightleftarrows}} A^{+} + C^{3+} \tag{10.1}$$

Table 10.1 Chemical reaction equations that occur in hybrid flow batteries (zinc–polyiodide), vanadium RFBs, and iron–chromium flow batteries [2].

Battery type	**Redox reaction**
Hybrid Flow Batteries	$Zn + I_3^- \underset{Charge}{\overset{Discharge}{\rightleftarrows}} Zn^{2+} + 3I^-$
Vanadium RFB	$VO_2^+ + V^{2+} \underset{Charge}{\overset{Discharge}{\rightleftarrows}} V^{3+} + VO^{2+}$
Iron–Chromium Flow Battery	$Fe^{3+} + Cr^{2+} \underset{Charge}{\overset{Discharge}{\rightleftarrows}} Cr^{3+} + Fe^{2+}$

10.1.1 Solar Flow Battery Concept

In the last few decades, with increasing energy demand, scientific research has focused on developments aimed at improving energy efficiency (EE), storing energy safely, and being environmentally friendly. Besides, it is important to switch from fossil fuels to renewable energy systems. Consequently, a significant amount of electricity can be obtained from renewable energy sources such as wind, biomass, solar, and hydroelectric power. The increase in the use of renewables is highly linked to the developments in energy storage technologies to attain stability of the electricity grid, high efficiency, and reliability. Especially, the intermittent nature of solar energy makes energy conversion and storage systems mandatory. Photoelectrochemical (PEC) storage systems that enable the conversion of solar energy into chemical energy have been proposed, and integrated photovoltaic (PV) systems have been developed [3]. Consequently, solar flow batteries (SFBs) have emerged as a new idea that offers a more cost-effective, compact approach, and efficiency has gained importance [4–6].

In 1976, the first PEC storage system resembling a solar RFB (SRFB) was developed by Hodes *et al.* using 1M sodium alkali solution and n-type CdSe crystal as the electrolyte and photoelectrode, respectively [7]. Between 1976 and 1987, various organic couples combined with photoelectrodes were investigated, mostly on a selenide basis [7]. Many redox couples used in PEC storage cells in the 1980s have recently been investigated as redox couples for RFBs [8]. In 1987, a 2-day open-air study of an SRFB with a CdSe photoanode demonstrated efficiency and energy density [4, 7, 8]. In 2016, a vanadium redox flow cell (CdS(s)|V^{3+},VO^{2+}||V^{3+},V^{2+}|Carbon Felt(s), E_0=0.6 V_{NHE}) was constructed using a CdS photoelectrode, providing a sufficient photovoltage to charge the vanadium battery up to 75% [9]. In the same year, SRFB integrated with catalytic H_2 production reaction based on $H_4[SiW_{12}{}^{VI}O_{40}]/H_6[SiW_{10}{}^{VI}W_2{}^{V}O_{40}]$ and $Fe^{2+/}Fe^{3+}$ and Fe^{2+}/Fe^{3+} redox species and perovskite solar cells were developed, and storage in redox species was achieved with a solar–chemical energy conversion efficiency of more than 15.2% during charging [10]. In 2020, they designed an all-in-one solar rechargeable flow cell (ASRFC) containing Fe/Quinone redox couples and an amorphous silicon (Si) based photoanode. They showed that 93% of the solar energy converted by the Si-based photoanode can be stored in the redox couples [11]. In 2021, they designed a Zn–Mn hybrid RFB (HRFB) and achieved an EE of 75.45% [12]. In 2022, an integrated SRFB device was fabricated with Si photoelectrode and 4-OH-TEMPO/ferricyanide redox couples and Si photoelectrode. The SRFB device can be photo-charged, and the electrical efficiency reached 4.5% [13]. In 2023,

SRFB with molybdenum disulfide on titanium dioxide (MoS_2@TiO_2) was designed, and an energy density of 87.45% was achieved [14]. In 2024, an SFB with nickel-doped titanium dioxide photoanode was intended, and the solar–chemical energy output efficiency reached 2.51% [15]. The recently reported SFB studies are summarized in Figure 10.1.

Yu *et al.* designed an SFB using TiO_2-based photoelectrode and aqueous lithium–iodine (Li–I) [16]. This system enables the simultaneous conversion and storage of solar energy. Li–I was charged under SFB AM1.5 illumination at a potential of 2.9 V. Upon a charging voltage below 3.3 V, energy is saved by approximately 20% compared to traditional Li–I batteries. Fu *et al.* designed a high-performance SFB based on a back-illuminated GaAs photoelectrode with an n-p-n sandwich design. The SFB also consists of a GaAs photoanode and two inert carbon felts in two electrolyte compartments separated by a Selemion DSV anion exchange membrane (AEM). The illumination is from the GaAs photoelectrode. The SFB device in that study can be configured in three variable operating modes. To enhance the solar-to-output electricity efficiencies (SOEEs) of SFB, numerical simulations were employed to ascertain the correlation between SOEE and the operational state of charge (SOC) range. As a result of these studies, a record SOEE of 15.4% was achieved among single-junction SFB devices. In addition, the TiO_2 protection layer and robust redox couples in neutral pH electrolytes enable SFBs to achieve stable cycling for 408 hours (150 cycles) [17]. Kumtepe *et al.* studied the RFB system with V^{+3}/VO^{+2} electrolyte to investigate the activity of hybrid TiO_2–In_2S_3 photoelectrode for the first time. The heterojunction photoelectrodes were fabricated using the ultrasonic spray pyrolysis technique, and the highest charge separation and quantum efficiencies of 46% and 20% were achieved, respectively [18].

10.1.2 Energy Conversion Equations

The EE (Equation 10.2), voltaic efficiency (VE) (Equation 10.3), and Coulombic efficiency (CE%) (Equation 10.4) calculations for the SFB are evaluated using the equations given below [14].

$$EE = CE\% \times VE = \frac{\int V_d I_d dt}{\int V_C I_C dt} x100\% \tag{10.2}$$

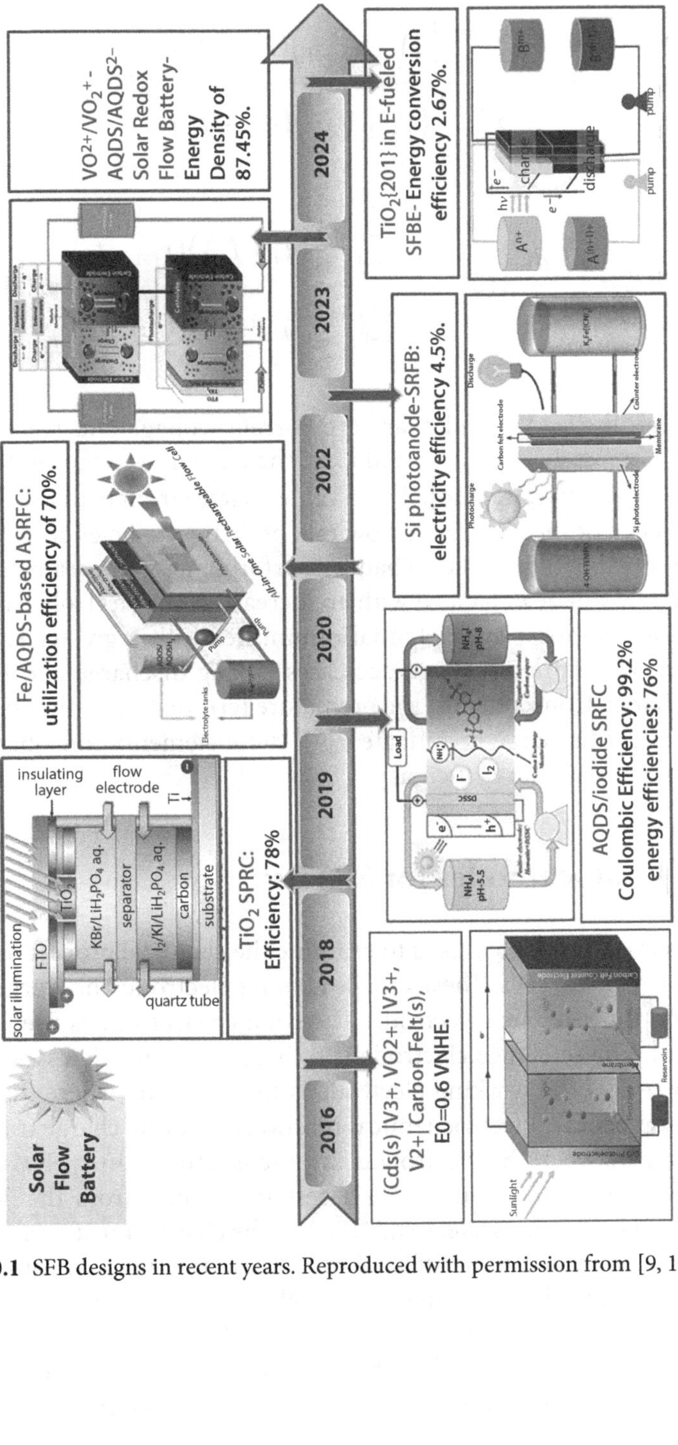

Figure 10.1 SFB designs in recent years. Reproduced with permission from [9, 11, 13, 14, 19–21].

$$VE = \frac{\sum_0^{m-1} \frac{1}{2}(V_j + V_{j+1})(t_{j+1} - t_j)}{\sum_0^{n-1} \frac{1}{2}(V_k + V_{k+1})(t_{k+1} - t_k)} x100\% \tag{10.3}$$

$$CE\% = \frac{\int I_d dt}{\int I_c dt} x100\% = \frac{\sum_0^{m-1} \frac{1}{2}(I_j + I_{j+1})(t_{j+1} - t_j)}{\sum_0^{n-1} \frac{1}{2}(I_k + I_{k+1})(t_{k+1} - t_k)} x100\% \tag{10.4}$$

Here, I_c, V_c, I_k, V_k, and t_k are the current–voltage and reaction time during charging. I_d, V_d, I_j, V_j, and t_j are the current, voltage, and reaction time during discharging [14]. These values relate strongly to the flow rate. The influence of the electrolyte flow rate on battery efficiency across various current densities is vital to battery performance. All parameters mentioned above will be enhanced with an increase in the electrolyte flow rate. As the concentration overpotential diminishes with a greater electrolyte flow rate, the terminal voltage decreases during discharge and increases during charge. Consequently, the premature termination at the conclusion of the charge/discharge is mitigated. All these elements contribute to the enhancement of battery efficiency.

10.2 Photocathodes for SRFBs

Three basic electrodes are used to examine the chemical process occurring in electrochemical cells. These are the counter electrode, the working electrode, and the reference electrode. The counter electrode is used to close the circuit. The reference electrode is utilized to accurately assess the circuit's potential. The working electrode is the electrode where the chemical reaction occurs. Electrochemical reactions occur by charge transfer on the surface between the electrolyte and the electrode. In typical electron transport, the reactant must be at a certain distance from the electrode surface on a molecular scale. In addition to the charge transfer process, the overall reaction rate can be affected by the reaction material coming onto the electrode and the removal of the product. Mass transfer can occur by electromigration, diffusion, and convection [22].

A proper electrode should have high chemical resistance, be cost-effective, have high electrical conductivity, and have good mechanical strength.

To improve the electrode material, it is necessary to increase the electrochemical reaction activity and optimize the electrode dissolution during charging and discharging. This can be done by changing the surface area, quantity, and quality between the electrode and the electrolyte. In addition, efficiency can be increased by increasing the selectivity on the electrode surface. Electrodes for RFBs are carbon-, metal-, or composite-based [22, 23].

Carbon-based electrodes: Carbon-based electrodes are often used in RFBs. Although carbon has lower conductivity than metals, it exhibits good electrical conductivity for flow battery applications. Carbon electrodes tend to decompose in highly oxidizing environments such as V^{5+} and Ce^{4+}, in sulfuric acid concentrations, and at high temperatures. Materials such as carbon felt, glassy carbon, and graphite stand out among carbon-based electrodes due to their charge transfer capabilities. Glassy carbon has a disadvantage due to its inflexibility and cannot be applied on different surfaces. Graphite vanadium has a higher reversibility than pure carbon in reactions, but it is not stable in highly oxidizing environments. It has been shown that graphite-based electrodes perform well in zinc–cerium systems. Carbon-based electrodes are constantly being developed to exhibit conductivity as good as metals. To develop electrodes with conductivity as good as metals, surface activation has been suggested to increase the amount of electrochemical field that the electrolyte can reach for redox reactions in zinc/bromine batteries. For carbon-based electrodes, it has been suggested to coat the weak carbon with more active carbon powder. The use of highly conductive activated carbon powder enhances the electrochemically active surface area accessible for redox processes [22, 23].

Metal-based electrodes: Metal-based electrodes are not preferred in RFBs due to corrosion problems unless they are coated with precious metals. Precious metals with high conductivity and chemical stability are not preferred as electrodes for large-scale storage systems because of their high cost. Sometimes, precious metals may not provide excellent electrochemical performance. Platinum may form non-conductive oxides during cerium oxidation, or gold is not reversible in vanadium redox reactions. Passivation processes have frequently been reported in different metal electrodes, including lead and titanium. Nevertheless, certain redox pairs exhibited greater reversibility on metal-based electrodes. For instance, Cr^{2+}/Cr^{3+} and Ti^{3+}/TiO^{2+} exhibited irreversibility on carbon and graphite electrodes, whereas composite lead and tungsten–rhenium alloys served as electrodes in chromium and titanium systems, respectively [22].

Composite electrodes: Carbon-composite materials typically consist of carbon particles and polymer binders. Carbon–polymer composites offer significant advantages over solid carbon, such as being lighter, lower cost, and having good mechanical properties. Carbon materials have the disadvantage of being brittle when applied to large areas on a large scale, while polymer–carbon composites offer advantages. Carbon–polymer composites are used in zinc–bromine, lead-acid, and all vanadium RFBs [22]. Wang *et al.* designed a SRFB combining $BiVO_4$ as a photoanode or Mo–$BiVO_4$ and polyterthiophene (pTTh) as photocathodes with suitable redox pairs. FESEM analysis conducted on the surface of Mo–$BiVO_4$ surface exhibited curved nanorods of approximately 200 nm length. PEC reactions occurring at the photoanode and photocathode with solar energy are given below [6].

Photocathode $2Fe^{3+} + 2e^{-} \rightarrow 2Fe^{2+}$

Photoanode $3Br^{-} + 2h^{+} \rightarrow Br_3^{-}$

Overall reaction $2Fe^{3+} + 3Br^{-} + \xrightarrow{h\nu} 2Fe^{2+} + Br_3^{-}$

The redox couples in the flow battery created fast redox kinetics that enabled charge transfer. This device was charged under simulated sunlight to store energy and convert the stored energy into electrical energy in a controlled manner. Research has shown that the open circuit voltage of Mo–$BiVO_4$ photoanode and pTTh photocathode is 0.34 V, and the short circuit current density is 0.38 mA cm^{-2}. Stable results have been obtained in constant current discharge tests. The results show that the flow battery that drives reversible redox reactions is low cost and stable and can be applied to large-scale areas.

In the flow battery, solar energy can be continuously converted into liquid fuel. However, the chemical imbalance between the photoanode and the electrolyte causes low efficiency. To solve this, Lu *et al.* developed an E-fuel solar flow cell with TiO_2 doped for the photoanode. The Mn-doped photoanode showed a stable current at 1.8 μA. With Mn doping, the band gap can be narrowed and the probability of electrons passing to the conduction band can be increased. However, as the doping rate increases, the PEC performance may not increase. This may be due to the adverse effect of oxidation on the photoanode side, increasing the electron–hole recombination rate. In addition, excessive doping can affect the stability of the photoanode and cause impurities. In short, the electrode and electrolyte interface was tried to be optimized by doping Mn into TiO_2. TiO_2 {201} is a spherical particle, and in the SEM analysis, Mn-doped TiO_2 {201} exhibited spherical particle size, and the particle size was found to be 93.1 nm.

Based on TiO_2 {201}, the flow battery achieved an energy conversion efficiency of 2.67% [15, 19].

The primary efforts to enhance the efficiency and stability of photoelectrodes are surface functionalization and the implementation of heterojunctions. Tian *et al.* developed a Nafion™-protected flow cell to improve solar redox flow cells. They fabricated a redox flow cell with MoS_2@TiO_2 nanomaterial on an FTO substrate. TiO_2 nanorods and molybdenum disulfide (MoS_2) nanoflowers were prepared on an FTO substrate using a hydrothermal method. SEM images show molybdenum disulfide MoS_2 nanoflowers on TiO_2 nanorod structures. UV–Vis results show that the photoanode can collect more light energy by adding MoS_2 and Nafion™. By coating the MoS_2@TiO_2 photoanode with Nafion™, corrosion is prevented, effective light absorption is maintained, and it is protected from strong acidic conditions. Thus, it shows long hours of stability under light since there is no current loss. The redox active species are VO^{2+} / VO_2^+ and AQDS / $AQDS^{2-}$, and when the sun comes to the electrode, the photogenerated holes VO^{2+} are oxidized to VO_2^+, and the photogenerated electrons are reduced to $AQDS^{2-}$ at the counter electrode. Chemical energy is conserved due to the oxidation and reduction forms. In addition, the EE is more than 17.5% with the addition of NH_4CI to the electrolyte, which shows that vanadium redox species support mass transport and charge transfer of AQDS in the electrolyte. The remaining energy density after 10 photocharge/discharge cycles with Nafion™ protection was found to be 87.45% [14].

10.3 Configuration

It is of crucial importance to make SRFBs capable of energy storage that is both efficient and effective. Efficient collection of solar energy is often accomplished through the utilization of PV and PEC systems [24]. In semiconductor–liquid junctions, the energy levels of the chosen semiconductor and redox couples must align to provide effective photovoltage generation, which is why conventional solar cell devices are often employed in these systems. Conversely, PV systems produce photovoltage *via* solid-state connections, thereby obviating the necessity for meticulous energy level alignment. PV RFB devices are advantageous over semiconductor–liquid junctions since they provide enhanced flexibility in material choices. However, PV RFB configurations require control systems with DC–DC converters for charging RFB systems. This requirement leads to an escalation in both the overall system's expenses and its complexity [25].

In situations where solar panel production is inadequate, the energy stored in the flow battery system can be used, hence improving the total EE. In these instances, the capacity of the RFB system plays a crucial role in determining system performance [1, 26].

In certain applications, electricity-assisted systems are also employed in RFB devices. These systems apply an external bias to reduce the potential difference between the redox couples and the photoelectrode [27]. This method guarantees optimal performance while simultaneously mitigating constraints in photovoltage production. This chapter will concentrate on fully integrated SRFBs, which have numerous substantial advantages in this domain, including advanced cell design options. The efficacy of SRFBs is affected by the efficiency of the system used. The effectiveness of the photoelectrode is a primary factor influencing system performance [28]. Semiconductor materials are extensively utilized as photoelectrodes, especially in PEC water-splitting systems, seen as viable alternatives for solar energy storage [29].

When choosing the kind and class of semiconductors for these systems, characteristics like band-gap energy are essential. The band gap is a crucial factor for photoelectrodes, as the energy difference between valence and conduction bands excites electrons. This excitation produces electron–hole pairs that can later dissociate, facilitating processes like water splitting [5]. The electrode produces free charge carriers and becomes photocharged. The photocharge mechanism denotes the transformation of solar energy into chemical energy [30]. The functionality of these systems relies on the selection of electrode materials, electrolytes, and separators.

This section will examine the operational principles of systems configured in diverse arrangements. Semiconductors are categorized into three types: n-type, p-type, and intrinsic. In p-type semiconductors, the density of positive charge carriers (holes) is elevated, while in n-type semiconductors, the density of negative charge carriers (electrons) is predominant. In intrinsic semiconductors, the concentrations of positive and negative charge carriers are equivalent [31]. Electrolytes in these systems generally have redox couplings in both the negative and positive half-cells, requiring the selection of a suitable membrane to enable ion transfer during operation. The selection of separators is crucial, since it relies on the electrode and electrolyte materials employed, since effective ion transfer is vital for optimal system performance. As of now, three primary designs of PEC RFB systems have been highlighted: (i) single-photoelectrode systems, (ii) dual-electrode systems, and (iii) systems utilizing metallic lithium anodes. Initially, PEC systems and RFB systems were created and examined separately. Nonetheless, following years of independent progress, the amalgamation of these technologies under the term "Solar-RFB" has established a new study field. This domain emphasizes the integration of *in situ* energy conversion functionalities of PEC systems (solar to chemical energy) with the beneficial attributes of RFB systems, including adaptable design, extended lifespans, and superior reliability. This integration has markedly improved the efficiency of hybrid SRFB systems [32]. In PEC RFBs, the system can be directly charged *via* a photoelectrode, facilitating the integration of the operational principles of both devices into a singular system. This integration enhances overall efficiency, with the photoelectrode being crucial in determining system performance. The semiconductor type, either n-type or p-type, significantly impacts the system's operational principles, while the configuration, whether single or dual photoelectrodes, influences solar energy absorption. The separator selection must correspond with the counter electrode and electrolyte to guarantee optimal functionality of all components. Furthermore, the energy levels of the photoelectrode and the redox couplings must align for optimal performance. In n-type semiconductors, when the Fermi energy level (E_F) exceeds the energy levels of electrolytes, electrons transfer from the conduction band of the semiconductor to the redox molecules in the electrolyte. This mechanism channels the electric field from the bulk to the surface of the semiconductor [33]. When exposed to light with energy exceeding the band gap, electron–hole pairs are produced and dissociated. Electrons travel toward the interior of the semiconductor, whereas holes advance to its surface. Redox reactions transpire at the surface according to their individual redox potentials, leading to photo-oxidation in n-type semiconductors [34]. In contrast, when the redox electrolyte's energy levels surpass the E_F, holes go toward the bulk of the semiconductor and electrons to its surface in p-type semiconductors. This mechanism inverts the electric field direction from the surface to the bulk. Comparable redox reactions transpire according to redox potentials [35]. Under sufficient illumination, oxidation occurs on the surface of n-type semiconductors, whereas reduction transpires on p-type semiconductors. While p-type photocathode materials can theoretically promote oxidation, only a few exhibit sufficiently high valence band edge energies to accomplish this in practice. The integration of p-type and n-type semiconductors in a dual-system architecture creates dual band gaps, hence improving efficiency. This configuration enables certain electrons to engage in reduction reactions through the conduction band of the p-type semiconductor, while others facilitate oxidation. Nonetheless, dual-electrode systems necessitate greater solar energy than single-electrode systems to produce equivalent photocurrent due to the recombination of certain electron–hole pairs that do not engage in redox processes [36]. In PEC systems, the characteristics of the semiconductor are paramount, resulting in the creation of more efficient designs that integrate sophisticated semiconductor topologies. In the selection of semiconductors, it is essential to account for energy loss results from concentration gradients and kinetic overpotential at the semiconductor–liquid interface. In single-photoelectrode systems, the material's band gap must be sufficiently large to offset these energy losses. Advanced configurations, including dual electrodes, tandem systems, n-type/p-type combinations, and PV electrode designs, can diminish the necessary band-gap energy while enhancing light-harvesting efficiency [28]. The cell voltage in SRFB systems depends on two key parameters: i) photovoltage, generated by the photoelectrode, and ii) the electrochemical potential variation among redox pairs. These parameters vary based on the system configuration, enabling tailored designs to optimize energy conversion and storage efficiency.

10.3.1 Single Photoelectrode with RFB Systems

The choice of photoelectrode is crucial for efficiency in single photoelectrode systems. Tandem structures are frequently utilized to attain elevated current densities and enhance EE, as they reduce recombination losses [37]. Tandem cells, especially those composed of Group III–V semiconductors, are extensively utilized because of their superior optical characteristics. These structures facilitate the efficient collection of solar energy and its conversion into chemical energy, which is then stored in RFBs. Employing the tandem cells as solar cells integrates solar energy conversion, usually conducted by PVs, with energy storage by RFBs into a cohesive, compact apparatus. In these systems, photocharging transpires between the photoelectrode and the counter electrode. The photoelectrode captures solar energy, producing electron–hole pairs that facilitate reduction and oxidation reactions based on the employed redox couple materials [14, 18, 38]. During discharge, the photoelectrode is detached, and the process transpires between two conductive current collectors. The photoelectrode must exhibit corrosion resistance due to its direct interaction with electrolytes containing redox couples separated by a membrane. Attaining elevated photovoltage is difficult for semiconductors with broadband gaps that predominantly absorb the short-wavelength portion of the sun spectrum. This constraint can be alleviated by employing several photoelectrodes. Nonetheless, individual photoelectrode systems are more economical and utilize readily accessible materials [30, 39–42]. Advances in material modification and the development of new materials hold promises for improving the efficiency of these systems. As indicated in Figure 10.2a, the locations of the components can be changed as long as the connection between the electrodes and the operating principle of the system does not change.

10.3.2 Dual Photoelectrode Systems with RFB

Numerous studies in the literature highlight the efficiency improvements achieved through dual photoelectrode systems [6, 28, 40, 43, 44]. The systems depicted in Figure 10.2b utilize photoelectrodes: a photoanode and a photocathode, between which photonic charge transpires.

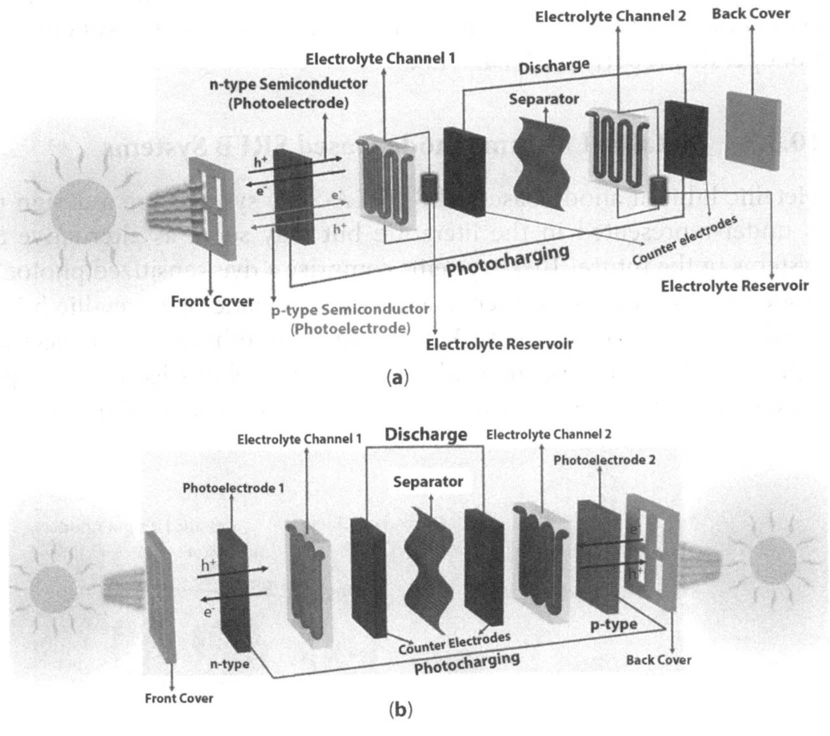

Figure 10.2 (a) The integrated PEC-RFB device comprises an n-type semiconductor single photoelectrode electron–hole movements shown as dark and a p-type semiconductor single photoelectrode movements shown as light, two counter electrodes, a separator, and redox couplings in the electrolyte. All of the electrodes are submerged in the electrolyte. (b) An integrated PEC-RFB device using dual-photoelectrode, two counter electrodes, a separator, and redox couples in electrolyte with two different reservoirs for each electrolyte. All of the electrodes are submerged in the electrolyte.

Distinct electrolytes are employed on the photoanode and photocathode sides, facilitating separate oxidation and reduction reactions. The two electrolytes are divided by a membrane that promotes charge equilibrium, while conductive surfaces on both sides allow for discharge. The high photocurrent is detected on the photoanode side, with the created holes promoting the oxidation of the redox partners in the electrolyte. Light absorption and charge separation transpire on the photocathode side, with the resultant electrons lowering the redox partners in the electrolyte. Each electrode independently creates photovoltage; yet, their collective function yields adequate photovoltage for the system. Consequently, dual photoelectrode systems markedly improve efficiency. Nonetheless, these systems exhibit greater complexity than individual photoelectrode systems, resulting in elevated prices and extended setup durations. Moreover, the working voltages of the two electrodes must be congruent; otherwise, the system would not operate properly [6, 40, 43–45].

10.3.3 Metallic Lithium Anode-Based SRFB Systems

Metallic lithium anode-based RFB (MLA-SFB) systems are a design that is under-represented in the literature but may serve as alternative SFB systems in the future. These systems comprise a dye-sensitized photoelectrode that captures solar energy, a device schematic of a metallic lithium anode as shown in Figure 10.3, and a metallic lithium anode electrode. Upon illumination, dye molecules on the TiO_2 photoelectrode are photoexcited, resulting in an oxidation reaction occurring in the electrolyte.

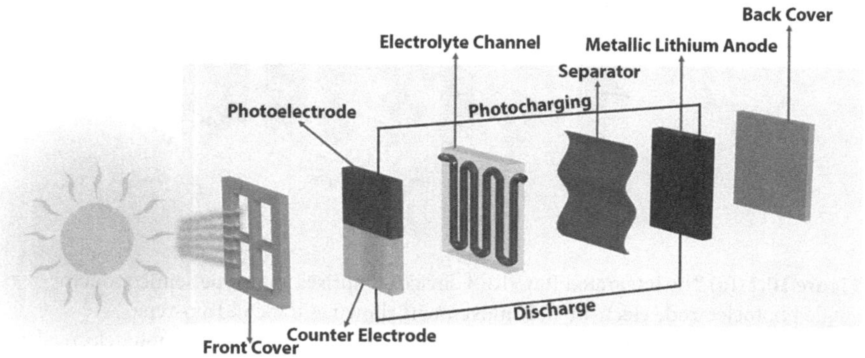

Figure 10.3 The schematic representation of an integrated MLA-SFB device using photoelectrode, the counter electrode, redox couples in the electrolyte (the electrolyte connects to a reservoir), a separator, and a metallic lithium-based anode electrode. All of the electrodes are submerged in the electrolyte.

Simultaneously, Li ions traverse to the surface of the metallic Li anode, where they undergo reduction, thus completing the photocharging process. Unlike conventional batteries, SFB systems offer energy savings due to a reduced charging voltage. Enhancing the performance of MLA-SFB depends on the quality of the components utilized in the system, while additional metallic materials-based SFBs, such as sodium-based SFB, are currently being developed as anode materials [16, 46, 47].

10.4 Counter Electrodes

A counter electrode is crucial in electrochemical systems, particularly in applications like solar cells, batteries, and electrochromic devices. The counter electrode operates in a manner that is different from that of the working electrode, as per the operational principle [48]. Electrochromic systems with two electrodes utilize a counter electrode that functions either as a passive ion storage layer visually or as a secondary electrode to achieve various colors. Counter electrodes improve color efficiency and device stability [49]. Color efficiency pertains to the charge transfers occurring in the redox reactions of the electrodes [50, 51]. The requirements such as high catalytic activity, large internal surface area, chemical stability, and high conductivity for high-performance counter electrodes were described in the literature [52]. The counter electrode's primary role is to gather electrons from the external circuit and facilitate the reduction of triiodide in the electrolyte to iodide. The decrease in the rate of the electrolyte at the counter electrode should typically align with the dye regeneration rate by the electrolyte at the photoanode in a dye-sensitized solar cell (DSSC) to sustain a low overvoltage and minimize energy loss, as shown by the short-circuit photocurrent density, Jsc [53].

Diverse characterization techniques have been established to elucidate the pertinent reaction process mechanisms behind counter electrodes, including charge creation, charge transfer, and charge recombination, and to improve the PV efficiency of DSSCs. Electrochemical and PEC approaches are regarded as the most effective techniques for analyzing component interactions and assessing the performance of DSSCs. The methodologies encompass current–voltage (I–V) characterization, cyclic voltammetry (CV), electrochemical impedance spectroscopy (EIS), intensity-modulated photocurrent spectroscopy, and intensity-modulated photovoltage spectroscopy. The counter electrode significantly affects the PV properties of DSSCs. The theoretical maximum photovoltage of the DSSC is determined by the energy difference between the redox potential of the mediator and the Fermi level of the metal oxide semiconductor on the photoanode, attainable only at zero current. Under load conditions, the output voltage frequently falls below the open-circuit value. The voltage loss arises from the aggregate overpotential of the counter electrode, induced by the current passing through the electrolyte and the electrolyte/counter electrode interface. Numerous counter electrodes have been produced nowadays. The most significant among these are platinized transparent counter electrodes, carbon counter electrodes, and conductive polymer counter electrodes. Platinum counter electrodes are the most commonly utilized. Platinum-coated conductive glass has been recognized as the standard material for counter electrodes in DSSCs. The processes at the counter electrode are affected by the particular redox shuttle that enables charge transfer between the photoelectrode and the counter electrode [54–58].

A study synthesized and evaluated metal oxides and sulfides, carbon-based materials, and conducting polymers with various morphologies as counter electrodes in DSSCs. Despite certain materials demonstrating commendable photoelectric conversion efficiency (PCE) and photoelectric responsiveness, they exhibit inadequate performance regarding photocatalytic activity, low charge transfer resistance, PCE, and stability in comparison to Pt-based counter electrodes. Among conducting polymers, poly-3,4 ethylenedioxythiothene (PEDOT) exhibits a significant photoelectric response to triiodide reduction; however, its expense precludes it from being a feasible substitute for Pt. Various metal oxides and sulfides have elevated catalytic activity and conductivity owing to their substantial surface area and little charge transfer resistance. Test results and investigations indicate that conducting polymers and carbon-based materials with diverse structural morphologies exhibit exceptional efficiency [59].

10.4.1 Semiconductors as Counter Electrodes

When the conventional counter electrodes are inadequate, semiconductor counter electrodes are specifically employed in PV and electrochemical cells. Semiconductor counter electrodes operate distinctively from conventional metal counter electrodes and are favored in several situations for their capacity to provide more efficient outcomes. The primary attribute of semiconductor counter electrodes is their capacity to regulate electrical conductivity *via* external influences like light or voltage. Upon exposure to light or an electrical stimulus, these counter electrodes can modify their conductivity or influence the flow of electric current. Common semiconductor materials utilized as counter electrodes include zinc oxide (ZnO), titanium dioxide (TiO_2), copper sulfide (Cu_2S), carbon nanotubes

(CNTs), and silicon (Si) [60–62]. While semiconductor and metal counter electrodes are favored for specific applications, they both possess distinct advantages and disadvantages. Several of the most significant items are presented below in Table 10.2.

10.4.2 Carbon-Based Counter Electrodes

Carbon is an abundantly present element in multiple forms. Carbon materials can be categorized based on morphology into carbon fiber, carbon sphere, carbon nanotubes, mesoporous carbon, graphene, and others. Moreover, diverse carbon materials have been employed in domains such as energy conversion and storage, electrocatalysis, medicine, delivery systems, and wastewater treatment, contingent upon the distinct characteristics of each carbon material type. Diverse carbon compounds have been employed as counter electrodes (CE) in DSSCs. Carbon is regarded as the most potential circular economy material owing to its affordability, elevated catalytic activity, and commendable stability. The incorporation of carbon counter electrodes in DSSCs has garnered considerable attention owing to their affordability, remarkable catalytic efficiency, and enhanced chemical stability, positioning them as a viable substitute for costly platinum electrodes in extensive DSSC applications. The catalytic activity of

Table 10.2 Comparison of semiconductor and metal counter electrodes [52, 63].

Parameter	Semiconductor counter electrodes	Metal counter electrodes
Efficiency	High efficiency in PV devices.	They are not active under light, and their efficiency is fixed.
Electrical conductivity	Conductivity depends on external factors.	Electrical conductivity is high.
Chemical durability	High chemical durability.	They may corrode over time.
Electrical properties	Can change depending on external factors.	It is fixed.
Thermal sensitivity	Performance decreases at high temperatures.	Stable at high temperatures.

the carbon counter electrode (CCE) can be significantly modified by engineering a high specific surface area, excellent conductivity, a distinctive structure, and functional surface groups. Nonetheless, the drawbacks of the CCE persist, encompassing inadequate mechanical stability, low opacity, and restricted charge transfer and mass transport phenomena at the CE/electrolyte interface [63, 64].

Based on their morphology, carbon materials can be classified as carbon fibres, carbon spheres, carbon nanotubes, mesoporous carbon, graphene, and a variety of other materials. Furthermore, depending on the characteristics that are unique to each form of carbon material, various carbon materials have been utilized in a variety of fields, including energy conversion and storage, electrocatalysis, medicine, delivery systems, and wastewater treatment. There have been a variety of carbon compounds utilized in DSSCs as counter electrodes (Figure 10.4). As a result of its low cost, high catalytic activity, and impressive stability, carbon is often considered to be the most promising material for carbonaceous electrodes. The utilization of carbon counter electrodes in DSSCs has garnered a significant amount of attention due to their low cost, remarkable catalytic efficiency, and enhanced chemical stability. These characteristics position carbon counter electrodes as a viable alternative to expensive platinum electrodes in a wide range of DSSC applications. It is possible to make large adjustments to the catalytic activity of the CCE through the engineering of a

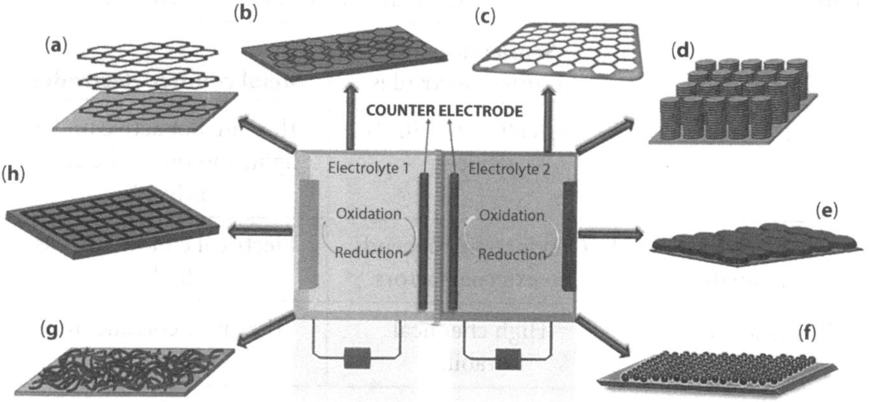

Figure 10.4 The following are some examples of the several types of carbon materials that are utilized in counter electrodes: (a) graphite, (b) CNTs that are arbitrarily orientated graphite, (c) graphene, (d) unidirectional CNTs, (e) activated carbon, (f) carbon black, (g) carbon nanofibers (CNFs), and (h) hollow active CNFs [52].

substantial specific surface area, high conductivity, a unique structure, and functional surface groups. Nevertheless, the drawbacks of the CCE persist. The disadvantages encompass inadequate mechanical stability, low opacity, and limited charge transfer and mass transport dynamics at the interface between the counter electrode and the electrolyte [64].

The application of activated carbon coated on a flexible graphite sheet led to the creation of a pure CCE for DSSCs. The electrode exhibited minimal series resistance (R_s) and charge transfer resistance (R_{ct}) due to the superior conductivity of flexible graphite and the enhanced catalytic activity of activated carbon. This was attributable to the exceedingly low levels of both features. Upon comparison of the platinum-coated fluorine-doped tin oxide (Pt/FTO) electrode with the pure CCE, it was found that the resistances R_s and R_{ct} of the pure CCE were reduced by one-fourth and two-thirds, respectively. The solar-to-electric conversion efficiencies of the pure CCE were determined to be 6.46% and 5.0%, respectively, when employed in DSSCs with active cell areas of 0.15 and 1 cm^2, respectively. The low sheet resistance of the pure carbon counter electrode significantly enhanced both the fill factor (FF) and efficiency (η). The efficiency of the pure carbon counter electrode was about double that of the platinum/fluorine-doped tin oxide counter electrode for the 1 cm^2 cell area. The pure carbon counter electrode is a crucial element for DSSCs due to its lightweight, flexible design, and affordability. This renders it a desirable element as it decreases manufacturing and transportation costs [65].

10.5 Electrolyte

In an SRFB, the electrolyte plays a crucial role. It acts as a connection inside the battery, transferring energy from one part to another *via* the photoelectrode so that the battery can convert and store energy. The electrolyte is shared across the two different types of batteries, which allows for their integration. The longevity of the battery is directly proportional to the electrolyte's ability to maintain its stability. Also, the battery's performance, notably its energy storage density, is affected by the electrolyte selection. Therefore, among the many electrolyte options, the SRFB is a good one since it can convert and store solar energy while also providing unilateral or even bilateral photo-driving power [66]. The chemical nature of the electrolyte in SRFBs is critical to the overall efficiency and durability of the system RFB devices; the active components comprise redox pairs dissolved in the electrolyte solutions. Therefore, the redox couple in the electrolyte must satisfy a number of requirements for SRFB applications. Features such as high solubility, adequate formal potential window, fast kinetic rates, good ionic conductivity, lack of interaction with solvent, atmosphere, photoelectrode materials, low viscosity, and abundant availability are all crucial. Other advantages include high energy density and current density operation, low light absorption within the semiconductor function range, and safety [36, 67–76]. The energy density of the RFB will be affected by the concentration of the electrolyte pairs. Equation 10.5 articulates the complete definition of energy density. In this equation, n represents the number of electrons transferred during the reaction, C indicates the concentration of active materials in the electrolyte expressed in molarity, and F specifies the Faraday constant. Optimal energy densities are attained at elevated concentrations; however, this is often limited by the solubilities of redox compounds generated while charging or discharging within the system's operational temperature limit. The water activity of semiconductors in aqueous solutions can be reduced only by a high concentration of redox species. The photogenerated carriers in aqueous SRFBs face direct competition from water. The photoelectrodes may deteriorate due to this competition. A larger concentration of redox species in the water may inhibit or halt corrosion, hence facilitating the transfer of photogenerated carriers.

Furthermore, a concentration gradient of reactants will be established by the surface reaction at the electrode, resulting in diffusion processes governed by Fick's first and second laws (Equations 10.6 and 10.7). The equations provide a flux, $J_{(x,t)}$ ($mol^{-1}cm^{-2}$), a concentration, $C_{(x,t)}$, at a distance x from the electrode surface at time t, and a diffusion coefficient, D. An additional potential, termed the concentration overpotential η_{conc}, is required to maintain the reactions at the specified current density when the process transitions from diffusion control to mass transport control. In Equation 10.8, this potential can be written as where C_0 represents the reactant concentration at the electrode surface and C_b denotes the reactant concentration in the bulk solution. As electroactive species are exhausted and reactants are limited at a constant flow rate, the concentration overpotential is typically maximal toward the end of the charging or discharging operation. To optimize the usage of pumping energy effectively, a mechanism that regulates the flow rate dynamically is crucial. Fick's first law, articulated in Equation 10.9, indicates that the concentration gradient of active species across the separator significantly influences the crossover rates. In this equation, J denotes the diffusion flux of the active species, C_B and C_A represent the concentrations of the redox components on each side of the membrane, and L indicates the thickness of the membrane. It is important to note that when the pumps are in standby mode, they may self-discharge and generate heat when diffusion rates over the membrane are excessively high [77–80].

$$\left(AhL^{-1}\right) = \frac{nCF}{3600} \tag{10.5}$$

$$-J(x,t) = D\frac{\partial C(x,t)}{\partial x} \tag{10.6}$$

$$\frac{\partial C(x,t)}{\partial t} = D\left(\frac{\partial^2 C(x,t)}{\partial x^2}\right) \tag{10.7}$$

$$\eta_{conc} = \frac{RT}{nF} ln \frac{C_o}{C_b} \tag{10.8}$$

$$J = -D\frac{\left(C_B - C_A\right)}{L} \tag{10.9}$$

By assessing the solubility of the redox agents in the supporting solution, one may ascertain the maximum concentration of these species in the electrolyte. The precipitation of species in the electrolyte may obstruct the batteries, potentially damaging the pumps or electrodes. Therefore, it is essential to acknowledge that solubility is contingent upon the supporting electrolyte, pH, temperature, and additives, and this relationship cannot be overlooked. Solar charging for an extended period can raise the temperature of the solution by as much as 15°C. This underscores the need for temperature stability, especially with SRFBs [81–90]. There is a relationship between the amount of ionic dissociation, the concentration, the temperature, and the conductivity of the electrolyte. To improve energy and power efficiency, high conductivity is sought after (Table 10.3). When it comes to aqueous solutions, the supporting electrolyte is often composed of strong electrolytes. When compared to organic solvents, organic solvents have a lower ionic conductivity; nevertheless, the reliability of the photoelectrode or electrode is expected to be higher. It is possible to improve ion mobility by increasing the temperature, which will result in a reduction in solution resistance and an increase in operating efficiency [91].

Table 10.3 Some charge carrier ions and their conductivities [36].

Ion	Molar conductivity $10^{-4} \cdot S \cdot m^2 \cdot mol^{-1}$	Ion	Molar conductivity $10^{-4} \cdot S \cdot m^2 \cdot mol^{-1}$
H^+	349.8	OH^-	198
Ca^{2+}	119	SO_4^{2-}	159.6
Mg^{2+}	106.1	Br-	78.4
K^+	73.5	I^-	76.8
Na^+	50.1	Cl^-	76.3
Li^+	38.6	NO_3^-	71.4

The viscosity of the electrolyte is a critical factor in RFBs since it influences the battery's EE by affecting flow patterns, flow rate, and mass transfer resistance. It is common for a high concentration of electrolyte pairs to increase the attraction between solvents and solutes, which ultimately leads to an increase in viscosity. Inversely, solute–solvent interactions are reduced as temperature increases, leading to a decrease in viscosity. The features of the solution in terms of its color and its ability to absorb light are determined by the electron configuration of the electrolyte. The light absorption spectrum of the electrolyte in SRFB systems must not correspond with that of the semiconductor being used; otherwise, this would result in a decrease in the efficiency of the photoelectrodes. The adsorption spectrum of the electrolyte can be altered by modifying the concentration, complexation, and hydration levels of the electrolyte simultaneously. Even though there has been a significant amount of study conducted on redox light-activated semiconductor/electrolyte junctions (Table 10.4), researchers have only recently started merging PEC cells with RFBs.

The electrolyte acts as a bridge for energy conversion and storage, affecting battery longevity and performance. The choice of electrolyte affects energy storage density, enabling simultaneous photo-driving power and solar energy conversion. Consequently, the type of electrolyte is of paramount significance for performance [66].

10.5.1 Inorganic–Inorganic

The redox couples of vanadium have two different impacts. One is the rapid redox reaction kinetics of vanadium, which enable it to swiftly eliminate carriers and reduce recombination, thus increasing the photocurrent.

Table 10.4 Redox half-cell reactions of different electrolytes.

Negative half-cell reaction	Potential(E_0)	Reference
$2Br^{-}+Cl^{-} - 2e^{-} \rightarrow ClBr_2^{-}$	1.4	[92]
$Cr^{3+} - e^{-} \rightarrow CrO_4^{2-}$	1.35	[36]
$VO^{2+} + H_2O - e^{-} \rightarrow VO_2^{+} + 2H^{+}$	1	[85]
$Fe^{2+} - e^{-} \rightarrow Fe^{3+}$	0.77	[36]
$2I^{-} - 2e^{-} \rightarrow I_2$	0.54	[93]
$Np^{4+} + e^{-} \rightarrow Np^{3+}$	0.15	[94]
$2Cl^{-} - 2e^{-} \rightarrow Cl_2$	0.14	[95]
$Pb^{2+} + 2e^{-} \rightarrow Pb$	-0.1262	[96]
$Cu^{2+} + e^{-} \rightarrow Cu^{+}$	-0.153	[93]
$V^{3+} + e^{-} \rightarrow V^{2+}$	-0.26	[97]
$Cd^{2+} + 2e^{-} \rightarrow Cd$	-0.403	[98]
$Cr^{3+} + 2e^{-} \rightarrow Cr^{+}$	-0.42	[36]
$Fe^{2+} + 2e^{-} \rightarrow Fe$	-0.447	[99]
$Zn^{2+} + 2e^{-} \rightarrow Zn$	-0.76	[67]

Additionally, it can change the dynamic equilibrium between carrier recombination and redox reaction toward the latter. An approach utilizing vanadium redox couples was initially proposed to enhance the photoresponse of regenerative solar cells and decrease the frequency of charge carrier recombination. The experimental results indicate that the addition of a minor quantity of vanadium oxide to an acidic electrolyte can markedly enhance the photocurrent of all-vanadium PEC batteries during photo-recharging and diminish the extent of photo etching in the photocatalyst during battery charging. The resistance ampere method and EIS were employed to provide preliminary evidence that solar energy can be stored and released even in the absence of light [100–102]. Later on, it was demonstrated that a continuous flow PEC storage battery could be developed using highly kinetic vanadium redox compounds. These chemicals can effectively prevent both charge recombination and side reactions. It was established that the forced convective motion of the electrolyte, devoid of external influence, led to a fivefold increase in photocurrent measurement compared to the static condition. The EIS results indicate a substantial reduction in charge transfer resistance within the electrolyte under forced convection flow. This results directly from the enhanced mass transfer efficiency attained with the incorporation of the serrated flow channel in the batteries. The findings offer more evidence that the electrolyte can be stimulated to flow convectively, resulting in a substantial decrease in charge transfer resistance on both battery sides.

10.5.2 Organic–Inorganic

A study was conducted using dual-phase electrolytes, specifically Li_2WO_4 as the anode and LiI as the cathode. The anode and cathode solutions contained LiI, while the electrolyte consisted of quinoxaline and its derivatives. Numerous investigations have evidenced the elevated redox potentials of quinoxaline and its derivatives. Enhancing the efficiency of SFBs can be accomplished by re-evaluating the design of organic molecules that serve as both electrolytes and electrochemically active species, with adjustable redox potentials [41, 103]. To substitute the V^{2+}/V^{3+} redox couple in the vanadium system and further demonstrate the proposed versatility of the electrolyte, an SFB utilizing V^{4+}/V^{5+} and AQDS/$AQDSH_2$ as active electrolytes was constructed [104]. In a different investigation, a form of SFBs was constructed, which utilized Br^-/Br_3^- and AQDS/$AQDSH_2$ as active electrolytes. These batteries were constructed using two distinct components. The process of photo-recharging involves the oxidation of Br^- on the surface of n-type silicon, while the reduction of p-type silicon AQDS occurs. When solar energy was converted into chemical energy, the efficiency of the conversion was approximately 5.9%, and when solar energy was converted into chemical energy, the efficiency of the conversion was 3.2%. In terms of discharge capacity, the battery was among the best available [105].

10.5.3 Organic–Organic

Additionally, organic–organic systems provide a mechanism for solar electrochemical storage that not only is effective but also has a significant potential capacity for storage. As an example, there is a type of SFB that is composed of 1,2-benzoquinone-3,5-disulfonic acid (BQDS) and 1,2-amino-2,5-disulfonic acid (AQDS) [37]. Electrolytes are crucial elements that significantly influence the energy density, potential, resistance, light absorption, and stability of the battery. A major obstacle hindering the advancement of solar cells is the need for a mix of semiconductors and electrolytes that is both efficient and stable. The identical photoelectrode and various redox electrolyte systems characterized by high solubility, rapid kinetics, low ionic resistance, and minimal light absorption were employed to perform multiple solar charge–discharge cycles to assess the stability of the semiconductor materials. Research is being undertaken to explore alternative semiconductor materials with suitable energy levels and stability techniques.

10.6 Membrane Separators

The separators can be categorized into four distinct groups: ionic exchange membranes, hybrid membranes, mesoporous separators, and solid ionic conductors. The ion exchange membrane (IEM) is crucial, preventing the mixing of electrolytes while facilitating necessary charge transfer between the electrodes of the device. Cation exchange membranes (CEM) facilitate the conduction of positively charged ions, while anion exchange membranes (AEM) enable the conduction of negatively charged ions.

Nonetheless, there exists a broad spectrum of choices within the two membrane groups. The predominant CEM membranes for SRFB applications are now Nafion™ 117 and 115, also Selemion AMV is a frequently utilized AEM membrane, albeit less commonly employed. The membrane's thickness is a critical aspect in attaining high-performance SRFBs, as an excessively thin membrane may lead to unintended electrolyte crossover. Conversely, an excessively thick membrane would lead to elevated resistance, impeding the viable charge transfer through the membrane. While values depend on measurement conditions like temperature and humidity, conductivities are often inversely correlated with thickness [106]. Typically, acidic or basic side chains are linked to a polymer backbone to form ionic exchange membranes. The organic acid or base's polar functional groups react with molecules of water or a dipolar solvent, causing the polymer to inflate. The process of effective swelling involves the transfer of ionic charges across a solvent molecular network. A charged stationary phase formed by the polymer's ionic groups allows for the selective transit of cations and anions across the membrane by the use of an electrostatic barrier.

A cationic ion-exchange membrane, an anionic ion-exchange membrane, and an amphoteric ion-exchange membrane are the three main types of ionic exchange membranes that are defined by the charge of their stationary phase [47]. Cationic ion-exchange membranes can be categorized into fluorinated and nonfluorinated subclasses according to the base polymer. Fluorinated cationic exchange membranes exhibit chemical stability against both oxidizing and reducing agents. They maintain stability in moderately concentrated acids and bases. Perfluorosulfonic acids (PFSA) are the major components of fluorinated exchange membranes. DuPont's Nafion™ serves as the quintessential illustration of this category of exchange material. This membrane product, designed for chlorine-alkali electrolysis, is widely recognized as a benchmark for CEMs. The sulfonic acid groups in Nafion™ are connected to the polymer backbone through completely fluorinated polypropylene glycol ether. The quantity of fluoropropyl ether groups dictates the polymer's chemical stability and swelling properties.

Presently, numerous enterprises provide PFSA membranes with differing side chain lengths. Two primary techniques exist for the production of PFSA membranes. They can be produced either by solution casting or by extruding the fluorosulfonic acid precursor followed by hydrolysis with alkali. The casting of membranes from a solution surpasses the extrusion process by promoting the formation of conductive domains as inverse-turned micelles, leading to the development of ionic channels. Solution casting facilitates the extraction of membranes from aqueous or alcoholic PFSA solutions for transfer onto a polymer substrate. It is feasible to amalgamate solutions with diverse polymers and cross-linkers to achieve the desired qualities in a singular membrane. The objectives for solution-derived modified membranes encompass heightened ionic conductivity, minimal swelling, improved selectivity, diminished electro-osmotic mobility, and low permeability to big cations, referred to as permselectivity. The buildup of vanadium species on the surface and within the matrix of PFSA membranes is a disadvantage of these membranes. The term for what is happening here is "fouling." Sulfonated polyether ether ketone is the usual component of nonfluorinated cation exchangers. Sulfonating polyimides or polyphenylene sulfone is another way to get nonfluorinated membranes. However, membranes composed of aromatic, nonfluorinated polymers often exhibit insufficient stability in vanadium electrolyte solutions and are susceptible to cracking [107–113].

When comparing anionic exchange membranes to cationic exchange membranes, it is evident that anionic exchange membranes exhibit superior permselectivity. This leads to improved current efficiency of the RFB. Triethylamine groups are commonly integrated into a base polymer by chloromethylation. These groupings exhibit cationic characteristics. Poly(vinylbenzylstyrene) copolymer cross-linked with adipic acid, poly(-fluorenyl ether), or poly(aryl ether ketone phthalazinone) (PAEKP) serves as an example of base polymers [114–118].

Amphoteric ion-exchange membranes contain equal quantities of cationic and anionic groups. Gamma rays facilitate the production of these membranes *via* radiation-graft copolymerization, which entails the modification of preformed films or powdered PVDF or ETFE. Cationic groups can be introduced by employing grafted polystyrene or α-methylstyrene, followed by sulfonation of the resultant molecule using chlorosulfuric acid. The copolymerization of dimethylaminoethyl methacrylate or trimethylaminoethyl methacrylate yields anionic groups from the resultant mixture [119–123].

The current porous separators in use are generally flexible polymer devices. Nanoporous materials are characterized by a pore size of less than 1 nm, microporous materials have a pore size ranging from 1 to 2 nm, mesoporous materials possess a pore size between 2 and 50 nm, and macroporous materials have a pore size beyond 50 nm. Nanofiltration membranes, ultrafiltration membranes, and microfiltration membranes are the classifications recognized for membrane separators. Nanofiltration membranes include a particle size of under 5 nm, ultrafiltration membranes range from 5 to 100 nm, and microfiltration membranes span from 100 nm to 10 μm.

Traditional battery separators, made from stretched polyolefin films, are frequently utilized in microfiltration membranes. The efficiency of a flow battery declines due to the migration of active ions through the pores. Therefore, these separators cannot be employed without extra adjustments. Historically, the use of commercial separators for lead batteries without prior treatment has been solely suggested for iron–vanadium redox-flow batteries. This is owing to the properties of lead batteries. Additionally, microfiltration or ultrafiltration membranes can be manufactured utilizing phase-inversion processes. Utilizing these technologies helps the manufacture of asymmetrical membranes, in which the pore radius is lowered to coincide with the membrane's thickness. Examples of these membranes include those manufactured from polyacrylonitrile (PAN), polyethersulfone in conjunction with sulfonated polyether ether ketone (PES/SPEEK), and polyvinylidene fluoride (PVDF). Furthermore, the pores may be packed and sealed utilizing inorganic fillers like methyl orthosilicates to improve pore selectivity and boost their water absorption capacity [124–128].

In the context of composite membranes, inorganic components are integrated into the matrix of polymers by the solution of the casting process, which extends from the sol–gel technique. The purpose is to lower the pore size in these membranes, thereby boosting the permselectivity of the panels. The types of membrane systems in this category include silicates in Nafion™ or partly fluorinated SPEEK, zirconium phosphate in partly fluorinated SPEEK, and polytungstate in sulfonated polyphenylene sulfide (PPS) [119, 129–133]. Ion-conducting crystals, ceramics, or glasses often only exhibit adequate conductivity at temperatures that are higher than

300°C, which is why solid-state conductors are employed in RFBs only under very constrained conditions. Compared to the conductivity of ionic exchange membranes, the conductivity of such devices is generally one to two orders of magnitude much lower.

Certain hybrid battery types incorporating solid metal electrodes or gas diffusion electrodes can technically operate without separators. To prevent short circuits resulting from dendritic metal deposition or the flooding of gas diffusion electrodes, hybrid RFBs often incorporate porous separators or membranes. In theory, any flow battery might function without separators. To achieve this, the electrolyte liquids of anolyte and catholyte within the cells must be perfectly co-laminar and devoid of turbulence. This state can be easily attained in a flow-through reactor at the low flow rate in the flow battery. Flow batteries devoid of membranes are usually employed as laminar microflow reactors. Porous flow electrodes in these structures may not contact the boundary layer, as this would lead to the mixing of electrolytes [134, 135]. Therefore, an optimal separator exhibits high ionic conductivity and little permeability to electrolyte active species. Exceptional mechanical, chemical, and electrochemical stability is required for membranes, especially for strong acids, bases, or oxidizing agents (VO_2^+, Br_3^-, Ce^{4+}). SRFBs require membrane stability under photon exposure or thermal conditions [136–143]. The ionic functional groups in the membrane can create electrostatic connections with oppositely charged ions in the solution, acting as exchange sites. Negatively charged groups like $-COO^-$, $C_6H_4O^-$, $-PO_3^{2-}$, $-PO_3H^-$, and $-SO_3^-$ are commonly used to enhance cation migration. Cationic groups, including $-NH^{3+}$, $-NRH_2^+$, $-NR_2H^+$, $-NR_3^+$, and $-SR_2^+$, can selectively enable anions to pass. Amphoteric, bipolar, or mosaic ion-exchange membranes have both positively and negatively charged functional groups organized in a random, layered, or domain pattern. The fundamental downside of ion-exchange membranes is their expensive cost, especially for Nafion™ along with other perfluorinated versions. Nonfluorinated ion-exchange membranes, such as those based on hydrocarbons, are significantly less expensive; however, their chemical stability in acidic and oxidative environments is relatively poor [112, 144, 145].

AEMs are attracting significantly less interest in RFB systems compared to CEMs, largely due to their poorer proton conductivity, diminished chemical durability, and excessive water transfer. The implementation of AEMs has been proved to limit the crossing of positively charged active species, including vanadium ions [116, 118, 137, 146]. The quantity of recorded investigations on amphoteric and other ion-exchange membranes in RFB systems is minimal, and stability remains a serious challenge [147–149].

A number of different flow battery structures have been built. Flow batteries can be classified as either symmetric or asymmetric flow batteries, depending on the degree of symmetry they possess in terms of the electrolyte composition. Symmetric flow batteries are dependent on the same parent molecule as the active species in both the catholyte and the anolyte. For instance, a vanadium flow battery is an example of such a parent molecule pair [47, 150–154]. For example, alkaline zinc–iron flow batteries are examples of asymmetric variations that make use of different active species in each electrolyte. The asymmetric arrangement can prevent capacity deterioration that is permanent. Certain active species, such as bromine and iodine, have a significant volumetric capacity, and they can be encapsulated within the electrode while simultaneously maintaining a distinct half-electrode with electrolyte circulation. Even though the single-flow architecture saves a pump and lowers the risk of electrolyte leakage, the capacity on one side is limited due to the size of the cell. When it comes to certain devices that employ solid-state active species on both sides, membranes are not necessary. It is vital to keep a gap between the negative and positive electrodes to avoid direct contact between the two poles. This will increase the resistance to mass transfer [155–157].

Enhanced ion selectivity, increased ion conductivity, and excellent durability are three essential properties of IEMs. Chemical stability, including resistance to strong oxidizing agents, very acidic or basic electrolytes, and mechanical integrity are the primary components of membrane reliability. Ion conductivity also refers to the ability to transport ions in the absence of active species. Reducing the ohmic polarization and improving voltage performance in a battery are both made possible by increased ion conductivity. To keep the capacity from degrading too much during cycling and to prevent active species from crossing over, an electrolyte must be ion-selective, meaning it can only conduct some ions while rejecting others. A membrane's selectivity and conductivity are important properties because they determine the associated flow battery's Coulombic efficiency and voltage efficiency or CE% and VE, respectively. Therefore, it is necessary to know the ion transport process in membranes for the development of better-performing devices [142, 151]. The ion transport process in ion-conducting membranes is intricate due to the variability of charge carriers associated with different types of flow batteries, including vanadium, alkaline zinc–iron, zinc–bromine, neutral zinc–iodine, and organic. The electrolyte solutions are essentially grouped as acidic, alkaline, and neutral solutions. In acidic electrolytes, HCl and/or H_2SO_4, in basic electrolyte NaOH and/or KOH, and, in neutral electrolytes, NaCl and/or KCl solutions are employed.

Various fuel cells necessitate distinct types of electrolytes and specific membrane characteristics. The membranes must exhibit exceptional stability owing to the oxidizing and acidic electrolytes including in vanadium flow batteries. Neutral flow batteries, such as zinc–bromine and neutral zinc–iodine, demand the distinction between active species and supporting ions. Acidic or alkaline flow batteries commonly employ H^+ or OH^- as the principal charge carriers, but in a neutral system, alkali metal ions and halide ions are moved over a membrane, employing various ion transport techniques [150, 158]. The polymer composition of IEMs significantly influences both the morphology of the generated membranes and their future performance. The elevated ionic conductivity of a membrane will be guaranteed by a greater quantity of ion exchange groups, whereas the hydrophobic chains will contribute to the membrane's enhanced stability and selectivity.

In the preparation process, the hydrophobic and hydrophilic components of the polymer tend to self-assemble and segregate from each other. This results in a microphase-separated structure featuring hydrophobic matrices and hydrophilic clusters within the membranes. The ion transport channels of perfluorosulfonate membranes typified as Nafion™ series membranes are the microphase separation structures that have received the most study interest. The water-channel model has been the predominant characterization of ion channels in perfluorosulfonate membranes to date. This model demonstrates tubular waterways, where the walls include ion exchange groups and the core contains randomly scattered adsorbed water within a crystalline and non-crystalline hydrophobic substrate [159–161]. The configuration of the water channel can respond to the external environment. The accumulation of sulfonic acid groups facilitates the creation of a hydrophilic channel in the absence of water. The passage would inundated with water from the initial hydration shell as the water level ascended. When the water volume increases, the channel center is occupied by the second hydration shell, which comprises additional water molecules that can flow freely. This results in a proportional rise in water diffusivity. K^+, for example, can significantly decrease the dimensions of the water channel. Small-angle X-ray scattering can reveal the reduced channel size in Nafion™ 212 following treatment with a potassium hydroxide solution. This would lead to an augmentation in the membrane's area resistance and a decline in the efficacy of alkaline zinc–iron flow lithium batteries. To align the water channels in the stretching direction, mechanical stretching of Nafion™ is effective. Ion conductivity grows in the stretching-perpendicular direction and shrinks in the stretching-perpendicular one. There is an increase in the resistance to the permeability of vanadium ions in the vertical orientation of the water channels that are aligned with the stretching direction [32, 162–165].

In addition, an ion-solvating membrane is available. An electrolyte is employed to expand the membrane, leading to the creation of a uniform polymer/water/solute system. This obviates the necessity for phase separation and ion channels. Polybenzimidazole exemplifies a membrane within this specific type. The polybenzimidazole membrane infused with potassium hydroxide exhibits commendable performance in alkaline electrolyzer cells. Polyvinylpyrrolidone was employed in the fabrication of ion-solvating membranes with polybenzimidazole. This classification remains contentious, and the mechanism responsible for ion transport in this membrane type is still deemed unclear [166–170].

Under different conditions, membranes are subject to a variety of restrictions concerning their chemical stability. Cell membranes are vulnerable to three processes: the nucleophilic assault, the Wittig reaction initiated by hydroxyl radicals, and the Hoffmann elimination. Proton exchange membrane fuel cells and flow batteries necessitate more stable membranes due to the presence of oxidizing agents, pentavalent vanadium ions, and exceedingly acidic electrolytes [112, 171–174].

While cracks or leaks can occur when the mechanical properties and structure of a membrane are compromised due to degradation of the polymer backbone, cracks or leaks may not occur when functional groups are lost due to degradation of the side chains. Enhancing the longevity of membranes necessitates prioritizing the stability of the backbone. The primary chain architectures of polymers encompass polyolefin, polyphenyl ether, polyaryletherketone, polyphenyl, and imidazole, among others. Because π–π bonds have a high bond dissociation energy, aromatic polymers are more stable compared to polyolefins. Aromatic polymers can increase the membrane's durability due to their rigid structure. This has led to a plethora of studies focusing on aromatic polymers. Usually, heteroatoms weaken the main chain's stability because their electrophilic or nucleophilic properties might throw off the even distribution of electrons, leading to the breakage of main chain segments by electrophilic or nucleophilic attack. Therefore, it is expected that the primary chains of aromatic polymers free of heteroatoms will be very stable. A reason to disturb this stability is introduction of ion exchange groups. Because the electron-withdrawing sulfonic acid groups increase the asymmetry of electron distribution and make nucleophilic assaults easier, sulfonation can reduce the stability of poly (ether ether ketone). Quaternary ammonium groups are equally problematic. Another approach to producing very stable polymers is to use chemical bonds that have a high bonding energy.

While fluorine atoms protect the C–C link against reactive species assaults, the bonding energy of the C–F bond in perfluorosulfonate IEMs is much higher than that of the C–H bond. To increase the longevity of the polymer materials, monomers containing fluorine atoms are added to the polymer framework utilizing the perfluorosulfonate ion exchange resin. There are other possible bonds outside the C–F bond, such as the inorganic Si–O bond [132].

Since Si–O is already an oxide, it shows a lot of resistance to substances that oxidize it. Because Si–O-based polymers do not have sufficient mechanical strength, composite membranes are preferred. Furthermore, augmenting the contact among polymer chains constitutes an excellent approach to improving the chemical stability of the membrane. Cross-linked membranes may mitigate the swelling effect and extend their longevity. The selection of cross-linking agents would also confer additional properties to the membrane.

An example involved the synthesis of cross-linked chloromethylated polysulfone with imidazole, which functions as a cross-linker, confers a positive charge to the membranes, and improves vanadium resistance *via* charge repulsion. A symmetrically structured polymer with a high level of crystallinity may improve membrane stability. Daramic demonstrates stability in vanadium flow batteries (VFB) owing to the crystalline PE polymer's efficient reinforcement of the membrane structure. Extensive conjugated structural polymers demonstrate considerable durability, with polybenzimidazole being a prominent instance [175, 176]. The functional group's steadiness is essential for maintaining the functionality of IEMs. Steric hindrance, which is created by initiating significant steady groups around the functional center, can improve the stability of functional groups. The stability of functional groups can be diminished when electron-withdrawing moieties are located close to them. Furthermore, the polymer's swelling, mechanical strength, and contact area with active species can all be negatively impacted by an overabundance of functionalization, which, in turn, increases the likelihood of breakdown. Thus, it is crucial to take into account a chemical structure that is stable and has a sufficient level of functionality while developing membranes [177, 178].

Because the backbone and functional groups work together, the polymer is stable. As mentioned before, the electron-withdrawing ion exchange groups may affect the main chain's stability. Stability could be greatly improved by increasing the distance between ion exchange groups and the main chain. An extended side chain connected anion exchange groups to the primary chain, which improves membrane stability and reduces the primary chain's electron-withdrawing action. The micro-phase structure and stability of the membrane are determined by the interaction between the main chain and functional groups. It is possible that active species can interact less with the primary chain due to a novel micro-phase separation, leading to improved membrane stability. Incorporating longer hydrophobic chains can enhance stability and facilitate micro-phase separation. A more flexible meta-terphenyl rather than para-terphenyl may improve micro-phase separation and increase membrane stability and ion conductivity. Most flow batteries operate in highly acidic, alkaline, and oxidative conditions, leading to restricted stability of the majority of membranes, particularly non-perfluorinated types, during cycling. A considerable body of research has investigated the breakdown mechanisms of IEMs, demonstrating that the ion exchange groups markedly affect membrane stability [112, 179, 180].

10.7 Electrochemical Characterization of an SFB

As research into renewable energy sources continues to expand, it is becoming increasingly important to explain the structure and composition of energy materials, as well as the response mechanisms that govern them, to create more advanced energy storage methods. Having comprehensive information with immediate measuring systems during the operation and storage of energy system components and materials has become increasingly crucial in recent years. There is vital information contained within characterization procedures that can be utilized for optimization under long-term operational conditions. The electrochemical techniques of characterization like CV and EIS do not do any damage to the material while it is being measured. Furthermore, if evaluated with great care, these approaches can provide valuable information for devices [181–183].

10.7.1 Performance Evaluation

Integrating SFBs, which are akin to solar cells, can only be as effective as their light response capability. An example of a solar vanadium RFB utilized TiO_2 as the photo-active electrode, whereas the counter electrode was Pt metal in the solution containing VO_2^+/VO^{2+} and V^{3+}/V^{2+} as redox couples [66]. Upon exposure to sunlight, electron–hole pairs are generated on the TiO_2 photoelectrode, and the resulting holes migrate to the interface between the photoelectrode and the electrolyte to oxidize VO^{2+} in the electrolyte. Concurrently, electrons are sent to the anode by an external circuit, leading to the reduction of the anode material, V^{3+}.

Ions migrate across the membrane from within to maintain charge equilibrium on both sides. Throughout discharge, the cathode and the anode are connected by an external load. Electrons traverse from the anode to the cathode, resulting in the reduction of the electrochemically active material in the catholyte, while the active material in the anode or the anolyte undergoes oxidation, in contrast to the charging process. To enhance cell performance, the utilization of composite electrodes as photoelectrodes increases the absorption of light. The interaction between the band structures inside the composite electrode is altered in a way that facilitates the separation and transport of charges through the electrode. As a result of the reduction in charge recombination and the enhancement of charging current density, the battery's light response capabilities have been significantly improved [100–102].

Connecting the photoelectrodes in series is another approach that could be applied to achieve improved performance. Either the photoelectrode could lack sufficient photovoltage to significantly alter the redox potential, or it comprises a material with a substantial band gap that possesses considerable PV potential but fails to absorb a significant portion of the available light. An arrangement of two tandem photoelectrode cells, divided by a proton/cation exchange membrane, provides superior photoelectric efficiency compared to a singular semiconductor electrode, by absorbing solar light at higher intensities, thereby enhancing the efficiency of potential energy conversion. The tandem photoelectrodes RFBs, characterized by high efficiency and elevated photovoltage, are optimized to facilitate effective off-grid solar energy utilization and electrification [37, 184].

The requirements for light response capacity should encompass solar energy harvesting, spectrum response range, and illumination. The majority of individuals evaluate only one or two factors; nevertheless, it is feasible to formulate an equation for evaluating light response capability by incorporating all three dimensions, which may also be examined. The photoelectrode's vulnerability to electrolyte corrosion and the redox pair's instability are the primary causes of why integrated SFBs fail to last extensively. Establishing stable redox couples that avoid performance degradation of the redox couple and the integrated SFBs' photoanode and photocathode is vital for improving the battery life of integrated SFBs [45].

In an extended cycle test, the capacity utilization rate of the devices must be maintained to verify the achievement of long-life redox couples. To mitigate photoelectrode corrosion caused by the electrolyte, an alternate photoelectrode, such as TiO_2, can be utilized as a protective layer to avert direct contact between the electrolyte and the photoelectrode. The efficacy of solar energy diminishes during charging, necessitating that external devices be fully charged, hence precluding spontaneous charging. The band of photoanode generates a minimal discharge potential in the absence of bias, eliminating the need for additional apparatus [21].

To attain unbiased charging, a coating of dye molecules can be applied to the surface of the photoanode. A dye-sensitized titanium dioxide acts as the photoelectrode, while I^-/I_3^- and (AQDS)/($AQDSH_2$) serve as the redox couple in the electrolyte, thus forming a distinct category of integrated SFBs. The pH level influences the possible tolerance of AQDS. The typical battery voltage is about 0.73 V at a pH of 8.55. The average voltage in 1 mole of H_2SO_4 is a modest 0.28 V. Acquiring neutral photo spontaneous charging is linked to pH. Nonetheless, the photoelectrode exhibits instability, making it challenging to enhance the durability of dye molecules under prolonged sunlight exposure, and its ingredients are costly, particularly platinum and ruthenium [16, 185]. Cyclic testing of the photocharge–discharge sequence can assess the performance of SRFBs by capturing current, time, and voltage data for a duration. Multiple efficiencies can be derived from this test. Before analyzing the data, it is crucial to acknowledge that the highest overall conversion efficiency for the photo to chemical energy under solar illumination is limited by the properties of the semiconductor. The efficiency is calculated using Equations 10.10 and 10.11, where J_g denotes the absorbed photon flux, μ_{ex} represents the surplus chemical potential generated by light absorption (J), φ_{conv} indicates the quantum yield for absorbed photons (the frequency of a specific event per photon absorbed by the system), and P_{in} signifies the total incoming solar energy (Wcm^{-2}). The prospective solar conversion efficiency of an individual semiconductor possessing a band gap of 1.26 eV (984 nm) is roughly 40% [186].

If SRFBs require an external bias to facilitate reduction and oxidation, it is feasible to assess the efficiency of the batteries. Furthermore, it is feasible to estimate the average efficiency of solar to chemical energy conversion during the photocharge process. In Equation 10.12, J denotes the observed current density, E_0' signifies the reversible potential of the redox couplings, E_{app} indicates the applied voltage (V), and t represents the time of illumination. In the event that there is no requirement for an external bias, the efficiency of the photocharge process can be evaluated at E_{app} = 0V. It is possible to calculate the total EE (the conversion of photochemical energy to electrical energy) for a single photocharge–discharge cycle by utilizing Equation 10.13, where E_{dis} stands for the discharge voltage. Throughout the photocharge and discharge process, the CE% (η_c) acts as an indicator of Faradic loss, as described in Equation 10.14. The voltage efficiency (η_v) is determined by dividing the EE by the CE%.

$$\eta = \frac{J_g \mu_{ex} \phi_{conv}}{P_{in}} x100\% \tag{10.10}$$

$$\eta = \frac{J_g \left(E_o' - E_{app}\right)}{P_{in}} x100\% \tag{10.11}$$

$$\eta = \frac{\int j(t)dt\left(E_o' - E_{app}\right)}{P_{in}} x100\% \tag{10.12}$$

$$\eta = \frac{\int j(t)(E_o' - E_{app}(t))dt}{P_{in}} x \frac{\int j(t)E_{dis}(t)dt)}{\int j(t)dt(E^{o'} - E_{app})} x100\% \tag{10.13}$$

$$\eta_c = \frac{\int j_{discharge}(t)dt}{\int j_{photocharge}(t)dt} \tag{10.14}$$

SRFBs necessitate both elevated overall EE and substantial discharge power density for efficient practical deployment. There is a method that may be utilized for the purpose of measuring, and that is the discharge of the cell under varying current densities. To find the power density, take the average cell voltage within a specific SOC range and multiply it by the current density. In this way, power density can be quantified. The slope of the polarization curve, which depicts the voltage of the discharge with the current density, can be used to quantify the resistance of the discharge cell. The slope of the curve can be used for it.

10.7.2 State of Charge

A common method for quantifying the amount of charge present in an electrolyte is to utilize the state of charge, abbreviated as SOC. For instance, in the VRFB, the SOC of the battery is 100% if the electrolyte in the positive and negative tanks is entirely composed of VO_2^+ and V^{2+}, respectively. This indicates that the battery is fully charged. On the other hand, if the catholyte or the anolyte is composed of pure VO^{2+} or V^{3+}, the SOC is 0%, which indicates that the battery is not charged at all. The state of the current in SRFBs may be determined by using Equation 10.15, where A represents the area of the electrode. This equation can be used to compute the current after photo-charging for a variety of periods. Measuring the open-circuit potential (E_{oc}), which is derived from the Nernst equation, and the correlations between the current state and redox species concentrations, as shown in Equation 10.16, is further means of monitoring the current state. The formal cell potential of the RFB is denoted by the symbol E_0. This potential can be determined by measuring E_{oc} between the electrodes at a point in time where the concentration of the reduced species in the electrolyte is equal to the concentration of the oxidized species in each reservoir, denoted by 50% SOC.

$$SOC(\%) = \frac{\int j_{photocharge}(t)dt}{nCVF} x100\% \tag{10.15}$$

$$E_{oc} = E_o - \frac{RT}{F}\ln\left[\frac{c(VO_2^+)c(V^{2+})}{c(VO^{2+})c(V^{3+})}\right] = E_{oc@50\%SOC} + 2\mathrm{RT}\ln\left(\frac{SOC}{1-SOC}\right) \tag{10.16}$$

10.7.3 CV Measurements

The formal potential of a half-cell reaction could be readily ascertained by CV, provided that both the redox species exhibit chemical stability throughout the measurement period [187, 188]. Voltage ramps at a predetermined scan rate are executed between the working electrode and the reference electrode, with the resultant current densities (or current) documented as an expression of voltage. Current initiates when the potential is anodic enough and persists in increasing while the oxidation reaction transpires. As the reactant is depleted, the current attains a maximum value before declining until the limiting current is achieved, at which point the potential scan direction reverses. As the potential diminishes, the reduction process transpires, resulting in the emergence of a corresponding cathodic peak until the oxidized species at the electrode surface is exhausted. As a result, a CV curve reveals numerous critical parameters, including anodic peak potential (E_{pa}), anodic peak current (i_{pa}), cathodic peak potential (E_{pc}), cathodic peak current (i_{pc}), and half-wave potential ($E_{1/2}$), which is a few millivolts from the formal potential (E_0'). If $E_{pa} - E_{pc} = \frac{57}{n}$ mV and $|i_{pa}/i_{pc}|$ = 1, the redox process is categorized as reversible, with mass transfer being the exclusive limiting factor. While the scan rate has no effect on the characteristics of reversible processes, the concentration of electroactive species is the sole determinant of peak currents. The peak separation of the redox reaction typically increases with increased scan speeds in quasi-reversible processes affected by mass and charge transfer. Irreversible processes are essentially limited by electron transport, requiring a greater overpotential to produce any current. The values of rate constants, namely 10^{-1} and 10^{-5} $cm.s^{-1}$, are distinguished between reversible, quasi-reversible, and irreversible reactions. Theoretically, one can find the diffusion coefficient D at the disc electrode surface in a dilute electrolyte by plotting the slope of the peak current densities Jp versus the square root of the scan rate $v^{1/2}$, as specified in Equation 10.17, where n denotes the number of electrons involved in the redox reaction, F represents the Faraday constant, R indicates the gas constant, T signifies the temperature, and C_b refers to the concentration of reactants in the solution.

The reaction constant k_0 can be ascertained from the intercept of log j_p versus $(E_p - E_0')$, where α denotes the transfer coefficient, E_p indicates the peak potential, and E_0' represents the formal potential. The value of E_0' can be estimated by Equations 10.18 and 10.19. Nevertheless, CV measurement is imprecise in the quantitative assessment of kinetic parameters due to the challenges in accurately establishing the baseline of current densities. However, it remains a straightforward method for acquiring qualitative information regarding redox processes.

$$j_P = 0.452\left(\frac{n^{3/2}F^{3/2}}{R^{1/2}T^{1/2}}\right)D^{1/2}C_b v^{1/2} \tag{10.17}$$

$$\log j_P = \log\left(0.227nFC_b k_o\right) - \left(\frac{\alpha nF}{RT}\right)\left(E_P - E_o'\right) \tag{10.18}$$

$$E_o' = \frac{1}{n}\sum \frac{E_{pa} + E_{pc}}{2} \tag{10.19}$$

10.7.4 Electrochemical Impedance Spectroscopy

EIS refers to the resistance of a circuit to electric current and describes the approach where the impedance is graphed as a function of frequency. This methodology has evolved over approximately 60 years and is employed in supercapacitors, photodetectors, fuel cells, and flow batteries. EIS is an economical, extensively utilized measurement technique that preserves the integrity of the sample and delivers sensitive, rapid results using sinusoidal waves.

The characterization of flow batteries is essential for modeling, elucidating electrochemical mechanisms, assessing battery interface resistance, determining the SOC, evaluating battery body resistance, predicting battery lifespan, comprehending the diffusion layer, and analyzing reaction kinetics. It can also assist in identifying potential issues and faults inside the testing system [181, 182, 189].

To design an electrical circuit for an electrochemical system, it is essential to examine the characteristics of current, voltage, and resistance. Impedance is derived from the principle of resistance as articulated in Ohm's law. In Ohm's law (E=IR), E represents voltage and I denotes current. Impedance is the generalized concept of electrical resistance, representing the opposition a resistor presents when a potential current is applied. Ohm's law is a principle that omits the phase shift between sinusoidal voltage and current, which is contingent upon time and frequency. When an alternating current (AC) potential is applied to the circuit, the potential E_{ac0} sin(ωt) and the current I_{ac0} sin(ωt) are expressed in sinusoidal form, and their relationship is defined by Ohm's law ($R = E_{ac0} / I_{ac0}$). E_{ac0} denotes the potential amplitude, while I_{ac0} represents the current amplitude. When a voltage is applied to a resistor, the voltage across it and the resultant current are illustrated in Figure 10.5. When voltage is applied, the resulting current follows without delay, and the sinusoidal waveforms (v(t) and I(t)) are in phase (Φ=0).

When an AC potential is applied to a capacitor, a phase discrepancy arises between the current and frequency, as illustrated in the accompanying graphs of potential and current (Figures 10.5 and 10.6). When an AC potential is delivered to a capacitor, the current is expressed as follows [190];

$$I_{ac} = C\frac{dE}{dt} = C\frac{d\{E_{ac0}\sin(\omega t)\}}{dt} = C\omega E_{ac0}\cos(\omega t) \qquad (10.20)$$

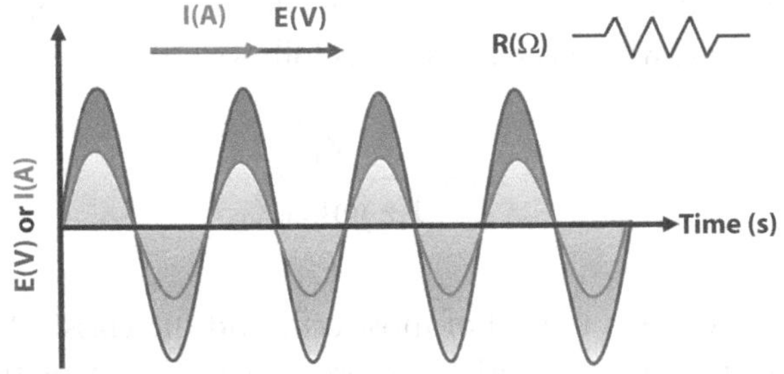

Figure 10.5 E/V and I/A graphs for resistance.

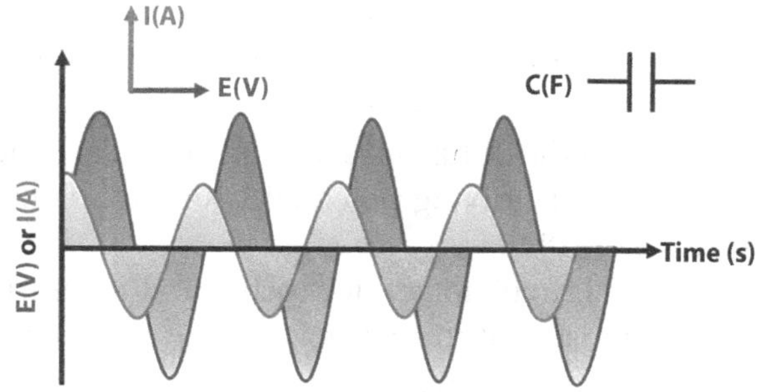

Figure 10.6 E/V and I/A graphs for the capacitor.

Numerous circuit models exist for the investigation of electrochemical impedance spectra, with the equivalent circuit model frequently employed. Electrochemical processes are estimated using the equivalent circuit model, which comprises circuit elements including resistors, capacitors, and inductors. These circuit components signify the physical and chemical processes occurring within the device. The impedance spectrum derived from experiments is achieved by correlating the circuit model with mathematical equations. The potentiostatic and galvanostatic modes remain stable provided the electrochemical characteristics of the system are unchanged during measurement [182]. The circuit with solely resistive elements is seen in Figure 10.7. The circuit's impedance is defined as the sum of the overall impedance (Z), the interface impedance (Z_0), and the electrolyte resistance (R_e) (Equation 10.21).

$$Z = R_e + Z_0 \tag{10.21}$$

The circuit comprising a resistor and a capacitor is illustrated in Figure 10.7. In this circuit, the total current is the sum of the current (I_C) flowing through the double-layer capacitor (C_{dl}) and the Faradic current (I_F). The interface impedance arises from the parallel double-layer capacitor and the Faradic impedance, which is expressed as follows [191]:

$$Z = R_e + \frac{Z_F}{1 + j\omega C_{dl} Z_F} \tag{10.22}$$

The kinetic processes typical of most cells and the classical impedance spectra of the interlayer material are presented. This depiction encompasses the charge transfer across each layer, incorporating ionic conduction

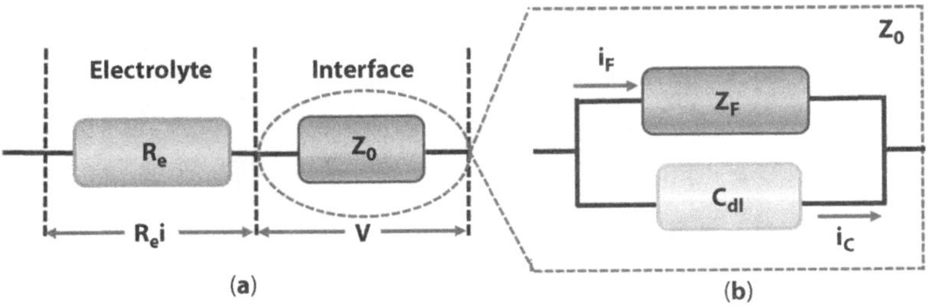

Figure 10.7 (a) Series configuration of an electrical circuit illustrating electrolyte resistance and interface impedance, and (b) parallel configuration of an electrical circuit depicting Faradaic impedance and double-layer capacitance that constitute the interface impedance [191].

via the electrolyte in the interstitial spaces between particles, as well as the resistance of the insulating layer. At frequencies below 1 mHz, activities like the creation of new crystals may become rate-limiting dynamics. Conduction is not included in the figure because, at elevated frequencies (exceeding 10 kHz), the inductance of the cables and the electrode connections become significant factors [192].

Impedance spectroscopy (EIS) study investigates three categories of resistance: mass transfer resistance, charge transfer resistance, and ohmic resistance. Ohmic resistance pertains to the resistance of the test system, charge transfer resistance denotes the processes occurring at the cathode and the anode, while mass transfer resistance arises during the manufacture of the electrolyte. In Nyquist plots, the intersection with the x-axis indicates the internal resistance (R_{in}) of the system, represented by the first resistor element in the circuit, while the semicircular arc illustrates the charge transfer resistance (R_{CT}). In flow batteries characterized by two semicircles, the initial semicircle, depicted by a secondary parallel circuit element, signifies the charge transfer resistance (R_{CT}) at the cathode, whereas the subsequent semicircle, illustrated by a tertiary parallel circuit element, denotes the charge transfer resistance (R_{CT}) at the anode [193].

The figure illustrates the Nyquist curves at 298 K with a flow rate of 15 ml min^{-1} for the electrolyte solution, corresponding to different current densities ranging from 33.33 to 122.22 mA cm^{-2}. The collection of AC impedance spectra indicates minimal alteration in the starting value of the Nyquist plot, yet the diameter of the current–capacitance arc expands. The equivalent elements and equivalent circuit are employed to characterize the AC impedance of the overloaded vanadium RFBs, as illustrated in Figure 10.8. In this context, R_0, R_1, and R_2 represent resistances, C_1 and C_2

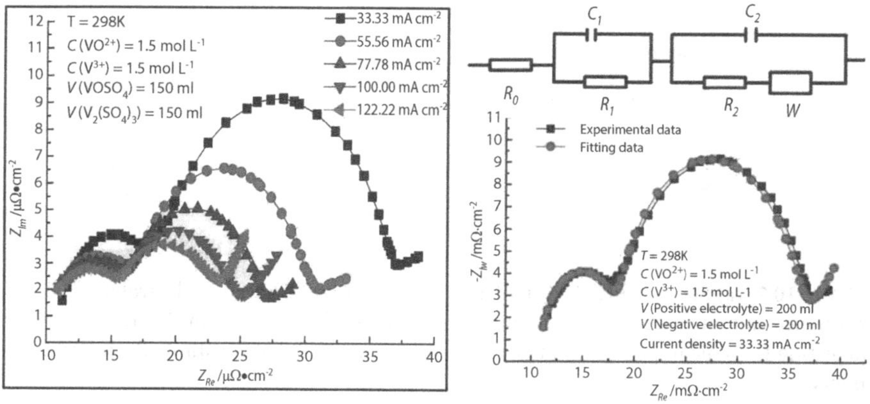

Figure 10.8 Vanadium RFBs (VRB). (a) Equivalent circuit for overcharged VRB. (b) The fitting and experimental data for the overcharged VRB [194].

denote capacitances, and W signifies the Warburg resistance at the positive electrode [194]. Comparison of the circuit's fitting data, as seen in Figure 10.8, with the experimental data indicates a near-exact correspondence between the two sets of data. The findings show that the comparable circuit is efficient for the overloaded vanadium redox battery [194].

10.7.5 Mott-Schottky Methods

Mott-Schottky analysis is widely employed to determine essential working parameters of semiconductor photoelectrodes, namely the flat-band potential (E_{FB}) and the donor concentration N_d (for n-type semiconductor photoanodes) or acceptor concentration N_a (for p-type photocathodes). Obtaining values for these parameters *via* a Mott-Schottky plot is relatively straightforward [29, 195, 196]. Applying EIS over a variety of input potentials and perturbation frequencies is the first step in the Mott-Schottky approach. It is common practice to fit the measured impedance to equivalent circuit models to calculate the capacitance of the semiconductor space-charge layer, normalized to the geometrical area of the photoelectrode-electrolyte interface; this capacitance represents the combined effects of various charge transport and transfer processes within the system. Constant phase components are sometimes employed instead of actual capacitors to cover a wide frequency range; nevertheless, they can be converted into capacitances using the Mott-Schottky method.

The Mott–Schottky equation concisely relates the semiconductor space–charge or depletion layer to the applied potential (E), considering its flat-band potential and other semiconductor characteristics, as represented in Equations 10.23 and 10.24 [197, 198]. In this context, Z_{im} is the imaginary resistance, and f is the frequency in Equation 10.23. ϵ and ϵ_0 represent the dielectric constant and vacuum permittivity, respectively; e denotes the electronic charge; k_B indicates the Boltzmann constant; and T refers to the temperature. This means that a graph of C_{SC}^{-2} vs E should reveal the flat-band potential as the linearly extrapolated intercept on the x-axis, while N_d will be inversely related to the slope. This connection arises from the Schottky-type junction that ideally forms at the interface between the semiconductor and the electrolyte. In reverse bias, when applied potentials surpass the flat-band potential for an n-type photoanode, a Schottky junction forms a space–charge or depletion layer with a width, W, approximately equal to Equation 10.25.

$$C_{SC} = \frac{-i}{2\pi f Z_{im}} \tag{10.23}$$

$$\frac{1}{C_{SC}^2} = \frac{2}{\epsilon\epsilon_o e N_d}\left(E - E_{FB} - \frac{k_b T}{e}\right) \tag{10.24}$$

$$W = \sqrt{\frac{2\epsilon\epsilon_o\left(E - E_{FB}\right)}{eN_d}} \tag{10.25}$$

The substitution of W with E defines the attributes of the Mott–Schottky plot. Similar to a parallel-plate capacitor, the capacitance within the space–charge layer decreases as the distance between the plates rises. Equations 10.23 and 10.24 signify the semiconductor–electrode junction having a planar geometry. This estimation was appropriate in the early semiconductor photoelectrochemistry research, which predominantly employed single crystals. However, optimal Schottky junctions are rarely detected in current studies based on nanostructured or catalyst-functionalized semiconductor photoelectrodes. An extensive aggregation of surface states or a textured nanostructured semiconductor may elicit nonlinearities in the Mott–Schottky plot. Even if a credible Mott–Schottky plot displaying a linear region is generated, one should not presume that valid conclusions may be drawn from Equation 10.23 [199–201].

10.8 Conclusion

SRFBs represent a cost-efficient and practical approach to solar energy storage, integrating the *in situ* energy conversion functions of PVs with redox flow selection of electrode battery technology. RFB system capacity is vital for performance, and the materials, electrolytes, and separators are critical. PEC RFBs integrate the operational principles of two devices into a singular system, with the photoelectrode serving a crucial function. Progress in material modification and the development of novel materials hold significant potential for enhancing efficiency.

Flow batteries store energy by moving electrolytes from tanks to electrochemical cells. They have increased safety, long cycle life, and low self-discharge. The battery stack and external tanks make up RFB systems. Power and energy can be scaled separately. Research has focused on EE, energy storage, and sustainability. SFBs are a new, compact, cost-effective, and efficient option.

Single-junction SFBs have achieved a record 15.4% sun-to-output electricity efficiency, according to recent studies. Electrochemical cells are studied using counter, working, and counter electrodes. RFBs use carbon, metal, or composite electrodes. Many RFBs employ carbon-based electrodes because of their strong electrical conductivity; however, they disintegrate in highly oxidizing situations. Carbon particles and polymer binders make composite electrodes lighter, cheaper, and more mechanically efficient than solid carbon. The flow battery's redox couplings provided fast redox kinetics for charge transfer, and it was charged under simulated sunshine to store and convert energy into electrical energy. Photoelectrode efficiency affects SRFB efficiency.

The performance of the RFB system depends on its capacity. In some applications, electricity-assisted systems are used in RFB devices to lower the potential difference between the redox couplings and photoelectrode. PEC RFBs charge directly from a photoelectrode, improving efficiency. The photoelectrode and redox coupling energy levels must match for maximum performance. Using p-type and n-type semiconductors to create dual band gaps improves efficiency but requires more solar energy. Dual electrodes, tandem systems, n-type/p-type combinations, and PV electrode designs reduce band-gap energy and increase light-harvesting efficiency.

Tandem structures increase current densities and EE in single photoelectrode with RFB systems. Tandem cells, especially Group III–V semiconductors, efficiently convert solar energy into chemical energy for RFBs. Dual photoelectrode systems are more efficient but more complicated, increasing cost and setup time. Electrochemical systems including solar cells, batteries, and electrochromic devices require counter electrodes. As passive ion storage layers or secondary electrodes, they boost color efficiency and device stability.

Carbon materials like carbon fibers, carbon spheres, carbon nanotubes, mesoporous carbon, and graphene are used in many fields due to their low cost, strong catalytic activity, and stability. Energy conversion and storage depend on the electrolyte, affecting battery performance and longevity. Organic–inorganic systems may efficiently store solar electrochemical energy and convert it into chemical energy. Electrolytes affect battery energy density, potential, resistance, light absorption, and stability. Alternative semiconductor materials with adequate energy levels and stability methods are being researched.

In flow battery systems, ionic exchange membranes, hybrid membranes, mesoporous separators, and solid ionic conductors are employed. IEMs prevent electrolyte mixing and enable electrode charge transfer. Flexible polymer porous separators are utilized in RFBs. Hybrid batteries with solid metal or gas diffusion electrodes can run without separators but require superb mechanical, chemical, and electrochemical stability. High expense and low chemical stability in acidic and oxidative settings make ion-exchange membranes less common. Flow batteries are symmetric or asymmetric depending on their electrolyte. IEMs must be stable, conduct ions well, and select ions well to reduce ohmic polarization and improve voltage performance. Ion transport channels in perfluorosulfonate membranes have been explored. Ion-solvating membranes in alkaline electrolyzer cells must be more stable due to oxidizing agents, pentavalent vanadium ions, and acidic electrolytes.

To create enhanced energy storage systems, energy materials' structure and reaction processes must be understood as renewable energy sources become more relevant. Long-term operational devices can benefit from electrochemical methods like CV and EIS. Increase light absorption and charge transport to improve SFBs and cells. Composite electrodes and series photoelectrodes boost performance. Redox couples must be stable to extend battery life.

CV measures the formal potential of a half-cell reaction if reduced and oxidized species are chemically stable. EIS is a cheap, widely used testing method that uses sinusoidal waves to protect sample integrity and produce sensitive, fast results. The equivalent circuit model is used to study impedance derived from Ohm's law. Mott–Schottky analysis is commonly used to evaluate semiconductor photoelectrode functioning parameters such as flat-band potential and donor/acceptor concentration.

Flow batteries have long cycle life, low self-discharge, and increased safety. As renewable energy sources become more popular, energy storage technology research has concentrated on efficiency, storage, and sustainability.

References

1. Sadeghi, S., Study using the flow battery in combination with solar panels and solid oxide fuel cell for power generation. *Sol. Energy*, 170, 732–740, Aug. 2018. doi: 10.1016/j.solener.2018.05.091.
2. Zhang, H., Zhang, H., Li, X., Zhang, J., *Redox Flow Batteries*, CRC Press, Boca Raton, 2017, doi: 10.1201/9781315152684.
3. Altaf, C.T., Yolacan, D., Sankir, N.D., Decoration of 3D ZnO nanoelectrodes with CuInS2 for solar water splitting. *Mater. Lett.*, 236, 710–714, 2019. doi: 10.1016/j.matlet.2018.11.056.
4. Wedege, K., Bae, D., Smith, W.A.A., Mendes, A., Bentien, A., Solar Redox Flow Batteries with Organic Redox Couples in Aqueous Electrolytes: A Minireview. *J. Phys. Chem. C*, 122, 45, 25729–25740, Nov. 2018. doi: 10.1021/acs.jpcc.8b04914.
5. Li, W. and Jin, S., Design Principles and Developments of Integrated Solar Flow Batteries. *Acc. Chem. Res.*, 53, 11, 2611–2621, Nov. 2020. doi: 10.1021/acs.accounts.0c00373.
6. Wang, J., *et al.*, Dual photoelectrode-drived Fe–Br rechargeable flow battery for solar energy conversion and storage: A cost-effective approach. *J. Power Sources*, 618, 235163, Oct. 2024. doi: 10.1016/j.jpowsour.2024.235163.
7. Hodes, G., Manassen, J., Cahen, D., Photoelectrochemical energy conversion and storage using polycrystalline chalcogenide electrodes. *Nature*, 261, 5559, 403–404, Jun. 1976. doi: 10.1038/261403a0.
8. Sharon, M., Veluchamy, P., Natarajan, C., Kumar, D., Solar rechargeable battery-principle and materials. *Electrochim. Acta*, 36, 7, 1107–1126, Jan. 1991. doi: 10.1016/0013-4686(91)85097-Q.
9. Azevedo, J., *et al.*, Unbiased solar energy storage: Photoelectrochemical redox flow battery. *Nano Energy*, 22, 396–405, Apr. 2016. doi: 10.1016/j.nanoen.2016.02.029.
10. Ma, W., *et al.*, High-Performance Solar Redox Flow Battery toward Efficient Overall Splitting of Hydrogen Sulfide. *ACS Energy Lett.*, 5, 2, 597–603, Feb. 2020. doi: 10.1021/acsenergylett.9b02206.
11. Liu, M., *et al.*, Iron/Quinone-based all-in-one solar rechargeable flow cell for highly efficient solar energy conversion and storage. *Nano Energy*, 76, 104907, Oct. 2020. doi: 10.1016/j.nanoen.2020.104907.
12. Pandiyan Naresh, R., Mariyappan, K., Dixon, D., Ulaganathan, M., Ragupathy, P., Investigations on New Electrolyte Composition and Modified

Membrane for High Voltage Zinc–Manganese Hybrid Redox Flow Batteries. *Batter. Supercaps*, 4, 9, 1464–1472, Sep. 2021. doi: 10.1002/batt.202100071.
13. Liu, F., Ma, Z., Liu, Q., Wang, Z., He, C., An integrated solar redox flow battery using a single Si photoanode and near-neutral electrolytes. *J. Power Sources*, 549, 231987, Nov. 2022. doi: 10.1016/j.jpowsour.2022.231987.
14. Tian, G., Jervis, R., Sobrido, A.J., Influence of NH4Cl additive in a VO2+/VO2+ - AQDS/AQDS2– solar redox flow battery. *Electrochim. Acta*, 461, May, 142671, Sep. 2023. doi: 10.1016/j.electacta.2023.142671.
15. Lu, P., *et al.*, Outside-to-inside: Efficacy comparation of Mn bulk and surface-doped TiO2{201} in E-fueled solar flow battery system. *Surf. Interfaces*, 46, 104174, Mar. 2024. doi: 10.1016/j.surfin.2024.104174.
16. Yu, M., McCulloch, W.D., Beauchamp, D.R., Huang, Z., Ren, X., Wu, Y., Aqueous Lithium–Iodine Solar Flow Battery for the Simultaneous Conversion and Storage of Solar Energy. *J. Am. Chem. Soc.*, 137, 26, 8332–8335, Jul. 2015. doi: 10.1021/jacs.5b03626.
17. Fu, H.C., *et al.*, An efficient and stable solar flow battery enabled by a single-junction GaAs photoelectrode. *Nat. Commun.*, 12, 156, 2021. doi: 10.1038/s41467-020-20287-w.
18. Kumtepe, A., *et al.*, Indium Sulfide Based Photoelectrodes for All-Vanadium Photoelectrochemical Redox Flow Batteries. *ACS Appl. Energy Mater.*, 3, 4, 3127–3133, Apr. 2020. doi: 10.1021/acsaem.9b02034.
19. Lu, P., *et al.*, Study on performance enhancement of electro-fueled solar flow battery system by nickel-doped titanium dioxide photoanode. *Energy Convers. Manag.*, 301, August 2023, 117997, Feb. 2024. doi: 10.1016/j.enconman.2023.117997.
20. Khataee, A., *et al.*, Integrated design of hematite and dye-sensitized solar cell for unbiased solar charging of an organic-inorganic redox flow battery. *Nano Energy*, 62, April, 832–843, Aug. 2019. doi: 10.1016/j.nanoen.2019.06.001.
21. Zhang, S., *et al.*, TiO 2 -Photoanode-Assisted Direct-Solar-Energy Harvesting and Storage in a Solar-Powered Redox Cell Using Halides as Active Materials. *ACS Appl. Mater. Interfaces*, 10, 27, 23048–23054, Jul. 2018. doi: 10.1021/acsami.8b04314.
22. Shah, A.A., Leung, P., Xu, Q., Sui, P.-C., Xing, W., *New Paradigms in Flow Battery Modelling*, vol. 16, Springer Nature Singapore, Singapore, 2023, doi: 10.1007/978-981-99-2524-7.
23. Rajarathnam, G.P. and Vassallo, A.M., *The Zinc/Bromine Flow Battery*, Springer Singapore, Singapore, 2016, doi: 10.1007/978-981-287-646-1.
24. Bae, D., Kanellos, G., Wedege, K., Draževič, E., Bentien, A., Smith, W.A., Tailored energy level alignment at MoOX/GaP interface for solar-driven redox flow battery application. *J. Chem. Phys.*, 152, 12, Mar. 2020. doi: 10.1063/1.5136252.
25. Bhattacharjee, A., Samanta, H., Banerjee, N., Saha, H., Development and validation of a real time flow control integrated MPPT charger for solar PV

applications of vanadium redox flow battery. *Energy Convers. Manag.*, 171, 1449–1462, Sep. 2018. doi: 10.1016/j.enconman.2018.06.088.
26. Millán, M., *et al.*, Towards the sustainable powering of the electrocoagulation of wastewater through the use of solar-vanadium redox flow battery: A first approach. *Electrochim. Acta*, 270, 14–21, Apr. 2018. doi: 10.1016/j.electacta.2018.03.055.
27. Zhou, Y., *et al.*, Efficient Solar Energy Harvesting and Storage through a Robust Photocatalyst Driving Reversible Redox Reactions. *Adv. Mater.*, 30, 31, Aug. 2018. doi: 10.1002/adma.201802294.
28. Liao, S., *et al.*, Integrating a dual-silicon photoelectrochemical cell into a redox flow battery for unassisted photocharging. *Nat. Commun.*, 7, 1, 11474, May 2016. doi: 10.1038/ncomms11474.
29. Gerischer, H., Electrochemical photo and solar cells principles and some experiments. *J. Electroanal. Chem. Interfacial Electrochem.*, 58, 1, 263–274, Jan. 1975. doi: 10.1016/S0022-0728(75)80359-7.
30. Wedege, K., Azevedo, J., Khataee, A., Bentien, A., Mendes, A., Direct Solar Charging of an Organic–Inorganic, Stable, and Aqueous Alkaline Redox Flow Battery with a Hematite Photoanode. *Angew. Chem. Int. Ed.*, 55, 25, 7142–7147, Jun. 2016. doi: 10.1002/anie.201602451.
31. Hisatomi, T., Kubota, J., Domen, K., Recent advances in semiconductors for photocatalytic and photoelectrochemical water splitting. *Chem. Soc. Rev.*, 43, 22, 7520–7535, Jan. 2014. doi: 10.1039/C3CS60378D.
32. So, S., Cha, M.S., Jo, S.-W., Kim, T.-H., Lee, J.Y., Hong, Y.T., Hydrophilic Channel Alignment of Perfluoronated Sulfonic-Acid Ionomers for Vanadium Redox Flow Batteries. *ACS Appl. Mater. Interfaces*, 10, 23, 19689–19696, Jun. 2018. doi: 10.1021/acsami.8b03985.
33. Guijarro, N., Prévot, M.S., Sivula, K., Surface modification of semiconductor photoelectrodes. *Phys. Chem. Chem. Phys.*, 17, 24, 15655–15674, 2015. doi: 10.1039/C5CP01992C.
34. Bard, A.J., Photoelectrochemistry. *Sci. (80-.).*, 207, 4427, 139–144, Jan. 1980. doi: 10.1126/science.207.4427.139.
35. Nozik, A.J., Photochemical diodes. *Appl. Phys. Lett.*, 30, 11, 567–569, Jun. 1977. doi: 10.1063/1.89262.
36. Cao, L., Skyllas-Kazacos, M., Wang, D., Solar Redox Flow Batteries: Mechanism, Design, and Measurement. *Adv. Sustain. Syst.*, 2, 8–9, Aug. 2018. doi: 10.1002/adsu.201800031.
37. Li, W., Fu, H.-C., Zhao, Y., He, J.-H., Jin, S., 14.1% Efficient Monolithically Integrated Solar Flow Battery. *Chem*, 4, 11, 2644–2657, Nov. 2018. doi: 10.1016/j.chempr.2018.08.023.
38. Li, W., *et al.*, High-performance solar flow battery powered by a perovskite/silicon tandem solar cell. *Nat. Mater.*, 19, 12, 1326–1331, Dec. 2020. doi: 10.1038/s41563-020-0720-x.

39. Ma, J., Pan, Z., Tagliabue, G., High-Performance Hematite Photoanodes for Unassisted Recharging of Solar Redox Flow Battery. *Sol. RRL*, 8, 17, Sep. 2024. doi: 10.1002/solr.202400477.
40. Cheng, Q., *et al.*, Photorechargeable High Voltage Redox Battery Enabled by Ta 3 N 5 and GaN/Si Dual-Photoelectrode. *Adv. Mater.*, 29, 26, Jul. 2017. doi: 10.1002/adma.201700312.
41. Yan, N.F., Li, G.R., Gao, X.P., Solar rechargeable redox flow battery based on Li2WO4/LiI couples in dual-phase electrolytes. *J. Mater. Chem. A*, 1, 24, 7012, 2013. doi: 10.1039/c3ta11360d.
42. Varadhan, P., Fu, H.-C., Kao, Y.-C., Horng, R.-H., He, J.-H., An efficient and stable photoelectrochemical system with 9% solar-to-hydrogen conversion efficiency *via* InGaP/GaAs double junction. *Nat. Commun.*, 10, 1, 5282, Nov. 2019. doi: 10.1038/s41467-019-12977-x.
43. Wedege, K., Bae, D., Dražević, E., Mendes, A., Vesborg, P.C.K., Bentien, A., Unbiased, complete solar charging of a neutral flow battery by a single Si photocathode. *RSC Adv.*, 82018, 12, 6331–6340, 2018. doi: 10.1039/C8RA00319J.
44. Li, W., Fu, H., Li, L., Cabán-Acevedo, M., He, J., Jin, S., Integrated Photoelectrochemical Solar Energy Conversion and Organic Redox Flow Battery Devices. *Angew. Chem. Int. Ed.*, 55, 42, 13104–13108, Oct. 2016. doi: 10.1002/anie.201606986.
45. Li, W., *et al.*, A Long Lifetime Aqueous Organic Solar Flow Battery. *Adv. Energy Mater.*, 9, 31, Aug. 2019. doi: 10.1002/aenm.201900918.
46. Yu, M., Ren, X., Ma, L., Wu, Y., Integrating a redox-coupled dye-sensitized photoelectrode into a lithium–oxygen battery for photoassisted charging. *Nat. Commun.*, 5, 1, 5111, Oct. 2014. doi: 10.1038/ncomms6111.
47. Park, M., Ryu, J., Wang, W., Cho, J., Material design and engineering of next-generation flow-battery technologies. *Nat. Rev. Mater.*, 2, 1, 16080, Nov. 2016. doi: 10.1038/natrevmats.2016.80.
48. Hwang, I. and Yong, K., Counter Electrodes for Quantum-Dot-Sensitized Solar Cells. *ChemElectroChem*, 2, 5, 634–653, May 2015. doi: 10.1002/celc.201402405.
49. Kim, C.-Y., Cho, S.-G., Lim, T.-Y., Cycle test and degradation analysis of WO3/PC+LiClO4/CeO2·TiO2 electrochromic device. *Sol. Energy Mater. Sol. Cells*, 93, 12, 2056–2061, Dec. 2009. doi: 10.1016/j.solmat.2009.07.007.
50. Ahmed, U., Alizadeh, M., Rahim, N.A., Shahabuddin, S., Ahmed, M.S., Pandey, A.K., A comprehensive review on counter electrodes for dye sensitized solar cells: A special focus on Pt-TCO free counter electrodes. *Sol. Energy*, 174, 1097–1125, Nov. 2018. doi: 10.1016/j.solener.2018.10.010.
51. Moulki, H., *et al.*, Electrochromic performances of nonstoichiometric NiO thin films. *Thin Solid Films*, 553, 63–66, Feb. 2014. doi: 10.1016/j.tsf.2013.10.154.

52. Samantaray, M.R., *et al.*, Synergetic Effects of Hybrid Carbon Nanostructured Counter Electrodes for Dye-Sensitized Solar Cells: A Review. *Mater. (Basel).*, 13, 12, 2779, Jun. 2020. doi: 10.3390/ma13122779.
53. Kouhnavard, M., Ludin, N.A., Ghaffari, B.V., Sopian, K., Ikeda, S., Carbonaceous Materials and Their Advances as a Counter Electrode in Dye-Sensitized Solar Cells: Challenges and Prospects. *ChemSusChem*, 8, 9, 1510–1533, May 2015. doi: 10.1002/cssc.201500004.
54. Dhillon, K.K., Patyal, M., Bhogal, S., Gupta, N., A review on advanced counter electrode materials for high-efficiency dye sensitized solar cells. *J. Coord. Chem.*, 77, 17–19, 1933–1968, Oct. 2024. doi: 10.1080/00958972.2024.2404070.
55. Wu, J., *et al.*, Counter electrodes in dye-sensitized solar cells. *Chem. Soc. Rev.*, 46, 19, 5975–6023, 2017. doi: 10.1039/C6CS00752J.
56. Ondersma, J.W. and Hamann, T.W., Recombination and redox couples in dye-sensitized solar cells. *Coord. Chem. Rev.*, 257, 9–10, 1533–1543, May 2013. doi: 10.1016/j.ccr.2012.09.010.
57. Hagfeldt, A., Boschloo, G., Sun, L., Kloo, L., Pettersson, H., Dye-Sensitized Solar Cells. *Chem. Rev.*, 110, 11, 6595–6663, Nov. 2010. doi: 10.1021/cr900356p.
58. Murakami, T.N. and Grätzel, M., Counter electrodes for DSC: Application of functional materials as catalysts. *Inorg. Chim. Acta*, 361, 3, 572–580, Feb. 2008. doi: 10.1016/j.ica.2007.09.025.
59. Papageorgiou, N., Counter-electrode function in nanocrystalline photoelectrochemical cell configurations. *Coord. Chem. Rev.*, 248, 13–14, 1421–1446, Jul. 2004. doi: 10.1016/j.ccr.2004.03.028.
60. Poudel, P. and Qiao, Q., Carbon nanostructure counter electrodes for low cost and stable dye-sensitized solar cells. *Nano Energy*, 4, 157–175, Mar. 2014. doi: 10.1016/j.nanoen.2013.10.012.
61. Kusmierek, E., Semiconductor Electrode Materials Applied in Photoelectrocatalytic Wastewater Treatment—an Overview. *Catalysts*, 10, 4, 439, Apr. 2020. doi: 10.3390/catal10040439.
62. Lin, C.-Y., Teng, C.-Y., Li, T.-L., Lee, Y.-L., Teng, H., Photoactive p-type PbS as a counter electrode for quantum dot-sensitized solar cells. *J. Mater. Chem. A*, 1, 4, 1155–1162, 2013. doi: 10.1039/C2TA00251E.
63. Chen, M. and Shao, L.-L., Review on the recent progress of carbon counter electrodes for dye-sensitized solar cells. *Chem. Eng. J.*, 304, 629–645, Nov. 2016. doi: 10.1016/j.cej.2016.07.001.
64. Wu, M., Sun, M., Zhou, H., Ma, J., Ma, T., Carbon Counter Electrodes in Dye-Sensitized and Perovskite Solar Cells. *Adv. Funct. Mater.*, 30, 7, Feb. 2020. doi: 10.1002/adfm.201906451.
65. Chen, J., *et al.*, A flexible carbon counter electrode for dye-sensitized solar cells. *Carbon N. Y.*, 47, 11, 2704–2708, Sep. 2009. doi: 10.1016/j.carbon.2009.05.028.
66. Lu, P., Leung, P., Su, H., Yang, W., Xu, Q., Materials, performance, and system design for integrated solar flow batteries – A mini review. *Appl. Energy*, 282, 116210, Jan. 2021. doi: 10.1016/j.apenergy.2020.116210.

67. Barnartt, S. and Forejt, D.A., Bromine-Zinc Secondary Cells. *J. Electrochem. Soc.*, 111, 11, 1201, 1964. doi: 10.1149/1.2425960.
68. Alotto, P., Guarnieri, M., Moro, F., Redox flow batteries for the storage of renewable energy: A review. *Renew. Sustain. Energy Rev.*, 29, 325–335, Jan. 2014. doi: 10.1016/j.rser.2013.08.001.
69. Pan, F. and Wang, Q., Redox Species of Redox Flow Batteries: A Review. *Molecules*, 20, 11, 20499–20517, Nov. 2015. doi: 10.3390/molecules201119711.
70. Arenas, L.F., Ponce de León, C., Walsh, F.C., Engineering aspects of the design, construction and performance of modular redox flow batteries for energy storage. *J. Energy Storage*, 11, 119–153, Jun. 2017. doi: 10.1016/j.est.2017.02.007.
71. Noack, J., Roznyatovskaya, N., Herr, T., Fischer, P., The Chemistry of Redox-Flow Batteries. *Angew. Chem. Int. Ed.*, 54, 34, 9776–9809, Aug. 2015. doi: 10.1002/anie.201410823.
72. Chalamala, B.R., Soundappan, T., Fisher, G.R., Anstey, M.R., Viswanathan, V.V., Perry, M.L., Redox Flow Batteries: An Engineering Perspective. *Proc. IEEE*, 102, 6, 976–999, Jun. 2014. doi: 10.1109/JPROC.2014.2320317.
73. Chakrabarti, M.H., *et al.*, Prospects of applying ionic liquids and deep eutectic solvents for renewable energy storage by means of redox flow batteries. *Renew. Sustain. Energy Rev.*, 30, 254–270, Feb. 2014. doi: 10.1016/j.rser.2013.10.004.
74. Weber, A.Z., Mench, M.M., Meyers, J.P., Ross, P.N., Gostick, J.T., Liu, Q., Redox flow batteries: a review. *J. Appl. Electrochem.*, 41, 10, 1137–1164, Oct. 2011. doi: 10.1007/s10800-011-0348-2.
75. Ponce de León, C., Frías-Ferrer, A., González-García, J., Szánto, D.A., Walsh, F.C., Redox flow cells for energy conversion. *J. Power Sources*, 160, 1, 716–732, Sep. 2006. doi: 10.1016/j.jpowsour.2006.02.095.
76. Skyllas-Kazacos, M., Chakrabarti, M.H., Hajimolana, S.A., Mjalli, F.S., Saleem, M., Progress in Flow Battery Research and Development. *J. Electrochem. Soc.*, 158, 8, R55, 2011. doi: 10.1149/1.3599565.
77. Tang, A., Bao, J., Skyllas-Kazacos, M., Dynamic modelling of the effects of ion diffusion and side reactions on the capacity loss for vanadium redox flow battery. *J. Power Sources*, 196, 24, 10737–10747, Dec. 2011. doi: 10.1016/j.jpowsour.2011.09.003.
78. Tang, A., Bao, J., Skyllas-Kazacos, M., Thermal modelling of battery configuration and self-discharge reactions in vanadium redox flow battery. *J. Power Sources*, 216, 489–501, Oct. 2012. doi: 10.1016/j.jpowsour.2012.06.052.
79. Tang, A., Bao, J., Skyllas-Kazacos, M., Studies on pressure losses and flow rate optimization in vanadium redox flow battery. *J. Power Sources*, 248, 154–162, Feb. 2014. doi: 10.1016/j.jpowsour.2013.09.071.
80. Gerischer, H. and Lübke, M., Competition between Photocorrosion and Photooxidation of Redox Systems at n-Type Semiconductor Electrodes. *Ber. Der Bunsengesellschaft für Phys. Chem.*, 87, 2, 123–128, Feb. 1983. doi: 10.1002/bbpc.19830870211.

81. Roe, S., Menictas, C., Skyllas-Kazacos, M., A High Energy Density Vanadium Redox Flow Battery with 3 M Vanadium Electrolyte. *J. Electrochem. Soc.*, 163, 1, A5023–A5028, Jul. 2016. doi: 10.1149/2.0041601jes.
82. Chen, Y.D., Santhanam, K.S.V., Bard, A.J., Solution Redox Couples for Electrochemical Energy Storage: I . Iron (III)-Iron (II) Complexes with O-Phenanthroline and Related Ligands. *J. Electrochem. Soc.*, 128, 7, 1460–1467, Jul. 1981. doi: 10.1149/1.2127663.
83. Lu, X., Spectroscopic study of vanadium(V) precipitation in the vanadium redox cell electrolyte. *Electrochim. Acta*, 46, 28, 4281–4287, Aug. 2001. doi: 10.1016/S0013-4686(01)00639-9.
84. Skyllas-Kazacos, M. and Milne, N., Evaluation of iodide and titanium halide redox couple combinations for common electrolyte redox flow cell systems. *J. Appl. Electrochem.*, 41, 10, 1233–1243, Oct. 2011. doi: 10.1007/s10800-011-0287-y.
85. Skyllas-Kazacos, M., Cao, L., Kazacos, M., Kausar, N., Mousa, A., Vanadium Electrolyte Studies for the Vanadium Redox Battery—A Review. *ChemSusChem*, 9, 13, 1521–1543, Jul. 2016. doi: 10.1002/cssc.201600102.
86. Kim, S., *et al.*, 1 kW/1 kWh advanced vanadium redox flow battery utilizing mixed acid electrolytes. *J. Power Sources*, 237, 300–309, Sep. 2013. doi: 10.1016/j.jpowsour.2013.02.045.
87. Vijayakumar, M., *et al.*, Towards understanding the poor thermal stability of V5+ electrolyte solution in Vanadium Redox Flow Batteries. *J. Power Sources*, 196, 7, 3669–3672, Apr. 2011. doi: 10.1016/j.jpowsour.2010.11.126.
88. Zhang, J., *et al.*, Effects of additives on the stability of electrolytes for all-vanadium redox flow batteries. *J. Appl. Electrochem.*, 41, 10, 1215–1221, Oct. 2011. doi: 10.1007/s10800-011-0312-1.
89. Mohammadi, F., Timbrell, P., Zhong, S., Padeste, C., Skyllas-Kazacos, M., Overcharge in the vanadium redox battery and changes in electrical resistivity and surface functionality of graphite-felt electrodes. *J. Power Sources*, 52, 1, 61–68, Nov. 1994. doi: 10.1016/0378-7753(94)01938-X.
90. Tang, A., Ting, S., Bao, J., Skyllas-Kazacos, M., Thermal modelling and simulation of the all-vanadium redox flow battery. *J. Power Sources*, 203, 165–176, Apr. 2012. doi: 10.1016/j.jpowsour.2011.11.079.
91. Kennedy, J.H. and Frese, K.W., Photo-oxidation of Water at Barium Titanate Electrodes. *J. Electrochem. Soc.*, 123, 11, 1683–1686, Nov. 1976. doi: 10.1149/1.2132667.
92. Skyllas-Kazacos, M., Novel vanadium chloride/polyhalide redox flow battery. *J. Power Sources*, 124, 1, 299–302, Oct. 2003. doi: 10.1016/S0378-7753(03)00621-9.
93. Sanz, L., Lloyd, D., Magdalena, E., Palma, J., Kontturi, K., Description and performance of a novel aqueous all-copper redox flow battery. *J. Power Sources*, 268, 121–128, Dec. 2014. doi: 10.1016/j.jpowsour.2014.06.008.
94. Shiokawa, Y., Yamana, H., Moriyama, H., An Application of Actinide Elements for a Redox Flow Battery. *J. Nucl. Sci. Technol.*, 37, 3, 253–256, Mar. 2000. doi: 10.1080/18811248.2000.9714891.

95. Jorn, J., Kim, J.T., Kralik, D., The zinc-chlorine battery: half-cell overpotential measurements. *J. Appl. Electrochem.*, 9, 5, 573–579, Sep. 1979. doi: 10.1007/BF00610944.
96. Pletcher, D. and Wills, R., A novel flow battery: A lead acid battery based on an electrolyte with soluble lead(ii). *Phys. Chem. Chem. Phys.*, 6, 8, 1779, 2004. doi: 10.1039/b401116c.
97. Sum, E. and Skyllas-Kazacos, M., A study of the V(II)/V(III) redox couple for redox flow cell applications. *J. Power Sources*, 15, 2–3, 179–190, Jun. 1985. doi: 10.1016/0378-7753(85)80071-9.
98. Cheng, J., Zhang, H.M., Wen, Y.H., Cao, G.P., Yang, Y.S., Study on a Cd-Fe Redox Flow Battery in a Sulphuric Acid Electrolyte. *Adv. Mater. Res.*, 399–401, 1519–1523, Nov. 2011. doi: 10.4028/www.scientific.net/AMR.399-401.1519.
99. Hruska, L.W. and Savinell, R.F., Investigation of Factors Affecting Performance of the Iron-Redox Battery. *J. Electrochem. Soc.*, 128, 1, 18–25, Jan. 1981. doi: 10.1149/1.2127366.
100. Liu, D., *et al.*, Reversible Electron Storage in an All-Vanadium Photoelectrochemical Storage Cell: Synergy between Vanadium Redox and Hybrid Photocatalyst. *ACS Catal.*, 5, 4, 2632–2639, Apr. 2015. doi: 10.1021/cs502024k.
101. Liu, D., Wei, Z., Hsu, C., Shen, Y., Liu, F., Efficient Solar Energy Storage Using A TiO2/WO3 Tandem Photoelectrode in An All-vanadium Photoelectrochemical Cell. *Electrochim. Acta*, 136, 435–441, Aug. 2014. doi: 10.1016/j.electacta.2014.05.129.
102. Wei, Z., Liu, D., Hsu, C., Liu, F., All-vanadium redox photoelectrochemical cell: An approach to store solar energy. *Electrochem. Commun.*, 45, 79–82, Aug. 2014. doi: 10.1016/j.elecom.2014.05.018.
103. Yan, N.F., Li, G.R., Gao, X.P., Electroactive Organic Compounds as Anode-Active Materials for Solar Rechargeable Redox Flow Battery in Dual-Phase Electrolytes. *J. Electrochem. Soc.*, 161, 5, A736–A741, Mar. 2014. doi: 10.1149/2.065405jes.
104. Wei, Z., Almakrami, H., Lin, G., Agar, E., Liu, F., An organic-inorganic hybrid photoelectrochemical storage cell for improved solar energy storage. *Electrochim. Acta*, 263, 570–575, Feb. 2018. doi: 10.1016/j.electacta.2017.11.087.
105. Bertani, R., Geothermal power generation in the world 2005–2010 update report. *Geothermics*, 41, 1–29, Jan. 2012. doi: 10.1016/j.geothermics.2011.10.001.
106. Whitley, S. and Bae, D., Perspective—Insights into Solar-Rechargeable Redox Flow Cell Design: A Practical Perspective for Lab-Scale Experiments. *J. Electrochem. Soc.*, 168, 12, 120517, Dec. 2021. doi: 10.1149/1945-7111/ac3ab3.
107. Chen, D. and Hickner, M.A., V5+ degradation of sulfonated Radel membranes for vanadium redox flow batteries. *Phys. Chem. Chem. Phys.*, 15, 27, 11299, 2013. doi: 10.1039/c3cp52035h.
108. Zhang, H., Ding, C., Cao, J., Xu, W., Li, X., Zhang, H., A novel solvent-template method to manufacture nano-scale porous membranes for

vanadium flow battery applications. *J. Mater. Chem. A*, 2, 25, 9524, 2014. doi: 10.1039/c4ta00917g.
109. Vijayakumar, M., *et al.*, Spectroscopic investigations of the fouling process on Nafion membranes in vanadium redox flow batteries. *J. Memb. Sci.*, 366, 1–2, 325–334, Jan. 2011. doi: 10.1016/j.memsci.2010.10.018.
110. Yeager, H.L. and Steck, A., Cation and Water Diffusion in Nafion Ion Exchange Membranes: Influence of Polymer Structure. *J. Electrochem. Soc.*, 128, 9, 1880–1884, Sep. 1981. doi: 10.1149/1.2127757.
111. Kim, S., *et al.*, Cycling performance and efficiency of sulfonated poly(sulfone) membranes in vanadium redox flow batteries. *Electrochem. Commun.*, 12, 11, 1650–1653, Nov. 2010. doi: 10.1016/j.elecom.2010.09.018.
112. Kim, S., *et al.*, Chemical and mechanical degradation of sulfonated poly(sulfone) membranes in vanadium redox flow batteries. *J. Appl. Electrochem.*, 41, 10, 1201–1213, Oct. 2011. doi: 10.1007/s10800-011-0313-0.
113. Li, J., Zhang, Y., Zhang, S., Huang, X., Wang, L., Novel sulfonated polyimide/ZrO 2 composite membrane as a separator of vanadium redox flow battery. *Polym. Adv. Technol.*, 25, 12, 1610–1615, Dec. 2014. doi: 10.1002/pat.3411.
114. Zhang, S., Zhang, B., Zhao, G., Jian, X., Anion exchange membranes from brominated poly(aryl ether ketone) containing 3,5-dimethyl phthalazinone moieties for vanadium redox flow batteries. *J. Mater. Chem. A*, 2, 9, 3083, 2014. doi: 10.1039/c3ta14503d.
115. Zhang, B., Zhang, S., Xing, D., Han, R., Yin, C., Jian, X., Quaternized poly(phthalazinone ether ketone ketone) anion exchange membrane with low permeability of vanadium ions for vanadium redox flow battery application. *J. Power Sources*, 217, 296–302, Nov. 2012. doi: 10.1016/j.jpowsour.2012.06.027.
116. Zhang, S., Yin, C., Xing, D., Yang, D., Jian, X., Preparation of chloromethylated/quaternized poly(phthalazinone ether ketone) anion exchange membrane materials for vanadium redox flow battery applications. *J. Memb. Sci.*, 363, 1–2, 243–249, Nov. 2010. doi: 10.1016/j.memsci.2010.07.046.
117. Chen, D., Hickner, M.A., Agar, E., Kumbur, E.C., Selective anion exchange membranes for high coulombic efficiency vanadium redox flow batteries. *Electrochem. Commun.*, 26, 37–40, Jan. 2013. doi: 10.1016/j.elecom.2012.10.007.
118. Xing, D., Zhang, S., Yin, C., Zhang, B., Jian, X., Effect of amination agent on the properties of quaternized poly(phthalazinone ether sulfone) anion exchange membrane for vanadium redox flow battery application. *J. Memb. Sci.*, 354, 1–2, 68–73, May 2010. doi: 10.1016/j.memsci.2010.02.064.
119. Qiu, J., *et al.*, Performance of vanadium redox flow battery with a novel amphoteric ion exchange membrane synthesized by two-step grafting method. *J. Memb. Sci.*, 342, 1–2, 215–220, Oct. 2009. doi: 10.1016/j.memsci.2009.06.043.
120. Qiu, J., *et al.*, Amphoteric ion exchange membrane synthesized by radiation-induced graft copolymerization of styrene and dimethylaminoethyl methacrylate into PVDF film for vanadium redox flow battery applications. *J. Memb. Sci.*, 334, 1–2, 9–15, May 2009. doi: 10.1016/j.memsci.2009.02.009.

121. Hu, G., *et al.*, A novel amphoteric ion exchange membrane synthesized by radiation-induced grafting α-methylstyrene and N,N-dimethylaminoethyl methacrylate for vanadium redox flow battery application. *J. Memb. Sci.*, 407–408, 184–192, Jul. 2012. doi: 10.1016/j.memsci.2012.03.042.
122. Yuan, J., *et al.*, Facile synthesis of amphoteric ion exchange membrane by radiation grafting of sodium styrene sulfonate and N,N -dimethylamino-ethyl methacrylate for vanadium redox flow battery. *J. Polym. Sci. Part A Polym. Chem.*, 51, 24, 5194–5202, Dec. 2013. doi: 10.1002/pola.26949.
123. Ma, J., *et al.*, Designing a new process to prepare amphoteric ion exchange membrane with well-distributed grafted chains for vanadium redox flow battery. *J. Memb. Sci.*, 419–420, 1–8, Nov. 2012. doi: 10.1016/j.memsci.2012.04.034.
124. Zhang, H., Zhang, H., Li, X., Mai, Z., Wei, W., Silica modified nanofiltration membranes with improved selectivity for redox flow battery application. *Energy Environ. Sci.*, 5, 4, 6299–6303, 2012. doi: 10.1039/C1EE02571F.
125. Wei, X., *et al.*, Microporous separators for Fe/V redox flow batteries. *J. Power Sources*, 218, 39–45, Nov. 2012. doi: 10.1016/j.jpowsour.2012.06.073.
126. Zhang, H., Zhang, H., Li, X., Mai, Z., Zhang, J., Nanofiltration (NF) membranes: the next generation separators for all vanadium redox flow batteries (VRBs)? *Energy Environ. Sci.*, 4, 5, 1676, 2011. doi: 10.1039/c1ee01117k.
127. Xu, W., Li, X., Cao, J., Yuan, Z., Zhang, H., Morphology and performance of poly(ether sulfone)/sulfonated poly(ether ether ketone) blend porous membranes for vanadium flow battery application. *RSC Adv.*, 4, 76, 40400–40406, Aug. 2014. doi: 10.1039/C4RA05083E.
128. Wei, W., Zhang, H., Li, X., Zhang, H., Li, Y., Vankelecom, I., Hydrophobic asymmetric ultrafiltration PVDF membranes: an alternative separator for VFB with excellent stability. *Phys. Chem. Chem. Phys.*, 15, 6, 1766–1771, 2013. doi: 10.1039/C2CP43761A.
129. Pan, J., Wang, S., Xiao, M., Hickner, M., Meng, Y., Layered zirconium phosphate sulfophenylphosphonates reinforced sulfonated poly (fluorenyl ether ketone) hybrid membranes with high proton conductivity and low vanadium ion permeability. *J. Memb. Sci.*, 443, 19–27, Sep. 2013. doi: 10.1016/j.memsci.2013.04.068.
130. Chen, D., Wang, S., Xiao, M., Han, D., Meng, Y., Sulfonated poly (fluorenyl ether ketone) membrane with embedded silica rich layer and enhanced proton selectivity for vanadium redox flow battery. *J. Power Sources*, 195, 22, 7701–7708, Nov. 2010. doi: 10.1016/j.jpowsour.2010.05.026.
131. Kim, J., Jeon, J.-D., Kwak, S.-Y., Nafion-based composite membrane with a permselective layered silicate layer for vanadium redox flow battery. *Electrochem. Commun.*, 38, 68–70, Jan. 2014. doi: 10.1016/j.elecom.2013.11.002.
132. Xi, J., Wu, Z., Qiu, X., Chen, L., Nafion/SiO2 hybrid membrane for vanadium redox flow battery. *J. Power Sources*, 166, 2, 531–536, Apr. 2007. doi: 10.1016/j.jpowsour.2007.01.069.

133. Teng, X., Zhao, Y., Xi, J., Wu, Z., Qiu, X., Chen, L., Nafion/organically modified silicate hybrids membrane for vanadium redox flow battery. *J. Power Sources*, 189, 2, 1240–1246, Apr. 2009. doi: 10.1016/j.jpowsour.2008.12.040.
134. Ferrigno, R., Stroock, A.D., Clark, T.D., Mayer, M., Whitesides, G.M., Membraneless Vanadium Redox Fuel Cell Using Laminar Flow. *J. Am. Chem. Soc.*, 124, 44, 12930–12931, Nov. 2002. doi: 10.1021/ja020812q.
135. Braff, W.A., Bazant, M.Z., Buie, C.R., Membrane-less hydrogen bromine flow battery. *Nat. Commun.*, 4, 1, 2346, Aug. 2013. doi: 10.1038/ncomms3346.
136. Wei, X., Li, B., Wang, W., Porous Polymeric Composite Separators for Redox Flow Batteries. *Polym. Rev.*, 55, 2, 247–272, Apr. 2015. doi: 10.1080/15583724.2015.1011276.
137. Maurya, S., Shin, S.-H., Kim, Y., Moon, S.-H., A review on recent developments of anion exchange membranes for fuel cells and redox flow batteries. *RSC Adv.*, 5, 47, 37206–37230, 2015. doi: 10.1039/C5RA04741B.
138. Doan, T.N.L., Hoang, T.K.A., Chen, P., Recent development of polymer membranes as separators for all-vanadium redox flow batteries. *RSC Adv.*, 5, 89, 72805–72815, 2015. doi: 10.1039/C5RA05914C.
139. Wu, X., *et al.*, Ion exchange membranes for vanadium redox flow batteries. *Pure Appl. Chem.*, 86, 5, 633–649, May 2014. doi: 10.1515/pac-2014-0101.
140. Tanaka, Y., Redox Flow Battery, in: *Ion Exchange Membranes*, pp. 471–485, Elsevier, Amsterdam, 2015, doi: 10.1016/B978-0-444-63319-4.00023-7.
141. Prifti, H., Parasuraman, A., Winardi, S., Lim, T.M., Skyllas-Kazacos, M., Membranes for Redox Flow Battery Applications. *Membr. (Basel).*, 2, 2, 275–306, Jun. 2012. doi: 10.3390/membranes2020275.
142. Schwenzer, B., Zhang, J., Kim, S., Li, L., Liu, J., Yang, Z., Membrane Development for Vanadium Redox Flow Batteries. *ChemSusChem*, 4, 10, 1388–1406, Oct. 2011. doi: 10.1002/cssc.201100068.
143. Parasuraman, A., Lim, T.M., Menictas, C., Skyllas-Kazacos, M., Review of material research and development for vanadium redox flow battery applications. *Electrochim. Acta*, 101, 27–40, Jul. 2013. doi: 10.1016/j.electacta.2012.09.067.
144. Sukkar, T. and Skyllas-Kazacos, M., Membrane stability studies for vanadium redox cell applications. *J. Appl. Electrochem.*, 34, 2, 137–145, Feb. 2004. doi: 10.1023/B:JACH.0000009931.83368.dc.
145. Tanaka, Y., Preparation of Ion Exchange Membranes, in: *Ion Exchange Membranes*, pp. 3–28, Elsevier, Amsterdam, 2015, doi: 10.1016/B978-0-444-63319-4.00001-8.
146. Hwang, G., Crosslinking of anion exchange membrane by accelerated electron radiation as a separator for the all-vanadium redox flow battery. *J. Memb. Sci.*, 132, 1, 55–61, Aug. 1997. doi: 10.1016/S0376-7388(97)00040-9.
147. Xi, J., Wu, Z., Teng, X., Zhao, Y., Chen, L., Qiu, X., Self-assembled polyelectrolyte multilayer modified Nafion membrane with suppressed vanadium ion crossover for vanadium redox flow batteries. *J. Mater. Chem.*, 18, 11, 1232, 2008. doi: 10.1039/b718526j.

148. Sukkar, T. and Skyllas-Kazacos, M., Modification of membranes using polyelectrolytes to improve water transfer properties in the vanadium redox battery. *J. Memb. Sci.*, 222, 1–2, 249–264, Sep. 2003. doi: 10.1016/S0376-7388(03)00316-8.
149. Mohammadi, T. and Skyllas-Kazacos, M., Use of polyelectrolyte for incorporation of ion-exchange groups in composite membranes for vanadium redox flow battery applications. *J. Power Sources*, 56, 1, 91–96, Jul. 1995. doi: 10.1016/0378-7753(95)80014-8.
150. Yang, Z., *et al.*, Electrochemical Energy Storage for Green Grid. *Chem. Rev.*, 111, 5, 3577–3613, May 2011. doi: 10.1021/cr100290v.
151. Li, X., Zhang, H.H., Mai, Z., Zhang, H.H., Vankelecom, I., Ion exchange membranes for vanadium redox flow battery (VRB) applications. *Energy Environ. Sci.*, 4, 4, 1147, 2011. doi: 10.1039/c0ee00770f.
152. Potash, R.A., McKone, J.R., Conte, S., Abruña, H.D., On the Benefits of a Symmetric Redox Flow Battery. *J. Electrochem. Soc.*, 163, 3, A338–A344, Dec. 2016. doi: 10.1149/2.0971602jes.
153. Dai, Q., Zhao, Z., Shi, M., Deng, C., Zhang, H., Li, X., Ion conductive membranes for flow batteries: Design and ions transport mechanism. *J. Memb. Sci.*, 632, 119355, Aug. 2021. doi: 10.1016/j.memsci.2021.119355.
154. Dunn, B., Kamath, H., Tarascon, J.-M., Electrical Energy Storage for the Grid: A Battery of Choices. *Sci. (80-.)*, 334, 6058, 928–935, Nov. 2011. doi: 10.1126/science.1212741.
155. Yuan, Z., Liu, X., Xu, W., Duan, Y., Zhang, H., Li, X., Negatively charged nanoporous membrane for a dendrite-free alkaline zinc-based flow battery with long cycle life. *Nat. Commun.*, 9, 1, 3731, Sep. 2018. doi: 10.1038/s41467-018-06209-x.
156. Yuan, Z., Duan, Y., Liu, T., Zhang, H., Li, X., Toward a Low-Cost Alkaline Zinc-Iron Flow Battery with a Polybenzimidazole Custom Membrane for Stationary Energy Storage. *iScience*, 3, 40–49, May 2018. doi: 10.1016/j.isci.2018.04.006.
157. Yuan, Z., Yin, Y., Xie, C., Zhang, H., Yao, Y., Li, X., Advanced Materials for Zinc-Based Flow Battery: Development and Challenge. *Adv. Mater.*, 31, 50, Dec. 2019. doi: 10.1002/adma.201902025.
158. Xie, C., Duan, Y., Xu, W., Zhang, H., Li, X., A Low-Cost Neutral Zinc–Iron Flow Battery with High Energy Density for Stationary Energy Storage. *Angew. Chem. Int. Ed.*, 56, 47, 14953–14957, Nov. 2017. doi: 10.1002/anie.201708664.
159. Gierke, T.D., Munn, G.E., Wilson, F.C., The morphology in nafion perfluorinated membrane products, as determined by wide- and small-angle x-ray studies. *J. Polym. Sci. Polym. Phys. Ed.*, 19, 11, 1687–1704, Nov. 1981. doi: 10.1002/pol.1981.180191103.
160. Schmidt-Rohr, K. and Chen, Q., Parallel cylindrical water nanochannels in Nafion fuel-cell membranes. *Nat. Mater.*, 7, 1, 75–83, Jan. 2008. doi: 10.1038/nmat2074.

161. Hsu, W.Y. and Gierke, T.D., Ion transport and clustering in nafion perfluorinated membranes. *J. Memb. Sci.*, 13, 3, 307–326, Feb. 1983. doi: 10.1016/S0376-7388(00)81563-X.
162. Mendil-Jakani, H., Pouget, S., Gebel, G., Pintauro, P.N., Insight into the multiscale structure of pre-stretched recast Nafion® membranes: Focus on the crystallinity features. *Polym. (Guildf)*, 63, 99–107, Apr. 2015. doi: 10.1016/j.polymer.2015.02.046.
163. Li, J., Park, J.K., Moore, R.B., Madsen, L.A., Linear coupling of alignment with transport in a polymer electrolyte membrane. *Nat. Mater.*, 10, 7, 507–511, Jul. 2011. doi: 10.1038/nmat3048.
164. Hu, J., Zhang, H., Xu, W., Yuan, Z., Li, X., Mechanism and transfer behavior of ions in Nafion membranes under alkaline media. *J. Memb. Sci.*, 566, 8–14, Nov. 2018. doi: 10.1016/j.memsci.2018.08.057.
165. Zhao, Q., Majsztrik, P., Benziger, J., Diffusion and Interfacial Transport of Water in Nafion. *J. Phys. Chem. B*, 115, 12, 2717–2727, Mar. 2011. doi: 10.1021/jp1112125.
166. Aili, D., Kraglund, M.R., Tavacoli, J., Chatzichristodoulou, C., Jensen, J.O., Polysulfone-polyvinylpyrrolidone blend membranes as electrolytes in alkaline water electrolysis. *J. Memb. Sci.*, 598, 117674, Mar. 2020. doi: 10.1016/j.memsci.2019.117674.
167. Aili, D., *et al.*, Heterogeneous anion conducting membranes based on linear and crosslinked KOH doped polybenzimidazole for alkaline water electrolysis. *J. Memb. Sci.*, 447, 424–432, Nov. 2013. doi: 10.1016/j.memsci.2013.07.054.
168. Aili, D., Wright, A.G., Kraglund, M.R., Jankova, K., Holdcroft, S., Jensen, J.O., Towards a stable ion-solvating polymer electrolyte for advanced alkaline water electrolysis. *J. Mater. Chem. A*, 5, 10, 5055–5066, 2017. doi: 10.1039/C6TA10680C.
169. Jensen, J.O., Aili, D., Hansen, M.K., Li, Q., Bjerrum, N.J., Christensen, E., (Invited) A Stability Study of Alkali Doped PBI Membranes for Alkaline Electrolyzer Cells. *ECS Trans.*, 64, 3, 1175–1184, Aug. 2014. doi: 10.1149/06403.1175ecst.
170. Kraglund, M.R., *et al.*, Ion-solvating membranes as a new approach towards high rate alkaline electrolyzers. *Energy Environ. Sci.*, 12, 11, 3313–3318, 2019. doi: 10.1039/C9EE00832B.
171. Yuan, Z., Li, X., Hu, J., Xu, W., Cao, J., Zhang, H., Degradation mechanism of sulfonated poly(ether ether ketone) (SPEEK) ion exchange membranes under vanadium flow battery medium. *Phys. Chem. Chem. Phys.*, 16, 37, 19841–19847, 2014. doi: 10.1039/C4CP03329A.
172. Zhao, Y., Tsuchida, E., Choe, Y.-K., Wang, J., Ikeshoji, T., Ohira, A., Theoretical studies on the degradation of hydrocarbon copolymer ionomers used in fuel cells. *J. Memb. Sci.*, 487, 229–239, Aug. 2015. doi: 10.1016/j.memsci.2015.04.005.
173. Sun, Z., Pan, J., Guo, J., Yan, F., The Alkaline Stability of Anion Exchange Membrane for Fuel Cell Applications: The Effects of Alkaline Media. *Adv. Sci.*, 5, 8, Aug. 2018. doi: 10.1002/advs.201800065.

174. Chempath, S., *et al.*, Mechanism of Tetraalkylammonium Headgroup Degradation in Alkaline Fuel Cell Membranes. *J. Phys. Chem. C*, 112, 9, 3179–3182, Mar. 2008. doi: 10.1021/jp7115577.
175. Yuan, Z., Duan, Y., Zhang, H., Li, X., Zhang, H., Vankelecom, I., Advanced porous membranes with ultra-high selectivity and stability for vanadium flow batteries. *Energy Environ. Sci.*, 9, 2, 441–447, 2016. doi: 10.1039/C5EE02896E.
176. Yuyue Zhao, Y.Z., Lu, W., Yuan, Z., Qiao, L., Li, X., Zhang, H., Advanced charged porous membranes with flexible internal crosslinking structures for vanadium flow batteries. *J. Mater. Chem. A*, 5, 13, 6193–6199, 2017. doi: 10.1039/C7TA00117G.
177. Wang, Y., Chen, K.S., Mishler, J., Cho, S.C., Adroher, X.C., A review of polymer electrolyte membrane fuel cells: Technology, applications, and needs on fundamental research. *Appl. Energy*, 88, 4, 981–1007, 2011. doi: 10.1016/j.apenergy.2010.09.030.
178. Fan, J., *et al.*, Poly(bis-arylimidazoliums) possessing high hydroxide ion exchange capacity and high alkaline stability. *Nat. Commun.*, 10, 1, 2306, May 2019. doi: 10.1038/s41467-019-10292-z.
179. Li, N., Yan, T., Li, Z., Thurn-Albrecht, T., Binder, W.H., Comb-shaped polymers to enhance hydroxide transport in anion exchange membranes. *Energy Environ. Sci.*, 5, 7, 7888, 2012. doi: 10.1039/c2ee22050d.
180. Nuñez, S.A., Capparelli, C., Hickner, M.A., N -Alkyl Interstitial Spacers and Terminal Pendants Influence the Alkaline Stability of Tetraalkylammonium Cations for Anion Exchange Membrane Fuel Cells. *Chem. Mater.*, 28, 8, 2589–2598, Apr. 2016. doi: 10.1021/acs.chemmater.5b04767.
181. Alhammadi, A., Fetyan, A., Agung Susantyoko, R., Mustafa, I., Bamgbopa, M.O., Understanding characteristic electrochemical impedance spectral data of redox flow batteries with multiphysics modeling. *J. Energy Chem.*, 102, 329–339, 2025. doi: 10.1016/j.jechem.2024.11.007.
182. Krakowiak, J., Bącalski, W., Lentka, G., Peljo, P., Ślepski, P., Three modes of electrochemical impedance spectroscopy measurements performed on vanadium redox flow battery. *Sustain. Mater. Technol.*, 40, April, 2024. doi: 10.1016/j.susmat.2024.e00957.
183. Roth, C., Noack, J., Skyllas-Kazacos, M. (Eds.), *Flow Batteries*, Wiley, Weinheim, Germany, 2023, doi: 10.1002/9783527832767.
184. Durant, B.K., She, Y., Wang, P., Kraus, T., Parkinson, B.A., Solar Charged Redox Flow Battery: A Tandem Photoelectrochemical Approach. *J. Electrochem. Soc.*, 166, 5, H3001–H3008, Sep. 2019. doi: 10.1149/2.0011905jes.
185. McCulloch, W.D., Yu, M., Wu, Y., pH-Tuning a Solar Redox Flow Battery for Integrated Energy Conversion and Storage. *ACS Energy Lett.*, 1, 3, 578–582, Sep. 2016. doi: 10.1021/acsenergylett.6b00296.
186. Walter, M.G., *et al.*, Solar Water Splitting Cells. *Chem. Rev.*, 110, 11, 6446–6473, Nov. 2010. doi: 10.1021/cr1002326.

187. Heinze, J., Cyclic Voltammetry—'Electrochemical Spectroscopy'. New Analytical Methods (25). *Angew. Chem. Int. Ed. English*, 23, 11, 831–847, Nov. 1984. doi: 10.1002/anie.198408313.
188. Evans, D.H., O'Connell, K.M., Petersen, R.A., Kelly, M.J., Cyclic voltammetry. *J. Chem. Educ.*, 60, 4, 290, Apr. 1983. doi: 10.1021/ed060p290.
189. Trovò, A., Zamboni, W., Guarnieri, M., Multichannel Electrochemical Impedance Spectroscopy and equivalent circuit synthesis of a large-scale vanadium redox flow battery. *J. Power Sources*, 493, December 2020, 229703, 2021. doi: 10.1016/j.jpowsour.2021.229703.
190. Srinivasan, R. and Fasmin, F., *An Introduction to Electrochemical Impedance Spectroscopy*, First edition, CRC Press, Boca Raton, 2021, doi: 10.1201/9781003127932.
191. Orazem, M.E. and Tribollet, B., *Electrochemical Impedance Spectroscopy*, Wiley, Pennington, NJ, 2017, doi: 10.1002/9781119363682.
192. Barsoukov, E. and Macdonald, J.R. (Eds.), *Impedance Spectroscopy*, Wiley, New Jersey, USA, 2018, doi: 10.1002/9781119381860.
193. Chitvuttichot, S., Yeetsorn, R., Tuantranont, A., Investigating impact of charging parameters on discharge efficiency of sodium polysulfide-bromide flow battery. *J. Energy Storage*, 102, PA, 114070, 2024. doi: 10.1016/j.est.2024.114070.
194. Sun, H., *et al.*, Characteristics of charge/discharge and alternating current impedance in all-vanadium redox flow batteries. *Energy*, 168, 693–701, 2019. doi: 10.1016/j.energy.2018.11.130.
195. Sivula, K., Mott–Schottky Analysis of Photoelectrodes: Sanity Checks Are Needed. *ACS Energy Lett.*, 6, 7, 2549–2551, Jul. 2021. doi: 10.1021/acsenergylett.1c01245.
196. Gelderman, K., Lee, L., Donne, S.W., Flat-Band Potential of a Semiconductor: Using the Mott–Schottky Equation. *J. Chem. Educ.*, 84, 4, 685, Apr. 2007. doi: 10.1021/ed084p685.
197. Hsu, C.H. and Mansfeld, F., Technical Note: Concerning the Conversion of the Constant Phase Element Parameter Y 0 into a Capacitance. *CORROSION*, 57, 9, 747–748, Sep. 2001. doi: 10.5006/1.3280607.
198. Bisquert, J., Giménez, S., Bertoluzzi, L., Herraiz-Cardona, I., Analysis of Photoelectrochemical Systems by Impedance Spectroscopy, in: *Photoelectrochemical Solar Fuel Production*, pp. 281–321, Springer International Publishing, Cham, 2016, doi: 10.1007/978-3-319-29641-8_6.
199. Peter, L.M., Gurudayal, L.H., Abdi, F.F., Understanding the role of nanostructuring in photoelectrode performance for light-driven water splitting. *J. Electroanal. Chem.*, 819, 447–458, Jun. 2018. doi: 10.1016/j.jelechem.2017.12.031.
200. Cesar, I., Sivula, K., Kay, A., Zboril, R., Grätzel, M., Influence of Feature Size, Film Thickness, and Silicon Doping on the Performance of Nanostructured Hematite Photoanodes for Solar Water Splitting. *J. Phys. Chem. C*, 113, 2, 772–782, Jan. 2009. doi: 10.1021/jp809060p.
201. Klahr, B., Gimenez, S., Fabregat-Santiago, F., Hamann, T., Bisquert, J., Water Oxidation at Hematite Photoelectrodes: The Role of Surface States. *J. Am. Chem. Soc.*, 134, 9, 4294–4302, Mar. 2012. doi: 10.1021/ja210755h.

11

Perovskite-Based Photo-Supercapacitors as Self-Charging and Energy Storage Devices

Muhamad Yudatama Perdana[1], Abdurrahman Imam[1], Mohammed Ashraf Gondal[1,2*], Ahmar Ali[1] and Mohamed Jaffer Sadiq Mohamed[1]

[1]*Laser Research Group, Department of Physics & Interdisciplinary Research Center for Hydrogen Technologies and Carbon Management (IRC-HTCM), King Fahd University of Petroleum and Minerals (KFUPM), Dhahran, Saudi Arabia*
[2]*K. A. CARE Energy Research and Innovation Center, King Fahd University of Petroleum and Minerals, Dhahran, Saudi Arabia*

Abstract
The demand of self-charging supercapacitors is growing rapidly as portable electronic devices, which are also economically efficient. Owing to its abundant availability, solar energy is considered a green energy resource. Thus, integrating the energy harvesting and energy storage as a single device has attracted researchers worldwide, and such a device is known as a photo-supercapacitor. Many efforts have been made to assemble high-performance photo-supercapacitors. Perovskite-based oxides and halide perovskite materials have recently been used for supercapacitor applications. Several studies have used perovskites as dual-functional material that acts as a light absorber to generate electrons and the electrodes for supercapacitor function. In such an application, perovskite should have good ionic and electron conductivity properties because ion migration and electron charge transfer play important roles for dual-functional applications. It is worth mentioning that photoinduced charge transfer at the interface of perovskite/electrode will affect additional photocapacitance and ion migration to contribute to the total specific capacitance. This book chapter will shed light on different types of perovskite-based materials, perovskite as a light absorber, studies on perovskite materials for supercapacitor applications, and the evaluation parameters of photo-supercapacitors.

**Corresponding author*: magondal@kfupm.edu.sa

Nurdan Demirci Sankir and Mehmet Sankir (eds.) Solar Capacitors and Batteries, (351–388)

Keywords: Supercapacitors, perovskite, photo-supercapacitors, charge transfer, conductivity

11.1 Introduction

Solar energy has a major role in energy sources and sustaining the environment. For solar energy harvesting, solar cells of various types have been established. Electronic devices in a portable manner integrated with solar cells become famous, but they have a drawback such as power output instability of the solar cells under solar illumination [1]. The difference between the timings of energy supply and demand is able to be managed by integrating energy conversions and storage in portable and off-grid power systems [2, 3]. Merging the mechanism of photovoltaic and energy storage devices has an obstacle, such as the device structures with the complexity of multilayers, which correlates with the transport of electrons and ions. The bifunctional properties of the advanced photoelectrodes can unravel the problem of combining energy storage with energy harvesting [4]. There are several shortcomings with this system, such as a low power density, a brief lifespan, and a sluggish charging/discharging reaction time [5].

In general, a supercapacitor (SC) comprises a porous electrode with conducting characteristics, and the electrodes are interfaced with the electrolyte solution. The diffusion of ions from the electrolyte into the porous electrode, or electric double-layer capacitance, or both of these processes, will be used to store the charge at the interface between the conductor and the electrolyte. In the energy-harvesting device, photoelectrochemical cells, which involve semiconductors and electrolytes, cannot store the energy [6]. Energy generation and energy storage within the same device can be achieved by using material that has multifunctional properties of electronic and ionic conductivity in the semiconducting electrodes, which have porous sites and interface with the suitable electrolyte [5]. Photocapacitor was discovered by Miyasaka and Murakami (2004) by using dye-sensitized nanocrystalline film to absorb visible light. The sandwich configuration consists of a photoelectrode, which acts as a light absorber, an electrolyte with redox-free solution, and a counter electrode. For the capacitor part, activated carbon, a solution of propylene carbonate with 15 wt% $(CH_3CH_2)NBF_4$, and porous resin film were used for the active material, electrolyte, and separator, respectively [7].

Several applications in perovskite-based light-harvesting devices can be found in dye-sensitized solar cell (DSSC) using perovskite oxide [8], perovskite solar cells (PSCs) [9], and photodetector [10]. Energy storage applications, such as batteries and SCs, have used halide-based perovskites as they have good stability of their ionic conductivity properties. In battery, intercalation/de-intercalation with the electrodes fabricated with halide-based perovskite can be done by the metal ions [11–13]. In addition, metal halide perovskites (MHPs) have exceptional light absorbance [14].

In this discussion, we will reveal several types of perovskite material. Work mechanisms of photo-supercapacitors will be delivered. Also, several studies on photo-supercapacitors will be brought to the discussion.

11.2 Perovskite Materials

11.2.1 Classification of MHPs

MHPs can be classified in various ways depending on their properties and characteristics of interest. Here are a few possible classifications, along with references to further reading.

11.2.1.1 By Composition

MHPs can be classified based on the specific combination of metal, halide, and organic/inorganic cations used in their synthesis. For example, some common MHPs include cesium lead bromide ($CsPbBr_3$), formamidinium lead iodide ($FAPbI_3$), and methylammonium lead iodide ($MAPbI_3$). This classification is often used in the context of materials synthesis and characterization [15].

There are three categories of perovskite. They could be all-inorganic, all-organic, or a combination of organic–inorganic (hybrid) materials. The categorization is based on cations used in the crystal structure of ABX_3, where ionic permutations between the sites at A and B occur (Table 11.1) [16]. Among these categories, organic-based perovskites, which involve organic cations at A and B sites, are the least commonly studied. A recent study by A. Stoddart focused on all-organic perovskites for ferroelectric applications, utilizing a divalent organic cation. For instance, at the A site, N-methyl-N'-diazabicyclo [2.2.2] octonium can be employed. For the B-site, ammonium ion (NH^{4+}) and the X site can be filled by halides [17]. Because there are not sufficient metallic ions within the crystal structure, organic-based perovskites are called metal-free perovskites.

On the other hand, all-inorganic perovskites consist of metallic cations occupying the A and B sites. Inorganic-based-perovskite can be found in $BaTiO_3$ [18], $CsPbI_3$ [19], $LaYS_3$ [20], and $LaWN_3$ [21]. Schouwink *et al.*

Table 11.1 Perovskite materials classifications according to their x-site.

Class based on x site	Properties	Applications	Ref.
Oxides	Exhibits properties such as ferromagnetism, piezoelectricity, magnetism, and superconductivity	Commonly used in photocatalyst, ferroelectric diode, integrated memory, ceramic capacitors, devices, spin electronic devices, sensors, photovoltaic, and laser	[18, 24–26]
Sulfides	Exhibits semiconducting and optoelectronic properties	Particularly suitable for tandem solar cells, photoelectrodes, photovoltaic applications, and electrochemical water splitting application with tandem system	[20, 27–29]
Telluride	Demonstrates semiconducting and optoelectronic characteristics	Primarily used in photodetector, tandem solar cells, photovoltaic applications, and gamma-ray detector	[30]
Nitrides	Possesses properties like semiconductivity, electrical conductivity, ferromagnetism, magnetism, and catalytic activity	Widely utilized in photovoltaic applications, photocatalysis for hydrogen generation, capacitors, and integrated memory devices	[21, 31]

(*Continued*)

Table 11.1 Perovskite materials classifications according to their x-site. (*Conitnued*)

Class based on x site	Properties	Applications	Ref.
Selenides	Possesses semiconducting and optoelectronic properties	Commonly employed in photovoltaic applications and photodetectors	[32]
Hydrides	Known for permanent nanoporosity ferroelectric behavior and rapid ion conduction	Utilized in various applications such as thermal neutron detectors, thermal neutron detectors, pressure sensors, light-emitting diodes, and hydrogen storage	[22, 23, 33]
Halides	Exhibits photovoltaic and optoelectronic behavior, ionically conducting nature, catalytic properties, and luminescence	Widely used in light-emitting diodes, solar cells, SCs, phototransistors, photodiodes, photosensors, and membrane materials	[34–37]

conducted a study of a material that is potential for ferroelectric applications where the X site was replaced with the borohydride (BH_4) anionic radical, resulting in a series of stable $AB(BH_4)_3$ perovskites [22]. Some materials of such inorganic-based perovskite are $CsCa(BH_4)_3$, $RbCa(BH_4)_3$, and $KCa(BH_4)_3$ [22, 23].

Likewise, there exist organic–inorganic hybrid perovskites where organic cations occupy the A site and inorganic cations occupy the B site, or in the reverse system. Examples of perovskites in this type comprise $CH_3NH_3PbX_3$, $(CH_3NH_3)_3Bi_3X_9$, and $CH(NH_2)_2PbX_3$, where halides predominately occupy the X sites [38, 39]. Based on the sort of anions present at the X site, perovskite can also be divided into several categories. This leads to the categorization of perovskites as halide-, selenide-, telluride-, sulfide-, hydride-, nitride-, or oxide-based, depending on the specific anions that occupy the X site. For instance, $BaTiO_3$ is an example of perovskite oxide,

$LaYS_3$ is a perovskite sulfide, and $LaWN_3$ is a nitride-based perovskite. Furthermore, $CH_3NH_3PbX_3$ is a combination of organic and inorganic MHP, sometimes referred to as an organometallic halide or hybrid halide perovskite (HHP). In certain cases, the X site may be occupied by a combination of different halides, resulting in mixed halide perovskites, such as ABI_xCl_{3-x}, ABI_xBr_{3-x}, or $ABBr_xCl_{3-x}$, where A and B represent predefined cations. Table 11.1 summarizes the characteristics and uses of several perovskite materials, categorized based on their anionic (-X) sites.

In the 12-fold cuboctahedral structure, carrier thermalization generates heat within the solar cell. This heat enhances the vibration of $CH_3NH_3^+$ within the $[PbI_6]^{2-}$ octahedral framework, which can lead to the expulsion of the $CH_3NH_3^+$ group from the perovskite lattice. Once separated, $CH_3NH_3^+$ readily degrades through photochemical catalysis. In the work by Li *et al.* (2018), a strategy was proposed to mitigate the issue of cation volatility and lattice instability in 3D perovskite matrices. Gallium(III) acetylacetonate (GAA_3) was incorporated into the perovskite precursor for the *in-situ* formation of a core–shell structure. The resulting shell, composed of $[GAA_3]_4$, serves to protect the 3D $Cs_xFA_{1-x}PbI_3$ crystallites from ambient moisture and to passivate defect states. This approach significantly improves the environmental stability of the perovskite films while preserving high power conversion efficiency [39].

Additionally, some perovskites incorporate double anions, where combinations such as halide–sulfide, halide–nitride, oxide–halide, or similar compositions occupy the X site, forming double-anion perovskites [40]. Laboratory-synthesized examples of double-anion perovskites include $SrFeO_{3-0.05-0.10}F_{0.05}$ and $SrFe_{0.9}Ti_{0.1O3-0.05-0.10}F_{0.05}$ [41]. Furthermore, using density functional theory simulations, other examples in a similar category, such as $CH_3NH_3BiI_2Te$, $CH_3NH_3BiI_2Se$, and $CH_3NH_3BiI_2S$, have been predicted.

Double perovskites, represented by the general formula A_2BCX_6, have been successfully synthesized in laboratory settings. Examples of such double perovskites include Cs_2InAgX_6, Rb_2CuInX_6, Cs_2AgBiX_6, and more [42]. These double perovskites can be viewed as a combination of ABX_3 and ACX_3 perovskites. Similarly, the A_3BCDX_9 triple perovskite can be understood as a combination of ABX_3, ACX_3, and ADX_3 perovskite, while the A_4BCDX_{12} quadruple perovskite consists of ABX_3, ACX_3, ADX_3, and AEX_3 perovskite components. Laboratory synthesis has successfully demonstrated the formation of the $Sr_3CrFeMoO_9$ triple perovskite for ferroelectric applications and the $CaCu_3Ru_4O_{12}$ quadruple perovskite for water oxidation applications [43, 44]. Moreover, triple-cation double-halide perovskites like $Cs_{0.15}FA_{0.85}Pb(I_{0.9}Br_{0.1})_3$ and $(MAPbBr_3)_{0.15}(FAPbI_3)_{0.85}$ have been synthesized and utilized in optoelectronic applications [45, 46]. Patil J *et al.* have also synthesized quadruple-cation double-halide perovskite materials for solar cell fabrications, as exemplified by $Cs_{0.05}(MA_{0.17}FA_{0.83})_{0.95}Pb(I_{0.83}Br_{0.17})_3$ [47, 48].

The most popular among all the classes of perovskite materials mentioned above is a class of semiconductors known as MHPs. MHPs have the formula ABX_3 where X is a halide, A is a cation, and B is a divalent metal ion, and also known with the potential to provide more affordable and effective photovoltaics and optoelectronics applications than silicon-based technology [49–55]. Due to their distinct optical and electrical characteristics, the last 10 years have seen a significant increase in scientific interest in MHP. They have so far been utilized in sectors including solar energy conversion technology, light-emitting diodes, and photovoltaics. Particularly, a significant increase in photovoltaic efficiency has been followed by a better understanding of the underlying physical properties of the materials and working of devices [52]. Additionally, recent developments in the generation and characterization of single crystals at both the macro- and nanoscales necessitate a rationalization of the various perovskite semiconductor forms outside of the popular polycrystalline thin films. Compared to crystalline Si and GaAs, two other common semiconductors, MHPs stand out due to their remarkable crystallinity despite being processed at low temperatures (around 200°C) [56–58]. This makes it possible to process on a variety of substrates and makes it simple to produce perovskite materials in a variety of shapes, from nanocrystals to macroscopic single crystals [49, 59, 60].

MHPs can also be classified based on their dimensional structure. For example, three-dimensional (3D) MHPs have a bulk crystal structure that is 3D, while 2D MHPs have a layered structure that is only two-dimensional. Similarly, 1D MHPs have a linear structure, while 0D MHPs are nanoparticles or quantum dots [61]. This classification is often used in the context of materials physics and optoelectronics. Details will follow in subsequent sections.

11.2.1.2 By Dimensionality

MHPs can also be classified based on their dimensional structure. For example, 3D MHPs have a bulk crystal structure that is 3D, while 2D MHPs have a layered structure that is only two-dimensional. Similarly, 1D MHPs have a linear structure, while 0D MHPs are nanoparticles or quantum dots [61]. This classification is often used in the context of materials physics and optoelectronics. Details will follow in subsequent sections.

11.2.1.3 By Crystal Symmetry

MHPs can also be classified based on their crystal symmetry. The most common crystal systems for MHPs are cubic, tetragonal, and orthorhombic, but some materials may also exhibit other crystal symmetries. This classification is often used in the context of crystallography and materials science [62].

11.2.1.4 By Stability

MHPs can also be classified based on their stability, which is a critical property for their use in practical applications. Some researchers have focused on developing stable MHPs that can resist degradation over time, while others have studied the causes of instability and the factors that can lead to degradation. This classification is often used in the context of materials engineering and device fabrication [63, 64].

These are just a few examples of the many possible ways to classify MHPs. The references provided should serve as a starting point for further reading and exploration.

11.2.2 Perovskite Properties as a Light Absorber

11.2.2.1 High Absorption Coefficients

The absorption coefficient is a measure of how efficiently a material can absorb incident light. A high absorption coefficient means that the material can absorb a larger fraction of incident photons, leading to improved light-harvesting capability and higher sensitivity in photodetection devices. MHPs such as $MAPbI_3$ (methylammonium lead iodide) and $MAPbBr_3$ (methylammonium lead bromide) exhibit unique optical properties that contribute to their high absorption coefficients. These materials have a direct bandgap, which means that the energy difference between the valence and conduction bands corresponds to a specific wavelength of light. This direct bandgap allows MHPs to efficiently absorb photons with energies matching their bandgap, covering a broad range of the electromagnetic spectrum from the visible to the near-infrared (NIR) region [65]. The high absorption coefficient of MHPs can be attributed to several factors. First, their crystal structure allows for strong light–matter interactions. MHPs have a well-defined lattice structure with a long-range order, which promotes efficient absorption and exciton generation. Additionally, the presence of heavy elements, such as lead (Pb) and iodine (I) or bromine (Br), leads to strong spin–orbit coupling, which enhances the absorption of light [66, 67]. Furthermore, MHPs exhibit a high degree of crystallinity and low defect density, which minimize non-radiative recombination processes and enhance the efficiency of photogenerated charge carriers [68]. This property is crucial for achieving high photoconversion efficiency in photodetection devices.

Compared to traditional photodetection materials such as cadmium telluride (CdTe), copper indium gallium selenide (CIGS), gallium arsenide (GaAs), amorphous silicon (a-Si), and crystalline silicon (c-Si), MHPs exhibit much higher absorption coefficients as shown in Figure 11.7 [69]. The high absorption coefficient of MHPs is advantageous in photodetection applications because it enables the detection of even low levels of light. This is particularly important in low-light conditions or when working with weak light sources. The efficient light absorption of MHPs allows for enhanced sensitivity and improved signal-to-noise ratios in photodetectors.

11.2.2.2 Bandgap Tunability

Tunable bandgaps refer to the ability to adjust the energy range in which a material can absorb light or conduct electricity. The energy difference between a material's conduction band, which has the lowest energy level suitable for electron conduction, and its lowest energy level, the valence band, which has the highest energy level occupied by electrons, is known as the bandgap. In photodetection, the bandgap determines the range of wavelengths that a material can absorb light effectively [70, 71]. One of the key advantages of MHPs is the ability to modify their bandgaps by changing their composition and structure. This tunability allows for the customization of the absorption spectrum to match specific photodetection requirements, spanning from visible to NIR wavelengths. MHPs, such as $MAPbX_3$, $FAPbX_3$, and $CsPbX_3$, offer tunable bandgaps due to their unique crystal structure and composition [72]. The bandgap of a perovskite material can be adjusted by changing the halide composition, the size of the organic cation [73–75], or the arrangement of atoms within the crystal lattice [76].

By engineering the bandgap of MHPs, it is possible to tailor their absorption properties to specific wavelengths or energy ranges. This tunability allows for the optimization of perovskite photodetectors for different applications. For example, in visible light photodetection, the bandgap can be tuned to match the energy of photons in the visible spectrum. In other cases, perovskites can be modified to absorb light in the ultraviolet (UV) or NIR regions.

11.2.2.3 High Charge Carrier Mobility

The capacity of charge carriers, such as electrons or holes, to flow freely through a material in response to an electric field is referred to as charge carrier mobility. It is a critical parameter in various electronic and optoelectronic devices, including photodetectors. The physics behind high charge carrier mobility involves several factors that contribute to efficient charge transport. Charge carriers in an MHP material move in response to an applied electric field. The relationship between the electric field *E* and the velocity *v* of charge carriers is described by the mobility, μ:$v = \mu E$.

The ratio of the applied electric field to the charge carrier drift velocity is known as mobility. A high mobility implies that charge carriers can move quickly in the presence of an electric field, resulting in efficient charge transport.

High charge carrier mobility is one of the remarkable properties of MHPs that has garnered significant attention in the field of optoelectronics [77, 78]. MHPs, such as $CH_3NH_3PbI_3$ and $HC(NH_2)_2PbI_3$, exhibit exceptionally high charge carrier mobilities compared to many other semiconductor materials [79]. The exceptional performance of MHP-based devices, such as photodetectors, light-emitting diodes, and solar cells, is largely due to their high mobility. The high charge carrier mobility in MHPs can be attributed to several underlying physics principles such as the following.

11.2.2.3.1 Crystalline Structure

MHPs possess a unique crystal structure that facilitates efficient charge transport. They exhibit a 3D network of metal halide octahedra connected by cations. This structure provides a long-range order and enables smooth pathways for charge carriers to move through the material.

11.2.2.3.2 Low Defect Density

MHPs can be synthesized with a low defect density, which minimizes the scattering of charge carriers and enhances their mobility. Defects, such as vacancies, impurities, or grain boundaries, can act as scattering centers that hinder charge carrier motion [80]. Minimizing defect densities through precise fabrication and processing techniques helps maintain high mobility in MHPs.

11.2.2.3.3 Long Electron–Hole Diffusion Lengths

MHPs exhibit long electron and hole diffusion lengths, which indicate the distance that charges carriers can travel without significant scattering or recombination. The long diffusion lengths suggest efficient transport and low charge carrier capture that leads to high mobility. Experimental studies have reported the lengths of electron diffusion between several micrometers and tens of micrometers and the lengths of hole diffusion between hundreds of nanometers and several micrometers in MHPs. Detailed discussion on this subtopic will follow in the next section.

The high charge carrier mobility in MHPs has been experimentally demonstrated [77, 78]. For example, studies have reported electron mobilities exceeding 100 cm^2/Vs and hole mobilities surpassing 10 cm^2/Vs (see Table 11.1) in certain MHP compositions and device configurations [81]. These values are significantly higher than those observed in traditional semiconductors, such as amorphous silicon (a-Si), which typically exhibit mobilities on the order of 1 cm^2/Vs (see Table 11.1). These desirable properties in MHPs offer several advantages for optoelectronic devices. It enables efficient collection of photo-generated charge carriers, reduces recombination losses, and enhances device performance. Additionally, high mobility contributes to rapid response times and high sensitivity in photodetectors.

Table 11.1 offers a summary of the research on experimentally established charge carrier mobilities for ABX_3 perovskites where A can be Cs, formamidinium, methylammonium, then B can be Pb or Sn, and X can construct from I, Br, Cl, and alloys thereof. Perovskite with layered structure (quasi-2D) examples is also given. However, the discrepancies in the reported charge carrier mobilities of MHPs have been a subject of debate and ongoing research within the scientific community [79]. While there is a general consensus that MHPs exhibit relatively high charge carrier mobilities compared to some other semiconductor materials, there are variations in reported values and discrepancies among different studies. The debate mainly revolves around measurement techniques, device architectures, and material quality.

One aspect of the debate relates to the distinction between intrinsic and extrinsic charge carrier mobilities. Intrinsic mobility refers to the charge transport properties within the bulk of the perovskite material, while extrinsic mobility takes into account the influence of defects, grain boundaries, and interface effects on the charge transport. It is generally accepted that intrinsic mobilities of MHPs are relatively high, between 1 and 100 $cm^2/(V{\cdot}s)$ for both electrons and holes, as demonstrated in various studies [1–3].

Similarly, there are challenges in accurately measuring intrinsic mobility due to the presence of defects and grain boundaries, which can significantly affect the transport properties. The presence of traps, such as surface defects or grain boundaries, can act as charge carrier recombination centers, leading to lower effective mobilities in practical devices. Additionally, the mobility can vary depending on the film quality, crystal structure, doping, and processing conditions.

Another aspect of the debate is related to the choice of measurement techniques. Different methods, such as space-charge-limited current measurements, field-effect transistor measurements, time-of-flight measurements, and others, can yield different mobility values depending on the specific experimental conditions and assumptions. Each technique has its advantages and limitations, and the interpretation of the results requires careful consideration of the device architecture and underlying assumptions. However, to resolve the debate and establish a comprehensive understanding of charge carrier mobilities in MHPs, further research is needed to improve measurement techniques, reduce material defects, and develop more accurate models for charge transport in these complex materials.

11.2.2.4 Long Carrier Diffusion Lengths (L_{diff})

Carrier diffusion length refers to the maximum distance that electrons or holes can traverse within a material before undergoing recombination, a phenomenon pivotal for achieving efficient photovoltaic and optoelectronic devices. The noteworthy extended carrier diffusion lengths witnessed in MHPs underscore their remarkable power conversion efficiencies and have ignited fervent research attention. This attribute bears paramount importance in photodetection devices as it curtails charge recombination, thereby bolstering the collection of photoinduced carriers. This robust charge transport capability contributes to heightened responsivity, rapid response rates, and reduced noise levels, all of which synergistically enhance the overall performance of photodetectors [82].

In the context of MHPs materials, L_{diff} signifies how far charge carriers can move before they encounter a recombination event, which could yield in the loss of the carriers and hinder device operation. Mathematically, the length of carrier diffusion L_{diff} is related to the mobility of charge carrier μ and the lifetime of carrier τ through the following equation:

$$\mathrm{L} = \sqrt{\mu D \tau}$$

where μ is the carrier mobility, which represents the carrier's ability to move under the influence of an electric field; D is the diffusion coefficient, which describes how quickly charge carriers spread out in response to a concentration gradient; and τ is the carrier lifetime, denoting the average time a carrier exists before recombination. The parameter L_{diff} is particularly important due to its impact on the performance of photovoltaic and optoelectronic devices. Long L_{diff} suggests efficient carrier transport and minimal recombination, leading to higher device efficiency.

MHPs have been observed to possess long L_{diff}, which can be attributed to factors such as low defect density, balanced electron and hole mobilities, and unique crystal structures. These properties allow charge carriers to travel relatively long distances before encountering recombination events, contributing to the impressive performance of perovskite-based devices.

It is important to note that L_{diff} is not a fixed value but rather a material-dependent parameter that can be influenced by various factors including crystal quality, temperature, and doping.

Numerous investigations corroborate the substantial carrier diffusion lengths in MHPs. For instance, by using the measurements of time-resolved photoluminescence, in the study of Stranks *et al.* (2013), they delved into carrier dynamics within methylammonium lead iodide ($MAPbI_3$) perovskite thin films. The outcome revealed the lengths of carrier diffusion reaching several micrometers in these materials, indicative of proficient charge transport and diminished recombination probabilities [83].

Furthermore, the research by Peng *et al.* (2017) scrutinized L_{diff} across diverse types of MHPs, encompassing single crystals and polycrystalline thin films. Employing time-resolved microwave conductivity measurements, the L_{diff} spanned from hundreds of nanometers to several micrometers, thereby showcasing these materials' remarkable charge transport attributes [84].

The origins of these elongated L_{diff} in MHPs trace back to various factors. Firstly, their elevated tolerance to defects and minimal defect density curtails the odds of charge carrier entrapment, affording carriers the ability to traverse longer distances without recombination. Secondly, the harmonized electron and hole mobilities within these materials expedite efficient charge transport [85–87]. Moreover, inherent structural traits of MHPs, including substantial dielectric constants and low effective masses of charge carriers, contribute to their extended diffusion lengths [88].

In summation, the conspicuous elongated L_{diff} observed in MHPs, notably MAPbI3, fundamentally underpins their high-performance photovoltaic characteristics. These extensive diffusion lengths exemplify the materials' commendable charge transport capabilities, minimal carrier entrapment tendencies, and well-balanced electron–hole mobilities. As the scientific community perseveres in exploring and optimizing these materials, comprehending and harnessing their prolonged L_{diff} will persist as a

Table 11.2 Comparison between the carrier diffusion lengths of metal halides and some other materials.

Material	Carrier diffusion lengths	Ref.
Metal Halides Perovskite		
$CH_3NH_3PbI_3$ ($MAPbI_3$)	1–10	[83, 87, 89]
$CsPbI_3$	1–10	[90, 91]
$FAPbI_3$	~0.1–1	[92]
Other Materials		
Silicon (Si)	~100–150	[93]
Organic semiconductors	~10–50	[94, 95]
Gallium arsenide (GaAs)	~0.1–1	[96, 97]
Polymers	~1–10	[98, 99]

pivotal facet in propelling the realm of PSCs. Table 11.2 offers a comparative overview of L_{diff} in MHPs and other prevalent materials.

It should be noted that L_{diff} can vary significantly depending on factors such as material quality, crystal structure, defects, and measurement techniques. The values provided in the table are approximate ranges and can vary from study to study. References are provided for each material to guide the reader to further information.

11.3 Perovskite Materials for Photo-Supercapacitor

A study done by Liu *et al.* (2017) fabricated a photocapacitor based on PSC with $CH_3NH_3PbI_3$ ($MAPBI_3$) and a polyaniline/carbon nanotube (PANI/CNT) for the SC part. In that study, perovskite was used as an absorber, and CNT bridge was used to integrate PSC and SC. When the light was turned on, photo-excited electron–hole pairs would develop in the $MAPbI_3$ absorber. While holes traveled to the MWCNT electrode and the common electrode, electrons would migrate to the front electrodes made of TiO_2 and FTO [100].

Perovskite-based photocapacitor was also built by the study of Gao *et al.* (2019). The counter electrode for PSC and the working electrode of the SC part used the same CGP film; furthermore, it was called the common electrode. Then, the counter electrode of the SC used another CGP film. During the charging and discharging procedure, the SC, PSC, and load were linked. The photo-charge component of the photo-charge process uses the PSC to transform light energy into electrical energy, which it then stores in the SC. The SC can power the load in total darkness while the discharge is taking place [101].

HHP of methylammonium lead tribromide ($MAPbBr_3$) was used as porous perovskite electrodes in the study of Kumar *et al.* (2022). Carbon black quantity was varied in this study. Electrons and holes are generated through a photoexcited event, and ion migration comes from the light-induced event. Electron–hole pairs and the ion migration give rise to significant boost to the perovskite-based photoelectrode specific capacitance. In this device, the ions' intercalation and deintercalation from the liquid electrolyte into perovskite sites make the diffusion mechanism. A maximum capacitance that can be achieved is ~85 F/g with an electrode that contains 15 mg of carbon black and with illumination of 20,000 *lux* or ~20 $mWcm^{-2}$ from white LED [5].

Heterostructure material comprising activated carbon (AC) combined with Ag_2CrO_4 and SnS (AC/Ag_2CrO_4/SnS) was used as an electrode for photocapacitor application. Ag_2CrO_4/SnS acts as the main component, and Ag_2CrO_4 was used to absorb light, which is the photoactive part to produce electron–hole pairs. Then, the SnO_x layer, which serves as the capacitive part, has a role to store to produced holes. Excellent visible absorption of Ag_2CrO_4 makes it suitable for photocapacitor application. In their work, three-electrode system was used for the electrochemical analysis. From the obtained cyclic voltammetry (CV) curve, a pair of redox peaks was present, and it indicates the behavior of the Faradaic redox that leads to the pseudocapacitive manner of the electrode [102]. The reversible redox reaction that occurred might be $Sn(II) \rightarrow Sn(IV) + 2e^-$. The electrode of FTO/AC/Ag_2CrO_4/SnS gave photoelectrochemical capacitance, which is due to the mechanism of quasi reversible electron transfer process. Sn^{2+}/Sn^{4+} redox couple was taken into account in the mechanism, and the buffer electrolyte contains OH^- ions, which might also facilitate through the system. The highest specific capacitance that can be attained is 4782 mF/g, and it is recorded at a scan rate of 10 mV/s. The electrode's pore size, pore distribution, and surface area give significant contribution on the specific capacitance performance [103].

11.4 Some Studies on Light-Induced SC

Methylammonium bismuth triiodide (MBI) has been used for the electrode material, and a new polymer gel electrolyte was used; furthermore, the device is denoted as MBI:CPH-G photo-supercapacitor. The polymer gel electrolyte has non-aqueous property in this study. Two different conditions were applied for the electrochemical investigations, under dark and light environment. The illustrations for working mechanism without light illumination and under illumination are shown in Figures 11.1a and b [14].

Polymer gel electrolyte has been used as one electrolyte for photo-supercapacitor. In this investigation, bismuth-based organometallic halide perovskite was used as the electrode. This electrolyte has a good surface reaction and may diffuse into the photocapacitive electrode. More ion supply penetrates to the electrode due to the time independence of electrode and electrolyte interactions at a low potential scan rate. For a higher scan rate, there are only a few ions that move toward the electrode since the time for the electrode and electrolyte interaction is not sufficient [14].

11.4.1 Work Mechanism under Dark Condition

The capacitive and diffusion behavior can be separated from the CV current. Power's law can be used for this aim. The formalization illustrates the relationship between the CV current and the scan rate. Power's law is expressed in Equation 11.1a, and Equation 11.1b may be used to express the logarithmic form, where v, i, a, and b present the scan rate, the current at a certain potential, and the parametric variables, respectively [14, 104].

$$i = av^{b} \tag{11.1a}$$

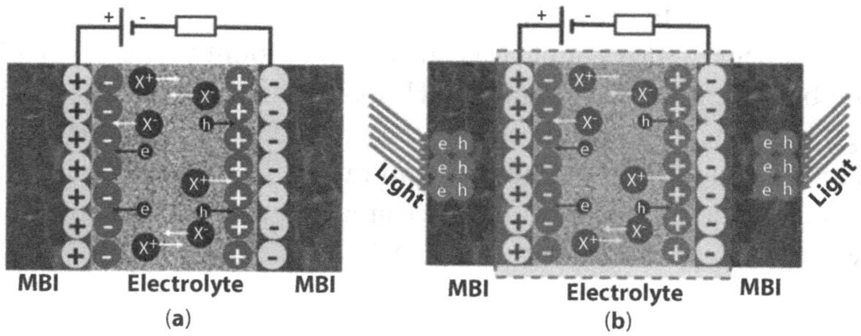

Figure 11.1 (a) Work mechanism of photo-assisted SC under dark-working environment. (b) Mechanism of the device under illumination [14].

$$\log i = b(\log v) + \log a \tag{11.1b}$$

From the log i vs. log v plot, we can get b-value, which corresponds to the slope value. If the b-value is 0.5, it indicates that the mechanism is diffusion-controlled, while if the b-value is 1.0, it shows capacitive controlled mechanism. In other case, capacitive charge storage mechanism and the hybrid diffusion is indicated by a b-value between 0.5 and 1.0 [104, 105].

For the capacitive currents and the diffusion depicted in the CV curve, Dunn's equation applies. The equation, as expressed in Equation 11.2a, corresponds between the total current when it reaches certain potential and the scan rate. From the equation, $c_1(v)$ is the capacitive controlled current, while $c_2(v^{1/2})$ is the diffusion-controlled current [106, 107].

$$i_D(V) = c_1(v) + c_2(v^{1/2}) \tag{11.2a}$$

Taking the modification from Equation 11.2a, c_1 and c_2 can be obtained from the plot as given in Equation 11.2b. From the plot, the values of c_1 and c_2 are shown by the slope and the intersection, respectively.

$$i_D(V) / \left(v^{\frac{1}{2}}\right) = c_1\left(v^{\frac{1}{2}}\right) + c_2 \tag{11.2b}$$

11.4.2 Working Principle under Light Environment

MBI perovskite-based electrodes yield photocurrent, which shows photocapacitive behavior. The photocurrent is contributed by electron–hole pairs supply. From the CV analysis, the onset of electron–hole pairs recombination results in decreasing the photocurrent, and it might occur as the scan rates rise [14].

When light working condition is applied, the CV current can be resolved qualitatively to its three mechanisms, which comprise photo-induced event, capacitive behavior, and diffusion-controlled elements by using Power's law. Under the light-environment condition, the b-value are below 0.5. This corresponds to the further ejection of generating electron–hole pairs (excitons) from the electrodes, which have light-induced capacitive property when illumination is applied. Under the light-working condition, Dunn's equation is not applicable. However, it can be modified such that we can differ the contribution of those three mechanisms, and the modified equations can be expressed in Equation 11.3. From Equation 11.3a, i_{mD}, i_D, and χ are the increased CV current under illumination, the CV current without illumination, and the total effective electron–hole pairs, which is produced per time under illumination with constant intensity. χ can represent the effective light-induced current. While from Equation 11.3b, n, e^-, h^+, and t correspond to the produced excitons number, electrons number, the holes number, and time. Scan rate also affects electron–hole recombination within the electrodes [14].

$$i_{mD}(V) = i_D(V) + \chi(v) \tag{11.3a}$$

$$i_{mD}(V) = c_1(v) + c_2(v^{1/2}) + \chi(v) \tag{11.3b}$$

$$\chi(v) = \frac{n(e^- - h^+)}{t} \tag{11.3c}$$

11.4.3 Electrochemical Analysis on Photo-Supercapacitor

Deviated-rectangular form, which is shown from the CV curves of methylammonium bismuth triiodide-based photo-supercapacitor, indicates the presence of combined mechanism of diffusion and capacitive behavior [108, 109]. Shown in Figure 11.2a is the CV plot under dark environment. The current obtained from the CV increased as the scan rate increased, and there was no significant change of CV curves' shape. Those are indications of capacitive behavior, good stability, and rate capability [110]. Recorded at 0.5 V/s scan rate, the peak current obtained from MBI-based photo-supercapacitor was 10.27 μA [14].

For light-working environment, 100 mW/cm^2, Figure 11.3a shows the CV plot under light-working environment, with the potential windows at 0 V and 1.0 and at scan rates between 0.001 V/s and 0.5 V/s [14].

The formula that was used to approximate the capacitance of the photo-supercapacitor can be expressed in Equation 11.4. From the equation, i, V_0, dt, and v are the current (A), maximum voltage (V), differential time (s), and applied scan rate (V/s), respectively [14].

$$C = \frac{\int_0^{\frac{2V_0}{v}} |i| dt}{2V_0} \tag{11.4}$$

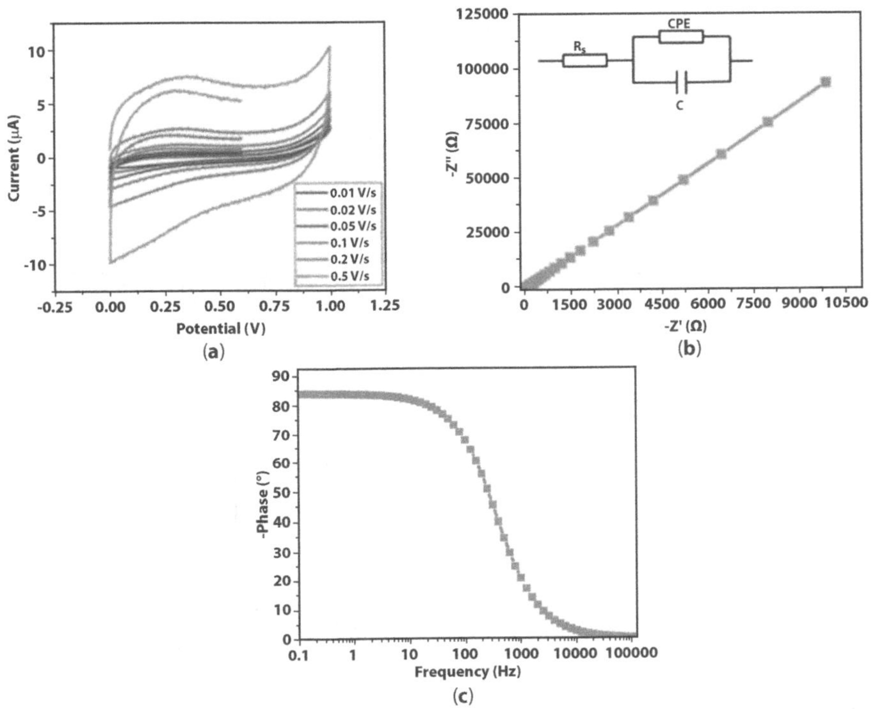

Figure 11.2 (a) Electrochemical analysis of CV at various scan rates under dark-operating condition. (b) Nyquist plot under dark-operating condition. (c) Bode plot under dark-operating condition [14].

The formula for the gravimetric capacitance and areal capacitance obtained from the CV-plot calculation can be expressed in Equation 11.5. From Equation 11.4, Π is the mass of the photocapacitive electrode for gravimetric capacitance and surface area for areal capacitance [14].

$$C_s = \frac{C}{\Pi} \tag{11.5}$$

The calculations to obtain the photo-supercapacitors' estimated energy density (E_D) and power density (P_D) are given in Equations 11.6 and 11.7, respectively [111, 112].

$$E_D = \frac{C(\Delta V)^2}{7200\Pi} \tag{11.6}$$

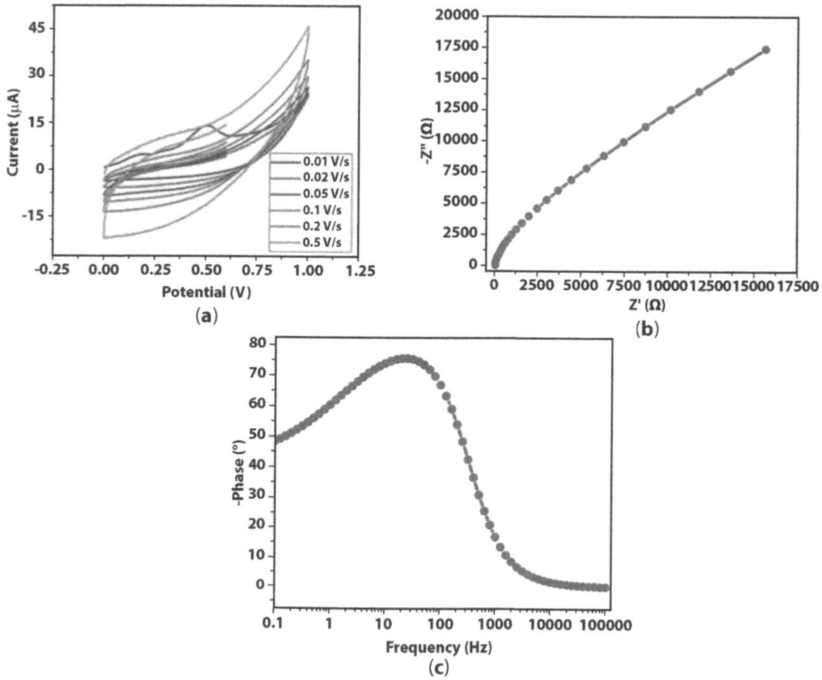

Figure 11.3 (a) CV curves at different scan rates for photo-supercapacitor. (b) Nyquist plot under illuminating-operation condition. (c) Bode plot under illuminated-operation condition [14].

$$P_D = \frac{V_0^2}{4(ESR)\Pi} \tag{11.7}$$

For dark-environment investigation, the areal capacitance at 0.01 *V/s* scan rate is 0.28 mF/cm^2. The corresponding specific capacitance is 0.35 *F/g* that can be obtained at a 0.01 *V/s* scan rate. There is a significant result of the light-induced SC performance while light condition was applied. Under light-working condition, the specific capacitance and the areal capacitance values, taken at a 0.01 *V/s* scan rate, are ~4.96 *F/g*~ and 3.96 mF/cm^2, respectively [14].

Figure 11.2b shows the EIS analysis result for the methylammonium bismuth triiodide-based photo-supercapacitor under non-illumination environment. The inset shows the circuit that corresponds with the Nyquist plot. The semicircle is absence on the Nyquist plot at large frequency. It suggests that there is little any charge transfer at the device's electrolyte–electrode contact [114, 115]. The photo-supercapacitor's charge storage mechanism exhibits ascending capacitive behavior, as seen by the straight line of the Nyquist plot. The combination between the electrolyte ionic resistance, electrode intrinsic resistance, and resistance at the interface between the electrode and FTO glass as the current collector gives contribution to equivalent series resistance (ESR) [113, 114, 116]. The comparatively low ESR value of 54.92 in this study was attained as a result of the electrode's wide surface area that needed to be supplied with electrolyte [117].

The photo-induced SC's bode plot under dark-operating condition is shown in Figure 11.2c to reveal how the phase angle depends on the frequency. Capacitive behavior is indicated by the phase angle at 90Y, while diffusion mechanism is indicated by the phase angle at 45Y [104]. In this study, the phase angle at 78.35Y was obtained. Both capacitive behavior and a diffusion mechanism are indicated by the value that lies between 45Y and 90Y.

The Nyquist plot for illuminated photo-supercapacitor is shown in Figure 11.3b. At high frequency, there is no semicircle present in the plot and non-linear behavior appeared. Capacitive-controlled behavior can be more favorable than diffusion-controlled behavior, and it is indicated by the straight line at a low frequency range. Under light environment, the ESR value of the photo-supercapacitor obtained from the Nyquist plot is 42.83 Ω. Lowered intrinsic electrodes' resistance makes the ESR value under light-operating condition smaller than that of the condition under dark environment. Photocurrent from the electron–hole pairs within the electrodes due to light-induced mechanism gives contribution to this reduced resistance.

The Bode plot for MBI-based photo-supercapacitor that worked under light environment is depicted in Figure 11.3c. The obtained phase angle is 48.19Y. The mechanism of light-induced electron–hole pairs production gives more contribution rather than the capacitive controlled charge storage mechanism since the phase angle is largely deviated from 90Y [14].

Cu_2SbI_6-based rudorffite material has been used as hybrid energy storage and energy-harvesting device [11]. Polymer-gel electrolyte was utilized in this study. The device was denoted as CHP photo-responsive electrochemical capacitor (EC). Shown in Figure 11.4a are CV curves for the device when the dark-operating condition was applied. From the CV analysis, it shows that diffusion-controlled mechanism dominates within the EC as the redox peaks appeared. Migration between cations and anions within the electrode gives contribution to the pseudocapacitive behavior. From the CV study, the scan rate affects the CV current.

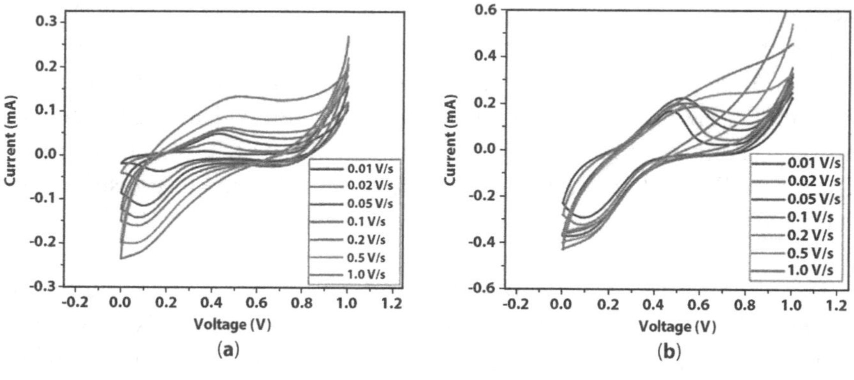

Figure 11.4 The CHP-based photoelectrode EC's CV plot at various scan rates: (a) with non-illumination and (b) under illumination [11].

It shows the rate capability of the EC, which has photo-responsive property. The CV current will grow as the scan rate does, which means that the EC has high rate capability. The stability and reversibility of the device can be considered from the alternation level of the CV curve shape. If there are small changes of the shape as the scan rate increases, it indicates good stability and reversibility. Under the dark-operating condition, the maximum CV current that can be obtained is -0.23 mA at the 1 *V/s* scan rate [11].

For the device that worked under light-operating condition at 100 mW/cm^2, the CV current increased as shown in Figure 11.4b. It indicates that light-induced electron–hole pairs are present within the device. The Cu_2SbI_6-based electrodes produce those electron–hole pairs. Under illumination, the maximum CV current was -0.43 mA [11].

Comparison of the CV curves under dark- and illuminated-operating condition with the scan rate variation can be shown in Figure 11.5. The augmentation is obtained from the CV curve *via* photo-induced electron–hole pairs. As shown in Figure 11.5a–d, at lower scan rates, the augmentation presents more under light-operating condition compared to dark-operating condition. At a higher scan rate, the augmentation presents less. The recombination onset of electron–hole pairs will increase as the augmentation reduces. The decline of augmentation occurs when the scan rate rises. The recombination leads to less production of the light-induced current. The maximum quantity of stored charges is caused by recombination at the electrode–electrolyte interface and in the active electrode material. Due to the elevation of the scan rate, the light-induced electron–hole pairs recombination will be accelerated [11].

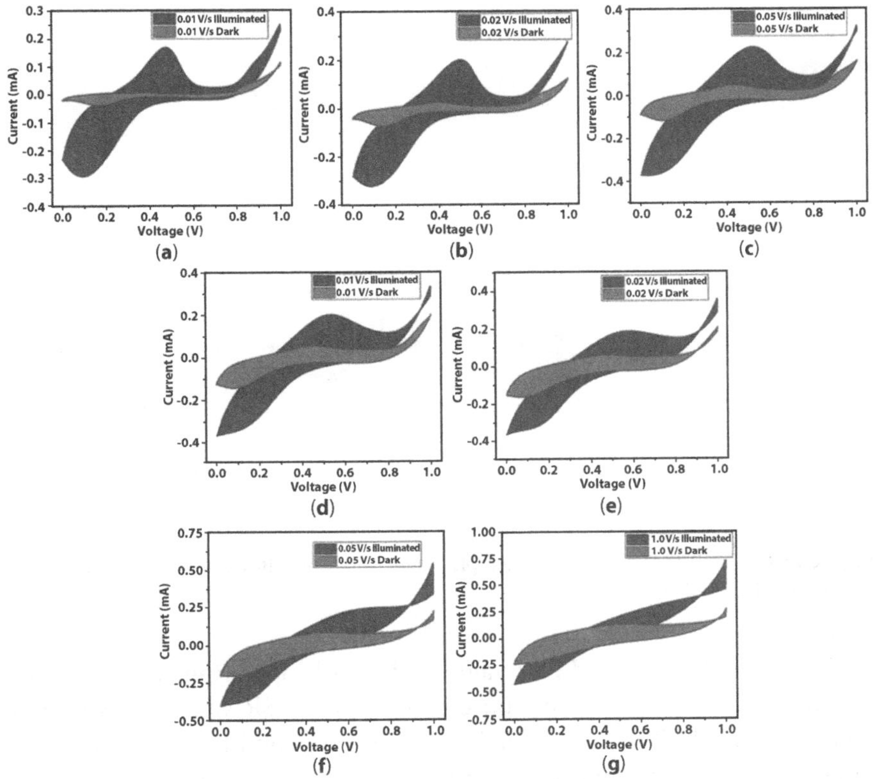

Figure 11.5 Comparisons of the CV plot at various scan rates, including of (a) 0.001 V/s scan rate; (b) 0.02 V/s scan rate; (c) 0.05 V/s scan rate; (d) 0.01 V/s scan rate; (e) 0.02 V/s scan rate; (f) 0.05 V/s scan rate; and g) 1 V/s scan rate [11].

The Nyquist plot for the light-induced EC is represented in Figure 11.6 under dark- and light-operating condition. The applied perturbation potential is 10 mV, while the direct current value was adjusted to 0.5 V. There is no significant difference between those two conditions. A vertex is exhibited in the lowest frequency range, followed by a wide curve at the moderate frequency range and a small curve at the high frequency range. The intercalation/de-intercalation redox process of the EC is dominant, as seen by the Nyquist plot form. Under dark and illuminated treatment, tremendous charge-transfer resistances are indicated by the small curve. ESR value is influenced by the interfacial resistance between the electrode, the internal resistance of the electrode and the counter electrode, the ionic resistance of the gel electrolyte, and FTO glass [118–120], and by measuring the real impedance axis intercept, one may determine the ESR value.

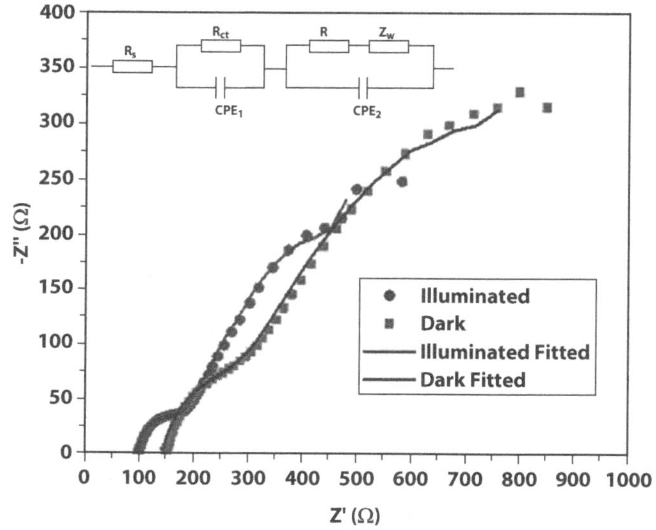

Figure 11.6 Nyquist plot for CHP-based photoelectrode capacitor. Shown in the inset is equivalent circuit diagram [11].

In this work, the obtained ESR value under dark-operating condition was 149 Ω, while under illumination condition, the ESR value of 98 Ω was obtained. The ESR value under light-operating condition is lower than under the dark condition due to the presence of light-induced excitons (electron–hole pairs). The R_{ct} value for the device under dark condition is 190 Ω; however, the R_{ct} value under illumination is 134 Ω. The lower R_{ct} value for illuminated photo-responsive EC gives evidence that illuminated device has better charge transfer. CPE_1 (constant phase element) values for the device are 0.02 μF with n = 0.53 and 0.01 μF with n=0.84 under non-illumination and illumination condition, respectively. The CPE_2 value under non-illumination and illumination condition is 36.9 μF with n = 0.89, while for device under illumination condition, the CPE_2 value is 22.3 μF with n=0.57, obtained from the Nyquist plot. Warburg diffusion impedance (Z_w) can be obtained from the EIS analysis. The obtained values are 0.05 Ω and 0.04 Ω under dark- and illuminated-operating condition, respectively. Lower Z_w under illuminated-operating condition confirms the photo-induced effect within the active electrode. Under dark-operating environment, the leakage resistance is 226 Ω, while for the illuminated device, the leakage resistance of 215 Ω was obtained [11].

Lead-free perovskite $Cu_3Bi_2I_9$ has also been used for photocapacitive electrodes [121]. CV curves under non-illumination condition at the various scan rates are shown in Figure 11.7a. Under non-illumination condition,

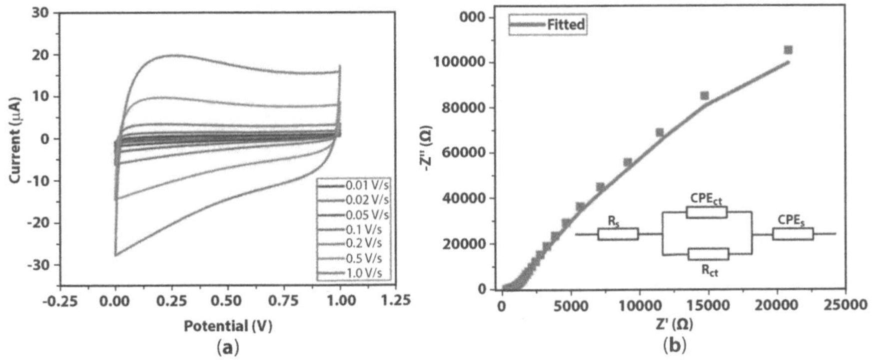

Figure 11.7 (a) CV electrochemical analysis for $Cu_3Bi_2I_9$ perovskite photo-supercapacitor at different rates. (b) Nyquist plot with its equivalent circuit, under dark-working condition [121].

the CV curves show the shape of a partial rectangle. The form suggests that the SC has capacitive and diffusion mechanism [122]. Tremendous stability and high rate capability of the photocapacitive electrode with copper-based perovskite are shown through no significant change in the CV curves as the scan rate increases. Recorded at a scan rate of 1.0 *V/s*, the value of −27.7 *μA* CV current was obtained. The specific capacitance was 2.73 *mF/g* at a scan rate 0.01 *V/s*. At a 1 *V/s* scan rate, the specific capacitance of 158 *mF/g* can be achieved. There was sufficient time for the interaction between Cu-based perovskite photocapacitive electrode and the gel electrolyte of H_3PO_4/polyvinylpyrrolidone/acetonitrile. Shown in Figure 11.7b is the Nyquist plot for the device without illumination. The plot exhibits a small curve at high frequency, and no semicircle appeared at low frequency, which indicates there were diffusion and capacitive mechanism. The small arc that appeared means there is a reasonable charge transfer of the device [123, 124]. Shown in the inset of Figure 11.7b is the corresponding circuit of the Nyquist plot. The R_s value of 264 Ω can be obtained for the photo-supercapacitor. The R_{ct} value of 1004 Ω was also attained in this work. Double-layer capacitance of 3.78 *mF* was obtained from the photocapacitive constant phase element (CPE_{CT}). The CPE_s value of 1 *mF* is a dominant capacitive characteristic of the photo-supercapacitor [121].

The *b*-values obtained from the slope in Equation 11.1b are 1.0 with potential of 0.4 *V*, 0.93 with potential of 0.6 *V*, 0.83 with potential of 0.8 *V*, and 0.76 with potential of 1.0 *V*. These b-values represent the dominance mechanism of capacitive-controlled. The values of c_1 and c_2 can be obtained from the plot in Figure 11.8a by using Equation 11.2b. Figure 11.5c depicts the capacitive and diffusion-current contribution within Cu-based perovskite photo-supercapacitor. Diffusion and capacitive current are shown

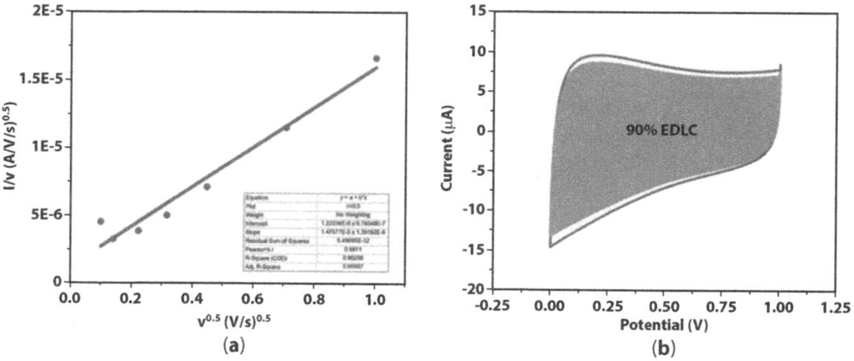

Figure 11.8 (a) $i/v^{0.5}$ vs. $v^{0.5}$ graph at 0.8 v. (b) The role of capacitive-controlled mechanism to the $Cu_3Bi_2I_9$ photo-supercapacitor, recorded at 0.5 *V/s* [121].

by shaded and unblocked area of the graph, respectively. While running the electrochemical analysis at a 0.5 *V/s* scan rate, the capacitive current can contribute 90% with Cu-based perovskite photo-supercapacitor. At entirely scan rates, capacitive mechanism gives more contribution. An increment of capacitive current appeared, and the diffusion current decreases as the scan rate increases [121].

Shown in Figure 11.9a are CV curves under illumination of 100 mW/cm^2 Oriel lamp solar simulator intensity at various scan rates under illumination condition. Partial rectangular shapes are shown by the CV curves during this condition. As the scan rate goes up, CV current rises. At a 1.0 *V/s* scan rate, a CV current peak is found at −26.6 μA. The optoionic production occurred within the Cu-perovskite-based photocapacitive electrode. The positrode and negatrode can produce excitons. The applied bias separates these light-induced excitons to prevent the start of recombination. Then, charge accumulation will occur at the electrodes.

Under light-working operation, the maximum specific capacitance is 621 *mF/g* taken at the 0.01 *V/s* scan rate. The specific capacitance is higher if it is compared to the device under non-illumination condition. It indicates the contribution of optoionic that gives better photo-supercapacitor performance.

The Nyquist plot for the device under illumination is depicted in Figure 11.9b. The curve appeared at low frequency, while at high frequency, the behavior is not completely linear. Diffusion-controlled mechanism is evidenced by this non-linear shape. The value of R_s, R_{CT}, CPE_{CT}, and CPE_s are 101 Ω, 436 Ω, 9.90 *mF*, and 2.29 *mF*, respectively [121].

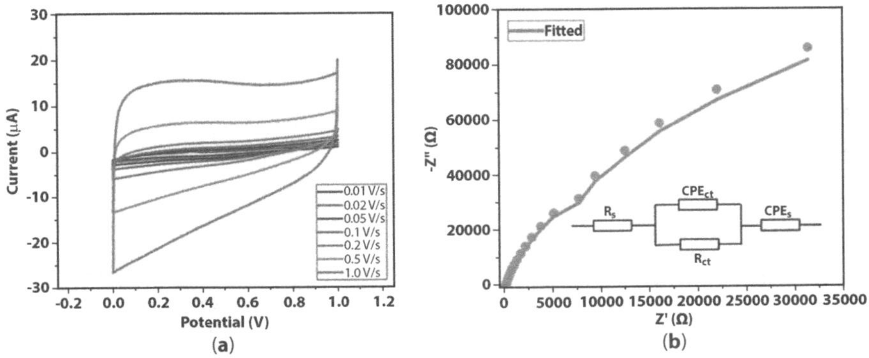

Figure 11.9 (a) CV curves for $Cu_3Bi_2I_9$-based photo-supercapacitor at different rates under light-working condition. (b) Nyquist plot for the device under light-working condition [121].

Diffusion- and capacitive-controlled mechanism under light-operating environment can be correlated with Power's law. The logarithmic plot that corresponds with Equation 11.1b can be shown through Figure 11.10a. The *b*-value that can be obtained from the plot are 1.0 with potential of 0.4 *V*, 0.94 with potential of 0.6 *V*, 0.69 with potential of 0.8 *V*, and 0.61 with potential of 1.0 *V*. The diffusion-controlled mechanism can be indicated by the *b*-value of 0.69 and 0.61. Diffusion-controlled behavior results from CV current's optoionic role. The distribution of the diffusion and capacitive roles can be associated with Dunn's equation by plotting $i/v^{0.5}$ vs. $v^{0.5}$ with v as the scan rate. The plot can be shown in Figure 11.10b. At 0.5 *V/s* scan rate, the capacitive current contributes 67% for the light-induced Cu-based perovskite SC, as illustrated in Figure 11.10c (shaded area). Under illumination, the diffusion mechanism role is found at 33% (unshaded area) due to the optoionic role from the electrodes. In this work, the energy density of Cu-based perovskite light-induced SC obtained at 0.01 *V/s* scan rate is ~86.4 mWh/kg under light-working environment and ~38.0 mWh/kg under dark-working condition [121].

Excellent stability of the device was evidenced after 10,000 charge–discharge cycles that the device was able to maintain 93.8% of the capacitance The Nyquist plot after 10,000 charge–discharge cycles is shown in Figure 11.10d, which is similar to that of Figure 11.7b [121].

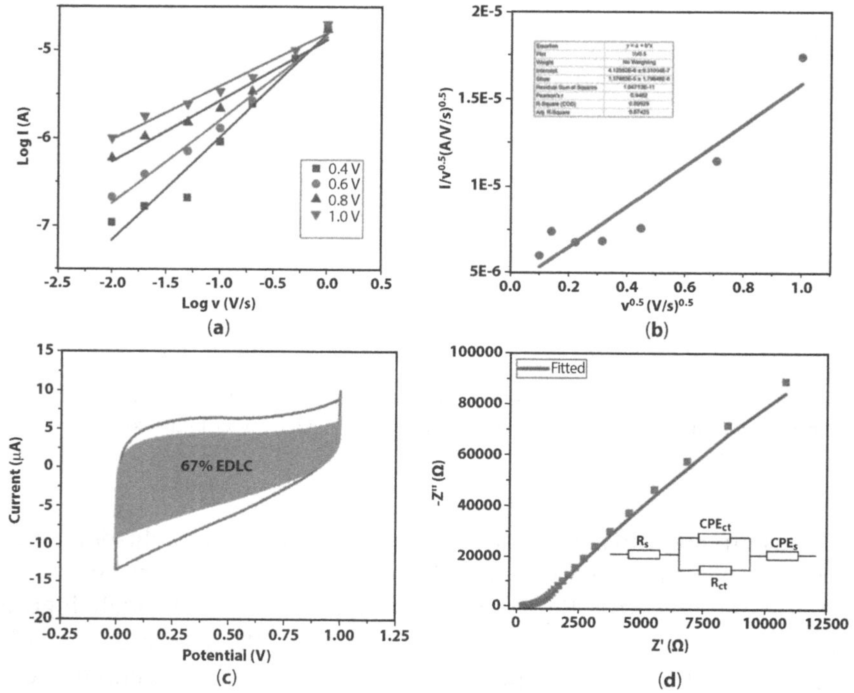

Figure 11.10 (a) Power's law logarithmic plot for $Cu_3Bi_2I_9$ perovskite photo-supercapacitor with illumination. (b) $i/v^{0.5}$ vs. $v^{0.5}$ plot under light-working condition. (c) Ratio of capacitive and diffusion mechanism for the device. (d) Nyquist plot after 10,000 CV cycles [121].

11.5 Conclusions

Photo-supercapacitors have tremendous advantages due to their ability combining the light-harvesting part and the capacitor part to store the energy. Thus, these self-charging devices are crucial to be one part in renewable energy sustainability. The need of long lifetime use makes them suitable to be employed in a real environment as they have a massive number of lifetime cycles. The fluctuation of input power with the changes of weather condition can be resolved by the typical feature of SC due to their high-power density. Another advantage of the SC is due to their safety in relatively extreme condition compared to other energy storage devices.

Perovskite materials, especially halide perovskites, have been proven in solar cell applications as they attain high conversion efficiency. Easy for tuning the bandgap is one of the essential properties for light-harvesting devices. In some previous studies, methylammonium bismuth triiodide and $Cu_3Bi_2I_9$ are among the perovskite materials that succeed in photo-supercapacitor applications. It was evidenced from their behavior in dark- and light-operating conditions. In brief, the working mechanism can be divided into capacitive-controlled and diffusion-controlled mechanism. From Power's law, the *b*-value can be obtained, which further indicates whether the mechanism is capacitive- or diffusion-controlled. A *b*-value of 0.5 indicates diffusion-controlled behavior, while a *b*-value of 1.0 indicates capacitive-controlled behavior. In addition, *b*-value between 0.5 and 1.0 represent hybrid mechanism. In case of *b*-value below 0.5, they correspond to having more ejections of electron–hole pairs production.

Acknowledgment

We thank the Deanship of Research Oversight and Coordination through the Renewable Energy Technical Incubator (RETI) by the Interdisciplinary Research Center for Renewable Energy and Power Systems, King Fahd University of Petroleum and Minerals under project No.#CREP2522 funded by the National Industrial Development and Logistics Program (NIDLP).

References

1. Lu, Y., Chen, M., Zhu, G., Zhang, Y., Recent progress in the study of integrated solar cell-energy storage systems. *Nanoscale*, 16, 8778–8790, 2024.
2. Dong, L., Xu, C., Li, Y., Huang, Z.-H., Kang, F., Yang, Q.-H., Zhao, X., Flexible electrodes and supercapacitors for wearable energy storage: a review by category. *J. Mater. Chem. A*, 4, 4659–85, 2016.
3. Kanaujiya, N., Kumar, N., Singh, M., Sharma, Y., Varma, G.D., CoMn2O4 nanoparticles decorated on 2D MoS2 frame: a synergetic energy storage composite material for practical supercapacitor applications. *J. Energy Storage*, 35, 102302, 2021.
4. Boruah, B.D. and De Volder, M., Vanadium dioxide–zinc oxide stacked photocathodes for photo-rechargeable zinc-ion batteries. *J. Mater. Chem. A*, 9, 23199–205, 2021.
5. Kumar, R., Kumar, A., Shukla, P.S., Das, V.G., Venkataraman, D., Bag, M., Photorechargeable Hybrid Halide Perovskite Supercapacitors. *ACS Appl. Mater. Interfaces*, 14, 35592–9, 2022.
6. Grätzel, M., Photoelectrochemical cells. *Nature*, 414, 338–44, 2001.
7. Miyasaka, T. and Murakami, T.N., The photocapacitor: An efficient self-charging capacitor for direct storage of solar energy. *Appl. Phys. Lett.*, 85, 3932–4, 2004.

8. Xu, M., Wang, W., Zhong, Y., Xu, X., Wang, J., Zhou, W., Shao, Z., Enhancing the triiodide reduction activity of a perovskite-based electrocatalyst for dye-sensitized solar cells through exsolved silver nanoparticles. *J. Mater. Chem. A*, 7, 17489–97, 2019.
9. Balaji, G., Joshi, P.H., Abbas, H.A., Zhang, L., Kottokkaran, R., Samiee, M., Noack, M., Dalal, V.L., CH 3 NH 3 PbI 3 from non-iodide lead salts for perovskite solar cells *via* the formation of PbI 2. *Phys. Chem. Chem. Phys.*, 17, 10369–72, 2015.
10. Popoola, A., Gondal, M.A., Popoola, I.K., Oloore, L.E., Bakr, O.M., Fabrication of bifacial sandwiched heterojunction photoconductor–type and MAI passivated photodiode–type perovskite photodetectors. *Org. Electron.*, 84, 105730, 2020.
11. Popoola, I.K., Gondal, M.A., Oloore, L.E., Popoola, A., Inorganic antimony-based rudorffite photo-responsive electrochemical capacitor utilizing non-aqueous polyvinylpyrrolidone polymer gel electrolyte for hybrid energy harvesting and storage applications. *Mater. Sci. Eng. B*, 291, 116373, 2023.
12. Tathavadekar, M., Krishnamurthy, S., Banerjee, A., Nagane, S., Gawli, Y., Suryawanshi, A., Bhat, S., Puthusseri, D., Mohite, A.D., Ogale, S., Low-dimensional hybrid perovskites as high performance anodes for alkali-ion batteries. *J. Mater. Chem. A*, 5, 18634–42, 2017.
13. Vicente, N. and Garcia-Belmonte, G., Methylammonium lead bromide perovskite battery anodes reversibly host high Li-ion concentrations. *J. Phys. Chem. Lett.*, 8, 1371–4, 2017.
14. Popoola, I.K., Gondal, M.A., Popoola, A., Oloore, L.E., Bismuth-based organometallic-halide perovskite photo-supercapacitor utilizing novel polymer gel electrolyte for hybrid energy harvesting and storage applications. *J. Energy Storage*, 53, 105167, 2022.
15. Zhang, W., Eperon, G.E., Snaith, H.J., Metal halide perovskites for energy applications. *Nat. Energy*, 1, 1–8, 20162016.
16. Popoola, A., Oloore, L.E., Popoola, I.K., A Demystified Guide on Fabrication and Characterization of Perovskites-Based Optoelectronic Devices. *Adv. Photonics Res.*, 3, 2100122, 2022.
17. Stoddart, A., Purely organic perovskites. *Nat. Rev. Mater.*, 3, 226, 2018.
18. Zhu, J., Li, H., Zhong, L., Xiao, P., Xu, X., Yang, X., Zhao, Z., Li, J., Perovskite oxides: Preparation, characterizations, and applications in heterogeneous catalysis. *ACS Catal.*, 4, 2917–40, 2014.
19. Ha, S.T., Su, R., Xing, J., Zhang, Q., Xiong, Q., Metal halide perovskite nanomaterials: synthesis and applications. *Chem. Sci.*, 8, 2522–36, 2017.
20. Bystrický, R., Tiwari, S.K., Hutár, P., Vančo, L., Sýkora, M., Synthesis of Sulfide Perovskites by Sulfurization with Boron Sulfides. *Inorg. Chem.*, 61, 18823–7, 2022.
21. Geng, S. and Xiao, Z., Can Nitride Perovskites Provide the Same Superior Optoelectronic Properties as Lead Halide Perovskites? *ACS Energy Lett.*, 8, 2051–7, 2023.

22. Schouwink, P., Ley, M.B., Tissot, A., Hagemann, H., Jensen, T.R., Smrčok, L., Černý, R., Structure and properties of complex hydride perovskite materials. *Nat. Comm.*, 5, 1–10, 2014.
23. Schouwink, P., D'Anna, V., Ley, M.B., Lawson Daku, L.M., Richter, B., Jensen, T.R., Hagemann, H., Černý, R., Bimetallic Borohydrides in the System M(BH4)2–KBH4 (M = Mg, Mn): On the Structural Diversity. *J. Phys. Chem. C*, 116, 10829–40, 2012.
24. Peña, M.A. and Fierro, J.L.G., Chemical structures and performance of perovskite oxides. *Chem. Rev.*, 101, 1981–2017, 2001.
25. Wang, Y., Ren, J., Wang, Y., Zhang, F., Liu, X., Guo, Y., Lu, G., Nanocasted synthesis of mesoporous LaCoO3 perovskite with extremely high surface area and excellent activity in methane combustion. *J. Phys. Chem. C*, 112, 15293–8, 2008.
26. Sadakane, M., Asanuma, T., Kubo, J., Ueda, W., Facile procedure to prepare three-dimensionally ordered macroporous (3DOM) perovskite-type mixed metal oxides by colloidal crystal templating method. *Chem. Mater.*, 17, 3546–51, 2005.
27. Comparotto, C., Ström, P., Donzel-Gargand, O., Kubart, T., Scragg, J.J.S., Synthesis of BaZrS3 Perovskite Thin Films at a Moderate Temperature on Conductive Substrates. *ACS Appl. Energy Mater.*, 5, 5, 6335–43, 2022 Apr 22.
28. Pilania, G., Ghosh, A., Hartman, S.T., Mishra, R., Stanek, C.R., Uberuaga, B.P., Anion order in oxysulfide perovskites: origins and implications. *npj Comput. Mater.*, 6, 1, 1–11, 2020.
29. Peng, H., Sun, W., Li, Y., Yan, W., Yu, P., Zhou, H., Bian, Z., Huang, C., High-performance cadmium sulphide-based planar perovskite solar cell and the cadmium sulphide/perovskite interfaces. *J. Photonics Energy*, 6, 022002, 2016.
30. Fu, L., Nie, Y., Li, B., Li, N., Cao, B., Yin, L., Bismuth Telluride Interlayer for All-Inorganic Perovskite Solar Cells with Enhanced Efficiency and Stability. *Sol. RRL*, 3, 1900233, 2019.
31. Sarmiento-Pérez, R., Cerqueira, T.F.T., Körbel, S., Botti, S., Marques, M.A.L., Prediction of Stable Nitride Perovskites. *Chem. Mater.*, 27, 5957–63, 2015.
32. Zhang, H., Wu, X., Ding, K., Xie, L., Yang, K., Ming, C., Bai, S., Zeng, H., Zhang, S., Sun, Y.Y., Prediction and Synthesis of a Selenide Perovskite for Optoelectronics. *Chem. Mater.*, 35, 4135, 2022.
33. Li, H.W., Yan, Y., Orimo, S.I., Züttel, A., Jensen, C.M., Recent progress in metal borohydrides for hydrogen storage. *Energies*, 4, 185–214, 2011.
34. He, C. and Liu, X., The rise of halide perovskite semiconductors. *Light Sci. Appl.*, 12, 1–2, 2023.
35. Huang, Y.T., Kavanagh, S.R., Scanlon, D.O., Walsh, A., Hoye, R.L.Z., Perovskite-inspired materials for photovoltaics and beyond-from design to devices. *Nanotechnology*, 32, 13, 132004, 2021.
36. Liu, X.K., Xu, W., Bai, S., Jin, Y., Wang, J., Friend, R.H., Gao, F., Metal halide perovskites for light-emitting diodes. *Nat. Mater.*, *20*, 1, 10–21, 2020.

37. Seok, S., II and Guo, T.F., Halide perovskite materials and devices. *MRS Bull.*, 45, 427–30, 2020.
38. Li, K., Dong, L.Y., Xu, H.X., Qin, Y., Li, Z.G., Azeem, M., Li, W., Bu, X.H., Electronic structures and elastic properties of a family of metal-free perovskites. *Mater. Chem. Front.*, 3, 1678–85, 2019.
39. Li, W., Zhang, C., Ma, Y., Liu, C., Fan, J., Mai, Y., Schropp, R.E., II, *In situ* induced core/shell stabilized hybrid perovskites *via* gallium(III) acetylacetonate intermediate towards highly efficient and stable solar cells. *Energy Environ. Sci.*, 11, 286–93, 2018.
40. Wang, X., Tang, R., Wu, C., Zhu, C., Chen, T., Development of antimony sulfide–selenide Sb2(S, Se)3-based solar cells. *J. Energy Chem.*, 27, 713–21, 2018.
41. Zhang, Z., Zhu, Y., Zhong, Y., Zhou, W., Shao, Z., Anion Doping: A New Strategy for Developing High-Performance Perovskite-Type Cathode Materials of Solid Oxide Fuel Cells. *Adv. Energy Mater.*, 7, 1700242, 2017.
42. Zhang, P., Yang, J., Wei, S.H., Manipulation of cation combinations and configurations of halide double perovskites for solar cell absorbers. *J. Mater. Chem. A*, 6, 1809–15, 2018.
43. Sahu, M., Hajra, S., Bijelic, J., Oh, D., Djerdj, I., Kim, H.J., Triple perovskite-based triboelectric nanogenerator: a facile method of energy harvesting and self-powered information generator. *Mater. Today Energy*, 20, 100639, 2021.
44. Kim, N., II, Sa, Y.J., Yoo, T.S., Choi, S.R., Afzal, R.A., Choi, T., Seo, Y.S., Lee, K.S., Hwang, J.Y., Choi, W.S., Joo, S.H., Park, J.Y., Oxygen-deficient triple perovskites as highly active and durable bifunctional electrocatalysts for oxygen electrode reactions. *Sci. Adv.*, 4, eapp9360, 2018.
45. Reyna, Y., Salado, M., Kazim, S., Pérez-Tomas, A., Ahmad, S., Lira-Cantu, M., Performance and stability of mixed FAPbI3(0.85)MAPbBr3(0.15) halide perovskite solar cells under outdoor conditions and the effect of low light irradiation. *Nano Energy*, 30, 570–9, 2016.
46. Yang, H., Cai, T., Dube, L., Chen, O., Synthesis of double perovskite and quadruple perovskite nanocrystals through post-synthetic transformation reactions. *Chem. Sci.*, 13, 4874–83, 2022.
47. Patil, J.V., Mali, S.S., Hong, C.K., A thiourea additive-based quadruple cation lead halide perovskite with an ultra-large grain size for efficient perovskite solar cells. *Nanoscale*, 11, 21824–33, 2019.
48. Susic, I., Gil-Escrig, L., Palazon, F., Sessolo, M., Bolink, H.J., Quadruple-Cation Wide-Bandgap Perovskite Solar Cells with Enhanced Thermal Stability Enabled by Vacuum Deposition. *ACS Energy Lett.*, 7, 1355–63, 2022.
49. Zhang, W., Eperon, G.E., Snaith, H.J., Metal Halide Perovskites for Energy Applications. *Nat. Energy*, 1, 1–8, 2016.
50. Zhou, Y. and Saliba, M., Zooming in on Metal Halide Perovskites: New Energy Frontiers Emerge. *ACS Energy Lett.*, 6, 2750–4, 2021.

51. Hao, J. and Xiao, X., Recent Development of Optoelectronic Application Based on Metal Halide Perovskite Nanocrystals. *Front. Chem.*, 9, 822106, 2022.
52. Dong, H., Ran, C., Gao, W., Li, M., Xia, Y., Huang, W., Metal Halide Perovskite for next-generation optoelectronics: progresses and prospects. *eLight*, *3*, 1, 1–16, 2023.
53. Mohapatra, A., Kar, M.R., Bhaumik, S., Recent Progress and Prospects on Metal Halide Perovskite Nanocrystals as Color Converters in the Fabrication of White Light-Emitting Diodes. *Front. Electron. Mater.*, 2, 891983, 2022.
54. Manser, J.S., Christians, J.A., Kamat, P.V., Intriguing optoelectronic properties of metal halide perovskites. *Chem. Rev.*, 116, 12956–13008, 2016.
55. Duan, D., Ge, C., Rahaman, M.Z., Lin, C.H., Shi, Y., Lin, H., Hu, H., Wu, T., Recent progress with one-dimensional metal halide perovskites: from rational synthesis to optoelectronic applications. *NPG Asia Mater.*, *15*, 1, 1–22, 2023.
56. Meng, Q., Chen, Y., Xiao, Y.Y., Sun, J., Zhang, X., Han, C.B., Gao, H., Zhang, Y., Yan, H., Effect of temperature on the performance of perovskite solar cells. *J. Mater. Sci.: Mater. Electron.*, 32, 12784–92, 2021.
57. Reddy, S.H., Di Giacomo, F., Di Carlo, A., Low-Temperature-Processed Stable Perovskite Solar Cells and Modules: A Comprehensive Review. *Adv. Energy Mater.*, 12, 2103534, 2022.
58. Gupta, R., Korukonda, T.B., Gupta, S.K., Dhamaniya, B.P., Chhillar, P., Datt, R., Vashishtha, P., Gupta, G., Gupta, V., Srivastava, R., Pathak, S., Room temperature synthesis of perovskite (MAPbI3) single crystal by anti-solvent assisted inverse temperature crystallization method. *J. Cryst. Growth*, 537, 125598, 2020.
59. Sunny, F., Varghese, L.M., Kalarikkal, N., Subila, K.B., Sunny, F., Varghese, L.M., Kalarikkal, N., Subila, K.B., Metal Halide Hybrid Perovskites, in: *Recent Advances in Multifunctional Perovskite Materials*, 2022.
60. Yan, J., Li, H., Aldamasy, M.H., Frasca, C., Abate, A., Zhao, K., Hu, Y., Advances in the Synthesis of Halide Perovskite Single Crystals for Optoelectronic Applications. *Chem. Mater.*, 35, 2683–712, 2022.
61. Gao, P., Bin Mohd Yusoff, A.R., Nazeeruddin, M.K., Dimensionality engineering of hybrid halide perovskite light absorbers. *Nat. Comm.*, *9*, 1, 1–14, 2018.
62. Wang, C., Ding, Y., Liu, B., Weng, B., Hofkens, J., Roeffaers, M.B.J., Crystal structure engineering of metal halide perovskites for photocatalytic organic synthesis. *Chem. Comm.*, 59, 3122–5, 2023.
63. Dipta, S.S. and Uddin, A., Stability Issues of Perovskite Solar Cells: A Critical Review. *Energy Tech.*, 9, 2100560, 2021.
64. Zhou, Y. and Zhao, Y., Chemical stability and instability of inorganic halide perovskites. *Energy Environ. Sci.*, 12, 1495–511, 2019.

65. Ou, Q., Bao, X., Zhang, Y., Shao, H., Xing, G., Li, X., Shao, L., Bao, Q., Band structure engineering in metal halide perovskite nanostructures for optoelectronic applications. *Nano Mater. Sci.*, 1, 268–87, 2019.
66. Bourelle, S.A., Camargo, F.V.A., Ghosh, S., Neumann, T., van de Goor, T.W.J., Shivanna, R., Winkler, T., Cerullo, G., Deschler, F., Optical control of exciton spin dynamics in layered metal halide perovskites *via* polaronic state formation. *Nat. Comm.*, *13*, 1, 1–8, 2022.
67. Protesescu, L., Yakunin, S., Bodnarchuk, M., II, Krieg, F., Caputo, R., Hendon, C.H., Yang, R.X., Walsh A and Kovalenko, M.V., Nanocrystals of Cesium Lead Halide Perovskites (CsPbX3, X = Cl, Br, and I): Novel Optoelectronic Materials Showing Bright Emission with Wide Color Gamut. *Nano Lett.*, 15, 3692–6, 2015.
68. Wang, F., Bai, S., Tress, W., Hagfeldt, A., Gao, F., Defects engineering for high-performance perovskite solar cells. *npj Flexible Electron.*, *2*, 1, 1–14, 2018.
69. De Wolf, S., Holovsky, J., Moon, S.J., Löper, P., Niesen, B., Ledinsky, M., Haug, F.J., Yum, J.H., Ballif, C., Organometallic halide perovskites: Sharp optical absorption edge and its relation to photovoltaic performance. *J. Phy. Chem. Lett.*, 5, 1035–9, 2014.
70. Alqahtani, S.M., Alsayoud, A.Q., Alharbi, F.H., Structures, band gaps, and formation energies of highly stable phases of inorganic ABX3 halides: A = Li, Na, K, Rb, Cs, Tl; B = Be, Mg, Ca, Ge C.OMMAS.R.X.X.X, Sn, Pb; and X = F, Cl, Br, I. *RSC Adv.*, 13, 9026–32, 2023.
71. Zhou, Y., Yang, M., Game, O.S., Wu, W., Kwun, J., Strauss, M.A., Yan, Y., Huang, J., Zhu, K., Padture, N.P., Manipulating Crystallization of Organolead Mixed-Halide Thin Films in Antisolvent Baths for Wide-Bandgap Perovskite Solar Cells. *ACS Appl. Mater. Interfaces*, 8, 2232–2237, 2016.
72. Tang, M.C., Dang, H.X., Lee, S., Barrit, D., Munir, R., Wang, K., Li, R., Smilgies, D.M., De Wolf, S., Kim, D.Y., Anthopoulos, T.D., Amassian, A., Wide and Tunable Bandgap MAPbBr3–xClx Hybrid Perovskites with Enhanced Phase Stability: *In Situ* Investigation and Photovoltaic Devices. *Sol. RRL*, 5, 2000718, 2021.
73. Senno, M. and Tinte, S., Mixed formamidinium-methylammonium lead iodide perovskite from first-principles: hydrogen-bonding impact on the electronic properties. *Phy. Chem. Chem. Phy.*, 23, 7376–85, 2021.
74. Liu, D., Li, Q., Wu, K., Ethylammonium as an alternative cation for efficient perovskite solar cells from first-principles calculations. *RSC Adv.*, 9, 7356–61, 2019.
75. Tang, Z.K., Xu, Z.F., Zhang, D.Y., Hu, S.X., Lau, W.M., Liu, L.M., Enhanced optical absorption *via* cation doping hybrid lead iodine perovskites. *Sci. Rep.*, 7, 7843, 2017.
76. Handley, C.M., Ward, R.E., Freeman, C.L., Reaney, I.M., Sinclair, D.C., Harding, J.H., Dynamic tilting in perovskites. *Acta. Crystallogr. A Found Adv.*, 79, 163–70, 2023.

77. Motta, C., El-Mellouhi, F., Sanvito, S., Charge carrier mobility in hybrid halide perovskites. *Sci. Rep.*, 5, 1, 1–8, 2015.
78. Liu, T., Tang, W., Luong, S., Fenwick, O., High charge carrier mobility in solution processed one-dimensional lead halide perovskite single crystals and their application as photodetectors. *Nanoscale*, 12, 9688–95, 2020.
79. Herz, L.M., Charge-Carrier Mobilities in Metal Halide Perovskites: Fundamental Mechanisms and Limits. *ACS Energy Lett.*, 2, 1539–48, 2017.
80. Zhang, Z., Kim, W., Ko, M.J., Li, Y., Perovskite single-crystal thin films: preparation, surface engineering, and application. *Nano. Converg.*, 10, 23, 2023.
81. Shi, D., Adinolfi, V., Comin, R., Yuan, M., Alarousu, E., Buin, A., Chen, Y., Hoogland, S., Rothenberger, A., Katsiev, K., Losovyj, Y., Zhang, X., Dowben, P.A., Mohammed, O.F., Sargent, E.H., Bakr, O.M., Solar cells. Low trap-state density and long carrier diffusion in organolead trihalide perovskite single crystals. *Science*, 347, 519–22, 2015.
82. Schleuning, M., Kölbach, M., Abdi, F.F., Schwarzburg, K., Stolterfoht, M., Eichberger, R., van de Krol, R., Friedrich, D., Hempel, H., Generalized Method to Extract Carrier Diffusion Length from Photoconductivity Transients: Cases of BiVO4, Halide Perovskites, and Amorphous and Crystalline Silicon. *PRX Energy*, 1, 023008, 2022.
83. Stranks, S.D., Eperon, G.E., Grancini, G., Menelaou, C., Alcocer, M.J.P., Leijtens, T., Herz, L.M., Petrozza, A., Snaith, H.J., Electron-hole diffusion lengths exceeding 1 micrometer in an organometal trihalide perovskite absorber. *Science*, 342, 341–4, 2013.
84. Peng, J., Chen, Y., Zheng, K., Pullerits, T., Liang, Z., Insights into charge carrier dynamics in organo-metal halide perovskites: from neat films to solar cells. *Chem. Soc. Rev.*, 46, 5714–29, 2017.
85. Hodes G and Kamat, P.V., Understanding the Implication of Carrier Diffusion Length in Photovoltaic Cells. *J. Phys. Chem. Lett.*, 6, 4090–2, 2015.
86. Dong, Q., Fang, Y., Shao, Y., Mulligan, P., Qiu, J., Cao, L., Huang, J., Electron-hole diffusion lengths > 175 μm in solution-grown CH3NH3PbI3 single crystals. *Sci. (1979)*, 347, 967–70, 2015.
87. Shi, D., Adinolfi, V., Comin, R., Yuan, M., Alarousu, E., Buin, A., Chen, Y., Hoogland, S., Rothenberger, A., Katsiev, K., Losovyj, Y., Zhang, X., Dowben, P.A., Mohammed, O.F., Sargent, E.H., Bakr, O.M., Low trap-state density and long carrier diffusion in organolead trihalide perovskite single crystals. *Sci. (1979)*, 347, 519–22, 2015.
88. Wilson, J.N., Frost, J.M., Wallace, S.K., Walsh, A., Dielectric and ferroic properties of metal halide perovskites. *APL Mater.*, 7, 010901, 2019.
89. Xing, G., Mathews, N., Sun, S., Lim, S.S., Lam, Y.M., Grätzel, M., Mhaisalkar, S., Sum, T.C., Long-range balanced electron-and hole-transport lengths in organic-inorganic CH3NH3PbI3. *Sci. (1979)*, 342, 344–7, 2013.

90. Hu, J., Zhao, C., He, S., Tian, W., Hao, C., Jin, S., Carrier dynamics in CsPbI3 perovskite microcrystals synthesized in solution phase. *Chin. Chem. Lett.*, 29, 699–702, 2018.
91. Li, B., Zhang, Y., Fu, L., Yu, T., Zhou, S., Zhang, L., Yin, L., Surface passivation engineering strategy to fully-inorganic cubic CsPbI3 perovskites for high-performance solar cells. *Nat. Commun.*, *9*, 1, 1–8, 2018.
92. Zhumekenov, A.A., Saidaminov, M., II, Haque, M.A., Alarousu, E., Sarmah, S.P., Murali, B., Dursun, I., Miao, X.H., Abdelhady, A.L., Wu, T., Mohammed, O.F., Bakr, O.M., Formamidinium Lead Halide Perovskite Crystals with Unprecedented Long Carrier Dynamics and Diffusion Length. *ACS Energy Lett.*, 1, 32–7, 2016.
93. Kato, S., Kurokawa, Y., Miyajima, S., Watanabe, Y., Yamada, A., Ohta, Y., Niwa, Y., Hirota, M., Improvement of carrier diffusion length in silicon nanowire arrays using atomic layer deposition. *Nanoscale Res. Lett.*, 8, 1–8, 2013.
94. Coropceanu, V., Cornil, J., da Silva Filho, D.A., Olivier, Y., Silbey, R., Brédas, J.L., Charge transport in organic semiconductors. *Chem. Rev.*, 107, 926–52, 2007.
95. Longeaud, C., Fath Allah, A., Schmidt, J., El Yaakoubi, M., Berson, S., Lemaitre, N., Determination of diffusion lengths in organic semiconductors: Correlation with solar cell performances. *Org. Electron.*, 31, 253–7, 2016.
96. Yadav, D. and Agarwal, S.C., Measurement of the diffusion length of minority carriers using a steady-state photocarrier grating. *Solid State Commun.*, 150, 321–4, 2010.
97. Ahrenkiel, R.K., Keyes, B.M., Shen, T.C., Chyi, J., II, Morkoc, H., Minority-carrier lifetime in AlxGa1-xAs grown by molecular-beam epitaxy. *J. Appl. Phys.*, 69, 3094–6, 1991.
98. McGehee, M.D. and Heeger, A.J., Semiconducting (conjugated) polymers as materials for solid-state lasers. *Adv. Mater.*, 12, 1655–1668, 2000.
99. Laquai, F., Andrienko, D., Deibel, C., Neher, D., Charge carrier generation, recombination, and extraction in polymer–fullerene bulk heterojunction organic solar cells. *Adv. Polym. Sci.*, 272, 267–91, 2017.
100. Liu, R., Liu, C., Fan, S., A photocapacitor based on organometal halide perovskite and PANI/CNT composites integrated using a CNT bridge. *J. Mater. Chem. A*, 5, 23078–84, 2017.
101. Gao, K., Ti, D., Zhang, Z., A photocapacitor with high working voltage and energy density. *Sustain. Energy Fuels*, 3, 1937–42, 2019.
102. Rani, B.J., Pradeepa, S.S., Hasan, Z.M., Ravi, G., Yuvakkumar, R., Hong, S., II, Supercapacitor and OER activity of transition metal (Mo, Co, Cu) sulphides. *J. Phys. Chem. Solids*, 138, 109240, 2020.
103. Kajana, T., Pirashanthan, A., Yuvapragasam, A., Velauthapillai, D., Ravirajan, P., Senthilnanthanan, M., Bimetallic AC/Ag2CrO4/SnS heterostructure photoanode for energy conversion and storage: A self-powered Photocapacitor. *J. Power Sources*, 520, 230883, 2022.

104. Sankar, K.V., Selvan, R.K., Meyrick, D., Electrochemical performances of CoFe 2 O 4 nanoparticles and a rGO based asymmetric supercapacitor. *RSC Adv.*, 5, 99959–67, 2015.
105. Sankar, K.V. and Selvan, R.K., The ternary MnFe2O4/graphene/polyaniline hybrid composite as negative electrode for supercapacitors. *J. Power Sources*, 275, 399–407, 2015.
106. Brezesinski, T., Wang, J., Tolbert, S.H., Dunn, B., Ordered mesoporous α-MoO3 with iso-oriented nanocrystalline walls for thin-film pseudocapacitors. *Nat. Mater.*, 9, 146–51, 2010.
107. Wang, J., Polleux, J., Lim, J., Dunn, B., Pseudocapacitive contributions to electrochemical energy storage in TiO2 (anatase) nanoparticles. *J. Phys. Chem. C*, 111, 14925–31, 2007.
108. Kunwar, R., Harilal, M., Krishnan, S.G., Pal, B., Misnon, I., II, Mariappan, C.R., Ezema, F., II, Elim, H., II, Yang, C.-C., Jose, R., Pseudocapacitive charge storage in thin nanobelts. *Adv. Fiber Mater.*, 1, 205–13, 2019.
109. Javed, M.S., Shah, S.S.A., Najam, T., Aslam, M.K., Li, J., Hussain, S., Ahmad, M.A., Ashfaq, M., Raza, R., Mai, W., Synthesis of mesoporous defective graphene-nanosheets in a space-confined self-assembled nanoreactor: Highly efficient capacitive energy storage. *Electrochim. Acta*, 305, 517–27, 2019.
110. Javed, M.S., Khan, A.J., Asim, S., Shah, S.S.A., Najam, T., Siyalg, S.H., Tahir, M.F., Zhao, Z., Mai, W., Insights to pseudocapacitive charge storage of binary metal-oxide nanobelts decorated activated carbon cloth for highly-flexible hybrid-supercapacitors. *J. Energy Storage*, 31, 101602, 2020.
111. Yan, J., Wang, Q., Wei, T., Fan, Z., Recent advances in design and fabrication of electrochemical supercapacitors with high energy densities. *Adv. Energy Mater.*, 4, 1300816, 2014.
112. Lee, K., Lee, H., Shin, Y., Yoon, Y., Kim, D., Lee, H., Highly transparent and flexible supercapacitors using graphene-graphene quantum dots chelate. *Nano Energy*, 26, 746–54, 2016.
113. Foo, C.Y., Lim, H.N., Mahdi, M.A., Wahid, M.H., Huang, N.M., Three-dimensional printed electrode and its novel applications in electronic devices. *Sci. Rep.*, 8, 7399, 2018.
114. Liu, W., Feng, Y., Yan, X., Chen, J., Xue, Q., Superior micro-supercapacitors based on graphene quantum dots. *Adv. Funct. Mater.*, 23, 4111–22, 2013.
115. Vijayakumar, V., Anothumakkool B, A.T.A.T., Nair, S.B., Badiger, M.V., Kurungot, S., An all-solid-state-supercapacitor possessing a non-aqueous gel polymer electrolyte prepared using a UV-assisted *in situ* polymerization strategy. *J. Mater. Chem. A Mater.*, 5, 8461–76, 2017.
116. Huang, F. and Chen, D., Towards the upper bound of electrochemical performance of ACNT@polyaniline arrays as supercapacitors. *Energy Environ. Sci.*, 5, 5833–5841, 2012.

117. Pious, J.K., Lekshmi, M.L., Muthu, C., Rakhi, R.B., Vijayakumar, C., Zero-dimensional methylammonium bismuth iodide-based lead-free perovskite capacitor. *ACS Omega*, 2, 5798–802, 2017.
118. Muslu, E., Eren, E., Oksuz, A.U., Research progress on flexible WO_3 based thin film electrodes for supercapacitor applications: a comprehensive review. *Emerg. Mater.*, 7, 2205–2236, 2024.
119. Wang, Y., Shi, Z., Huang, Y., Ma, Y., Wang, C., Chen, M., Chen, Y., Supercapacitor devices based on graphene materials. *J. Phys. Chem. C*, 113, 13103–7, 2009.
120. Stoller, M., Park, S., Yanwu, Z., An, J., Ruoff, R., Graphene-Based Ultracapacitors. *Nano Lett.*, 8, 3498–502, 2008.
121. Popoola, I.K., Gondal, M.A., Popoola, A., Oloore, L.E., Younas, M., Inorganic perovskite photo-assisted supercapacitor for single device energy harvesting and storage applications. *J. Energy Storage*, 73, 108828, 2023.
122. Aderyani, S., Flouda, P., Shah, S.A., Green, M.J., Lutkenhaus, J.L., Ardebili, H., Simulation of cyclic voltammetry in structural supercapacitors with pseudocapacitance behavior. *Electrochim. Acta*, 390, 138822, 2021.
123. Nayak, M.K., Sahoo, B.B., Thatoi, D.N., Nazari, S., Ali, R., Chamkha, A.J., Recent advances on supercapacitor electrode materials from biowastes – a review. *J. Sci.: Adv. Mater. Devices*, 9, 3, 100734, 2024.
124. Barzegar, F., Bello, A., Fashedemi, O.O., Dangbegnon, J.K., Momodu, D.Y., Taghizadeh, F., Manyala, N., Synthesis of 3D porous carbon based on cheap polymers and graphene foam for high-performance electrochemical capacitors. *Electrochim. Acta*, 180, 442–450, 2015.

12

Photo-Supercapacitors Based on Perovskite Materials

Tanuj Kumar[1]* and Monojit Bag[1,2]†

[1]*Advanced Research in Electrochemical Impedance Spectroscopy Laboratory, Indian Institute of Technology Roorkee, Roorkee, India*
[2]*Centre for Nanotechnology, Indian Institute of Technology Roorkee, Roorkee, India*

Abstract
Solar energy, a renewable and eco-friendly resource, offers an affordable substitute for traditional power sources. Nevertheless, its fluctuating nature requires effective energy storage systems to align production with consumption. Photovoltaic (PV) systems are essential in transforming solar energy into electrical power, and combining them with energy storage technologies provides a practical solution for addressing variability. This integration necessitates advancements in device architecture and innovative materials to improve efficiency and scalability. Metal halide perovskites have garnered considerable interest as a transformative class of materials in PV technology, providing exceptional power conversion efficiency, economic viability, and facile manufacturability for solar cell development. Their potential extends to photorechargeable electrochemical supercapacitors (SCs), wherein integrated perovskite-based PV devices and SCs can facilitate autonomous self-charging under illumination, ensuring a sustained power supply in off-grid and remote applications. Photorechargeable SCs have been engineered with both external (four-terminal) and internal (three-terminal) PV-SCs integration, with the latter being more favorable due to its reduced cost, lower ohmic losses, and minimized energy dissipation. In addition to these two types of integration, researchers have developed advanced photosensitive bifunctional electrodes capable of simultaneously harvesting solar energy and storing it within a single material. The optoelectronic properties of perovskites, combined with their electronic and ionic conductivity, make them ideal candidates for bifunctional SCs that can be efficiently charged under illumination.

**Corresponding author*: tanuj_k@ph.iitr.ac.in
†*Corresponding author*: monojit.bag@ph.iitr.ac.in

Nurdan Demirci Sankir and Mehmet Sankir (eds.) Solar Capacitors and Batteries, (389–422) © 2025 Scrivener Publishing LLC

Keywords: Perovskite, integrated devices, bifunctional material, photorechargeable supercapacitors

12.1 Introduction

Energy storage devices play a crucial part in numerous sectors of modern civilization including transportation and agriculture; thus, they are essential for humanity [1]. However, fast technological advancement of society and population growth have resulted in considerable energy consumption during the past decades. Worldwide energy consumption is expected to increase by about 60% in the coming years. Most of the countries are using fossil fuels and pumping out more toxic and greenhouse gases such as CO, CO_2, NO_x, etc., in the environment. This has caused significant environmental issues in recent years. Overreliance on fossil fuels results in their persistent loss and enhanced CO_2 emissions, a primary contributor to global warming and many other serious environmental problems [2–5]. To attain a more sustainable future, extracting energy from sustainable and green sources is considered the most viable option for addressing both the energy problem and polluted environment. Wind, biomass, geothermal, hydropower, tidal, and solar energy have garnered interest to meet carbon reduction and sustainability goals [2, 4]. Thus, wind, biomass, ocean, and sun energy are becoming more desirable. Solar energy, being an eco-friendly, economical, everlasting renewable resource and easy to harvest, may substantially aid in meeting global energy demands *via* its transformation into various energy forms [2, 4, 6–10]. Solar energy produces no emissions during consumption and is readily accessible, making it advantageous for environmental protection and providing significant ease for energy demands in distant regions [5, 11–14]. Despite extensive research on radiation from the sun across various fields, including photovoltaic (PV), it is not easy to use it practically as a consistent supply of electricity due to its intermittent nature and inevitable day and seasonal fluctuations due to weather conditions, which prevents it from being used in large-scale applications. Consequently, identifying viable storage methods and exploring new technology for solar energy has emerged as a prominent area of study for providing a steady energy supply. The utilization of renewable energy resources for energy storage by high-performance energy storage systems has a slight impact on the climate [15, 16]. Over the span of the last several decades, a significant amount of focus has been devoted to the advancement of the development of PVs. Currently, Si-solar cells have several drawbacks, such as significant consumption of electricity, expansive, restricted bandgap flexibility, and the poor theoretical efficiency of single-junction cells [17]. Identifying such materials is, therefore, of the highest priority to the researchers. Certain alloys and compounds, such as GaAs and CdTe, used in thin-film PVs, have been extensively analyzed by researchers, but due to their high cost, they cannot be employed in commercially viable solar panels for practical applications [18–20]. This enables individuals to shift to the efficient use of green energy supplies. Materials capable of responding to light by photon absorption, including ZnO, TiO_2, nickel sulfide (NS), nickel cobalt sulfide (NCS), organohalide perovskites (OHPs), and lead-free perovskite, are exceptional choices for the development of photorechargeable supercapacitors (PRSCs) [21–26]. Next-generation PVs must be cheap and efficient for the practical uses. Within the context of PV applications, perovskite, which is the most recent advancement in the next generation of PV materials, has a plethora of exciting potential and distinctive benefits that combine low cost and high efficiency. Their interesting photoelectric characteristics—including notable carrier mobility, elevated absorption coefficient, high photoluminescence quantum yield, ambipolar charge transport, and broadened diffusion length for photo-generated carriers—guarantee their recognized progress in recent years, positioning them as potential substitutes for Si-based solar cells and enabling them to establish an important portion of the market over a longer duration [21, 27–31]. Furthermore, these materials are widespread on the planet and can potentially be used in the production of flexible cells and devices that are low cost, are environmentally harmless, and have an enormous scope of application. In addition, the unusual structure of perovskite materials gives them specific features that make them very desirable and adaptable for many kinds of uses. These materials can be used in PVs, neuromorphic devices, lasers, memristors, photodetectors, LEDs, batteries, and supercapacitors (SCs) (Figure 12.1) [32–41].

Electrochemical techniques, such as batteries and capacitors, have important applications in renewable and clean energy technology owing to their easy operation, suitable efficiency, and eco-friendly features. As electrochemical energy storage devices can store power significantly, over the last several decades, SCs or electrochemical capacitors have been increasingly attractive. They are stable with extended cycle lifespan and excellent durability, which can be operated in a wide temperature range, as well as fast rate of charge and discharge capability [42, 43]. The growing usage of

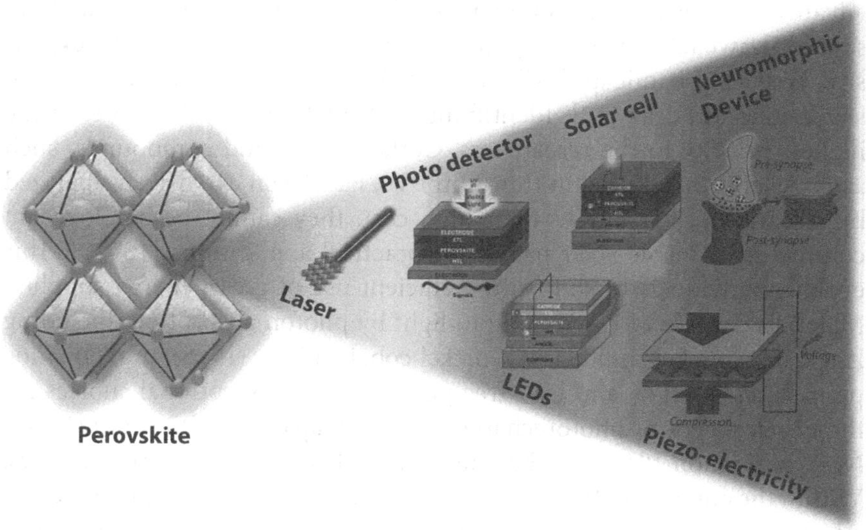

Figure 12.1 Schematic diagram of the various applications of perovskite materials in various fields.

wearable electronic devices is driving up the need for battery packs with high power density and improved energy storage capacity at high scan rates [44–46]. The possible application of SCs in electrical appliances and electric cars has made them an important category of storage devices for energy. In terms of the performance of the SCs, they lie in the middle of the batteries and the capacitors, having a power output of around 10,000 Wkg^{-1}, which is lower to capacitors but superior to batteries and fuel cells due to the minimum distance between the electrostatic charge in the electrochemical double-layer capacitor (EDLC) layer. Figure 12.2 illustrates the Ragone plot, a necessary tool to analyze the power and energy balances among the batteries, fuel cells, SCs, and traditional capacitors.

Recent efforts have focused on directly integrating solar cells based on perovskite materials and SCs as one unit facilitating harvest of the solar energy and storage of solar energy [47]. There are multiple efforts conducted to include energy conversion and storage devices into separate components for the purpose of developing devices that are highly efficient, compact, and portable. The integrated energy system converts and stores solar energy to stabilize power production and ensure continuous operation regardless of the intermittent nature of solar energy and its dependency on weather [12, 48, 49].

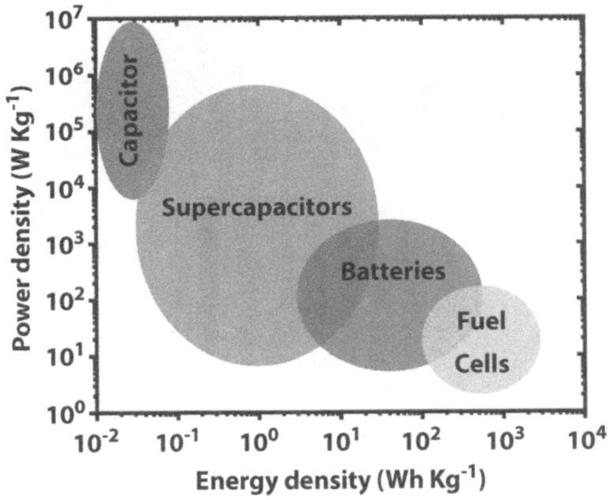

Figure 12.2 Ragone graphic illustrating the performance of electrical energy-storage devices, comparing power density and energy density.

12.2 Storage Mechanism of the SCs

SCs store energy by charging and discharging cycles at the interface of the electrode and the electrolyte. SCs store energy like common capacitors; however, they are better for fast release of energy and storing. SCs have electrodes with a larger effective surface area than normal capacitors, increasing capacitance fourfold. Through the suitable construction of SCs, the specific energy and specific power could vary by many orders of amount, causing SCs an adaptable storage solution. In recent years, research on SCs has significantly intensified to fulfil the need for applications demanding properties like increased specific energy, enhanced stability, and environmental safety with superior performance. The primary components of the SC construction are two electrodes, an electrolyte solution, an ion transporting membrane, and an electrode to collect the current. As shown in Figure 12.3, the storage mechanism of SCs depends on both the electrochemical charge storage through redox reaction taking place at the electrolyte/active material interface and the accumulation of ions on the electrode surface, forming an electric double layer capacitance (EDLC). It is possible to divide SCs into three basic categories, each of which is determined by the storage mechanism.

1. Electric double-layer capacitors (EDLCs)
2. Pseudo capacitors, and
3. Hybrid SCs

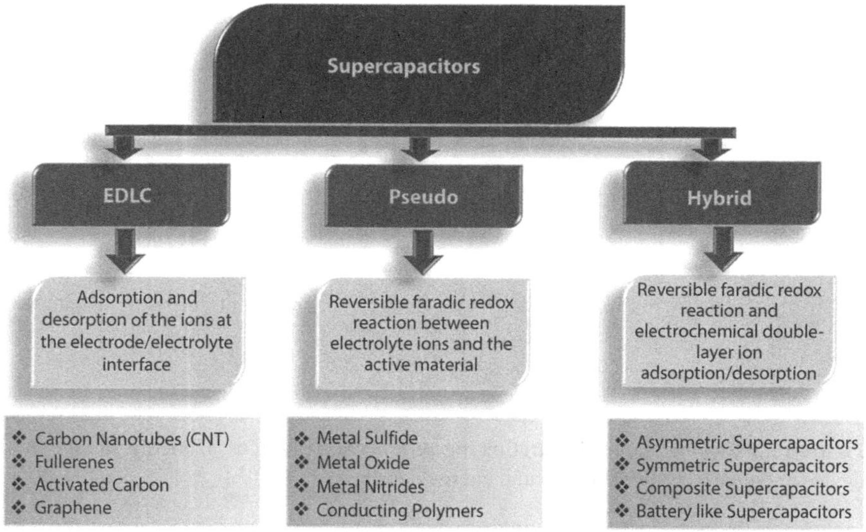

Figure 12.3 Overview of the categories and classifications of supercapacitors and their materials.

These various types of SCs each have their unique process for storing energy, which is illustrated in Figure 12.4.

12.2.1 Electric Double-Layer Capacitors

In terms of its capacitance mechanism, EDLC is analogous to the capacitance mechanism of a normal dielectric capacitor, which occurs between

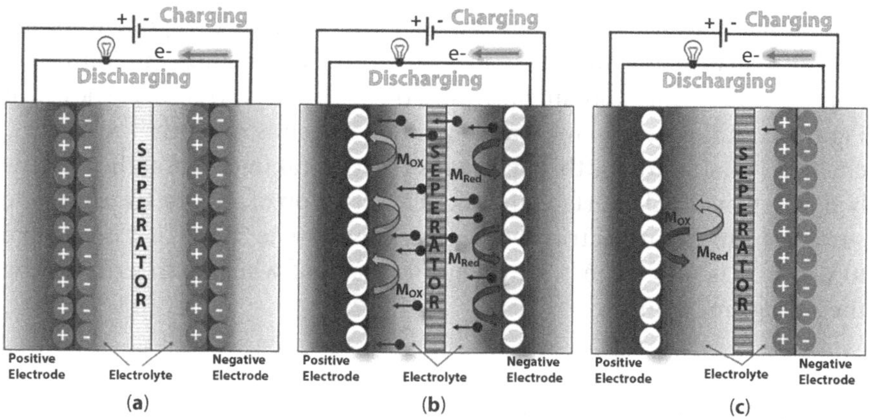

Figure 12.4 The charge storing mechanisms of several supercapacitor types. (a) EDL supercapacitor. (b) Pseudo capacitor. (c) Hybrid supercapacitor.

two charged metallic plates. This mechanism is responsible for the capacitance of the EDLC. EDLCs have an electrolyte, two electrodes, and a separator used between the electrolyte and the electrode interface to avoid the short circuit. EDLC may store charges electrostatically or with a non-Faradic mechanism by a reversible process by adsorbing the ions at the active electrode, which takes place without any electrostatic charge transfer [50]. These EDLC SCs have exceptional energy density and storage capability owing to the extensive specific interfacial area of the electrodes and the very short distance between the electrodes, which depends on the concentration of the ions and their size, which is around a few angstroms [51–53].

The excellent capacitive performance in the mechanism arises from charge separation resulting from polarized ions at the interface. To avoid ion recombination, a double layer of the electrostatic ions is formed in EDLCs. Additionally, since EDLCs have a unique storage mechanism, it is possible for them to rapidly soak up energy, distribute it, and produce an outstanding amount of power. Calculation of the capacitance of EDLC-type SCs can be done using the following equation:

$$C = \frac{A \times \in_0}{d}$$

In this equation, capacitance in farads is represented by C, the surface area of the electrode is represented by A, permittivity in free space is represented by ε_0, and the effective width of the EDLC layer is represented by d, which is also called as Debye length.

The choice of electrode material is crucial to determining electrical qualities since several EDLC materials have distinct attributes that affect device performance. Carbon is the most extensively studied material used as the electrode for the EDLC type capacitors because the surface area of this material is significantly high. Utilizing carbon commercially is advantageous due to its reasonable price and easy access. Carbon is available in numerous forms, including nanotubes, nanoparticles, conducting fibers and fabrics, conducting ink and paste, fullerenes, and porous conducting carbon foams, and various forms including carbon nanotubes, carbon nanoparticles, carbon fibers, carbon cloth, fullerenes, and foams. Since charge storage *via* the EDLC technique is a surface phenomenon, the surface characteristics of the electrode material influence capacitance.

12.2.2 Pseudocapacitors

The Faradaic nature, like oxidation–reduction processes, is responsible for charge storage. This process involves non-electrostatic transferring charge between the electrodes and the electrolyte, which reduces the energy density of these pseudocapacitors compared to the EDLC-type SCs [54]. Since pseudocapacitors store energy *via* redox processes, they behave like batteries. The pseudo electrodes primarily consist of metal oxides, metal sulfide, metal nitrides, carbon doped with metals, and conductive polymers. When a potential is applied to pseudocapacitor electrodes, a Faradaic current is induced due to processes such as electrosorption or redox reactions of the electroactive materials. This causes the capacitance to be produced in these SCs. Although pseudocapacitance is greater than EDLC by two orders of magnitude, its poor conductivity reduces cycle durability and power density [55, 56].

12.2.3 Hybrid SCs

Hybrid SCs show the properties of both types. Hybrid SCs are characterized by high capacitance and enhanced energy storage capacity compared to EDLC and pseudocapacitors [57]. Hybrid SCs can hold a substantial amount of charge at higher power rates than batteries that are rechargeable. Nonetheless, the charge accumulated on the outer layer limits their energy density or capacitance value to a lower level than that of batteries. Three distinct forms of HSCs can be distinguished according to the electrode configuration: the battery type, the asymmetric, and the composite [58].

12.3 Type of Integration of PV Unit with the SCs

Two primary methods exist in which PV cells and other types of solar cells can be combined with SCs. One of these methods is known as the traditional approach (conventional method). In contrast, the other is known as the monolithic integration (internal integration) method, which further separates into two specific types of integration, as shown in Figure 12.5.

12.3.1 Conventional Integration (External Integration, Isolated Integration)

The traditional approach for integrating energy to charge the SC unit, using energy generated by PV panels (PPVs), is referred to as an "isolated"

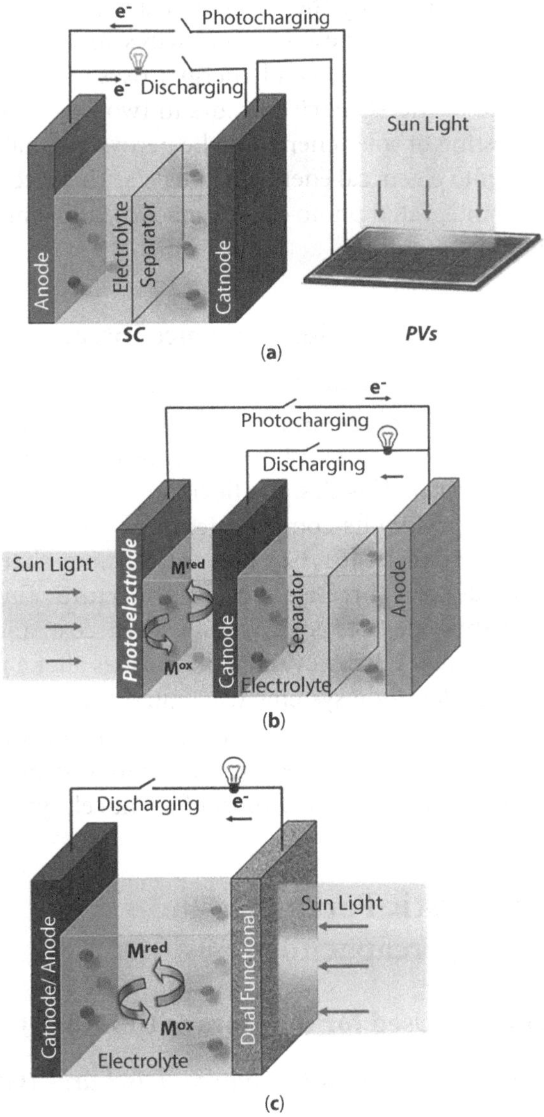

Figure 12.5 The type of integration of PVs and SCs. (a) External integration. (b) Three-electrode integration. (c) Two-electrode integration.

or "external integration" system. External integration demands that external electrical wires are commonly employed to connect solar cells for harvesting solar energy and rechargeable SCs that store this energy. However, the large volume of the whole device requires optimization for enhanced performance, making them unsuitable for flexible and smart electronic devices. Additionally, they are not flexible, and significant ohmic losses arise from the use of these external wires in the integration process. This also increases the costs and energy losses of these systems. The charging procedure of the SC device occurs in two stages. The first phase involves the harvesting of solar energy by the perovskite material, which is then transformed into electrical energy by the PVs. This electrical energy is then stored in the integrated photo-supercapacitor, allowing for discharge *via* a load for use [59, 60].

12.3.2 Monolithic Integration (Advance Integration)

There is also an alternative method that incorporates the internal integration of PVs and energy storage units, such as SCs, in which the separate modules are connected by a single photoelectrode monolithically by three- or two-electrode systems. This design's flexibility arises from the requirement for less cabling, while its compact design is particularly crucial for small-scale consumer devices [61, 62]. PVs and SC units share one positive or negative electrode in a three-electrode architecture. The shared electrode complicates these devices and increases their cost, and due to their dependence on PVs, their efficiency often diminishes after a few cycles. On the other hand, two-electrode systems use a dual-function electrode that concurrently functions as a harvesting material and stores charge storage simultaneously. These integrated devices have reduced prices, enhanced energy conversion efficiency, and the most efficient cell type [24, 62, 63].

12.4 Characteristic Parameters in Photo-Supercapacitors

12.4.1 Parameters Used for the Storage Unit (SCs)

12.4.1.1 Using the Galvanostatic Charge–Discharge (GCD) Cycles

The following formula is used to determine the areal capacitances (C_A) of SCs based on the potential window and discharging time extracted from the GCD curves at various current densities:

$$C_A = \frac{I \times t_{discharging}}{A \times \Delta V}$$

where the active surface area of the device is referred to as "A", $t_{discharging}$ is the discharge time extracted from the GCD curve, and ΔV is the potential window in which the device is operated.

To determine the output energy (E) of both the integrated device and the SC, apply the following formula:

$$E = \frac{1}{2} C_A (\Delta V)^2$$

The energy density can be calculated using the following formula:

$$E_d = \frac{1}{7200} C_A (\Delta V)^2$$

The unit of E_d is in Wh kg^{-1} when it is normalized by the mass loading and in Wh cm^{-2} when it is normalized by the active surface area.

The following equation can be utilized to determine the power density (P_d) of the combined device and SC based on the value of energy density (E_d):

$$P_d = \frac{E_d}{t_{discharging}}$$

The unit of P_d is in W kg^{-1} when it is normalized by the mass loading and in W cm^{-2} when it is normalized by the active surface area.

12.4.1.2 *Using the Cyclic Voltammetry (CV) Curve*

The specific capacitance (Cs) of the SC and integrated device can be determined *via* CV utilizing the following equation:

$$C_S = \frac{\int_{V_1}^{V_2} IdV}{v \times m \times \Delta V}$$

where $\int_{V_1}^{V_2} IdV$ is the enclosed area of the CV curve, v is the scan rate, m is the mass loading of the active electrode, and ΔV is the potential operating window.

12.4.1.3 Using the Electrochemical Impedance Spectroscopy (EIS)

The capacitance (C) by examining the Nyquist plot for both the SC and its integrated unit with solar cell can be determined by analyzing the EIS plot:

$$C = \frac{1}{2\pi f |Z|}$$

where $|Z|$ is the modulus of the impedance.

Round-trip efficiency (RTE)
The term "round-trip efficiency" refers to the capability of measuring the amount of energy that is lost during the charging and discharging operations. This efficiency may be determined by utilizing the equation that is provided below.

$$\eta_{\mathrm{RTE}} = \frac{\mathrm{E}_{\mathrm{discharging}}}{\mathrm{E}_{\mathrm{charging}}} \times 100\%$$

12.4.2 Parameters Used for the PVs

Solar cell's photoconversion efficiency (PCE) measures its capacity to convert solar radiation into electrical. The ratio of output energy to input light energy is defined as the PCE of the solar cell and can be calculated using the following equation:

$$\eta_{Solar\ cell} = PCE = \frac{FF \times J_{SC} \times V_{OC}}{P_{in}} \times 100$$

The fill factor (FF) denotes the quality of a solar cell's electrical output and is often expressed as a percentage; current density is referred to as J_{SG}, and the open-circuit voltage of the solar cells is denoted as V_{OC}.

According to the equation, the PVs' FF can be calculated directly using the following equation, which requires only the maximum power output, short-circuit current, and open-circuit voltage.

$$FF(in\ \%) = \frac{I_{Max} \times V_{Max}}{I_{SC} \times V_{OC}}$$

12.4.3 Parameters Used for the Integrated Device

Integrated device efficiency characteristics may be determined by analyzing the efficiency of PVs of the conversion of the solar energy and electrochemical energy storage devices using the $\eta_{overall}$ equation. The total value of η is determined by the ratio between the energy stored by the SC during the photo-charging process and the energy generated by the illumination resulting from the process.

$$\eta_{overall} = \frac{E_{Stored}}{E_{illuminated} \times A}$$

where A is the active area exposed in the light of the PVs.

By using the following equation, we can calculate the amount of energy that is produced by the integrated SC device with solar cells as a result of the illumination.

$$E_{illuminated} = P_{in} \times t_{charging}$$

When the light is illuminated, the power density is referred to as P_{in}, and $t_{charging}$ is the time of illumination.

For an integrated system, the energy storage efficiency of the combined device can be determined by using the overall efficiency and the efficiency of the PVs by the following equation.

$$\eta_{Storage} = \frac{\eta_{overall}}{\eta_{Solar\ cell}}$$

12.5 External Integration

The energy generated by the PVs may be stored *via* the external integration with the SC. Integrating a PV unit with an SC for energy storage enables continuous operation, independent of the intermittency of sunlight, hence addressing issues arising from discrepancies between solar production and energy consumption. Solar engineers are interested in this storage system because it can adjust to weather changes. The use of conducting FTO substrates, commonly employed in the fabrication of perovskite-based PVs, limits their application in smart electronic devices that require flexibility due to the heavy and non-flexible nature of FTO, rendering them rigid integrated devices and impeding their potential use. The use of photo-supercapacitors as a source of power in portable electronic devices is one of the potential applications of this technology in terms of future generations, where thin and flexible devices are greatly desirable. Compact flexible SCs made of all solid-state materials are better for mobile and wearable electronic devices than liquid-based ones [64]. Due to the high conductivity of electricity, better elastic properties, a huge area of coverage, and stable chemical properties make graphene films and graphene fibers, which are fabricated by numerous methods (hydrothermal method, chemical vapor deposition (CVD), and vacuum filtration), an attractive option for nano-structured carbon-based solid-state SCs fabricated on flexible substrates for the integrated system. Pengcheng *et al.* reported an SC constructed from self-stacked solvated graphene (SSG) films used as flexible substrate, which were produced *via* simple vacuum filtration as electrodes for the SC unit [65]. These SCs have a good storage performance and maintain 83% of their initial capacitance. Their specific capacitance is around 245 F g^{-1}, and their energy density is 8.0 Wh kg^{-1} as determined by the GCD curves at a current density of 1A g^{-1}. This SC has been coupled to a $CH_3NH_3PbI_{3-x}Cl_x$ perovskite-based solar cell unit, exhibiting a V_{OC} of 0.90 V and a PCE of 14.13%. These integrated devices may be charged under white light illumination, functioning as self-powered devices with significant potential in flexible electronics. The electron transport layer (ETL) in perovskite-based PVs (PPVs) demands temperatures that exceed 200°C, raising a significant demand for a flexible substrate capable of withstanding such high temperatures to produce high-performance flexible PVs. Yang *et al.* demonstrated a tungsten oxide (WO_x) layer, an inexpensive and reliable n-type semiconductor element with a large and tunable bandgap, deposited at room temperature on the ITO/PEN flexible substrate with varying thickness, as an ETL and $MAPbI_3$ perovskite layer as absorbing layer for PVs unit using electron-beam deposition, thereby addressing the challenges associated with the fabrication of flexible PPVs [59]. Carbon electrodes were employed for symmetrical SCs. The optimal champion PV device, including a WOx thickness of approximately 26 nm and a PCE of around 11%, exhibits a photo-to-electrical conversion efficiency of approximately 1.3% when integrated with a flexible SC, which is slightly lower than that of a rigid SC [59].

An integrated device using PVs based on MAPI perovskite to harvest the solar energy and polypyrrole SC that can store the solar energy was developed by Xiaobao *et al.* [60]. For this integrated device, inorganic–organic hybrid PVs were manufactured by using mesoporous TiO_2 (mp-TiO_2) of varying thicknesses as ETL (Figure 12.6a and b). These ETLs were then infiltrated with spin-coated Spiro-MeOTAD used as a hole transport layer (HTL) and with large-sized island-shaped MAPI perovskite, which enhanced the absorption of light as a result of the trapping effect. Using an improved m-TiO_2 layer of 250-nm thickness, the device has shown superior performance, achieving a V_{OC} of 0.974 and a PCE of 13.6% (Figure 12.6c). Despite the enhanced light absorption in a thicker layer, the current density diminishes owing to charge carrier recombination. The incident photo to electron conversion efficiency (IPCE) vs wavelength is shown in the curve for a device with a 250-nm-thick mp-TiO_2 layer, demonstrating the device's capability for effective solar light harvesting throughout a wide spectrum range (Figure 12.6d).

A new SC unit has been developed to store the harvested solar energy using symmetric SCs composed of two bacterial cellulose membrane/nanofiber made of polypyrrole/carbon nanotube using multi-walled architecture (MWCNT) films as electrodes. Based on the GCD curves, the calculated areal capacitance is around 570 mF cm^{-2}, using a redox mechanism for this SC. The specific capacitance value is measured at a discharge current of 1 mA cm^{-2} (Figure 12.6e). The capacitance retention of around 82% indicates higher electrical conductivity and effective ion transport.

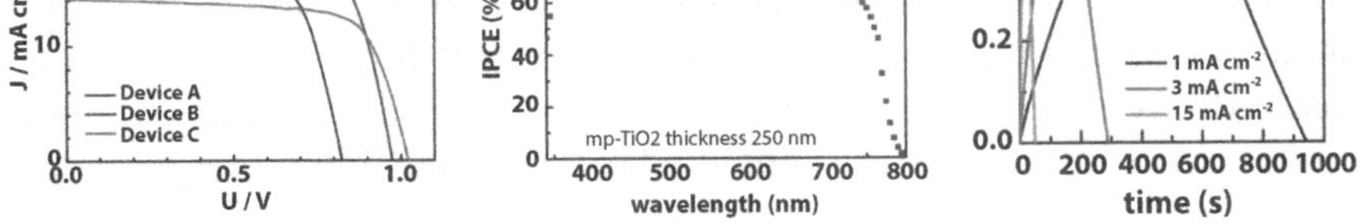

Figure 12.6 (a) Schematic representation of an integrated capacitor using polypyrrole (PPC) together with the $CH_3NH_3PbI_3$ (MAPI) perovskite solar cell (PSC). (b) Photograph of this integrated device. (c) The comparison of the J–V curves of photovoltaic (PVs) based MAPI based solar cells but varying the thickness of ETL layer (TiO_2 as ETL with thickness 300 nm (device A), 250 nm (device B, which has best performance), and 150 nm (device C). (d) Incident photon to electron conversion efficiency (IPEC) vs wavelength spectra for the optimal device B. (e) Galvanostatic charging and discharging curves (GCD curves) of the supercapacitor unit with various densities of current from 1 to 15 mA m^{-2}. (Reproduced with permission from ref [60]).

Under AM 1.5G illumination, the PV device with an area of approximately 0.06 cm^2 is subjected to exposure. Following the integration of this PPV with the SC, the voltage under illumination instantaneously rises to 0.3 V, subsequently increasing gradually over the next 200 seconds, ultimately reaching 0.71 V under continuous illumination of the exposed area. The solar cell and the SC unit were connected in series for the storage mode, and the value of the stored energy that was converted from the solar energy was around 10%, which is the greatest value that has been recorded [60].

12.6 Monolithic Integration

Monolithic integration of photorechargeable SCs can be classified into two categories based on the integration mechanism of energy-harvesting and storage devices. Unlike four-electrode integration, where solar PV and energy storage devices are connected externally through a battery management system (BMS), monolithic integration does not require external integration or the BMS.

12.6.1 Three-Electrode Integration

12.6.1.1 Organometal Halide Perovskite (OHP) Based PV Integration

Furthermore, considering the price and durability of PPVs, carbon-based electrode-based devices replace the HTL, establishing them as the most attractive renewable technologies [66–68]. SCs provide an excellent option for power storage as an emerging class of sustainable electronics, owing to their high power density, safety, prolonged life cycles, and rapid charge and discharge capabilities. Zhiyong *et al.* used a carbon layer, applied using the doctor blade technique, on a perovskite film that functions as a counter electrode to collect holes generated by the illumination inside the bulk of the MAPI perovskite film under solar irradiation [69]. The perovskite film for solar light absorption is deposited on the spin-coated compact and mesoscopic TiO_2 layer (600 nm) used as the ETL as shown in Figure 12.7a. They have examined the charging and discharging of an SC when combined with PV systems in both parallel (see Figure 12.7b and c) and series configurations (see Figure 12.7f and g). The curve between the current density and the voltage of the PV unit when illuminated with a conventional solar simulator (AM 1.5G with power density 100 mW cm^{-2}), shown in Figure 12.7d, indicates that the ideal cell reaches a power conversion efficiency of around ~7.8% with an open-circuit voltage of around 0.96 V *via* a reverse

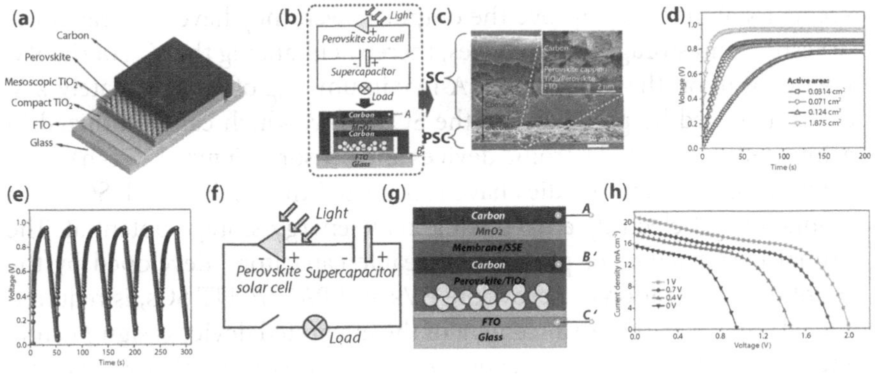

Figure 12.7 (a) Solar cell unit structure based on perovskite material (MAPI). (b) Circuit diagram and the schematic structure of the integration solar cell and supercapacitor when they are connected parallelly. (c) FESEM image of the device when they are connected parallelly shows the thickness of every layer. (d) The potential under illumination and the time when the device is charged under light. (The area of the device is varied from ~ 0.03 to ~1.8 cm^2.) (e) GCD curve under the illumination of the device, when illuminated with device area ~1.87 cm^2. The device is discharged at 2 mA cm^{-2}. (f) Circuit diagram and (g) schematic structure of the supercapacitor, which is connected with the solar cell in series combination. (h) J–V curve of the integrated power back when pre-charged by the solar cell to various potentials. (Reproduced with permission from ref [69]).

voltage scan. The negative electrode of the SC unit consists of a highly conductive and flexible graphite sheet, onto which a mesoporous carbon layer is applied *via* the screen-printing technique. In contrast, for the positive electrode, they used a MnO_2 layer with a significant surface area, exhibiting Faradic pseudocapacitive behavior through a redox mechanism. The GCD curves indicate that the C_A of this SC device reaches 61.01 F cm^{-2} at a constant value of the density of current 0.25 mA cm^{-2}, and it has a 65% capacitance retention rate surpassing values observed in comparable SCs. Upon exposure to AM1.0 sun irradiation, the voltage of the integrated device's PV considerably rises initially. The voltage after that rises somewhat to a steady level under continuous illumination. An increased surface area of the device will provide a greater photocurrent, charging the SC more rapidly and consequently generating a voltage that is higher than the capacitor. When it is charged under solar radiation, the overall conversion efficiency of the integrated devices first increases, reaching a maximum peak value before subsequently declining. The highest value is around 5.3%, with a maximum storage efficiency of 67% for a device with an area of roughly 0.07 cm^2 (Figure 12.7e). The integrated device's photocharge/galvanostatic discharge curve is shown in the figure, which illustrates its excellent stability. To improve the output power, they have integrated the harvesting and storage unit in series, thereby enhancing the V_{OC} of the PV, dependent upon the SC's charge. A maximum V_{OC} of approximately 2 V can be achieved by pre-charging the SC to 1 V, which can be utilized in various self-powered electronic devices and sensors (Figure 12.7f–h).

Numerous research studies have recognized that PANI/CNT SCs with an aqueous gel electrolyte have remarkable energy storage potential. The optimal performance of photorechargeable capacitors, developed by the monolithic integration of perovskite PVs and PANI/CNT SCs, is required [70, 71]. However, the advancement of the integrated device is significantly hindered by the perovskite's exposure to aqueous electrolytes; this promotes the diffusion of water molecules *via* the PANI/CNT layer, which is hydrophilic in nature, to the PSCs in a humid environment, thereby accelerating the degradation of the $MAPbI_3$ film employed in the PV unit. Liu *et al.* tried to solve the issue of damaging the $MAPbI_3$ perovskite absorbed layer by establishing a CNT bridge between the PV unit and the PANI/CNT-based SC, enhancing the system's efficiency. This system has a significant capacitance value of 422 mF cm^{-2} when normalized with the area and 103 F g^{-1} when normalized with the loaded mass. It is a reliable power-generating device under sophisticated illumination conditions, such as the intermittent nature of solar energy, exhibiting a coulombic efficiency of 95% and an overall energy storage efficiency of around 70% [24].

12.6.1.2 Mixed-Halide Mixed-Cation Perovskite (MCMHs) Based PV Integration: Impact of the ETL on the Overall Storage Conversion Efficiency

Perovskite crystals are sensitive to humidity, UV radiation, polar solvents, and other factors that degrade and prevent their use in practical applications. To avoid PSC decomposition, SCs with gel electrolytes and carbon-based electrodes with more thickness are needed. Perovskites' instability and the PSCs' low long-term stability provide significant challenges to developing a single, closely linked integrated device. Meng *et al.* recently employed an electrode based on double-sided TiO_2 nanotube arrays grown directly on Ti foil using electrochemical anodization to enhance charge transfer between the PV and the SC, which makes SC operations more flexible (Figure 12.8a) [72]. Figure 12.8b shows the FESEM image of the TiO_2 nanotube after the acid treatment. The TiO_2 nanotube deposited on the highly conductive Ti foil functions as the ETL in the PV unit and serves to store the charge created by the PVs. The high conductivity of Ti foil enables easy electron

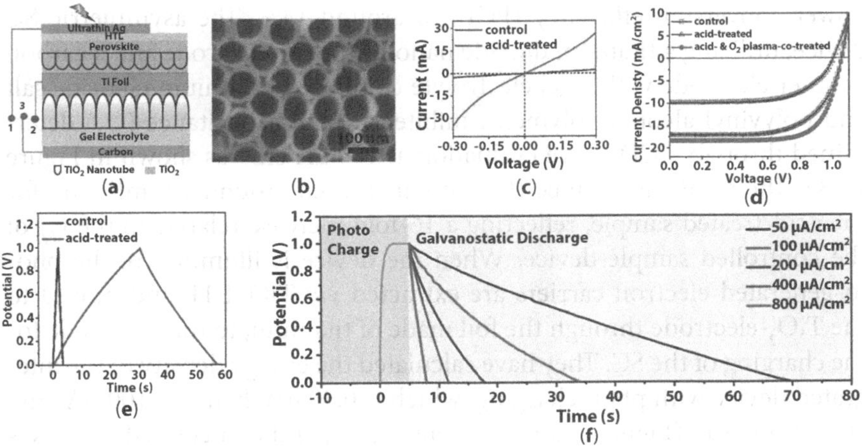

Figure 12.8 (a) Schematic structure of the monolithic system developed by the solar cell and the supercapacitor by connecting them using a double-sided anodized Ti foil, which forms TiO_2 nanotube array. it works as an electrode. (b) FESEM image of the TiO_2 nanotube after treatment by acid (scale bar—100 nm). (c) The linear sweep voltammetry curves of the film formed by TiO_2 nanotube array without treatment (controlled sample) and with acid treatment. (d) The current and voltage curves of the champion solar cell were subjected to various treatments on TiO_2 nanotube ETL. (e) GCD curves of supercapacitors using a control sample and TiO_2 nanotubes after acid treatment as electrodes have been performed at a current density of 100 μA/cm². (f) GCD curves are analyzed at different discharging current densities under illumination. (Reproduced with permission from ref [72]).

transfer between electrodes made by TiO_2 nanotubes bridged by it, so facilitating the charging under illumination or discharging the storage unit SC and helping to stabilize the perovskite film (Figure 12.8c). Using PTAA as the HTL, the MCMH perovskite was selected as a light-absorbing material because of its high efficiency and long-term stability. The device was coated with hydrophobic perylene film by vacuum deposition to protect it in a humid environment. It is treated with O_2 plasma to remove the oxygen vacancy on the surface while the bulk is left as it is, which enhances the bulk conductivity for charge transfer in PV operations.

Oxygen vacancies generate barriers to the movement of charge carriers, potentially trapping them and increasing the recombination rate, adversely affecting device performance. The oxygen vacancies in the TiO_2 nanotube arrays are detrimental to optimal device performance. Figure 12.8d illustrates the current and voltage characteristics of three distinct types of separate devices; among them, the device developed utilizing O2 plasma treatment and acid treatment has the highest performance, achieving a power conversion efficiency (PCE) of around 14%. The asymmetric SC device, utilizing a titanium dioxide nanotube array electrode and a carbon counter electrode with a gel electrolyte composed of lithium chloride salt and polyvinyl alcohol polymer, exhibited a areal capacitance (C_A) determined through GCD curve evaluation at 100 μA cm^{-2} as shown in Figure 12.8e. The areal capacitance (C_A) was measured around 2.8 mF cm^{-2} for the acid-treated sample, reflecting a 16-fold increase relative to the C_A of the controlled sample device. When the device is illuminated, the photo-generated electron carriers are extracted *via* TiO_2 ETL and moved to the TiO_2-electrode through the foil made of titanium, to the other side for the charging of the SC. They have calculated the energy density of the integrated device with photocharging, which is 0.40 μW hcm^{-2} at 100 μA cm^{-2} current density (Figure 12.8f). The value $\eta_{overall}$ of the integrated devices is 9.18%, resulting in an energy storage efficiency of 68.66%.

Chao *et al.* established a PPV-powered Li-ion capacitor (LIC), which can be charged under illumination that incorporates a PV unit of four MCMH perovskite-based units connected in series, each exhibiting a PCE of 14% and a V_{OC} of 1.05 V, utilizing a flexible ITO/PET substrate to enhance device flexibility. The flexible photo-LIC achieves an impressive 8.41% efficiency and 3-V output voltage at a constant current density of 0.1 A g^{-1}, making it the most efficient power source that can be charged under illumination [73]. Berestok *et al.* used an MCMH perovskite as the light-absorbing material for the p-i-n PV cells, achieving a high PCE of around 12.5% and a V_{OC} of 1.15 V. Mesopores must enable better penetration and surface access than micropores due to the large surface area so they use mesoporous nitrogen-doped carbon nanospheres with a distinct structure, uniform primary sizes of particles, and stable nature under operating conditions as active electrode material for energy storage, which provide the specific capacitance around 400 F g^{-1} and power density of ~117 W kg^{-1}. The integrated photo-supercapacitor can quickly charge up to 1 V due to its extremely high PV efficiency and outstanding electrochemical properties. The device's 11.5% photoelectrochemical energy conversion efficiency exceeds prior efforts [74].

12.6.1.3 All-Inorganic Perovskite (AIP) Based PV Integration

In particular, the V_{OC} of a solar cell ought to correspond with the voltage achieved by the photocharging of an electrochemical storage system for a photorechargeable SC. By combining dye-sensitized solar cell (DSSC) with a porous silicon wafer that was carbonized and titania, the group has developed a photorechargeable SC that has a PCE of about 4.8%. Despite its high efficiency, the DSSC's development for real-world uses is limited by electrolyte leakage. Furthermore, most DSSC-based photo-supercapacitors have poor energy density and low charging voltage. The immediate advancement of the PVs based on perovskite, particularly excellent light-absorbing material $MAPbI_3$, demonstrates exceptional optical and electronic characteristics and great potential in the integrated photo energy storage system (Figure 12.9a) [75]. However, the biggest barrier to commercialization remains the stability of this material over time in the humid environment, specifically. Chi *et al.* demonstrated a cesium-based halide perovskite film by substituting the MA cation in the organic perovskite and varying the halide ratio during fabrication. This approach exhibited enhanced stability for two weeks in humid conditions without encapsulation. Additionally, PVs based on this perovskite were integrated monolithically with an asymmetric SC, thereby minimizing ohmic losses due to the absence of external wiring. Compact and mesoporous TiO_2 was employed as ETL and spiro-OMeTAD as HTL, while the halide ratios in the inorganic halide perovskite were varied. The optimal performance of the PV was observed in the champion perovskite $CsPbBr_{2.9}I_{0.1}$, which is pinhole-free, thereby preventing degradation, and having a high open-circuit voltage of around ~0.65 V and PCE of around 0.5% (Figure 12.9b). The SC unit, which is integrated with the PV, includes PEDOT: PSS cast used a buffer cathode for energy storage, whereas PEDOT: PSS-assisted RZCo is used as the positive electrode, $LiClO_4$/acetonitrile as the electrolyte, and PyR negative as electrode resulting in a capacitance value around ~150 mF cm^{-2} at 0.1 mA cm^{-2}. The integrated device is charged under light until a cutoff potential of 0.07 V is achieved within 100 seconds and then discharged at 0.1 $mAcm^{-2}$ under dark, which shows the value of capacitance ~30 mF cm^{-2} for an integrated device (Figure 12.9c and d) [76].

Liang *et al.* provided an integrated AIP photo capacitor, which combines an all-inorganic PV unit employing $CsPbBr_3$ (band gap = 2.3eV) as the absorbing layer and a nanocarbon-based back electrode for hole collection, therefore enhancing device stability and achieving a high V_{OC} around 1.22 V and PCE around 6% under convention solar radiation. They constructed an energy storage unit using nanocarbon material as an electrode and silica-gel electrolyte operated between the potential window ranging from 0 to 1.3 V, which exhibits EDLC behavior. The areal capacitance determined using the GCD curves is around 40 mF cm^{-2} for this device, whereas the energy density is about 9 μWh cm^{-2}. The solar capacitor's total conversion efficiency (η) is determined to be 5.1% [62]. Ankur *et al.* recently fabricated a PV based on inorganic perovskite $CsPbBr_3$ layer as light-absorbing material that shows high stability in ambient conditions

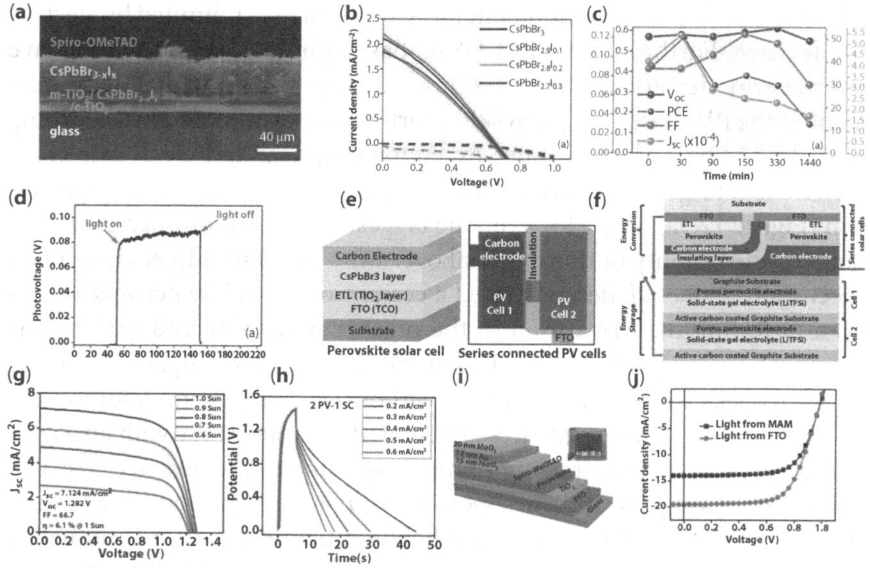

Figure 12.9 (a) FESEM cross-sectional view of the solar cell with the configuration ITO/compact-TiO_2/mesoporous-TiO_2/$CsPbBr_{3-x}I_x$/Hole transport layer. (b) The current and voltage (J–V) performances of $CsPbBr_{3-x}I_x$-sensitized solar cells with varying values of x under solar radiation (solid line) and in dark (dotted line). (c) The stability performance best photovoltaic device utilizing $CsPbBr_{2.9}I_{0.1}$ perovskite-sensitized material was evaluated during ultraviolet exposure and high humidity conditions (>80% RH) for a duration of 1450 minutes. (d) The photovoltage response of the most stable champion solar cell device based on $CsPbBr_{2.9}I_{0.1}$ perovskite under light and dark conditions. (Reproduced with permission from ref [76].) (e) Structure of the CsPbBr3 perovskite-based individual solar cell and series-connected solar cells with carbon-based HTL. (f) Schematic structure of two photovoltaic devices integrated with two supercapacitors coupled *via* a common electrode. (g) The I–V characteristic of isolated single solar cells under varying light intensities demonstrates a peak efficiency of approximately 6.1% at 1 sun illumination. (h) GCD curve of the integrated champion device with 2 PVs connected with 1 SC and discharged and various current densities ranging from 0.2 to 0.6 mA cm^{-2}. (Reproduced with permission from ref [77].) (i) Both structure of the separated solar cell unit and photo (inset) based on the transparent electrode made of MoO3/Au/MoO3 (MAM). (j) The J–V curves of a PPV based on MAM. when illuminated from the FTO side device shows a PCE of 12.5%, and it shows 9.4% efficiency when illuminated from the MAM side. (Reproduced with permission from ref [78]).

even without encapsulation compared to OHPs. Carbon paste is used as HTL and compact TiO_2 layer as ETL, achieving PCE 6.1% and V_{OC} 1.2 V under 1 sun illumination (Figure 12.9e–9g). The PV systems have been merged with SCs to create photorechargeable SCs in four distinct configurations. Among these, the champion devices exhibit a storage efficiency of around 87%, consisting of two PV systems connected in series integrated with one energy storage SC [77].

12.6.1.4 All Transparent Electrode-Based Integration for the Application of PVCC

As a result of the many potential uses of optically variable elements in architectural structures, automobile windows, airline windows, and roofs of buildings, windows with smart functionality that use electrochromic (EC), photoelectrochromic (PECC), and photovoltachromic cell (PVCC) technologies have attracted significant interest. The EC substances are used as electrodes in standard PECC and PVCC devices and have quick transmittance responses to light, switching between bleached and colored forms. Zhou *et al.* developed an all-transparent PV using a transparent electrode composed of MoO_3/Au/MoO_3 (MAM) at the top of the PV device structure, achieving 70% transmission in the visible spectrum for the integrated device structure (Figure 12.9i and j). The WO3 electrode they employed for the storage unit can store energy by changing the valence state of the electrode material. This is accomplished by the electrochemical reaction at the electrolyte/electrode interface. A novel PVCC structure can be developed by integrating WO3-based EC devices with PPVs. The integrated device, namely the co-cathode configuration, consists of a PV made of FTO/WO_3/PVA electrolyte/WO3/FTO/glass, which charges photovoltachromic SCs up to 0.68 V in 85 seconds. The color alteration effect begins at 0.50 V and achieves a color state after 56 seconds of continuous charging. The specific capacitance of PVCS under illuminated conditions has been determined to be 430 F m^{-2}, whereas the energy density is calculated to be around 25 mWh m^{-2} using the GCD curves [78].

12.7 Two-Electrode Integration

12.7.1 Non-Flexible Integrated Device

Jing *et al.* describe a device that brings the characteristics of solar energy conversion and storage into a single cell. This is accomplished by using the mesoscopic cell framework and the simple manufacturing of PPVs, which can be developed by printing. In the initial construction of the PV, the FTO substrate was screen-printed with the TiO_2 block layer, TiO_2 nanocrystalline layer as the ETL, ZrO_2 as a spacer layer, and mesoporous carbon electrode sequentially. The perovskite material is introduced into the porous carbon structure by applying the solution of $MAPbI_3$ over the coated carbon layer. They have used a PEDOT-carbon electrode in an integrated photo-supercapacitor (IPSC) system to link the PSC and SC components. MAI has been used as a stabilizer in the production of the electrolyte in IPA to enhance the stability of the perovskite layer and protect it from solvent corrosion. Prior to the integration of the PV unit with the SC unit, the PV exhibited an efficiency of 6.21% and an open-circuit voltage of 0.72 V. Following the integration, there was no significant change in performance, which remained consistent with that of the individual PV. The SC's areal capacitance was computed using GCD curves, ranging from 12 to 8.5 mF cm^{-2}. For the photocharging, when the integrated device is exposed to visible light, within a few seconds, the capacitor is charged to 0.70 V, nearing the V_{OC} of the manufactured PPVs (0.71 V), which signifies a substantial storage capacity of the PEDOT-carbon electrodes in the PSC device. The energy density was estimated to be 7.83×10^{-7} Wh cm^{-2}, while the power density was determined to be 7.4 mW cm^{-2}. The total efficiency with photocharge attains a peak of 4.70% within 1 second, while the related energy storage efficiency reaches about 74% [63].

12.7.2 Flexible Integrated Device

Chao *et al.* presented an all-solid-state ribbon that incorporates a very flexible and thin PV system fabricated using a solvent-assisted perovskite growth approach, achieving a PCE of 10%. These PVs are integrated with symmetric SCs, using a copper ribbon that collects photo-generated electrons from the PVs and functions as a shared electrode for the charge transfer. The exceptional flexibility and weave ability ribbon, which can harvest and store energy simultaneously, produces a fabric/matrix attractive option for supplying power to wearable electronic devices. The other side of this copper ribbon serves as a substrate for synthesizing copper hydroxide nanotubes (CuOHNTs) to develop a PVA/KOH electrolyte-based SC. The electrochemical performance is improved following the deposition of MnO_2 on the electrode, attributed to the increase in active surface area relative to the bare CuOHNT electrodes. This change exhibits a pseudocapacitive energy storage capacity, enabling the storage of substantial charge within the nanotubes of the electrodes, thereby enhancing the capacitance by a factor of ten, achieving a value of 145.15 mF cm^{-2} at a current density of 1 mA cm^{-2} from the GCD curve, for MnO_2-deposited CuOHNT electrode. In the photo-charging period, electrons generated in the perovskite under illumination are harvested and stored in the SCs, allowing the integrated device to achieve a potential of 0.8 V under illumination and discharge through a load for practical applications at varying current densities in the dark [79].

12.8 Photorechargeable SC (Dual Functional Electrode)

The conversion of energy and consumption could be more efficient with separate conversion and storage devices. As stated in the monolithic integration system, where the PV unit is integrated with the energy storage unit, it is gaining importance. Integration challenges involve controlling electrical and ionic transport in complicated multilayer monolithic devices and providing repeatability and dependability. Researchers have invented improved photosensitive bifunctional electrodes that harvest solar energy and store energy in just one substance to address these challenges. The optoelectronic characteristics of perovskite and electronic/ionic conduction make them the perfect materials for bifunctional devices. They could be used to construct photorechargeable energy storage systems like PRSCs. Ramesh *et al.* developed a photorechargeable SC by incorporating carbon black into halide perovskite porous electrodes, varying the carbon black ratio in perovskite and PVDF binder. The champion device with 25% carbon black exhibited stable energy storage performance in dark conditions, achieving specific capacitance around 73 F g^{-1}, and the calculated energy density by GCD curves is about 30 W h kg^{-1} at current density 0.3A g^{-1}, attributed to the enhanced surface area. As the carbon percentage rises, both capacitance and stability improve simultaneously in the absence of light; however, photo-capacitance begins to decline when the carbon black percentage surpasses 15%, at which the maximum photo-capacitance is recorded around 85 F g^{-1} (Figure 12.10a). The PRSCs, when exposed to a white light source of 20 mW cm^{-2}, has an implied photo conversion efficiency of 0.02% (Figure 12.10c) [23].

Our group demonstrated two types of photo-supercapacitors based on two synthesized active materials: one obtained directly from a mixed-halide perovskite single crystal ($MAPbBr_2I$) and the other developed by blending two different halide perovskites ($MAPbBr_3$ and $MAPbI_3$) in a stoichiometric ratio 2:1. The GCD spectra of both electrodes in dark and light circumstances reveal that the porosity of the electrode of single crystal-based active material exhibits a photo-capacitance improvement of up to 15 F g^{-1}, whereas the electrode composed of the blended active material demonstrates a photo-capacitance reduction due to trapping of charge carriers in the lower bandgap iodide sandwiched by the bromide phase caused by the micro-scale phase segregation of two distinct phases formed

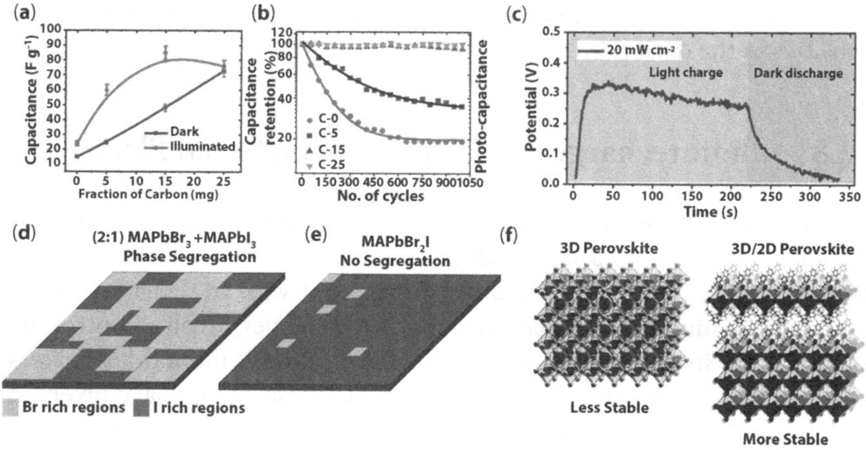

Figure 12.10 (a) The capacitance of the electrode vs the carbon percentage from 5% to 25% under light and illumination. (b) Capacitance retention measured for 1000 CV cycles. Stability increases with the increase in carbon fraction in the active material. The photocharging of the supercapacitor under white light illumination and discharging under dark. (Reproduced with permission from ref [23].) (f) Schematic of the phase segregation of the two regions (Br-rich region and I-rich region) of the electrode based on blended two halide perovskites ($MAPbBr_3$ and $MAPbI_3$) and the no-phase segregation of the electrode-based $MAPbBr_2I$. (Reproduced with permission from ref [80].) (g) The schematic diagram of the pure 3D perovskite and physical mixture of the 2D perovskite in 3D perovskite (3D/2D perovskite), which is more stable than 3D perovskite. (Reproduced with permission from ref [81]).

by the different halide content (Figure 12.10d). The photoelectrode made by $MAPbBr_2I$ perovskite single crystal active material exhibits enhanced homogeneity without any phase segregation and with excellent stability during photocharging, achieving a photoenergy density of about 1.6 Wh kg^{-1} under blue light illumination at a power of 90 mW cm^{-2}. The PCE is determined to be the highest at 0.017% [80]. Recently, we physically combined two perovskites of distinct dimensionalities (3D and 2D) by adjusting the weight percentage of the 2D perovskite within the 3D perovskite. This modification enhanced the stability of the electrodes as the proportion of 2D perovskite increased; however, beyond 5% of 2D perovskite, the charge storage performance began to decline due to the larger cations acting as a barrier to ion movement, thereby reducing capacitance. We developed two new electrodes with pure 3D perovskite and 10% blended 2D perovskite in 3D perovskite. Nonetheless, we reduced the carbon content to 2% to avoid unnecessary light absorption due to its high absorption capability (Figure 12.10e). The photovoltage of pure 3D perovskite-based electrodes reaches 1.11 V, exhibiting a PCE of around 0.02%. The photo-voltage of 10% 2D mixed, 3D perovskite-based electrodes reached 1.19 V under equal light illumination, exhibiting a conversion efficiency of around 0.03% [81].

12.9 Applications of the Integrated Photo-Supercapacitors

Due to the high energy density of lithium-ion batteries (LIBs), they are the most extensively utilized energy storage technology. This is even though LIBs are expensive, there are environmental safety concerns associated with the extraction of lithium from mining, and there is a limited supply of lithium material. In regions that need quick, strong, safe solar energy storage, SC could be beneficial. SCs may charge or discharge in short periods due to their lower energy density than batteries. For short periods of time, the electricity generated can go higher. The SC unit is quickly charged by the perovskite-based efficient PV unit of the integrated device whenever it is illuminated by any accessible indoor lights. The photo-supercapacitor can supply electricity to the sensors used in the emergency door, power grids, and handheld devices, including the display of the device testing for pregnancy, biological sensors, mobile devices, Bluetooth-enabled devices, and photochromic applications.

12.10 Conclusion

In conclusion, the advancement of solar energy is anticipated to yield a fossil fuel-free energy industry in the near future. This chapter has analyzed the most recent primary methods developed to integrate perovskite-based PV units with electrochemical energy storage SCs. Perovskite-based PV (PPV) devices have garnered significant attention owing to their significant improvements in power conversion efficiency (PCE), enabling them to rival silicon-based PVs. The existing PSCs expect significant advancements in the future. PPVs' efficiency is significant for transforming solar energy into electrical power, ensuring optimal device operation and energy sustainability. Solar technology can harvest solar energy; but it cannot store this energy, necessitating a storage device to retain the energy converted by PV systems in electrical form. The potential utilization of SCs in electronics and electric vehicles has made them an important class of energy storage devices. The energy density of SCs is between that of batteries and capacitors, with a power output of approximately 10,000 Wkg^{-1}, which is comparable to capacitors but superior to batteries and fuel cells. The integration of PV systems with SCs can be achieved through two distinct approaches: the conventional or traditional method and the monolithic approach. The conventional technique necessitated extra wiring to connect the energy storage device with the solar cells, resulting in energy loss, increased costs, and larger device sizes, hence limiting its applicability in commercial application. In monolithic integration, both devices are interconnected by a shared electrode, which reduces costs, size, and energy losses of the device. We have examined the integration of three-electrode and two-electrode systems, with the two-electrode configuration being more practical. In this context, the perovskite thin film is essential for the generation and storage part. We have examined several types of PPVs and their integration with different SCs, as well as the methods that make the device flexible, stable, and cost-effective for commercial applications. Consequently, AIPs are expected to enhance the stability of PVs. We have also investigated the photorechargeable SC utilizing a perovskite-based dual-functional electrode.

References

1. Poompavai, T. and Kowsalya, M., Control and energy management strategies applied for solar photovoltaic and wind energy fed water pumping system: A review. *Renew. Sustain. Energy Rev.*, 107, 108–122, 2019.
2. Bogachuk, D., Zouhair, S., Wojciechowski, K., Yang, B., Babu, V., Wagner, L., Xu, B., Lim, J., Mastroianni, S., Pettersson, H., Hagfeldt, A., Hinsch, A., Low-temperature carbon-based electrodes in perovskite solar cells. *Energy Environ. Sci.*, 13, 11, 3880–3916, 2020.
3. Hashemi, S.A., Ramakrishna, S., Aberle, A.G., Recent progress in flexible–wearable solar cells for self-powered electronic devices. *Energy Environ. Sci.*, 13, 3, 685–743, 2020.
4. Chang, X., Wang, T., Gong, J., CO2 photo-reduction: insights into CO2 activation and reaction on surfaces of photocatalysts. *Energy Environ. Sci.*, 9, 7, 2177–2196, 2016.
5. Gong, J., Li, C., Wasielewski, M.R., Advances in solar energy conversion. *Chem. Soc. Rev.*, 48, 7, 1862–1864, 2019.
6. Huang, S. and Liu, J., Geothermal energy stuck between a rock and a hot place. *Nature*, 463, 7279, 293, 2010.
7. Sun, H., You, X., Yang, Z., Deng, J., Peng, H., Winding ultrathin, transparent, and electrically conductive carbon nanotube sheets into high-performance

fiber-shaped dye-sensitized solar cells. *J. Mater. Chem. A Mater.*, 1, 40, 12422–12425, 2013.
8. Bajwa, D.S., Peterson, T., Sharma, N., Shojaeiarani, J., Bajwa, S.G., A review of densified solid biomass for energy production. *Renew. Sustain. Energy Rev.*, 96, 296–305, 2018.
9. Jurado, U.T., Pu, S.H., White, N.M., Grid of hybrid nanogenerators for improving ocean wave impact energy harvesting self-powered applications. *Nano Energy*, 72, 104701, 2020.
10. DeCarolis, J.F. and Keith, D.W., The Real Cost of Wind Energy. *Sci. (1979)*, 294, 5544, 1000–1003, 2001.
11. Wang, L., Han, J., Feng, J., Wang, X., Su, D., Hou, X., Hou, J., Liang, J., Dou, S.X., Simultaneously efficient light absorption and charge transport of CdS/TiO2 nanotube array toward improved photoelectrochemical performance. *Int. J. Hydrogen Energy*, 44, 59, 30899–30909, 2019.
12. Vega-Garita, V., Ramirez-Elizondo, L., Narayan, N., Bauer, P., Integrating a photovoltaic storage system in one device: A critical review. *Prog. Photovoltaics Res. Appl.*, 27, 4, 346–370, 2019.
13. Luo, B., Ye, D., Wang, L., Recent Progress on Integrated Energy Conversion and Storage Systems. *Adv. Sci.*, 4, 9, 1700104, 2017.
14. Hayat, M.B., Ali, D., Monyake, K.C., Alagha, L., Ahmed, N., Solar energy—A look into power generation, challenges, and a solar-powered future. *Int. J. Energy Res.*, 43, 3, 1049–1067, 2019.
15. Lu, Y., Chen, M., Zhu, G., Zhang, Y., Recent progress in the study of integrated solar cell-energy storage systems. *Nanoscale*, 16, 18, 8778–8790, 2024.
16. Fagiolari, L., Sampò, M., Lamberti, A., Amici, J., Francia, C., Bodoardo, S., Bella, F., Integrated energy conversion and storage devices: Interfacing solar cells, batteries and supercapacitors. *Energy Storage Mater.*, 51, 400–434, 2022.
17. Battaglia, C., Cuevas, A., De Wolf, S., High-efficiency crystalline silicon solar cells: status and perspectives. *Energy Environ. Sci.*, 9, 5, 1552–1576, 2016.
18. Moon, S., Kim, K., Kim, Y., Heo, J., Lee, J., Highly efficient single-junction GaAs thin-film solar cell on flexible substrate. *Sci. Rep.*, 6, 1, 30107, 2016.
19. Scarpulla, M.A., McCandless, B., Phillips, A.B., Yan, Y., Heben, M.J., Wolden, C., Xiong, G., Metzger, W.K., Mao, D., Krasikov, D., Sankin, I., Grover, S., Munshi, A., Sampath, W., Sites, J.R., Bothwell, A., Albin, D., Reese, M.O., Romeo, A., Nardone, M., Klie, R., Walls, J.M., Fiducia, T., Abbas, A., Hayes, S.M., CdTe-based thin film photovoltaics: Recent advances, current challenges and future prospects. *Sol. Energy Mater. Sol. Cells*, 255, 112289, 2023.
20. Zhu, J., Li, T., Qin, L., GaAs thin-film solar cells for perfect absorption in the visible and near-infrared bands. *Opt. Mater. (Amst)*, 143, 114269, 2023.
21. Popoola, I.K., Gondal, M.A., Popoola, A., Oloore, L.E., Younas, M., Inorganic perovskite photo-assisted supercapacitor for single device energy harvesting and storage applications. *J. Energy Storage*, 73, 108828, 2023.
22. Liang, J., Zhu, G., Lu, Z., Zhao, P., Wang, C., Ma, Y., Xu, Z., Wang, Y., Hu, Y., Ma, L., Chen, T., Tie, Z., Liu, J., Jin, Z., Integrated perovskite solar capacitors

with high energy conversion efficiency and fast photo-charging rate. *J. Mater. Chem. A Mater.*, 6, 5, 2047–2052, 2018.

23. Kumar, R., Kumar, A., Shukla, P.S., Varma, G., Venkataraman, D., Bag, M., Photorechargeable Hybrid Halide Perovskite Supercapacitors. *ACS Appl. Mater. Interfaces*, 14, 31, 35592–35599, 2022.
24. Liu, R., Liu, C., Fan, S., A photocapacitor based on organometal halide perovskite and PANI/CNT composites integrated using a CNT bridge. *J. Mater. Chem. A Mater.*, 5, 44, 23078–23084, 2017.
25. Usui, H., Suzuki, S., Domi, Y., Sakaguchi, H., TiO2/MnO2 composite electrode enabling photoelectric conversion and energy storage as photoelectrochemical capacitor. *Mater. Today Energy*, 9, 229–234, 2018.
26. Altaf, C.T., Coskun, O., Kumtepe, A., Rostas, A.M., Iatsunskyi, I., Coy, E., Erdem, E., Sankir, M., Sankir, N.D., Photo-supercapacitors based on nanoscaled ZnO. *Sci. Rep.*, 12, 1, 11487, 2022.
27. Liang, P.-W., Liao, C.-Y., Chueh, C.-C., Zuo, F., Williams, S.T., Xin, X.-K., Lin, J., Jen, A.K.-Y., Additive Enhanced Crystallization of Solution-Processed Perovskite for Highly Efficient Planar-Heterojunction Solar Cells. *Adv. Mater.*, 26, 22, 3748–3754, 2014.
28. Xiao, Z., Bi, C., Shao, Y., Dong, Q., Wang, Q., Yuan, Y., Wang, C., Gao, Y., Huang, J., Efficient, high yield perovskite photovoltaic devices grown by interdiffusion of solution-processed precursor stacking layers. *Energy Environ. Sci.*, 7, 8, 2619–2623, 2014.
29. Xing, G., Mathews, N., Sun, S., Lim, S.S., Lam, Y.M., Grätzel, M., Mhaisalkar, S., Sum, T.C., Long-Range Balanced Electron- and Hole-Transport Lengths in Organic-Inorganic CH3NH3PbI3. *Sci. (1979)*, 342, 6156, 344–347, 2013.
30. Stranks, S.D., Eperon, G.E., Grancini, G., Menelaou, C., Alcocer, M.J.P., Leijtens, T., Herz, L.M., Petrozza, A., Snaith, H.J., Electron-Hole Diffusion Lengths Exceeding 1 Micrometer in an Organometal Trihalide Perovskite Absorber. *Sci. (1979)*, 342, 6156, 341–344, 2013.
31. Etgar, L., Gao, P., Xue, Z., Peng, Q., Chandiran, A.K., Liu, B., Nazeeruddin, MK., Grätzel, M., Mesoscopic CH3NH3PbI3/TiO2 Heterojunction Solar Cells. *J. Am. Chem. Soc.*, 134, 42, 17396–17399, 2012.
32. Salim, T., Sun, S., Abe, Y., Krishna, A., Grimsdale, A.C., Lam, Y.M., Perovskite-based solar cells: impact of morphology and device architecture on device performance. *J. Mater. Chem. A Mater.*, 3, 17, 8943–8969, 2015.
33. John, R.A., Demirağ, Y., Shynkarenko, Y., Berezovska, Y., Ohannessian, N., Payvand, M., Zeng, P., Bodnarchuk, M.I., Krumeich, F., Kara, G., Shorubalko, I., Nair, M.V., Cooke, G.A., Lippert, T., Indiveri, G., Kovalenko, M.V., Reconfigurable halide perovskite nanocrystal memristors for neuromorphic computing. *Nat. Commun.*, 13, 1, 2074, 2022.
34. Fakharuddin, A., Gangishetty, M.K., Abdi-Jalebi, M., Chin, S.-H., bin Mohd Yusoff, AR., Congreve, D.N., Tress, W., Deschler, F., Vasilopoulou, M., Bolink, H.J., Perovskite light-emitting diodes. *Nat. Electron.*, 5, 4, 203–216, 2022.

35. Sun, Y., Ge, L., Dai, L., Cho, C., Ferrer Orri, J., Ji, K., Zelewski, S.J., Liu, Y., Mirabelli, A.J., Zhang, Y., Huang, J.-Y., Wang, Y., Gong, K., Lai, M.C., Zhang, L., Yang, D., Lin, J., Tennyson, E.M., Ducati, C., Stranks, S.D., Cui, L.-S., Greenham, N.C., Bright and stable perovskite light-emitting diodes in the near-infrared range. *Nature*, 615, 7954, 830–835, 2023.
36. Tian, J., Tan, Q.Y., Wang, Y., Yang, Y., Yuan, G., Adamo, G., Soci, C., Perovskite quantum dot one-dimensional topological laser. *Nat. Commun.*, 14, 1, 1433, 2023.
37. Kim, S.-Y., Zhang, H., Rubio-Magnieto, J., Operating Mechanism Principles and Advancements for Halide Perovskite-Based Memristors and Neuromorphic Devices. *J. Phys. Chem. Lett.*, 15, 40, 10087–10103, 2024.
38. Tang, Y., Jin, P., Wang, Y., Li, D., Chen, Y., Ran, P., Fan, W., Liang, K., Ren, H., Xu, X., Wang, R., Yang, Y.(M)., Zhu, B., Enabling low-drift flexible perovskite photodetectors by electrical modulation for wearable health monitoring and weak light imaging. *Nat. Commun.*, 14, 1, 4961, 2023.
39. John, R.A., Shah, N., Vishwanath, S.K., Ng, S.E., Febriansyah, B., Jagadeeswararao, M., Chang, C.-H., Basu, A., Mathews, N., Halide perovskite memristors as flexible and reconfigurable physical unclonable functions. *Nat. Commun.*, 12, 1, 3681, 2021.
40. Park, Y. and Lee, J.-S., Metal Halide Perovskite-Based Memristors for Emerging Memory Applications. *J. Phys. Chem. Lett.*, 13, 24, 5638–5647, 2022.
41. Wu, D., Zhang, Y., Liu, C., Sun, Z., Wang, Z., Lin, Z., Qiu, M., Fu, D., Wang, K., Recent Progress of Narrowband Perovskite Photodetectors: Fundamental Physics and Strategies. *Adv. Devices Instrum.*, 4, 0006, 2024.
42. K., N. and Rout, C.S., Photo-powered integrated supercapacitors: a review on recent developments, challenges and future perspectives. *J. Mater. Chem. A Mater.*, 9, 13, 8248–8278, 2021.
43. Xu, J., Chen, Y., Dai, L., Efficiently photo-charging lithium-ion battery by perovskite solar cell. *Nat. Commun.*, 6, 1, 8103, 2015.
44. Liu, R., Liu, Y., Zou, H., Song, T., Sun, B., Integrated solar capacitors for energy conversion and storage. *Nano Res.*, 10, 5, 1545–1559, 2017.
45. Shu, X., Chen, S., Chen, S., Pan, W., Zhang, J., Cobalt nitride embedded holey N-doped graphene as advanced bifunctional electrocatalysts for Zn-Air batteries and overall water splitting. *Carbon N Y*, 157, 234–243, 2020.
46. Lau, D., Song, N., Hall, C., Jiang, Y., Lim, S., Perez-Wurfl, I., Ouyang, Z., Lennon, A., Hybrid solar energy harvesting and storage devices: The promises and challenges. *Mater. Today Energy*, 13, 22–44, 2019.
47. Bati, A.S.R., Zhong, Y.L., Burn, P.L., Nazeeruddin, M.K., Shaw, P.E., Batmunkh, M., Next-generation applications for integrated perovskite solar cells. *Commun. Mater.*, 4, 1, 2, 2023.
48. Tian, Y., Cong, S., Su, W., Chen, H., Li, Q., Geng, F., Zhao, Z., Synergy of W18O49 and Polyaniline for Smart Supercapacitor Electrode Integrated

with Energy Level Indicating Functionality. *Nano Lett.*, 14, 4, 2150–2156, 2014.

49. Chen, P., Li, T.-T., Yang, Y.-B., Li, G.-R., Gao, X.-P., Coupling aqueous zinc batteries and perovskite solar cells for simultaneous energy harvest, conversion and storage. *Nat. Commun.*, 13, 1, 64, 2022.
50. Zhang, X., Zhang, H., Li, C., Wang, K., Sun, X., Ma, Y., Recent advances in porous graphene materials for supercapacitor applications. *RSC Adv.*, 4, 86, 45862–45884, 2014.
51. Nawaz, T., Ahmad, M., Ansari, M.Z., Hussain, I., Chen, X., Liu, L.-J., Walia, R., Low, K.H., Zhuang, S., Zhang, K., He, J., Structural study of atomically precise doped Au38–xAgx NCs@ZIF-8 electrode material for energy storage application. *Chem. Eng. J.*, 468, 143575, 2023.
52. Barai, H.R., Lopa, N.S., Ahmed, F., Khan, N.A., Ansari, S.A., Joo, S.W., Rahman, MM., Synthesis of Cu-Doped Mn3O4@Mn-Doped CuO Nanostructured Electrode Materials by a Solution Process for High-Performance Electrochemical Pseudocapacitors. *ACS Omega*, 5, 35, 22356–22366, 2020.
53. Da Silva, L.M., Cesar, R., Moreira, C.M.R., Santos, J.H.M., De Souza, L.G., Pires, B.M., Vicentini, R., Nunes, W., Zanin, H., Reviewing the fundamentals of supercapacitors and the difficulties involving the analysis of the electrochemical findings obtained for porous electrode materials. *Energy Storage Mater.*, 27, 555–590, 2020.
54. Barai, H.R., Rahman, MM., Roy, M., Barai, P., Joo, S.W., A calcium doped binary strontium-copper oxide electrode material for high-performance supercapacitors. *Mater. Sci. Semicond. Process.*, 90, 245–251, 2019.
55. Hussain, I., Ansari, M.Z., Lamiel, C., Hussain, T., Javed, M.S., Kaewmaraya, T., Ahmad, M., Qin, N., Zhang, K., *In Situ* Grown Heterostructure Based on MOF-Derived Carbon Containing n-Type Zn–In–S and Dry-Oxidative p-Type CuO as Pseudocapacitive Electrode Materials. *ACS Energy Lett.*, 8, 4, 1887–1895, 2023.
56. Balasubramaniam, S., Mohanty, A., Balasingam, S.K., Kim, S.J., Ramadoss, A., Comprehensive Insight into the Mechanism, Material Selection and Performance Evaluation of Supercapatteries. *Nano Micro Lett.*, 12, 1, 85, 2020.
57. Chatterjee, D.P. and Nandi, A.K., A review on the recent advances in hybrid supercapacitors. *J. Mater. Chem. A Mater.*, 9, 29, 15880–15918, 2021.
58. Liu, H., Liu, X., Wang, S., Liu, H.-K., Li, L., Transition metal based battery-type electrodes in hybrid supercapacitors: A review. *Energy Storage Mater.*, 28, 122–145, 2020.
59. Yang, Y., Fan, L., Pham, N.D., Yao, D., Wang, T., Wang, Z., Wang, H., Self-charging flexible solar capacitors based on integrated perovskite solar cells and quasi-solid-state supercapacitors fabricated at low temperature. *J. Power Sources*, 479, 229046, 2020.

60. Xu, X., Li, S., Zhang, H., Shen, Y., Zakeeruddin, S.M., Graetzel, M., Cheng, Y.-B., Wang, M., A Power Pack Based on Organometallic Perovskite Solar Cell and Supercapacitor. *ACS Nano*, 9, 2, 1782–1787, 2015.
61. Song, Z., Wu, J., Sun, L., Zhu, T., Deng, C., Wang, X., Li, G., Du, Y., Chen, Q., Sun, W., Fan, L., Chen, H., Lin, J., Lan, Z., Photocapacitor integrating perovskite solar cell and symmetrical supercapacitor generating a conversion storage efficiency over 20 %. *Nano Energy*, 100, 107501, 2022.
62. Liang, J., Zhu, G., Wang, C., Zhao, P., Wang, Y., Hu, Y., Ma, L., Tie, Z., Liu, J., Jin, Z., An all-inorganic perovskite solar capacitor for efficient and stable spontaneous photocharging. *Nano Energy*, 52, 239–245, 2018.
63. Xu, J., Ku, Z., Zhang, Y., Chao, D., Fan, H.J., Integrated Photo-Supercapacitor Based on PEDOT Modified Printable Perovskite Solar Cell. *Adv. Mater. Technol.*, 1, 5, 1600074, 2016.
64. Liu, R., Takakuwa, M., Li, A., Inoue, D., Hashizume, D., Yu, K., Umezu, S., Fukuda, K., Someya, T., An Efficient Ultra-Flexible Photo-Charging System Integrating Organic Photovoltaics and Supercapacitors. *Adv. Energy Mater.*, 10, 20, 2000523, 2020.
65. Du, P., Hu, X., Yi, C., Liu, H.C., Liu, P., Zhang, H.-L., Gong, X., Self-Powered Electronics by Integration of Flexible Solid-State Graphene-Based Supercapacitors with High Performance Perovskite Hybrid Solar Cells. *Adv. Funct. Mater.*, 25, 16, 2420–2427, 2015.
66. Mei, A., Li, X., Liu, L., Ku, Z., Liu, T., Rong, Y., Xu, M., Hu, M., Chen, J., Yang, Y., Grätzel, M., Han, H., A hole-conductor–free, fully printable mesoscopic perovskite solar cell with high stability. *Sci. (1979)*, 345, 6194, 295–298, 2014.
67. Li, X., Tschumi, M., Han, H., Babkair, S.S., Alzubaydi, R.A., Ansari, A.A., Habib, S.S., Nazeeruddin, M.K., Zakeeruddin, S.M., Grätzel, M., Outdoor Performance and Stability under Elevated Temperatures and Long-Term Light Soaking of Triple-Layer Mesoporous Perovskite Photovoltaics. *Energy Technol.*, 3, 6, 551–555, 2015.
68. Ku, Z., Rong, Y., Xu, M., Liu, T., Han, H., Full Printable Processed Mesoscopic CH3NH3PbI3/TiO2 Heterojunction Solar Cells with Carbon Counter Electrode. *Sci. Rep.*, 3, 1, 3132, 2013.
69. Liu, Z., Zhong, Y., Sun, B., Liu, X., Han, J., Shi, T., Tang, Z., Liao, G., Novel Integration of Perovskite Solar Cell and Supercapacitor Based on Carbon Electrode for Hybridizing Energy Conversion and Storage. *ACS Appl. Mater. Interfaces*, 9, 27, 22361–22368, 2017.
70. Meng, C., Liu, C., Fan, S., Flexible carbon nanotube/polyaniline paper-like films and their enhanced electrochemical properties. *Electrochem. Commun.*, 11, 1, 186–189, 2009.
71. Zhang, N., Luan, P., Zhou, W., Zhang, Q., Cai, L., Zhang, X., Zhou, W., Fan, Q., Yang, F., Zhao, D., Wang, Y., Xie, S., Highly stretchable pseudocapacitors based on buckled reticulate hybrid electrodes. *Nano Res.*, 7, 11, 1680–1690, 2014.

72. Zhang, M. and Lin, Z., Monolithic Perovskite Solar Capacitor Enabled by Double-Sided TiO2 Nanotube Arrays. *ACS Energy Lett.*, 7, 4, 1260–1265, 2022.
73. Li, C., Cong, S., Tian, Z., Song, Y., Yu, L., Lu, C., Shao, Y., Li, J., Zou, G., Rümmeli, M.H., Dou, S., Sun, J., Liu, Z., Flexible perovskite solar cell-driven photo-rechargeable lithium-ion capacitor for self-powered wearable strain sensors. *Nano Energy*, 60, 247–256, 2019.
74. Berestok, T., Diestel, C., Ortlieb, N., Buettner, J., Matthews, J., Schulze, P.S.C., Goldschmidt, J.C., Glunz, S.W., Fischer, A., High-Efficiency Monolithic Photosupercapacitors: Smart Integration of a Perovskite Solar Cell with a Mesoporous Carbon Double-Layer Capacitor. *Sol. RRL*, 5, 11, 2100662, 2021.
75. Grätzel, M., The light and shade of perovskite solar cells. *Nat. Mater.*, 13, 9, 838–842, 2014.
76. Ng, C.H., Lim, H.N., Hayase, S., Zainal, Z., Shafie, S., Lee, H.W., Huang, N.M., Cesium Lead Halide Inorganic-Based Perovskite-Sensitized Solar Cell for Photo-Supercapacitor Application under High Humidity Condition. *ACS Appl. Energy Mater.*, 1, 2, 692–699, 2018.
77. Yadav, A., Saini, A., Bag, M., Air-Processed All-Inorganic Halide Perovskites Integrated Photorechargeable Supercapacitors. *ACS Appl. Mater. Interfaces*, 16, 25, 32232–32239, 2024.
78. Zhou, F., Ren, Z., Zhao, Y., Shen, X., Wang, A., Li, Y.Y., Surya, C., Chai, Y., Perovskite Photovoltachromic Supercapacitor with All-Transparent Electrodes. *ACS Nano*, 10, 6, 5900–5908, 2016.
79. Li, C., Islam, MM., Moore, J., Sleppy, J., Morrison, C., Konstantinov, K., Dou, S.X., Renduchintala, C., Thomas, J., Wearable energy-smart ribbons for synchronous energy harvest and storage. *Nat. Commun.*, 7, 1, 13319, 2016.
80. Kumar, T., Kumar, A., Beniwal, S., Kumar, R., Bag, M., Photo-generated Charge Trapping in Phase Segregated Halide Perovskites – A Comprehensive Approach towards Efficient Photo-Rechargeable Ion Capacitors. *Batter. Supercaps*, 6, 10, e202300213, 2023.
81. Kumar, T., Thakuria, R., Kumar, M., Kumar, A., Kumar, R., Bag, M., Dimensional engineering to simultaneously enhance energy density and stability of MAPbBr3-based photo-rechargeable ion capacitors. *Sustain. Energy Fuels*, 7, 20, 5018–5028, 2023.

Index

Also of Interest

Check out these published titles in "Advances in Solar Cell Materials and Storage" series from Scrivener Publishing

Solar Capacitors and Batteries
Edited by Nurdan Demirci Sankir and Mehmet Sankir
Published 2025. ISBN 978-1-394-23376-2

Solar Fuels
Edited by Nurdan Demirci Sankir and Mehmet Sankir
Published 2023. ISBN 978-1-119-75057-4

Photoelectrochemical Solar Cells
Edited by Nurdan Demirci Sankir and Mehmet Sankir
Published 2019. ISBN 978-1-119-45993-4

Printable Solar Cells
Edited by Nurdan Demirci Sankir and Mehmet Sankir
Published 2017. ISBN 978-1-119-28371-3

www.scrivenerpublishing.com